Parallel Programming

Thomas Rauber • Gudula Rünger

Parallel Programming

for Multicore and Cluster Systems

Third Edition

 Springer

Thomas Rauber
Lehrstuhl für Angewandte Informatik II
University of Bayreuth
Bayreuth, Bayern, Germany

Gudula Rünger
Fakultät für Informatik
Chemnitz University of Technology
Chemnitz, Sachsen, Germany

Second English Edition was a translation from the 3rd German language edition: Parallele Programmierung (3. Aufl. 2012) by T. Rauber and G. Rünger, Springer-Verlag Berlin Heidelberg 2000, 2007, 2012.

ISBN 978-3-031-28923-1   ISBN 978-3-031-28924-8 (eBook)
https://doi.org/10.1007/978-3-031-28924-8

This Springer imprint is published by the registered company Springer Nature Switzerland AG
The registered company address is: Gewerbestrasse 11, 6330 Cham, Switzerland

Preface

Innovations in hardware architecture, such as hyper-threading or multicore processors, make parallel computing resources available for computer systems in different areas, including desktop and laptop computers, mobile devices, and embedded systems. However, the efficient usage of the parallel computing resources requires parallel programming techniques. Today, many standard software products are already based on concepts of parallel programming to use the hardware resources of multicore processors efficiently. This trend will continue and the need for parallel programming will extend to all areas of software development. The application area will be much larger than the area of scientific computing, which used to be the main area for parallel computing for many years. The expansion of the application area for parallel computing will lead to an enormous need for software developers with parallel programming skills. Some chip manufacturers already demand to include parallel programming as a standard course in computer science curricula. A more recent trend is the use of Graphics Processing Units (GPUs), which may comprise several thousands of cores, for the execution of compute-intensive non-graphics applications.

This book covers the new development in processor architecture and parallel hardware. Moreover, important parallel programming techniques that are necessary for developing efficient programs for multicore processors as well as for parallel cluster systems or supercomputers are provided. Both shared and distributed address space architectures are covered. The main goal of the book is to present parallel programming techniques that can be used in many situations for many application areas and to enable the reader to develop correct and efficient parallel programs. Many example programs and exercises are provided to support this goal and to show how the techniques can be applied to further applications. The book can be used as a textbook for students as well as a reference book for professionals. The material of the book has been used for courses in parallel programming at different universities for many years.

This third edition of the English book on parallel programming is an updated and revised version based on the second edition of this book from 2013. The three earlier German editions appeared in 2000, 2007, and 2012, respectively. The update of this new English edition includes an extended update of the chapter on computer architecture and performance analysis taking new developments such as the aspect of energy consumption into consideration. The description of OpenMP has been extended and now also captures the task concept of OpenMP. The chapter on message-passing programming has been extended and updated to include new features of MPI such as extended reduction operations and non-blocking collective communication operations. The chapter on GPU programming also has been updated. All other chapters also have been revised carefully.

The content of the book consists of three main parts, covering all areas of parallel computing: the architecture of parallel systems, parallel programming models and environments, and the implementation of efficient application algorithms. The emphasis lies on parallel programming techniques needed for different architectures.

The first part contains an overview of the architecture of parallel systems, including cache and memory organization, interconnection networks, routing and switching techniques as well as technologies that are relevant for modern and future multicore processors. Issues of power and energy consumption are also covered.

The second part presents parallel programming models, performance models, and parallel programming environments for message passing and shared memory models, including the message passing interface (MPI), Pthreads, Java threads, and OpenMP. For each of these parallel programming environments, the book introduces basic concepts as well as more advanced programming methods and enables the reader to write and run semantically correct and computationally efficient parallel programs. Parallel design patterns, such as pipelining, client-server, or task pools are presented for different environments to illustrate parallel programming techniques and to facilitate the implementation of efficient parallel programs for a wide variety of application areas. Performance models and techniques for runtime analysis are described in detail, as they are a prerequisite for achieving efficiency and high performance. A chapter gives a detailed description of the architecture of GPUs and also contains an introduction into programming approaches for general purpose GPUs concentrating on CUDA and OpenCL. Programming examples are provided to demonstrate the use of the specific programming techniques introduced.

The third part applies the parallel programming techniques from the second part to representative algorithms from scientific computing. The emphasis lies on basic methods for solving linear equation systems, which play an important role for many scientific simulations. The focus of the presentation is the analysis of the algorithmic structure of the different algorithms, which is the basis for a parallelization, and not so much on mathematical properties of the solution methods. For each algorithm, the book discusses different parallelization variants, using different methods and strategies.

Many colleagues and students have helped to improve the quality of this book. We would like to thank all of them for their help and constructive criticisms. For numerous corrections we would like to thank Robert Dietze, Jörg Dümmler, Marvin Ferber, Michael Hofmann, Ralf Hoffmann, Sascha Hunold, Thomas Jakobs, Oliver Klöckner, Matthias Korch, Ronny Kramer, Raphael Kunis, Jens Lang, Isabel Mühlmann, John O'Donnell, Andreas Prell, Carsten Scholtes, Michael Schwind, and Jesper Träff. Many thanks to Thomas Jakobs, Matthias Korch, Carsten Scholtes and Michael Schwind for their help with the program examples and the exercises. We thank Monika Glaser and Luise Steinbach for their help and support with the LATEX typesetting of the book. We also thank all the people who have been involved in the writing of the three German versions of this book. It has been a pleasure working with the Springer Verlag in the development of this book. We especially thank Ralf Gerstner for his continuous support.

Bayreuth and Chemnitz, January 2023

Thomas Rauber
Gudula Rünger

Contents

Chapter 1
Introduction

About this Chapter

Parallel programming is increasingly important for the software development today and in the future. This introduction outlines the more classical use of parallelism in scientific computing using supercomputers as well as the parallelism available in today's hardware, which broadens the use of parallelism to a larger class of applications. The basic concepts of parallel programming are introduced on less than two pages by informally defining key definitions, such as task decomposition or potential parallelism, and bringing them into context. The content of this book on parallel programming is described and suggestions for course structures are given.

1.1 Classical Use of Parallelism

Parallel programming and the design of efficient parallel programs is well-established in high performance, scientific computing for many years. The simulation of scientific problems is an important area in natural and engineering sciences of growing importance. More precise simulations or the simulation of larger problems lead to an increasing demand for computing power and memory space. In the last decades, high performance research also included the development of new parallel hardware and software technologies, and steady progress in parallel high performance computing can be observed. Popular examples are simulations for weather forecast based on complex mathematical models involving partial differential equations or crash simulations from car industry based on finite element methods. Other examples include drug design and computer graphics applications for film and advertising industry.

Depending on the specific application, computer simulation is the main method to obtain the desired result or it is used to replace or enhance physical experiments.

© The Author(s), under exclusive license to Springer Nature Switzerland AG 2023
T. Rauber, G. Rünger, *Parallel Programming*, https://doi.org/10.1007/978-3-031-28924-8_1

A typical example for the first application area is weather forecasting where the future development in the atmosphere has to be predicted, which can only be obtained by simulations. In the second application area, computer simulations are used to obtain results that are more precise than results from practical experiments or that can be performed at lower cost. An example is the use of simulations to determine the air resistance of vehicles: Compared to a classical wind tunnel experiment, a computer simulation can get more precise results because the relative movement of the vehicle in relation to the ground can be included in the simulation. This is not possible in the wind tunnel, since the vehicle cannot be moved. Crash tests of vehicles are an obvious example where computer simulations can be performed with lower cost.

Computer simulations often require a large computational effort. Thus, a low performance of the computer system used can restrict the simulations and the accuracy of the results obtained significantly. Using a high-performance system allows larger simulations which lead to better results and therefore, parallel computers have usually been used to perform computer simulations. Today, cluster systems built up from server nodes are widely available and are now also often used for parallel simulations. Additionally, multicore processors within the nodes provide further parallelism, which can be exploited for a fast computation. To use parallel computers or cluster systems, the computations to be performed must be partitioned into several parts which are assigned to the parallel resources for execution. These computation parts should be independent of each other, and the algorithm performed must provide enough independent computations to be suitable for a parallel execution. This is normally the case for scientific simulation, which often use one- or multi-dimensional arrays as data structures and organize their computations in nested loops. To obtain a parallel program for parallel excution, the algorithm must be formulated in a suitable programming language. Parallel execution is often controlled by specific runtime libraries or compiler directives which are added to a standard programming language, such as C, Fortran, or Java. The programming techniques needed to obtain efficient parallel programs are described in this book. Popular runtime systems and environments are also presented.

1.2 Parallelism in Today's Hardware

Parallel programming is an important aspect of high performance scientific computing but it used to be a niche within the entire field of hardware and software products. However, more recently parallel programming has left this niche and will become the mainstream of software development techniques due to a radical change in hardware technology.

Major chip manufacturers have started to produce processors with several power-efficient computing units on one chip, which have an independent control and can access the same memory concurrently. Normally, the term core is used for single computing units and the term multicore is used for the entire processor having several cores. Thus, using multicore processors makes each desktop computer a small parallel system. The technological development toward multicore processors was forced by physical reasons, since the clock speed of chips with more and more transistors cannot be increased at the previous rate without overheating.

Multicore architectures in the form of single multicore processors, shared memory systems of several multicore processors, or clusters of multicore processors with a hierarchical interconnection network will have a large impact on software development. In 2022, quad-core and oct-core processors are standard for normal desktop computers, and chips with up to 64 cores are already available for a use in high-end systems. It can be predicted from Moore's law that the number of cores per processor chip will double every 18 – 24 months and in several years, a typical processor chip might consist of dozens up to hundreds of cores where some of the cores will be dedicated to specific purposes such as network management, encryption and decryption, or graphics [138]; the majority of the cores will be available for application programs, providing a huge performance potential. Another trend in parallel computing is the use of GPUs for compute-intensive applications. GPU architectures provide many hundreds of specialized processing cores that can perform computations in parallel.

The users of a computer system are interested in benefitting from the performance increase provided by multicore processors. If this can be achieved, they can expect their application programs to keep getting faster and keep getting more and more additional features that could not be integrated in previous versions of the software because they needed too much computing power. To ensure this, there should definitely be support from the operating system, e.g., by using dedicated cores for their intended purpose or by running multiple user programs in parallel, if they are available. But when a large number of cores is provided, which will be the case in the near future, there is also the need to execute a single application program on multiple cores. The best situation for the software developer would be that there is an automatic transformer that takes a sequential program as input and generates a parallel program that runs efficiently on the new architectures. If such a transformer were available, software development could proceed as before. But unfortunately, the experience of the research in parallelizing compilers during the last 20 years has shown that for many sequential programs it is not possible to extract enough parallelism automatically. Therefore, there must be some help from the programmer and application programs need to be restructured accordingly.

For the software developer, the new hardware development toward multicore architectures is a challenge, since existing software must be restructured toward parallel execution to take advantage of the additional computing resources. In particular, software developers can no longer expect that the increase of computing power can automatically be used by their software products. Instead, additional effort is required at the software level to take advantage of the increased computing power. If a software company is able to transform its software so that it runs efficiently on novel multicore architectures, it will likely have an advantage over its competitors.

There is much research going on in the area of parallel programming languages and environments with the goal of facilitating parallel programming by providing support at the right level of abstraction. But there are also many effective techniques and environments already available. We give an overview in this book and present important programming techniques, enabling the reader to develop efficient parallel programs. There are several aspects that must be considered when developing a parallel program, no matter which specific environment or system is used. We give a short overview in the following section.

1.3 Basic Concepts of parallel programming

A first step in parallel programming is the design of a parallel algorithm or program for a given application problem. The design starts with the decomposition of the computations of an application into several parts, called **tasks**, which can be computed in parallel on the cores or processors of the parallel hardware. The decomposition into tasks can be complicated and laborious, since there are usually many different possibilities of decomposition for the same application algorithm. The size of tasks (e.g. in terms of the number of instructions) is called **granularity** and there is typically the possibility of choosing tasks of different sizes. Defining the tasks of an application appropriately is one of the main intellectual challenges in the development of a parallel program and is difficult to automate. The **potential parallelism** is an inherent property of an application algorithm and influences how an application can be split into tasks.

The tasks of an application are coded in a parallel programming language or environment and are assigned to **processes** or **threads** which are then assigned to physical computation units for execution. The assignment of tasks to processes or threads is called **scheduling** and fixes the order in which the tasks are executed. Scheduling can be done by hand in the source code or by the programming environment, at compile time or dynamically at runtime. The assignment of processes or threads onto the physical units, processors or cores, is called **mapping** and is usually done by the runtime system but can sometimes be influenced by the programmer. The tasks of an application algorithm can be independent but can also depend on each other resulting in data or control dependencies of tasks. Data and control dependencies may require a specific execution order of the parallel tasks: If a task needs data produced by another task, the execution of the first task can start only after the other task has actually produced these data and provides the information. Thus, dependencies between tasks are constraints for the scheduling. In addition, parallel programs need **synchronization** and coordination of threads and processes in order to execute correctly. The methods of synchronization and coordination in parallel computing are strongly connected with the way in which information is exchanged between processes or threads, and this depends on the memory organization of the hardware.

A coarse classification of the memory organization distinguishes between **shared memory** machines and **distributed memory** machines. Often the term *thread* is connected with shared memory and the term *process* is connected with distributed memory. For shared memory machines, a global shared memory stores the data of an application and can be accessed by all processors or cores of the hardware systems. Information exchange between threads is done by shared variables written by one thread and read by another thread. The correct behavior of the entire program has to be achieved by synchronization between threads so that the access to shared data is coordinated, i.e., a thread must not read a data element before the write operation by another thread storing the data element has been finalized. Depending on the programming language or environment, synchronization is done by the runtime system or by the programmer. For distributed memory machines, there exists a private memory for each processor, which can only be accessed by this processor and no synchronization for memory access is needed. Information exchange is done by sending data from one processor to another processor via an interconnection network by explicit **communication operations**.

Specific **barrier operations** offer another form of coordination which is available for both shared memory and distributed memory machines. All processes or threads have to wait at a barrier synchronization point until all other processes or threads have also reached that point. Only after all processes or threads have executed the code before the barrier, they can continue their work with the subsequent code after the barrier.

An important aspect of parallel computing is the **parallel execution time** which consists of the time for the computation on processors or cores and the time for data exchange or synchronization. The parallel execution time should be smaller than the sequential execution time on one processor so that designing a parallel program is worth the effort. The parallel execution time is the time elapsed between the start of the application on the first processor and the end of the execution of the application on all processors. This time is influenced by the distribution of work to processors or cores, the time for information exchange or synchronization, and **idle times** in which a processor cannot do anything useful but waiting for an event to happen. In general, a smaller parallel execution time results when the work load is assigned equally to processors or cores, which is called **load balancing**, and when the overhead for information exchange, synchronization and idle times is small. Finding a specific scheduling and mapping strategy which leads to a good load balance and a small overhead is often difficult because of many interactions. For example, reducing the overhead for information exchange may lead to load imbalance whereas a good load balance may require more overhead for information exchange or synchronization.

For a quantitative evaluation of the execution time of parallel programs, cost measures like **speedup** and **efficiency** are used, which compare the resulting parallel execution time with the sequential execution time on one processor. There are different ways to measure the cost or runtime of a parallel program and a large variety of parallel cost models based on parallel programming models have been proposed and used. These models are meant to bridge the gap between specific parallel hardware and more abstract parallel programming languages and environments.

1.4 Overview of the Book

The rest of the book is structured as follows. Chapter 2 gives an overview of important aspects of the hardware of parallel computer systems and addresses new developments such as the trends toward multicore architectures. In particular, the chapter covers important aspects of memory organization with shared and distributed address spaces as well as popular interconnection networks with their topological properties. Since memory hierarchies with several levels of caches may have an important influence on the performance of (parallel) computer systems, they are covered in this chapter. The architecture of multicore processors is also described in detail. The main purpose of the chapter is to give a solid overview of the important aspects of parallel computer architectures that play a role for parallel programming and the development of efficient parallel programs.

Chapter 3 considers popular parallel programming models and paradigms and discusses how the inherent parallelism of algorithms can be presented to a parallel runtime environment to enable an efficient parallel execution. An important part of this chapter is the description of mechanisms for the coordination of parallel programs, including synchronization and communication operations. Moreover, mechanisms for exchanging information and data between computing resources for different memory models are described. Chapter 4 is devoted to the performance analysis of parallel programs. It introduces popular performance or cost measures that are also used for sequential programs, as well as performance measures that have been developed for parallel programs. Especially, popular communication patterns for distributed address space architectures are considered and their efficient implementations for specific interconnection structures are given.

Chapter 5 considers the development of parallel programs for distributed address spaces. In particular, a detailed description of MPI (Message Passing Interface) is given, which is by far the most popular programming environment for distributed address spaces. The chapter describes important features and library functions of MPI and shows which programming techniques must be used to obtain efficient MPI programs. Chapter 6 considers the development of parallel programs for shared address spaces. Popular programming environments are Pthreads, Java threads, and OpenMP. The chapter describes all three and considers programming techniques to obtain efficient parallel programs. Many examples help to understand the relevant concepts and to avoid common programming errors that may lead to low performance or may cause problems such as deadlocks or race conditions. Programming examples and parallel programming patterns are presented. Chapter 7 introduces programming approaches for the execution of non-graphics application programs, e.g., from the area of scientific computing, on GPUs. The chapter describes the architecture of GPUs and concentrates on the programming environment CUDA (Compute Unified Device Architecture) from NVIDIA. A short overview of OpenCL is also given in this chapter. Chapter 8 considers algorithms from numerical analysis as representative examples and shows how the sequential algorithms can be transferred into parallel programs in a systematic way.

The main emphasis of the book is to provide the reader with the programming techniques that are needed for developing efficient parallel programs for different architectures and to give enough examples to enable the reader to use these techniques for programs from other application areas. In particular, reading and using the book is a good training for software development for modern parallel architectures, including multicore architectures.

The content of the book can be used for courses in the area of parallel computing with different emphasis. All chapters are written in a self-contained way so that chapters of the book can be used in isolation; cross-references are given when material from other chapters might be useful. Thus, different courses in the area of parallel computing can be assembled from chapters of the book in a modular way. Exercises are provided for each chapter separately. For a course on the programming of multicore systems, Chapters 2, 3 and 6 should be covered. In particular Chapter 6 provides an overview of the relevant programming environments and techniques. For a general course on parallel programming, Chapters 2, 5, and 6 can be used. These chapters introduce programming techniques for both distributed and shared address space. For a course on parallel numerical algorithms, mainly Chapters 5 and 8 are suitable; Chapter 6 can be used additionally. These chapters consider the parallel algorithms used as well as the programming techniques required. For a general course on parallel computing, Chapters 2, 3, 4, 5, and 6 can be used with selected applications from Chapter 8. Depending on the emphasis, Chapter 7 on GPU programming can be included in each of the courses mentioned above. The following web page will be maintained for additional and new material:
ai2.inf.uni-bayreuth.de/ppbook3

Chapter 2
Parallel Computer Architecture

About this Chapter

The possibility for a parallel execution of computations strongly depends on the architecture of the execution platform, which determines how computations of a program can be mapped to the available resources, such that a parallel execution is supported. This chapter gives an overview of the general architecture of parallel computers, which includes the memory organization of parallel computers, thread-level parallelism and multicore processors, interconnection networks, routing and switching as well as caches and memory hierarchies. The issue of energy efficiency is also considered.

In more detail, Section 2.1 gives an overview of the use of parallelism within a single processor or processor core. Using the available resources within a single processor core at instruction level can lead to a significant performance increase. Section 2.2 focuses on important aspects of the power and energy consumption of processors. Section 2.3 addresses techniques that influence memory access times and play an important role for the performance of (parallel) programs. Sections 2.4 introduces Flynn's taxonomy and Section 2.5 addresses the memory organization of parallel platforms. Section 2.6 presents the architecture of multicore processors and describes the use of thread-based parallelism for simultaneous multithreading.

Section 2.7 describes interconnection networks which connect the resources of parallel platforms and are used to exchange data and information between these resources. Interconnection networks also play an important role for multicore processors for the connection between the cores of a processor chip. The section covers static and dynamic interconnection networks and discusses important characteristics, such as diameter, bisection bandwidth and connectivity of different network types as well as the embedding of networks into other networks. Section 2.8 addresses routing techniques for selecting paths through networks and switching techniques for message forwarding over a given path. Section 2.9 considers memory hierarchies of sequential and parallel platforms and discusses cache coherence and memory consistency for shared memory platforms. Section 2.10 shows examples for the use of parallelism in today's computer architectures by describing the architecture of the Intel Cascade Lake and Ice Lake processors on one hand and the Top500 list on the other hand.

© The Author(s), under exclusive license to Springer Nature Switzerland AG 2023   9
T. Rauber, G. Rünger, *Parallel Programming*, https://doi.org/10.1007/978-3-031-28924-8_2

2.1 Processor Architecture and Technology Trends

Processor chips are the key components of computers. Considering the trends that can be observed for processor chips during recent years, estimations for future developments can be deduced.

An important performance factor is the **clock frequency** (also called clock rate or clock speed) of the processor which is the number of clock cycles per second, measured in Hertz = 1/second, abbreviated as Hz = 1/s. The clock frequency f determines the **clock cycle time** t of the processor by $t = 1/f$, which is usually the time needed for the execution of one instruction. Thus, an increase of the clock frequency leads to a faster program execution and therefore a better performance.

Between 1987 and 2003, an average annual increase of the clock frequency of about 40% could be observed for desktop processors [103]. Since 2003, the clock frequency of desktop processors remains nearly unchanged and no significant increases can be expected in the near future [102, 134]. The reason for this development lies in the fact that an increase in clock frequency leads to an increase in power consumption, mainly due to leakage currents which are transformed into heat, which then requires a larger amount of cooling. Using current state-of-the-art cooling technology, processors with a clock rate significantly above 4 GHz cannot be cooled permanently without a large additional effort.

Another important influence on the processor development are technical improvements in processor manufacturing. Internally, processor chips consist of transistors. The number of transistors contained in a processor chip can be used as a rough estimate of its complexity and performance. **Moore's law** is an empirical observation which states that the number of transistors of a typical processor chip doubles every 18 to 24 months. This observation has first been made by Gordon Moore in 1965 and has been valid for more than 40 years. However, the transistor increase due to Moore's law has slowed down during the last years [104]. Nevertheless, the number of transistors still increases and the increasing number of transistors has been used for architectural improvements, such as additional functional units, more and larger caches, and more registers, as described in the following sections.

In 2022, a typical desktop processor chip contains between 5 and 20 billions transistors, depending on the specific configuration. For example, an AMD Ryzen 7 3700X with eight cores (introduced in 2019) comprises about 6 billion transistors, an AMD Ryzen 7 5800H with eight cores (introduced in 2021) contains 10.7 billion transistors, and a 10-core Apple M2 (introduced in 2022) consists of 20 billion transistors, using an ARM-based system-on-a-chip (SoC) design, for more information see en.wikipedia.org/wiki/Transistor_count. The manufacturer Intel does not disclose the number of transistors of its processors.

The increase of the number of transistors and the increase in clock speed has led to a significant increase in the performance of computer systems. Processor performance can be measured by specific benchmark programs that have been selected from different application areas to get a representative performance metric of computer systems. Often, the SPEC benchmarks (*System Performance and Evaluation Cooperative*) are used to measure the integer and floating-point performance of computer systems [113, 104, 206, 136], see www.spec.org. Measurements with these benchmarks show that different time periods with specific performance increases can be identified for desktop processors [104]: Between 1986 and 2003, an average annual performance increase of about 50% could be reached. The time period of this large performance increase corresponds to the time period in which the clock frequency has been increased significantly each year. Between 2003 and 2011, the average annual performance increase of desktop processors is about 23%. This is still a significant increase which has been reached although the clock frequency remained nearly constant, indicating that the annual increase in transistor count has been used for architectural improvements leading to a reduction of the average time for executing an instruction. Between 2011 and 2015, the performance increase slowed down to 12% per year and is currently only at about 3.5 % per year, mainly due to the limited degree of parallelism available in the benchmark programs.

In the following, a short overview of architectural improvements that contributed to performance increase during the last decades is given. Four phases of microprocessor design trends can be observed [45] which are mainly driven by an internal use of parallelism:

1. **Parallelism at bit level**: Up to about 1986, the word size used by the processors for operations increased stepwise from 4 bits to 32 bits. This trend has slowed down and ended with the adoption of 64-bit operations at the beginning of the 1990's. This development has been driven by demands for improved floating-point accuracy and a larger address space. The trend has stopped at a word size of 64 bits, since this gives sufficient accuracy for floating-point numbers and covers a sufficiently large address space of 2^{64} bytes.

2. **Parallelism by pipelining**: The idea of pipelining at instruction level is to overlap the execution of multiple instructions. For this purpose, the execution of each instruction is partitioned into several steps which are performed by dedicated hardware units (pipeline stages) one after another. A typical partitioning could result in the following steps:

   (a) *fetch*: the next instruction to be executed is fetched from memory;
   (b) *decode*: the instruction fetched in step (a) is decoded;
   (c) *execute*: the operands specified are loaded and the instruction is executed;
   (d) *write back*: the result is written into the destination register.

   An instruction pipeline is similar to an assembly line in automobile industry. The advantage is that the different pipeline stages can operate in parallel, if there are no control or data dependencies between the instructions to be executed, see Fig. 2.1 for an illustration. To avoid waiting times, the execution of the different pipeline stages should take about the same amount of time. In each clock cycle, the execution of one instruction is finished and the execution of another instruction is started, if there are no dependencies between the instructions. The number of instructions finished per time unit is called the **throughput** of the pipeline. Thus, in the absence of dependencies, the throughput is one instruction per clock cycle.

In the absence of dependencies, all pipeline stages work in parallel. Thus, the number of pipeline stages determines the **degree of parallelism** attainable by a pipelined computation. The number of pipeline stages used in practice depends on the specific instruction and its potential to be partitioned into stages. Typical numbers of pipeline stages lie between 2 and 26 stages. Processors which use pipelining to execute instructions are called **ILP processors** (*instruction level parallelism processors*). Processors with a relatively large number of pipeline stages are sometimes called **superpipelined**. Although the available degree of parallelism increases with the number of pipeline stages, this number cannot be arbitrarily increased, since it is not possible to partition the execution of the instruction into a very large number of steps of equal size. Moreover, data dependencies often inhibit a completely parallel use of the stages.

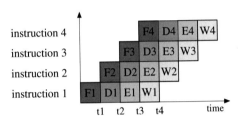

Fig. 2.1 Pipelined execution of four independent instructions 1–4 at times $t1,t2,t3,t4$. The execution of each instruction is split into four stages: *fetch* (F), *decode* (D), *execute* (E), and *write back* (W).

3. **Parallelism by multiple functional units**: Many processors are *multiple-issue processors*, which means that they use multiple, independent functional units, such as ALUs (*arithmetic logical unit*), FPUs (*floating-point unit*), load/store units, or branch units. These units can work in parallel, i.e., different independent instructions can be executed in parallel by different functional units, so that parallelism at instruction level is exploited. Therefore this technique is also referred to as **instruction-level parallelism (ILP)**. Using ILP, the average execution rate of instructions can be increased. Multiple-issue processors can be distinguished into **superscalar** processors and **VLIW** (*very long instruction word*) processors, see [104, 45] for a more detailed treatment.

The number of functional units that can efficiently be utilized is restricted because of data dependencies between neighboring instructions. For superscalar processors, these dependencies are dynamically determined at runtime by hardware, and decoded instructions are dispatched to the instruction units by hardware using dynamic scheduling. This may increase the complexity of the circuit significantly. But simulations have shown that superscalar processors with up to four functional units yield a substantial benefit over the use of a single functional unit. However, using a significantly larger number of functional units provides little additional gain [45, 123] because of dependencies between instructions and branching of control flow. In 2022, some server processors are able to submit up to ten independent instructions to functional units in one machine cycle per core, see Section 2.10.

4. **Parallelism at process or thread level**: The three techniques described so far assume a *single sequential* control flow which is provided by the compiler and which determines the execution order if there are dependencies between instructions. For the programmer, this has the advantage that a sequential programming language can be used nevertheless leading to a parallel execution of instructions due to ILP. However, the degree of parallelism obtained by pipelining and multiple functional units is limited. Thus, the increasing number of transistors available per processor chip according to Moore's law should be used for other techniques. One approach is to integrate larger caches on the chip. But the cache sizes cannot be arbitrarily increased either, as larger caches lead to a larger access time, see Section 2.9.

An alternative approach to use the increasing number of transistors on a chip is to put multiple, independent processor cores onto a single processor chip. This approach has been used for typical desktop processors since 2005. The resulting processor chips are called **multicore processors**. Each of the cores of a multicore processor must obtain a separate flow of control, i.e., parallel programming techniques must be used. The cores of a processor chip access the same memory and may even share caches. Therefore, memory accesses of the cores must be coordinated. The coordination and synchronization techniques required are described in later chapters.

A more detailed description of parallelism at hardware level using the four techniques described can be found in [45, 104, 171, 207]. Section 2.6 describes techniques such as simultaneous multithreading and multicore processors requiring an explicit specification of parallelism.

2.2 Power and Energy Consumption of Processors

Until 2003, a significant average annual increase of the clock frequency of processors could be observed. This trend has stopped in 2003 at a clock frequency of about 3.3 GHz and since then, only slight increases of the clock frequency could be observed. A further increase of the clock frequency is difficult because of the increased heat production due to leakage currents. Such leakage currents also occur if the processor is not performing computations. Therefore, the resulting power consumption is called **static power consumption**. The power consumption caused by computations is called **dynamic power consumption**. The overall power consumption is the sum of the static power consumption and the dynamic power consumption. In 2011, depending on the processor architecture, the static power consumption typically contributed between 25% and 50% to the total power consumption [104]. The heat produced by leakage currents must be carried away from the processor chip by using a sophisticated cooling technology.

An increase in clock frequency of a processor usually corresponds to a larger amount of leakage currents, leading to a larger power consumption and an increased heat production. Models to capture this phenomenon describe the dynamic power consumption P_{dyn} (measured in Watt, abbreviated as W) of a processor by

$$P_{dyn}(f) = \alpha \cdot C_L \cdot V^2 \cdot f \qquad (2.1)$$

where α is a switching probability, C_L is the load capacitance, V is the supply voltage (measured in Volt, abbreviated as V) and f is the clock frequency (measured in Hertz, abbreviated as Hz) [125, 186]. Since V depends linearly on f, a cubic dependence of the dynamic power consumption from the clock frequency results, i.e., the dynamic power consumption increases significantly if the clock frequency is increased. This can be confirmed by looking at the history of processor development: The first 32-bit microprocessors (such as the Intel 80386 processor) had a (fixed) clock frequency between 12 MHz and 40 MHz and a power consumption of about 2 W. A more recent 4.0 GHz Intel Core i7 6700K processor has a power consumption of about 95 W [104]. An increase in clock frequency which significantly exceeds 4.0 GHz would lead to an intolerable increase in dynamic power consumption. The cubic dependency of the power consumption on the clock frequency also explains why no significant increase in the clock frequency of desktop processors could be observed since 2003. Decreasing or increasing the clock frequency by a scaling factor is known as **frequency scaling**. There are several other factors that have an effect on the power consumption, including the computational intensity of an application program and the number of threads employed for the execution of the program. Using more threads usually leads to an increase of the power consumption. Moreover, the execution of floating-point operations usually leads to a larger power consumption than the use of integer operations [104].

The power consumption is normally not constant, but may vary during program execution, depending on the power consumption caused by the computations performed by the program and the clock frequency used during the execution. This can be seen in Figure 2.2 from [184]. The figure shows the power consumption of an application program for the numerical solution of ordinary differential equations using 10 threads (top) and 20 threads (bottom). Two different implementation versions of the solution methods are shown in the figure (Version 1 in red and Version 4 in green). Both versions exhibit an initialization phase with a small power consumption followed by a computation phase with the actual numerical solution using nine time steps, which can be clearly identified in the figure. During the individual time steps, there are large variations of the power consumption due to different types of computations. The figure also shows that during the computation phase, the use of 20 threads (bottom) leads to a larger power consumption than the use of 10 threads (top). Moreover, it can be seen that implementation Versions 1 and 4 have about the same execution time using 10 threads. However, when using 20 threads, implementation Version 4 is significantly faster than implementation Version 1.

To reduce energy consumption, modern microprocessors use several techniques such as shutting down inactive parts of the processor chip as well as **dynamic volt-**

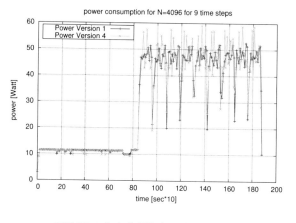

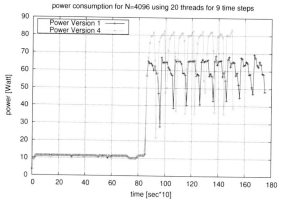

Fig. 2.2 Development of the power consumption during program execution for 10 threads (top) and 20 threads (bottom) on an Intel Broadwell processor when solving an ordinary differential equation [184].

age and frequency scaling (DVFS). The techniques are often controlled and co-ordinated by a special Power Management Unit (PMU). The idea of DVFS is to reduce the clock frequency of the processor chip to save energy during time periods with a small workload and to increase the clock frequency again if the workload increases again. Increasing the frequency reduces the cycle time of the processor, and in the same amount of time more instructions can be executed than when using a smaller frequency. An example of a desktop microprocessor with DVFS capability is the Intel Core i7 9700 processor (Coffee Lake architecture) for which 16 clock frequencies between 0.8 GHz and 3.0 GHz are available (3.0 GHz, 2.9 GHz, 2.7 GHz, 2.6 GHz, 2.4 GHz, 2.3 GHz, 2.1 GHz, 2.0 GHz, 1.8 GHz, 1.7 GHz, 1.5 GHz, 1.4GHz, 1.2 GHz, 1.1 GHz, 900 MHz, 800 MHz). The operating system can dynamically switch between these frequencies according to the current workload observed. Tools such as `cpufreq_set` can be used by the application programmer to adjust the clock frequency manually. For some processors, it is even possible to increase the frequency to a turbo-mode, exceeding the maximum frequency available for a short period of time. This is also called *overclocking* and allows an especially fast execution of computations during time periods with a heavy workload. Overclocking is typically restricted to about 10 % over the normal clock rate of the processor to avoid overheating [104].

The overall energy consumption of an application program depends on the execution time of the program obtained on a specific computer system and the power consumption of the computer system during program execution. The energy consumption E can be expressed as the product of the execution time T of the program and the average power consumption P_{av} during the execution:

$$E = P_{av} \cdot T. \tag{2.2}$$

Thus, the energy unit is $Watt \cdot sec = Ws = Joule$. The average power P_{av} captures the average static and dynamic power consumption during program execution. The clock frequency can be set at a fixed value or can be changed dynamically during program execution by the operating system. The clock frequency also has an effect on the execution time: decreasing the clock frequency leads to a larger machine cycle time and, thus, a larger execution time of computations. Hence, the overall effect of a reduction of the clock frequency on the energy consumption of a program execution is not a priori clear: reducing the energy decreases the power consumption, but increases the resulting execution time. Experiments with DVFS processor have shown that the smallest energy consumption does not necessarily correspond to the use of the smallest operational frequency available [186, 184]. Instead, using a small but not the smallest frequency often leads to the smallest energy consumption. The best frequency to be used depends strongly on the processor used and the application program executed, but also on the number of threads on parallel executions.

2.3 Memory access times

The access time to memory can have a large influence on the execution time of a program, which is referred to as the program's (runtime) performance. Reducing the memory access time can improve the performance. The amount of improvement depends on the memory access behavior of the program considered. Programs for which the amount of memory accesses is large compared to the number of computations performed may exhibit a significant benefit; these programs are called **memory-bound**. Programs for which the amount of memory accesses is small compared to the number of computations performed may exhibit a smaller benefit; these programs are called **compute-bound**.

The technological development with a steady reduction in the VLSI (Very-large-scale integration) feature size has led to significant improvements in processor performance. Since 1980, integer and floating performance on the SPEC benchmark suite has been increasing substantially per year, see Section 2.1. A significant contribution to these improvements comes from a reduction in processor cycle time. At the same time, the capacity of DRAM (Dynamic random-access memory) chips, which are used for building main memory of computer systems, increased significantly: Between 1986 and 2003, the storage capacity of DRAM chips increased by about 60% per year. Since 2003, the annual average increase lies between 25% and 40% [103]. In the following, a short overview of DRAM access times is given.

2.3.1 DRAM access times

Access to DRAM chips is standardized by the JEDEC Solid State Technology Association where JEDEC stands for Joint Electron Device Engineering Council, see jedec.org. The organization is responsible for the development of open industry standards for semiconductor technologies, including DRAM chips. All leading processor manufacturers are members of JEDEC.

For a performance evaluation of DRAM chips, the **latency** and the **bandwidth** are used. The latency of a DRAM chip is defined as the total amount of time that elapses between the point of time at which a memory access to a data block is issued by the CPU and the point in time when the first byte of the block of data arrives at the CPU. The latency is typically measured in micro-seconds (μs or nano-seconds (ns). The bandwidth denotes the number of data elements that can be read from a DRAM chip per time unit. The bandwidth is also denoted as **throughput**. The bandwidth is typically measured in megabytes per second (MB/s) or gigabytes per second (GB/s). For the latency of DRAM chips, an average decrease of about 5% per year could be observed between 1980 and 2005; since 2005 the improvement in access time has declined [104]. For the bandwidth of DRAM chips, an average annual increase of about 10% can be observed.

In 2022, the latency of the newest DRAM technology (DDR5, *Double Data Rate*) lies between 13.75 and 18 ns, depending on the specific JEDEC standard used. For the DDR5 technology, a bandwidth between 25.6 GB/s and 51.2 GB/s per DRAM chip is obtained. For example, the DDR5-3200 A specification leads to a peak bandwidth of 25.6 GB/s with a latency of 13.75 ns, the DDR5-6400 C specification has a peak bandwidth of 51.2 GB/s and a latency of 17.50 ns. Several DRAM chips (typically between 4 and 16) can be connected to DIMMs (*dual inline memory module*) to provide even larger bandwidths.

Considering DRAM latency, it can be observed that the average memory access time is significantly larger than the processor cycle time. The large gap between processor cycle time and memory access time makes a suitable organization of memory access more and more important to get good performance results at program level. Two important approaches have been proposed to reduce the average latency for memory access [14]: the simulation of **virtual processors** by each physical processor (multithreading) and the use of **local caches** to store data values that are accessed often. The next two subsections give a short overview of these approaches.

2.3.2 Multithreading for hiding memory access times

The idea of **interleaved multithreading** is to hide the latency of memory accesses by simulating a fixed number of virtual processors for each physical processor. The physical processor contains a separate program counter (PC) as well as a separate set of registers for each virtual processor. After the execution of a machine instruction, an implicit switch to the next virtual processor is performed, i.e. the virtual processors are simulated by the physical processor in a round-robin fashion. The number of virtual processors per physical processor should be selected such that the time between the execution of successive instructions of a specific virtual processor is sufficiently large to load required data from the global memory. Thus, the memory latency is hidden by executing instructions of other virtual processors. This approach does not reduce the amount of data loaded from the global memory via the network. Instead, instruction execution is organized such that a virtual processor does not access requested data until after its arrival. Therefore, from the point of view of a virtual processor, the memory latency cannot be observed. This approach is also called **fine-grained multithreading**, since a switch is performed after each instruction. An alternative approach is **coarse-grained multithreading** which switches between virtual processors only on costly stalls, such as level 2 cache misses [104]. For the programming of fine-grained multithreading architectures, a PRAM-like programming model can be used, see Section 4.6.1. There are two drawbacks of fine-grained multithreading:

- The programming must be based on a large number of virtual processors. Therefore, the algorithm executed must provide a sufficiently large potential of parallelism to employ all virtual processors.
- The physical processors must be especially designed for the simulation of virtual processors. A software-based simulation using standard microprocessors would be too slow.

There have been several examples for the use of fine-grained multithreading in the past, including Dencelor HEP (heterogeneous element processor) [202], NYU Ultracomputer [87], SB-PRAM [2], Tera MTA [45, 116], as well as the Oracle/Fujitsu T1 – T5 and M7/M8 multiprocessors. For example, each T5 processor supports 16 cores with 128 threads per core, acting as virtual processors. Graphics processing units (GPUs), such as NVIDIA GPUs, also use fine-grained multithreading to hide memory access latencies, see Chapter 7 for more information. Section 2.6.1 will describe another variation of multithreading which is simultaneous multithreading.

2.3.3 Caches for reducing the average memory access time

A **cache** is a small, but fast memory that is logically located between the processor and main memory. Physically, caches are located on the processor chip to ensure a fast access time. A cache can be used to store data that is often accessed by the processor, thus avoiding expensive main memory access. In most cases, the inclusion property is used, i.e., the data stored in the cache is a subset of the data stored in main memory. The management of the data elements in the cache is done by hardware, e.g. by employing a set-associative strategy, see [104] and Section 2.9.1 for a detailed treatment. For each memory access issued by the processor, it is first checked by hardware whether the memory address specified currently resides in the cache. If so, the data is loaded from the cache and no memory access is necessary. Therefore, memory accesses that go into the cache are significantly faster than memory accesses that require a load from the main memory. Since fast memory is expensive, several levels of caches are typically used, starting from a small, fast and expensive level 1 (L1) cache over several stages (L2, L3) to the large, but slower main memory. For a typical processor architecture, access to the L1 cache only takes 2-4 cycles whereas access to main memory can take up to several hundred cycles. The primary goal of cache organization is to reduce the average memory access time as far as possible and to achieve an access time as close as possible to that of the L1 cache. Whether this can be achieved depends on the memory access behavior of the program considered, see also Section 2.9.

Caches are used for nearly all processors, and they also play an important role for SMPs with a shared address space and parallel computers with distributed memory organization. If shared data is used by multiple processors or cores, it may be replicated in multiple caches to reduce access latency. Each processor or core should have a coherent view to the memory system, i.e., any read access should return the most recently written value no matter which processor or core has issued the corresponding write operation. A coherent view would be destroyed if a processor p changes the value in a memory address in its local cache without writing this value back to main memory. If another processor q would later read this memory address, it would not get the most recently written value. But even if p writes the value back to main memory, this may not be sufficient if q has a copy of the same memory location in its local cache. In this case, it is also necessary to update the copy in the local cache of q. The problem of providing a coherent view to the memory system is often referred to as **cache coherence problem**. To ensure cache coherency, a **cache coherency protocol** must be used, see Section 2.9.3 and [45, 104, 100] for a more detailed description.

2.4 Flynn's Taxonomy of Parallel Architectures

Parallel computers have been used for many years, and many different architectural alternatives have been proposed and implemented. In general, a parallel computer can be characterized as a collection of processing elements that can communicate and cooperate to solve large problems quickly [14]. This definition is intentionally quite vague to capture a large variety of parallel platforms. Many important details are not addressed by the definition, including the number and complexity of the processing elements, the structure of the interconnection network between the processing elements, the coordination of the work between the processing elements as well as important characteristics of the problem to be solved.

For a more detailed investigation, it is useful to introduce a classification according to important characteristics of a parallel computer. A simple model for such a classification is given by **Flynn's taxonomy** [64]. This taxonomy characterizes parallel computers according to the global control and the resulting data and control flows. Four categories are distinguished:

1. **Single-Instruction, Single-Data (SISD)**: There is one processing element which has access to a single program and a single data storage. In each step, the processing element loads an instruction and the corresponding data and executes the instruction. The result is stored back into the data storage. Thus, SISD describes a conventional sequential computer according to the *von Neumann model*.

2. **Multiple-Instruction, Single-Data (MISD)**: There are multiple processing elements each of which has a private program memory, but there is only one common access to a single global data memory. In each step, each processing element obtains the *same* data element from the data memory and loads an instruction from its private program memory. These possibly different instructions are then executed in parallel by the processing elements using the previously obtained (identical) data element as operand. This execution model is very restrictive and no commercial parallel computer of this type has ever been built.

3. **Single-Instruction, Multiple-Data (SIMD)**: There are multiple processing elements each of which has a private access to a (shared or distributed) data memory, see Section 2.5 for a discussion of shared and distributed address spaces. But there is only one program memory from which a special control processor fetches and dispatches instructions. In each step, each processing element obtains the *same* instruction from the control processor and loads a separate data element through its private data access on which the instruction is performed. Thus, the same instruction is synchronously applied in parallel by all processing elements to different data elements.

   For applications with a significant degree of data parallelism, the SIMD approach can be very efficient. Examples are multimedia applications or computer graphics algorithms which generate realistic three-dimensional views of computer-generated environments. Algorithms from scientific computing are often based on large arrays and can therefore also benefit from SIMD computations.

4. **Multiple-Instruction, Multiple-Data (MIMD)**: There are multiple processing elements each of which has a separate instruction and a separate data access to a (shared or distributed) program and data memory. In each step, each processing element loads a separate instruction and a separate data element, applies the instruction to the data element, and stores a possible result back into the data storage. The processing elements work asynchronously to each other. MIMD computers are the most general form of parallel computers in Flynn's taxonomy. Multicore processors or cluster systems are examples for the MIMD model.

Compared to MIMD computers, SIMD computers have the advantage that they are easy to program, since there is only one program flow, and the synchronous execution does not require synchronization at program level. But the synchronous execution is also a restriction, since conditional statements of the form

```
if (b==0) c=a; else c = a/b;
```

must be executed in two steps. In the first step, all processing elements whose local value of b is zero execute the then part. In the second step, all other processing elements execute the else part. Some processors support SIMD computations as additional possibility for processing large uniform data sets. An example is the x86 architecture which provides SIMD instructions in the form of SSE (Streaming SIMD Extensions) or AVX (Advanced Vector Extensions) instructions. AVX extensions have first been introduced in 2011 by Intel and AMD and are now supported in nearly all modern desktop and server processors. The features of AVX have been extended several times, and since 2017 AVX-512 is available. AVX-512 is based on a separate set of 512-bit registers. Each of these registers can store 16 single-precision 32-bit or eight double-precision 64-bit floating-point numbers, on which arithmetic operations can be executed in SIMD style, see Sect. 3.4 for a more detailed description. The computations of GPUs are also based on the SIMD concept, see Sect. 7.1 for a more detailed description.

MIMD computers are more flexible than SIMD computers, since each processing element can execute its own program flow. On the upper level, multicore processors as well as all parallel computers are based on the MIMD concept. Although Flynn's taxonomy only provides a coarse classification, it is useful to give an overview of the design space of parallel computers.

2.5 Memory Organization of Parallel Computers

Nearly all general-purpose parallel computers are based on the MIMD model. A further classification of MIMD computers can be done according to their memory organization. Two aspects can be distinguished: the physical memory organization and the view of the programmer to the memory. For the physical organization, computers with a physically shared memory (also called **multiprocessors**) and computers with a physically distributed memory (also called **multicomputers**) can be distinguished, see Fig. 2.3. But there exist also many hybrid organizations, for example providing a virtually shared memory on top of a physically distributed memory.

From the programmer's point of view, there are computers with a distributed address space and computers with a shared address space. This view does not necessarily correspond to the actual physical memory organization. For example, a parallel computer with a physically distributed memory may appear to the programmer as a computer with a shared address space when a corresponding programming environment is used. In the following, the physical organization of the memory is discussed in more detail.

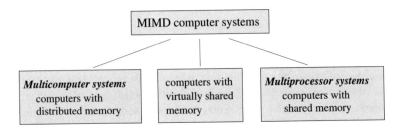

Fig. 2.3 Classification of the memory organization of MIMD computers.

2.5.1 Computers with Distributed Memory Organization

Computers with a physically distributed memory are also called **distributed memory machines** (DMM). They consist of a number of processing elements (called nodes) and an interconnection network which connects the nodes and supports the transfer of data between the nodes. A node is an independent unit, consisting of processor, local memory and, sometimes, peripheral elements, see Fig. 2.4 a).

The data of a program is stored in the local memory of one or several nodes. All local memory is *private* and only the local processor can access its own local memory directly. When a processor needs data from the local memory of other nodes to perform local computations, message-passing has to be performed via the interconnection network. Therefore, distributed memory machines are strongly connected with the message-passing programming model which is based on communication between cooperating sequential processes, see Chapters 3 and 5. To perform message-passing, two processes P_A and P_B on different nodes A and B issue corresponding send and receive operations. When P_B needs data from the local memory of node A, P_A performs a send operation containing the data for the destination process P_B. P_B performs a receive operation specifying a receive buffer to store the data from the source process P_A from which the data is expected.

The architecture of computers with a distributed memory has experienced many changes over the years, especially concerning the interconnection network and the coupling of network and nodes. The interconnection networks of earlier multicomputers were often based on **point-to-point connections** between nodes. A node is connected to a fixed set of other nodes by physical connections. The structure of the interconnection network can be represented as a graph structure. The nodes of the graph represent the processors, the edges represent the physical interconnections (also called *links*). Typically, the graph exhibits a regular structure. A typical network structure is the *hypercube* which is used in Fig. 2.4 b) to illustrate the node connections; a detailed description of interconnection structures is given in Section 2.7. In networks with point-to-point connections, the structure of the network determines the possible communications, since each node can only exchange data with its direct neighbors. To decouple send and receive operations, buffers can be used to store a message until the communication partner is ready. Point-to-point con-

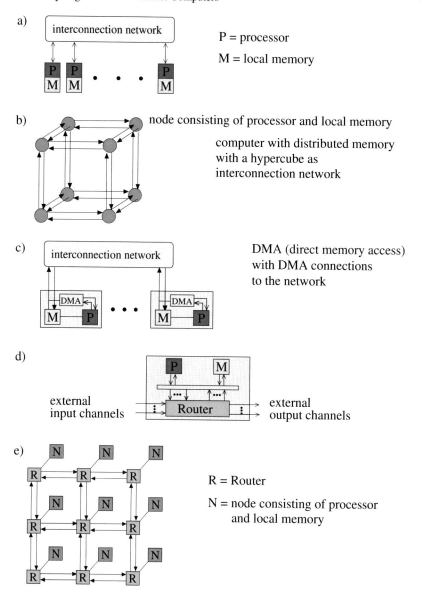

Fig. 2.4 Illustration of computers with distributed memory: a) abstract structure, b) computer with distributed memory and hypercube as interconnection structure, c) DMA (direct memory access), d) processor-memory node with router and e) interconnection network in form of a mesh to connect the routers of the different processor-memory nodes.

nections restrict parallel programming, since the network topology determines the possibilities for data exchange, and parallel algorithms have to be formulated such that their communication pattern fits to the given network structure [8, 144].

The execution of communication operations can be decoupled from the processor's operations by adding a **DMA controller** (DMA - direct memory access) to the nodes to control the data transfer between the local memory and the I/O controller. This enables data transfer from or to the local memory without participation of the processor (see Fig. 2.4 c) for an illustration) and allows asynchronous communication. A processor can issue a send operation to the DMA controller and can then continue local operations while the DMA controller executes the send operation. Messages are received at the destination node by its DMA controller which copies the enclosed data to a specific system location in local memory. When the processor then performs a receive operation, the data are copied from the system location to the specified receive buffer. Communication is still restricted to neighboring nodes in the network. Communication between nodes that do not have a direct connection must be controlled by software to send a message along a path of direct interconnections. Therefore, communication times between nodes that are not directly connected can be much larger than communication times between direct neighbors. Thus, it is still more efficient to use algorithms with a communication pattern according to the given network structure.

A further decoupling can be obtained by putting routers into the network, see Fig. 2.4 d). The routers form the actual network over which communication can be performed. The nodes are connected to the routers, see Fig. 2.4 e). Hardware-supported routing reduces communication times as messages for processors on remote nodes can be forwarded by the routers along a pre-selected path without interaction of the processors in the nodes along the path. With router support there is not a large difference in communication time between neighboring nodes and remote nodes, depending on the switching technique, see Sect. 2.8.3. Each physical I/O channel of a router can be used by one message only at a specific point in time. To decouple message forwarding, message buffers are used for each I/O channel to store messages and apply specific routing algorithms to avoid deadlocks, see also Sect. 2.8.1.

Technically, DMMs are quite easy to assemble since standard desktop computers or servers can be used as nodes. The programming of DMMs requires a careful data layout, since each processor can access only its local data directly. Non-local data must be accessed via message-passing, and the execution of the corresponding send and receive operations takes significantly longer than a local memory access. Depending on the interconnection network and the communication library used, the difference can be more than a factor of 100. Therefore, data layout may have a significant influence on the resulting parallel runtime of a program. The data layout should be selected such that the number of message transfers and the size of the data blocks exchanged are minimized.

The structure of DMMs has many similarities with networks of workstations (NOWs) in which standard workstations are connected by a fast local area network (LAN). An important difference is that interconnection networks of DMMs are typically more specialized and provide larger bandwidth and lower latency, thus leading to a faster message exchange.

Collections of complete computers with a dedicated interconnection network are often called **clusters**. Clusters are usually based on standard computers and even standard network topologies. The entire cluster is addressed and programmed as a single unit. The popularity of clusters as parallel machines comes from the availability of standard high-speed interconnections, such as FCS (Fibre Channel Standard), SCI (Scalable Coherent Interface), Switched Gigabit Ethernet, Myrinet, or Infini-Band, see [175, 104, 171]. A natural programming model of DMMs is the message-passing model that is supported by communication libraries, such as MPI or PVM, see Chapter 5 for a detailed treatment of MPI. These libraries are often based on standard protocols such as TCP/IP [139, 174].

The difference between cluster systems and **distributed systems** lies in the fact that the nodes in cluster systems use the same operating system and can usually not be addressed individually; instead a special job scheduler must be used. Several cluster systems can be connected to **grid systems** by using middleware software, such as the Globus Toolkit, see www.globus.org [72]. This allows a coordinated collaboration of several clusters. In Grid systems, the execution of application programs is controlled by the middleware software.

Cluster systems are also used for the provision of services in the area of **cloud computing**. Using cloud computing, each user can allocate virtual resources which are provided via the cloud infrastructure as part of the cluster system. The user can dynamically allocate and use resources according to his or her computational requirements. Depending on the allocation, a virtual resource can be a single cluster node or a collection of cluster nodes. Examples for cloud infrastructures are the Amazon Elastic Compute Cloud (EC2), Microsoft Azure, and Google Cloud. Amazon EC2 offers a variety of virtual machines with different performance and memory capacity that can be rented by users on an hourly or monthly basis. Amazon EC2 is part of Amazon Web Services (AWS), which provides many cloud computing services in different areas, such as computing, storage, database, network and content delivery, analytics, machine learning, and security, see aws.amazon.com. In the third quarter of 2022, AWS had a share of 34 % of the global cloud infrastructure service market, followed by Microsoft Azure with 21 % and Google Cloud with 11 %, see www.srgresearch.com.

2.5.2 Computers with Shared Memory Organization

Computers with a physically shared memory are also called **shared memory machines** (SMMs). The shared memory is also called **global memory**. SMMs consist of a number of processors or cores, a shared physical memory (global memory) and an interconnection network to connect the processors with the memory. The shared memory can be implemented as a set of memory modules. Data can be exchanged between processors via the global memory by reading or writing shared variables.

The cores of a multicore processor are an example for an SMM, see Sect. 2.6.2 for a more detailed description. Physically, the global memory usually consists of separate memory modules providing a common address space which can be accessed by all processors, see Fig. 2.5 for an illustration.

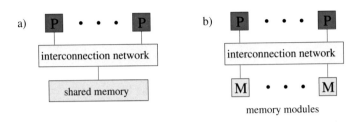

Fig. 2.5 Illustration of a computer with shared memory: a) abstract view and b) implementation of the shared memory with several memory modules.

A natural programming model for SMMs is the use of **shared variables** which can be accessed by all processors. Communication and cooperation between the processors is organized by writing and reading shared variables that are stored in the global memory. Accessing shared variables concurrently by several processors should be avoided, since **race conditions** with unpredictable effects can occur, see also Chapters 3 and 6.

The existence of a global memory is a significant advantage, since communication via shared variables is easy and since no data replication is necessary as it is sometimes the case for DMMs. But technically, the realization of SMMs requires a larger effort, in particular because the interconnection network must provide fast access to the global memory for each processor. This can be ensured for a small number of processors, but scaling beyond a few dozen processors is difficult.

A special variant of SMMs are symmetric multiprocessors (SMPs). SMPs have a single shared memory which provides a uniform access time from any processor for all memory locations, i.e., all memory locations are equidistant to all processors [45, 171]. SMPs usually have a small number of processors that are connected via a central bus or interconnection network, which also provides access to the shared memory. There are usually no private memories of processors or specific I/O processors, but each processor has a private cache hierarchy. As usual, access to a local cache is faster than access to the global memory. In the spirit of the definition from above, each multicore processor with several cores is an SMP system.

SMPs usually have only a small number of processors, since the central bus has to provide a constant bandwidth which is shared by all processors. When too many processors are connected, more and more access collisions may occur, thus increasing the effective memory access time. This can be alleviated by the use of caches and suitable cache coherence protocols, see Sect. 2.9.3. The maximum number of processors used in bus-based SMPs typically lies between 32 and 64. Interconnection schemes with a higher bandwidth are also used, such as parallel buses (IBM Power

8), ring interconnects (Intel Xeon E7) or crossbar connections (Fujitsu SPARC64 X+) [171].

Parallel programs for SMMs are often based on the execution of threads. A **thread** is a separate control flow which shares data with other threads via a global address space. It can be distinguished between **kernel threads** that are managed by the operating system, and **user threads** that are explicitly generated and controlled by the parallel program, see Section 3.8.2. The kernel threads are mapped by the operating system to processors or cores for execution. User threads are managed by the specific programming environment used and are mapped to kernel threads for execution. The mapping algorithms as well as the exact number of processors or cores can be hidden from the user by the operating system. The processors or cores are completely controlled by the operating system. The operating system can also start multiple sequential programs from several users on different processors or cores, when no parallel program is available. Small size SMP systems are often used as servers, because of their cost-effectiveness, see [45, 175] for a detailed description.

SMP systems can be used as nodes of a larger parallel computer by employing an interconnection network for data exchange between processors of different SMP nodes. For such systems, a shared address space can be defined by using a suitable cache coherence protocol, see Sect. 2.9.3. A coherence protocol provides the view of a shared address space, although the actual physical memory might be distributed. Such a protocol must ensure that any memory access returns the most recently written value for a specific memory address, no matter where this value is stored physically. The resulting systems are also called **distributed shared memory** (DSM) architectures. In contrast to single SMP systems, the access time in DSM systems depends on the location of a data value in the global memory, since an access to a data value in the local SMP memory is faster than an access to a data value in the memory of another SMP node via the coherence protocol. These systems are therefore also called NUMAs (non-uniform memory access), see Fig. 2.6 (b). Since single SMP systems have a uniform memory latency for all processors, they are also called UMAs (uniform memory access).

CC-NUMA (Cache-Coherent NUMA) systems are computers with a virtually shared address space for which cache coherence is ensured, see Fig. 2.6 (c). Thus, each processor's cache can store data not only from the processor's local memory but also from the shared address space. A suitable coherence protocol, see Sect. 2.9.3, ensures a consistent view by all processors. **COMA (Cache-Only Memory Architecture)** systems are variants of CC-NUMA in which the local memories of the different processors are used as caches, see Fig. 2.6 (d).

2.6 Thread-Level Parallelism

The architectural organization within a processor chip may require the use of explicitly parallel programs to efficiently use the resources provided. This is called **thread-level parallelism,** since the multiple control flows needed are often called

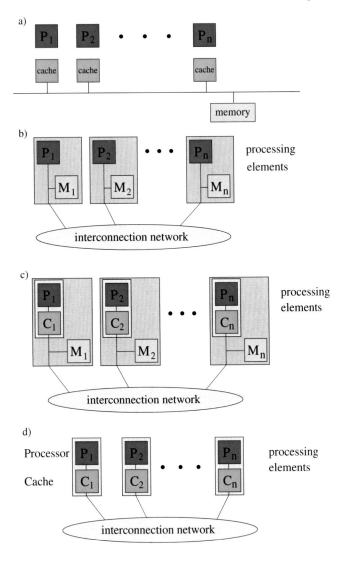

Fig. 2.6 Illustration of the architecture of computers with shared memory: a) SMP – symmetric multiprocessors, b) NUMA – non-uniform memory access, c) CC-NUMA – cache coherent NUMA and d) COMA – cache only memory access.

threads. The corresponding architectural organization is also called **chip multipro-cessing** (CMP). An example for CMP is the placement of multiple independent **exe-cution cores** with all execution resources onto a single processor chip. The resulting processors are called **multicore processors**, see Section 2.6.2.

An alternative approach is the use of *multithreading* to execute multiple threads simultaneously on a single processor by switching between the different threads when needed by the hardware. As described in Section 2.3, this can be obtained by fine-grained or coarse-grained multithreading. A variant of coarse-grained mul-tithreading is **timeslice multithreading** in which the processor switches between the threads after a predefined timeslice interval has elapsed. This can lead to situa-tions where the timeslices are not effectively used if a thread must wait for an event. If this happens in the middle of a timeslice, the processor may remain idle for the rest of the timeslice because of the waiting. Such unnecessary waiting times can be avoided by using **switch-on-event multithreading** [149] in which the processor can switch to the next thread if the current thread must wait for an event to occur as can happen for cache misses.

A variant of this technique is **simultaneous multithreading** (SMT) which will be described in the following. This technique is called **hyperthreading** for some Intel processors. The technique is based on the observation that a single thread of control often does not provide enough instruction-level parallelism to use all func-tional units of modern superscalar processors.

2.6.1 Simultaneous Multithreading

The idea of simultaneous multithreading (SMT) is to use several threads and to schedule executable instructions from different threads in the same cycle if neces-sary, thus using the functional units of a processor more effectively. This leads to a simultaneous execution of several threads which gives the technique its name. In each cycle, instructions from several threads compete for the functional units of a processor. Hardware support for simultaneous multithreading is based on the repli-cation of the chip area which is used to store the processor state. This includes the program counter (PC), user and control registers as well as the interrupt controller with the corresponding registers. Due to this replication, the processor appears to the operating system and the user program as a set of **logical processors** to which processes or threads can be assigned for execution. These processes or threads typ-ically come from a single user program, assuming their provision by this program using parallel programming techniques. The number of replications of the processor state determines the number of logical processors.

Each logical processor stores its processor state in a separate processor resource. This avoids overhead for saving and restoring processor states when switching to another logical processor. All other resources of the processor chip, such as caches, bus system, and function and control units, are shared by the logical processors. Therefore, the implementation of SMT only leads to a small increase in chip size.

For two logical processors, the required increase in chip area for an Intel Xeon processor is less than 5% [149, 225]. The shared resources are assigned to the logical processors for simultaneous use, thus leading to a simultaneous execution of logical processors. When a logical processor must wait for an event, the resources can be assigned to another logical processor. This leads to a continuous use of the resources from the view of the physical processor. Waiting times for logical processors can occur for cache misses, wrong branch predictions, dependencies between instructions, and pipeline hazards.

Investigations have shown that the simultaneous use of processor resources by two logical processors can lead to performance improvements between 15% and 30%, depending on the application program [149]. Since the processor resources are shared by the logical processors, it cannot be expected that the use of more than two logical processors can lead to a significant additional performance improvement. Therefore, SMT will likely be restricted to a small number of logical processors. In 2022, examples of processors that support SMT are the Intel Core i3, i5, and i7 processors (supporting two logical processors), the IBM Power9 and Power10 processors (four or eight logical processors, depending on the configuration), as well as the AMD Zen 3 processors (up to 24 threads per core).

To use SMT to obtain performance improvements, it is necessary that the operating system is able to control logical processors. From the point of view of the application program, it is necessary that there is a separate thread available for execution for each logical processor. Therefore, the application program must apply parallel programming techniques to get performance improvements for SMT processors.

2.6.2 Multicore Processors

The enormous annual increase of the number of transistors of a processor chip has enabled hardware manufacturers for many years to provide a significant performance increase for application programs, see also Section 2.1. Thus, a typical computer is considered old-fashioned and too slow after at most five years and customers buy new computers quite often. Hardware manufacturers are therefore trying to keep the obtained performance increase at least at the current level to avoid reduction in computer sales figures.

As discussed in Section 2.1, the most important factors for the performance increase per year have been an increase in clock speed and the internal use of parallel processing, such as pipelined execution of instructions and the use of multiple functional units. But these traditional techniques have mainly reached their limits:

- Although it is possible to put additional functional units on the processor chip, this would not increase performance for most application programs because dependencies between instructions of a single control thread inhibit their parallel execution. A single control flow does not provide enough instruction-level parallelism to keep a large number of functional units busy.

- There are two main reasons why the speed of processor clocks cannot be increased significantly [132]. First, the increase of the number of transistors on a chip is mainly achieved by increasing the transistor density. But this also increases the power density and heat production because of leakage current and power consumption, thus requiring an increased effort and more energy for cooling. Second, memory access times could not be reduced at the same rate as processor clock speed has been increased. This leads to an increased number of machine cycles for a memory access. For example, in 1990 main memory access has required between 6 and 8 machine cycles for a typical desktop computer system. In 2012, memory access time has significantly increased to 180 machine cycles for an Intel Core i7 processor. Since then, memory access time has increased further and in 2022, the memory latencies for an AMD EPYC Rome and an Intel Xeon Cascade Lake SP server processor are 220 cycles and 200 cycles, respectively [221]. Therefore, memory access times could become a limiting factor for further performance increase, and cache memories are used to prevent this, see Section 2.9 for a further discussion. In future, it can be expected that the number of cycles needed for a memory access will not change significantly.

There are more problems that processor designers have to face: Using the increased number of transistors to increase the complexity of the processor architecture may also lead to an increase in processor-internal wire length to transfer control and data between the functional units of the processor. Here, the speed of signal transfers within the wires could become a limiting factor. For example, a processor with a clock frequency of 3 GHz $= 3 \cdot 10^9 Hz$ has a cycle time of $1/(3 \cdot 10^9 Hz) = 0.33 \cdot 10^{-9} s = 0.33 ns$. Assuming a signal transfer at the speed of light (which is $0.3 \cdot 10^9$ m/s), a signal can cross a distance of $0.33 \cdot 10^{-9}$ s $\cdot 0.3 \cdot 10^9$ m/s $= 10$ cm in one processor cycle. This is not significantly larger than the typical size of a processor chip and wire lengths become an important issue when the clock frequency is increased further.

Another problem is the following: The physical size of a processor chip limits the number of pins that can be used, thus limiting the bandwidth between CPU and main memory. This may lead to a processor-to-memory performance gap which is sometimes referred to as *memory wall*. This makes the use of high-bandwidth memory architectures with an efficient cache hierarchy necessary [19].

All these reasons inhibit a processor performance increase at the previous rate when using the traditional techniques. Instead, new processor architectures have to be used, and the use of multiple cores on a single processor die is considered as the most promising approach. Instead of further increasing the complexity of the internal organization of a processor chip, this approach integrates multiple independent processing cores with a relatively simple architecture onto one processor chip. This has the additional advantage that the energy consumption of a processor chip can be reduced if necessary by switching off unused processor cores during idle times [102].

Multicore processors integrate multiple execution cores on a single processor chip. For the operating system, each execution core represents an independent logical processor with separate execution resources, such as functional units or execution pipelines. Each core has to be controlled separately, and the operating system can assign different application programs to the different cores to obtain a parallel execution. Background applications like virus checking, image compression, and encoding can run in parallel to application programs of the user. By using techniques of parallel programming, it is also possible to execute a computational-intensive application program (like computer games, computer vision, or scientific simulations) in parallel on a set of cores, thus reducing execution time compared to an execution on a single core or leading to more accurate results by performing more computations as in the sequential case. In the future, users of standard application programs as computer games will likely expect an efficient use of the execution cores of a processor chip. To achieve this, programmers have to use techniques from parallel programming.

The use of multiple cores on a single processor chip also enables standard programs, such as text processing, office applications, or computer games, to provide additional features that are computed in the background on a separate core so that the user does not notice any delay in the main application. But again, techniques of parallel programming have to be used for the implementation.

2.6.3 Architecture of Multicore Processors

There are many different design variants for multicore processors, differing in the number of cores, the structure and size of the caches, the access of cores to caches, and the use of heterogeneous components. From a high level view, [133] distinguished three main types of architectures: (a) a hierarchical design in which multiple cores share multiple caches that are organized in a tree-like configuration, (b) a pipelined design where multiple execution cores are arranged in a pipelined way, and (c) a network-based design where the cores are connected via an on-chip interconnection network, see Fig. 2.7 for an illustration. In 2022, most multicore processors are based on a network-based design using a fast on-chip interconnect to couple the different cores. Earlier multicore processors often relied on a hierarchical design.

2.6.3.1 Homogeneous Multicore Processors

As described in Section 2.1, the exploitation of parallelism by pipelining and ILP is limited due to dependencies between instructions. This limitation and the need for further performance improvement has led to the introduction of multicore designs starting in the year 2005 with the Intel Pentium D processor, with was a dual-core processor with two identical cores. The first quad-core processors were released in 2006, the first 12-core processors in 2010. This trend has continued, and in 2022 multicore with up to 96 cores are available, see Table 2.1. Processors with a large number of cores are typically used for server processors, whereas desktop computers or mobile computers normally use processors with a smaller number of cores, usu-

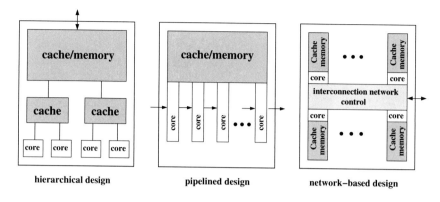

Fig. 2.7 Design choices for multicore chips according to [133].

ally between four and 16 cores. The reason for this usage lies in the fact that processors with a large number of cores provide a higher performance, but they also have a larger power consumption and a higher price than processors with a smaller number of cores. The larger performance can especially be exploited for server systems when executing jobs from different users. Desktop computers or mobile computers are used by a single user and the performance requirement is therefore smaller than for server systems. Hence, less expensive systems are sufficient with the positive effect that they are also more energy-efficient.

processor	number cores	number threads	clock GHz	L1 cache	L2 cache	L3 cache	year released
Intel Core i9-11900 "Rocket Lake"	8	16	2.5	8 x 64 KB	8 x 512 MB	16 MB	2021
Intel Xeon Platinum 8380 "Ice Lake"	40	80	2.3	40 x 48 KB	40 x 1.25 MB	60 MB	2021
Intel Mobil-Core i9-11950H "Tiger Lake-H"	8	16	2.6	8 x 48 KB	8 x 1.28 MB	24 MB	2020
AMD Ryzen 9 7950X "Zen 4"	16	32	4.5	1 MB	16 MB	64 MB	2022
AMD EPYC™ 9654P "Zen 4 Genoa"	96	192	2.4	6 MB	96 MB	384 MB	2022
IBM Power10	15	120	3.5	15 x 32 KB	15 x 2 MB	120 MB	2021
Ampere Altra Ma ARM v8	128	128	3.0	128 x 64 KB	128 x 1 MB	16 MB	2020

Table 2.1 Examples for multicore processors with a homogeneous design available in 2022.

Most processors for server and desktop systems rely on a homogeneous design, i.e., they contain identical cores where each core has the same computational performance and the same power consumption, see Figure 2.8 (left) for an illustration.

homogeneous   **heterogeneous**

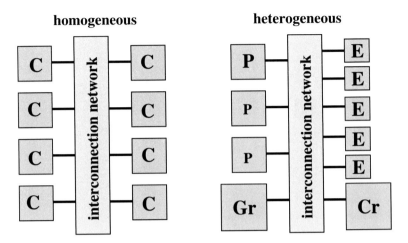

Fig. 2.8 Illustration of multicore architectures with a homogeneous design (left) and with a heterogeneous design (right). The homogeneous design contains eight identical cores C. The heterogeneous design has three performance cores P, five energy-efficient cores E, a cryptography core Cr, and a graphics unit Gr.

This has the advantage that the computational work can be distributed evenly among the different cores, such that each core gets about the same amount of work. Thus, the execution time of application programs can be reduced if suitable multithreading programming techniques are used.

Table 2.1 gives examples for multicore processors with a homogeneous design specifying the number of cores and threads, the base clock frequency in GHz as well as information about the size of the L1, L2, and L3 caches and the release year. The size for the L1 cache given in the table is the size of the L1 data cache. The Intel Core i9-11900 and the AMD Ryzen 9 7950X processors shown in the table are desktop processor, the Intel i9-11950H processor is a processor for notebooks. The other four processors are designed for a usage in server systems. These server processors provide a larger performance than the desktop or mobile processors, but they are also much more expensive. All processors shown have private L1 and L2 caches for each core and use shared L3 caches. The clock frequencies given are the base frequencies that can be increased to a higher turbo frequency for a short time period if required. Considering the power consumption, it can be observed that the server processors have a much larger power consumption that the desktop and mobile processors. This can be seen when considering the Thermal Design Power (TDP), which captures the maximum amount of heat that is generated by a processor. The TDP is measured in Watt and determines the cooling requirement for the processor and can be considered as a rough measure when comparing the average power consumption of processor chips. For example, the TDP is 270 W for the Intel Xeon Platinum 8380 server processor, 65 W for the Intel Core i9-11900 desktop processor, and 35 W for the Intel Mobil-Core i9-11950H mobile processor.

2.6.3.2 Heterogeneous Multicore Processors

Multicore processors with a heterogeneous design integrate cores with different architecture and different performance on a single chip, see Figure 2.8 (left) for an illustration. The idea for such a heterogeneous design is that the usage of the different cores can be adapted to the current workload of the processor. General-purpose cores with different performance and energy efficiency can be integrated as well as cores for specific applications such as encryption/decryption, multimedia or graphics. Heterogeneous designs are especially useful for mobile computers such as notebooks or smartphones, since unused cores can be deactivated, thus reducing the power consumption and increasing the battery lifetime of the mobile system.

processor	number cores	number threads	clock GHz	L1 cache	L2 cache	L3 cache	year released
Intel i9-12900K "Alder Lake"	16 8P+8E	24	3.2 P 2.4 E	P:8x48 KB E:8x32 KB	P:8x1.25 MB E:2x2 MB shared	30 MB	2021
Intel i7-1250U "Alder Lake"	10 2P+8E	12	1.1 P 0.8 E	P:2x48 KB E:8x32 KB	P:2x1.25 MB E:2x2 MB shared	12 MB	2022
Apple A15	8 2P+6E	8	3.24 P 2.01 E	8 x 64 KB	16 MB	32 MB	2021
Apple M2	8 4P+4E	8	3.5 P 2.4 E	P:4x128KB E:4x64KB	P: 16 MB E: 4 MB	8 MB	2022

Table 2.2 Examples for multicore processors with a heterogeneous design available in 2022. P denotes performance cores (P-cores), E denotes efficiency cores (E-cores). P-cores and E-cores have different characteristics such as different (base) operational frequencies and different cache sizes.

Table 2.2 shows examples for multicore desktop and mobile processors with a heterogeneous design. In the table, the main characteristics of the processors are shown. The operational frequency shown is the base frequency of the cores. The Intel Core i9-12900K processor with Intel's Alder Lake architecture is a desktop processor integrating eight so-called high-performance cores (P-cores) and eight so-called high-efficiency cores (E-cores). The P-cores are meant to support higher workloads due to their larger performance. The E-cores are more energy-efficient than the P-cores and are activated during times of lower workloads. Each P-core supports SMT (simultaneous multithreading) with two threads, the E-cores are single-threaded. The P-cores operate at 3.2 GHz, the E-cores at 2.3 GHz. The lower operational frequency reduces the power consumption significantly, see Equ. (2.1) in Section 2.2. Each P-core has a separate L1 data cache of size 48 KB, a separate L1 instruction cache of size 32 KB, and a separate L2 cache of size 1.25 MB. Each E-core has a separate L1 data cache of size 32 KB and a separate L1 instruction cache of size 64 KB. Four E-cores share an L2 cache of 2 MB. On the chip area, P-cores are much bigger than E-cores. The chip area of the Core i9-12900K processor contains ten segments. Eight of these segments contain single P-cores with their private L2 cache of size 1.25 MB. The remaining two segments are used for the E-cores: each of these two segments contains four E-cores along with an L2 cache of size 2 MB that is shared between these four E-cores, see anandtech.com for more information.

The Intel i7-1250U is also based on the Alder Lake architecture. It is especially designed for mobile devices: It integrates two P-cores with a base frequency of 1.1 GHz and eight E-cores with a base frequency of 0.8 GHz. Thus, the focus is on energy efficiency while still supporting higher workloads by the two P-cores. The low operational frequencies of both core types ensure a small power consumption: The TDP of the Intel i7-1250U is 9 W, whereas the Intel Core i9-12900K has a TDP of 125 W.

The Apple A15 processor is used for Apple iPhones and iPad Minis. The processor contains two high-performance cores called Avalanche running at 3.24 GHz and four energy-efficient cores called Blizzard running at 2.01 GHz. Both types of cores are single-threaded. The Apple M2 processor is used for Apple MacBooks and iPad Pro tablets. The processor contains four high-performance Avalanche cores and four energy-efficient Blizzard cores. Each performance core has an L1 data cache with 128 KB, an L1 instruction cache of size 192 KB. The performance cores share an L2 cache of size 16 MB. Each energy-efficient core has an L1 data cache with 64 KB, an L1 instruction cache of size 128 KB. The energy-efficient cores share an L2 cache of size 4 MB. As for the Apple A15 processor, SMT is not supported. Additionally, the M2 processor contains a ten-core GPU, where each core consists of 32 execution units. Each execution unit contains eight ALUs for arithmetic operations. The processor also contains a so-called Neural Engine and a so-called Media Engine to support processing of images and videos, signal processing, and security tasks. The TDP is 8.5 W for the Apple A15 and 22 W for the M2.

For a more detailed treatment of the architecture of multicore processors and further examples, we refer to [171, 104].

2.6.3.3 Future Trends and Developments

The potential of multicore processors has been realized by most processor manufacturers like Intel or AMD, and since about 2005, many manufacturers deliver processors with two or more cores. Since 2007, Intel and AMD provide quad-core processors (like the Quad-Core AMD Opteron and the Quad-Core Intel Xeon), and starting in 2010 the first oct-core processors were delivered. In 2022, all desktop and server processors are multicore processors, and server processors with up to 64 cores are available.

An important issue for the integration of a large number of cores in one processor chip is an efficient on-chip interconnection, which provides enough bandwidth for data transfers between the cores [102]. This interconnection should be *scalable* to support an increasing number of cores for future generations of processor designs and *robust* to tolerate failures of specific cores. If one or a few cores exhibit hardware failures, the rest of the cores should be able to continue operation. The interconnection should also support an efficient energy management which allows the scale-down of power consumption of individual cores by reducing the clock speed.

For an efficient use of processing cores, it is also important that the data to be processed can be transferred to the cores fast enough to avoid making the cores wait for data to be available. Therefore, an efficient memory system and I/O system are important. The memory system may use private first-level (L1) caches which can only be accessed by their associated cores, as well as shared second-level (L2) caches which can contain data of different cores. In addition, a shared third-level (L3) cache is often used. Processor chip with dozens or hundreds of cores will likely require an additional level of caches in the memory hierarchy to fulfill bandwidth requirements [102]. The I/O system must be able to provide enough bandwidth to keep all cores busy for typical application programs. At the physical layer, the I/O system must be able to bring hundreds of gigabits per second onto the chip. Such powerful I/O systems are under development since quite some time [102].

2.7 Interconnection Networks

A physical connection between the different components of a parallel system is provided by an **interconnection network**. Similar to control flow and data flow, see Section 2.4, or memory organization, see Section 2.5, the interconnection network can also be used for a classification of parallel systems. Internally, the network consists of links and switches which are arranged and connected in some regular way. In multicomputer systems, the interconnection network is used to connect the processors or nodes with each other. Interactions between the processors for coordination, synchronization, or exchange of data is obtained by communication through message-passing over the links of the interconnection network. In multiprocessor systems, the interconnection network is used to connect the processors with the memory modules. Thus, memory accesses of the processors are performed via the interconnection network. Interconnection networks are also used to connect the cores of multicore processors.

The main task of the interconnection network is to transfer a message from a specific processor or core to a specific destination. The message may contain data or a memory request. The destination may be another processor or a memory module. The requirement for the interconnection network is to perform the message transfer correctly and as fast as possible, even if several messages have to be transferred at the same time. Message transfer and memory accesses represent a significant part of operations of parallel systems independently of the organization of the address space. Therefore, the interconnection network used represents a significant part of the design of a parallel system and may have a large influence on its performance. Important design criteria of networks are:

- the **topology** describing the interconnection structure used to connect different processors or processors and memory modules and

- the **routing technique** describing the exact message transmission used within the network between processors or processors and memory modules.

The topology of an interconnection network describes the geometric structure used for the arrangement of switches and links to connect processors or processors and memory modules. The geometric structure can be described as a graph in which switches, processors or memory modules are represented as vertices and physical links are represented as edges. It can be distinguished between *static* and *dynamic* interconnection networks. **Static interconnection networks** connect nodes (processors or memory modules) *directly* with each other by fixed physical links. They are also called **direct networks** or **point-to-point networks**. The number of connections to or from a node may vary from only one in a star network to the total number of nodes in the network for a completely connected graph, see Section 2.7.2. Static networks are often used for systems with a distributed address space where a node comprises a processor and the corresponding memory module. The processor-internal interconnection in multicore processors is also based on static networks. **Dynamic interconnection networks** connect nodes *indirectly* via switches and links. They are also called **indirect networks**. Examples of indirect networks are *bus-based networks* or *switching networks* which consist of switches connected by links. Dynamic networks are used for both parallel systems with distributed and shared address space. Often, hybrid strategies are used [45].

The routing technique determines *how* and *along which path* messages are transferred in the network from a sender to a receiver. A path in the network is a series of nodes along which the message is transferred. Important aspects of the routing technique are the **routing algorithm** which determines the path to be used for the transmission and the **switching strategy** which determines whether and how messages are cut into pieces, how a routing path is assigned to a message, and how a message is forwarded along the processors or switches on the routing path.

The combination of routing algorithm, switching strategy and network topology determines the performance of a network. In Sections 2.7.2 and 2.7.4, important direct and indirect networks are described in more detail. Specific routing algorithms and switching strategies are presented in Sections 2.8.1 and 2.8.3. Efficient algorithms for the realization of common communication operations on different static networks are given in Chapter 4. A more detailed treatment of interconnection networks is given in [21, 45, 56, 89, 116, 144, 199].

2.7.1 Properties of Interconnection Networks

Static interconnection networks use fixed links between the nodes. They can be described by a connection graph $G = (V, E)$ where V is a set of nodes to be connected and E is a set of direct connection links between the nodes. If there is a direct physical connection in the network between the nodes $u \in V$ and $v \in V$, then it is $(u, v) \in E$. For most parallel systems, the interconnection network is *bidirectional*. This means that along a physical link messages can be transferred in both directions at the same time. Therefore, the connection graph is usually defined as an undirected graph. When a message must be transmitted from a node u to a node v and there is no direct connection between u and v in the network, a path from u to v must be selected which consists of several intermediate nodes along which the message is transferred. A sequence of nodes $(v_0, \ldots, v_k)$ is called *path* of length k between v_0 and v_k, if $(v_i, v_{i+1}) \in E$ for $0 \leq i < k$. For parallel systems, all interconnection networks fulfill the property that there is at least one path between any pair of nodes $u, v \in V$.

Static networks can be characterized by specific properties of the connection graph, including the following properties: number of nodes, diameter of the network, degree of the nodes, bisection bandwidth, node and edge connectivity of the network, and flexibility of embeddings into other networks as well as the embedding of other networks. In the following, a precise definition of these properties is given.

The **diameter** $\delta(G)$ of a network G is defined as the maximum distance between any pair of nodes:

$$\delta(G) = \max_{u,v \in V} \ \min_{\substack{\varphi \text{ path} \\ \text{from } u \text{ to } v}} \ \{k \mid k \text{ is the length of the path } \varphi \text{ from } u \text{ to } v\}.$$

The diameter of a network determines the length of the paths to be used for message transmission between any pair of nodes. The **degree** $d(G)$ of a network G is the maximum degree of a node of the network. The degree of a node n is the number of direct neighbor nodes of n:

$$g(G) = max\{g(v) \mid g(v) \text{ degree of } v \in V\}.$$

In the following, we assume that $|A|$ denotes the number of elements in a set A. The **bisection bandwidth** $B(G)$ of a network G is defined as the minimum number of edges that must be removed to partition the network into two parts of equal size without any connection between the two parts. For an uneven total number of nodes, the size of the parts may differ by 1. This leads to the following definition for $B(G)$:

$$B(G) = \min_{\substack{U_1, U_2 \text{ partition of } V \\ ||U_1| - |U_2|| \leq 1}} |\{(u, v) \in E \mid u \in U_1, v \in U_2\}|.$$

$B(G) + 1$ messages can saturate a network G, if these messages must be transferred at the same time over the corresponding edges. Thus, bisection bandwidth is a measure for the capacity of a network when transmitting messages simultaneously.

The **node** and **edge connectivity** of a network measure the number of nodes or edges that must fail to disconnect the network. A high connectivity value indicates a high reliability of the network and is therefore desirable. Formally, the node connectivity of a network is defined as the minimum number of nodes that must be deleted to disconnect the network, i.e., to obtain two unconnected network parts (which do not necessarily need to have the same size as it is required for the bisection bandwidth). For an exact definition, let $G_{V \setminus M}$ be the graph which is obtained by deleting all nodes in $M \subset V$ as well as all edges adjacent to these nodes. Thus, it is $G_{V \setminus M} = (V \setminus M, E \cap ((V \setminus M) \times (V \setminus M)))$. The **node connectivity** $nc(G)$ of G is then defined as

$$nc(G) = \min_{M \subset V} \{ |M| \mid \text{there exist } u, v \in V \setminus M \text{, such that there exists}$$

$$\text{no path in } G_{V \setminus M} \text{ from } u \text{ to } v \}.$$

Similarly, the edge connectivity of a network is defined as the minimum number of edges that must be deleted to disconnect the network. For an arbitrary subset $F \subset E$, let $G_{E \setminus F}$ be the graph which is obtained by deleting the edges in F, i.e., it is $G_{E \setminus F} = (V, E \setminus F)$. The **edge connectivity** $ec(G)$ of G is then defined as

$$ec(G) = \min_{F \subset E} \{ |F| \mid \text{there exist } u, v \in V \text{, such that there exists}$$

$$\text{no path in } G_{E \setminus F} \text{ from } u \text{ to } v \}.$$

The node and edge connectivity of a network is a measure for the number of independent paths between any pair of nodes. A high connectivity of a network is important for its availability and reliability, since many nodes or edges can fail before the network is disconnected. The minimum degree of a node in the network is an upper bound on the node or edge connectivity, since such a node can be completely separated from its neighboring nodes by deleting all incoming edges. Fig. 2.9 shows that the node connectivity of a network can be smaller than its edge connectivity.

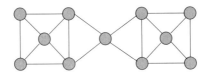

Fig. 2.9 Network with node connectivity 1, edge connectivity 2, and degree 4. The smallest degree of a node is 3.

The flexibility of a network can be captured by the notion of **embedding**. Let $G = (V, E)$ and $G' = (V', E')$ be two networks. An embedding of G' into G assigns each node of G' to a node of G such that different nodes of G' are mapped to different nodes of G and such that edges between two nodes in G' are also present between their associated nodes in G [21]. An embedding of G' into G can formally be described by a mapping function $\sigma : V' \rightarrow V$ such that the following holds:

- if $u \neq v$ for $u, v \in V'$, then $\sigma(u) \neq \sigma(v)$ and
- if $(u, v) \in E'$, then $(\sigma(u), \sigma(v)) \in E$.

If a network G' can be embedded into a network G, this means that G is at least as flexible as G', since any algorithm that is based on the network structure of G', e.g., by using edges between nodes for communication, can be re-formulated for G with the mapping function σ, thus using corresponding edges in G for communication.

The network of a parallel system should be designed to meet the requirements formulated for the architecture of the parallel system based on typical usage patterns. Generally, the following topological properties are desirable:

- a small diameter to ensure small distances for message transmission,
- a small node degree to reduce the hardware overhead for the nodes,
- a large bisection bandwidth to obtain large data throughputs,
- a large connectivity to ensure reliability of the network,
- embedding into a large number of networks to ensure flexibility, and
- easy extendability to a larger number of nodes.

Some of these properties are conflicting and there is no network that meets all demands in an optimal way. In the following, some popular direct networks are presented and analyzed. The topologies are illustrated in Fig. 2.10. The topological properties are summarized in Table 2.3.

2.7.2 Direct Interconnection Networks

Direct interconnection networks usually have a regular structure which is transferred to their graph representation $G = (V, E)$. In the following, we use $n = |V|$ for the number of nodes in the network and use this as a parameter of the network type considered. Thus, each network type captures an entire class of networks instead of a fixed network with a given number of nodes.

A **complete graph** is a network G in which each node is directly connected with every other node, see Fig. 2.10 (a). This results in diameter $\delta(G) = 1$ and degree $g(G) = n - 1$. The node and edge connectivity is $nc(G) = ec(G) = n - 1$, since a node can only be disconnected by deleting all $n - 1$ adjacent edges or neighboring nodes. For even values of n, the bisection bandwidth is $B(G) = n^2/4$: if two subsets of nodes of size $n/2$ each are built, there are $n/2$ edges from each of the nodes of one subset into the other subset, resulting in $n/2 \cdot n/2$ edges between the subsets. All other networks can be embedded into a complete graph, since there is a connection between any two nodes. Because of the large node degree, complete graph networks can only be built physically for a small number of nodes.

In a **linear array network**, nodes are arranged in a sequence and there is a bidirectional connection between any pair of neighboring nodes, see Fig. 2.10 (b), i.e., it is $V = \{v_1, \ldots, v_n\}$ and $E = \{(v_i, v_{i+1}) \mid 1 \leq i < n\}$. Since $n - 1$ edges have to be traversed to reach v_n starting from v_1, the diameter is $\delta(G) = n - 1$. The connectivity is $nc(G) = ec(G) = 1$, since the elimination of one node or edge disconnects the network. The network degree is $g(G) = 2$ because of the inner nodes, and the bisection bandwidth is $B(G) = 1$. A linear array network can be embedded in nearly all standard networks except a tree network, see below. Since there is a link only between neighboring nodes, a linear array network does not provide fault tolerance for message transmission.

In a **ring network**, nodes are arranged in ring order. Compared to the linear array network, there is one additional bidirectional edge from the first node to the last node, see Fig. 2.10 (c). The resulting diameter is $\delta(G) = \lfloor n/2 \rfloor$, the degree is $g(G) = 2$, the connectivity is $nc(G) = ec(G) = 2$, and the bisection bandwidth is also

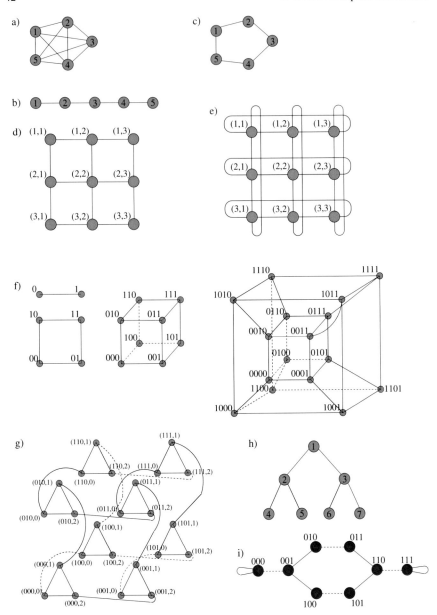

Fig. 2.10 Static interconnection networks: a) complete graph, b) linear array, c) ring, d) 2-dimensional mesh e) 2-dimensional torus, f) k-dimensional cube for k=1,2,3,4, g) Cube-connected-cycles network for k=3, h) complete binary tree, i) shuffle-exchange network with 8 nodes, where dashed edges represent exchange edges and straight edges represent shuffle edges.

$B(G) = 2$. In practice, ring networks can be used for small number of processors and as part of more complex networks.

A d-**dimensional mesh** (also called d-**dimensional array**) for $d \geq 1$ consists of $n = n_1 \cdot n_2 \cdot \ldots \cdot n_d$ nodes that are arranged as a d-dimensional mesh, see Fig. 2.10 (d). The parameter n_j denotes the extension of the mesh in dimension j for $j = 1, \ldots, d$. Each node in the mesh is represented by its position $(x_1, \ldots, x_d)$ in the mesh with $1 \leq x_j \leq n_j$ for $j = 1, \ldots, d$. There is an edge between node $(x_1, \ldots, x_d)$ and $(x'_1, \ldots x'_d)$, if there exists $\mu \in \{1, \ldots, d\}$ with

$$|x_\mu - x'_\mu| = 1 \text{ and } x_j = x'_j \text{ for all } j \neq \mu.$$

In the case that the mesh has the same extension in all dimensions (also called *symmetric mesh*), i.e., $n_j = r = \sqrt[d]{n}$ for all $j = 1, \ldots, d$, and therefore $n = r^d$, the network diameter is $\delta(G) = d \cdot (\sqrt[d]{n} - 1)$, resulting from the path length between nodes on opposite sides of the mesh. The node and edge connectivity is $nc(G) = ec(G) = d$, since the corner nodes of the mesh can be disconnected by deleting all d incoming edges or neighboring nodes. The network degree is $g(G) = 2d$, resulting from inner mesh nodes which have two neighbors in each dimension. A 2-dimensional mesh has been used for the Teraflop processor from Intel, see Section 2.6.3.

A d-**dimensional torus** is a variation of a d-dimensional mesh. The difference are additional edges between the first and the last node in each dimension, i.e., for each dimension $j = 1, \ldots, d$ there is an edge between node $(x_1, \ldots, x_{j-1}, 1, x_{j+1}, \ldots, x_d)$ and $(x_1, \ldots, x_{j-1}, n_j, x_{j+1}, \ldots, x_d)$, see Fig. 2.10 (e). For the symmetric case $n_j = \sqrt[d]{n}$ for all $j = 1, \ldots, d$, the diameter of the torus network is $\delta(G) = d \cdot \lfloor \sqrt[d]{n}/2 \rfloor$. The node degree is $2d$ for each node, i.e., $g(G) = 2d$. Therefore, node and edge connectivity are also $nc(G) = ec(G) = 2d$. Torus networks have often been used for the implementation of large parallel systems. Examples are the IBM BlueGene systems, where the BG/L and BG/P systems used a 3D torus and the newer BG/Q systems used a 5D torus network as central interconnect. Torus networks have also been used for the Cray XT3, XT4, and XT5 systems (3D torus) as well as for the Tofu interconnect of the Fujitsu Fugaku supercomputer (6D torus) [7].

A k-**dimensional cube** or **hypercube** consists of $n = 2^k$ nodes which are connected by edges according to a recursive construction, see Fig. 2.10 (f). Each node is represented by a binary word of length k, corresponding to the numbers $0, \ldots, 2^k - 1$. A 1-dimensional cube consists of two nodes with bit representations 0 and 1 which are connected by an edge. A k-dimensional cube is constructed from two given $(k-1)$-dimensional cubes, each using binary node representations $0, \ldots, 2^{k-1} - 1$. A k-dimensional cube results by adding edges between each pair of nodes with the same binary representation in the two $(k-1)$-dimensional cubes. The binary representations of the nodes in the resulting k-dimensional cube are obtained by adding a leading 0 to the previous representation of the first $(k-1)$-dimensional cube and adding a leading 1 to the previous representations of the second $(k-1)$-dimensional cube. Using the binary representations of the nodes $V = \{0, 1\}^k$, the recursive construction just mentioned implies that there is an edge between node $\alpha_0 \ldots \alpha_j \ldots \alpha_{k-1}$ and node $\alpha_0 \ldots \bar{\alpha}_j \ldots \alpha_{k-1}$ for $0 \leq j \leq k - 1$ where $\bar{\alpha}_j = 1$ for $\alpha_j = 0$ and $\bar{\alpha}_j = 0$

for $\alpha_j = 1$. Thus, there is an edge between every pair of nodes whose binary representation differs in exactly one bit position. This fact can also be captured by the Hamming distance.

The **Hamming distance** of two binary words of the same length is defined as the number of bit positions in which their binary representations differ. Thus, two nodes of a k-dimensional cube are directly connected, if their Hamming distance is 1. Between two nodes $v, w \in V$ with Hamming distance d, $1 \leq d \leq k$, there exists a path of length d connecting v and w. This path can be determined by traversing the bit representation of v bitwise from left to right and inverting the bits in which v and w differ. Each bit inversion corresponds to a traversal of the corresponding edge to a neighboring node. Since the bit representation of any two nodes can differ in at most k positions, there is a path of length $\leq k$ between any pair of nodes. Thus, the diameter of a k-dimensional cube is $\delta(G) = k$. The node degree is $g(G) = k$, since a binary representation of length k allows k bit inversions, i.e., each node has exactly k neighbors. The node and edge connectivity is $nc(G) = ec(G) = k$ as will be described in the following.

The connectivity of a hypercube is at most k, i.e., $nc(G) \leq k$, since each node can be completely disconnected from its neighbors by deleting all k neighbors or all k adjacent edges. To show that the connectivity is at least k, we show that there are exactly k independent paths between any pair of nodes v and w. Two paths are independent of each other if they do not share any edge, i.e., independent paths between v and w only share the two nodes v and w. The independent paths are constructed based on the binary representations of v and w, which are denoted by A and B, respectively, in the following. We assume that A and B differ in l positions, $1 \leq l \leq k$, and that these are the first l positions (which can be obtained by a renumbering). We can construct l paths of length l each between v and w by inverting the first l bits of A in different orders. For path i, $0 \leq i < l$, we stepwise invert bits $i, \ldots, l-1$ in this order first, and then invert bits $0, \ldots, i-1$ in this order. This results in l independent paths. Additional $k - l$ independent paths between v and w of length $l+2$ each can be constructed as follows: For i with $0 \leq i < k-l$, we first invert the bit $(l+i)$ of A and then the bits at positions $0, \ldots, l-1$ stepwise. Finally, we invert the bit $(l+i)$

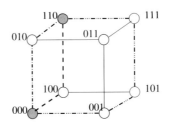

Fig. 2.11 In a 3-dimensional cube network, three independent paths (from node 000 to node 110) can be constructed. The Hamming distance between node 000 and node 110 is $l = 2$. There are two independent paths between 000 and 110 of length $l = 2$: path (000, 100, 110) and path (000, 010, 110). Additionally, there are $k - l = 1$ paths of length $l + 2 = 4$: path (000, 001, 101, 111, 110).

again, obtaining bit representation B. This is shown in Fig. 2.11 for an example. All k paths constructed are independent from each other, showing that $nc(G) \geq k$ holds.

A k-dimensional cube allows the embedding of many other networks as will be shown in the next subsection.

A **cube-connected cycles** (CCC) network results from a k-dimensional cube by replacing each node with a cycle of k nodes. Each of the nodes in the cycle has one off-cycle connection to one neighbor of the original node of the k-dimensional cube, thus covering all neighbors, see Fig. 2.10 (g). The nodes of a CCC network can be represented by $V = \{0,1\}^k \times \{0,\ldots,k-1\}$ where $\{0,1\}^k$ are the binary representations of the k-dimensional cube and $i \in \{0,\ldots,k-1\}$ represents the position in the cycle. It can be distinguished between cycle edges F and cube edges E:

$$F = \{((\alpha,i),(\alpha,(i+1) \bmod k)) \mid \alpha \in \{0,1\}^k, 0 \leq i < k\},$$

$$E = \{((\alpha,i),(\beta,i)) \mid \alpha_i \neq \beta_i \text{ and } \alpha_j = \beta_j \text{ for } j \neq i\}.$$

Each of the $k \cdot 2^k$ nodes of the CCC network has degree $g(G) = 3$, thus eliminating a drawback of the k-dimensional cube. The connectivity is $nc(G) = ec(G) = 3$ since each node can be disconnected by deleting its three neighboring nodes or edges. An upper bound for the diameter is $\delta(G) = 2k - 1 + \lfloor k/2 \rfloor$. To construct a path of this length, we consider two nodes in two different cycles with maximum hypercube distance k. These are nodes (α,i) and (β,j) for which α and β differ in all k bits. We construct a path from (α,i) to (β,j) by sequentially traversing a cube edge and a cycle edge for each bit position. The path starts with $(\alpha_0 \ldots \alpha_i \ldots \alpha_{k-1}, i)$ and reaches the next node by inverting α_i to $\bar{\alpha}_i = \beta_i$. From $(\alpha_0 \ldots \beta_i \ldots \alpha_{k-1}, i)$ the next node $(\alpha_0 \ldots \beta_i \ldots \alpha_{k-1}, (i+1) \bmod k)$ is reached by using a cycle edge. In the next steps, the bits $\alpha_{i+1}, \ldots, \alpha_{k-1}$ and $\alpha_0, \ldots, \alpha_{i-1}$ are successively inverted in this way, using a cycle edge between the steps. This results in $2k - 1$ edge traversals. Using at most $\lfloor k/2 \rfloor$ additional traversals of cycle edges starting from $(\beta, i+k-1 \bmod k)$ leads to the target node (β, j).

A **complete binary tree** network has $n = 2^k - 1$ nodes which are arranged as a binary tree in which all leaf nodes have the same depth, see Fig. 2.10 (h). The degree of inner nodes is 3, leading to a total degree of $g(G) = 3$. The diameter of the network is $\delta(G) = 2 \cdot \log \frac{n+1}{2}$ and is determined by the path length between two leaf nodes in different subtrees of the root node; the path consists of a subpath from the first leaf to the root followed by a subpath from the root to the second leaf. The connectivity of the network is $nc(G) = ec(G) = 1$, since the network can be disconnected by deleting the root or one of the edges to the root.

A k-dimensional **shuffle-exchange** network has $n = 2^k$ nodes and $3 \cdot 2^{k-1}$ edges [210]. The nodes can be represented by k-bit words. A node with bit representation α is connected with a node with bit representation β, if

- α and β differ in the last bit (*exchange edge*) or
- α results from β by a cyclic left shift or a cyclic right shift (*shuffle edge*).

Figure 2.10 (i) shows a shuffle-exchange network with 8 nodes. The permutation (α, β) where β results from α by a cyclic left shift is called **perfect shuffle**. The permutation (α, β) where β results from α by a cyclic right shift is called **inverse perfect shuffle**, see [144] for a detailed treatment of shuffle-exchange networks.

A k-**ary** d-**cube** with $k \geq 2$ is a generalization of the d-dimensional cube with $n = k^d$ nodes where each dimension i with $i = 0, \ldots, d-1$ contains k nodes. Each node can be represented by a word with d numbers $(a_0, \ldots, a_{d-1})$ with $0 \leq a_i \leq k-1$, where a_i represents the position of the node in dimension $i, i = 0, \ldots, d-1$. Two nodes $A = (a_0, \ldots, a_{d-1})$ and $B = (b_0, \ldots, b_{d-1})$ are connected by an edge if there is a dimension $j \in \{0, \ldots, d-1\}$ for which $a_j = (b_j \pm 1) \bmod k$ and $a_i = b_i$ for all other dimensions $i = 0, \ldots, d-1, i \neq j$. For $k = 2$, each node has one neighbor in each dimension, resulting in degree $g(G) = d$. For $k > 2$, each node has two neighbors in each dimension, resulting in degree $g(G) = 2d$. The k-ary d-cube captures some of the previously considered topologies as special case: A k-ary 1-cube is a ring with k nodes, a k-ary 2-cube is a torus with k^2 nodes, a 3-ary 3-cube is a 3-dimensional torus with $3 \times 3 \times 3$ nodes, and a 2-ary d-cube is a d-dimensional cube.

Table 2.3 summarizes important characteristics of the network topologies described.

Table 2.3 Key characteristics of static interconnection networks for selected topologies.

network G with n nodes	degree $g(G)$	diameter $\delta(G)$	edge-connectivity $ec(G)$	bisection bandwidth $B(G)$
complete graph	$n-1$	1	$n-1$	$\left(\frac{n}{2}\right)^2$
linear array	2	$n-1$	1	1
ring	2	$\lfloor \frac{n}{2} \rfloor$	2	2
d-dimensional mesh $(n = r^d)$	$2d$	$d(\sqrt[d]{n} - 1)$	d	$n^{\frac{d-1}{d}}$
d-dimensional torus $(n = r^d)$	$2d$	$d \lfloor \frac{\sqrt[d]{n}}{2} \rfloor$	$2d$	$2n^{\frac{d-1}{d}}$
k-dimensional hyper-cube $(n = 2^k)$	$\log n$	$\log n$	$\log n$	$\frac{n}{2}$
k-dimensional CCC-network $(n = k2^k$ for $k \geq 3)$	3	$2k - 1 + \lfloor k/2 \rfloor$	3	$\frac{n}{2k}$
complete binary tree $(n = 2^k - 1)$	3	$2\log \frac{n+1}{2}$	1	1
k-ary d-cube $(n = k^d)$	$2d$	$d \lfloor \frac{k}{2} \rfloor$	$2d$	$2k^{d-1}$

2.7.3 Embeddings

In this section, embeddings of several networks (ring, two-dimensional mesh and d-dimensional mesh) into a hypercube network are presented, demonstrating that the hypercube topology is versatile and flexible. The notion of embedding is introduced in Sect. 2.7.1. Each of the specific embeddings is described by a suitable mapping function σ.

2.7.3.1 Embedding a ring into a hypercube network

For an embedding of a ring network with $n = 2^k$ nodes represented by $V' = \{1, \ldots, n\}$ in a k-dimensional cube with nodes $V = \{0, 1\}^k$, a bijective function from V' to V is constructed such that a ring edge $(i, j) \in E'$ is mapped to a hypercube edge. In the ring, there are edges between neighboring nodes in the sequence $1, \ldots, n$. To construct the embedding, we have to arrange the hypercube nodes in V in a sequence such that there is also an edge between neighboring nodes in the sequence. The sequence is constructed as reflected Gray code (RGC) sequence which is defined as follows:

A k-bit RGC is a sequence with 2^k binary strings of length k such that two neighboring strings differ in exactly one bit position. The RGC sequence is constructed recursively, as follows:

- the 1-bit RGC sequence is $RGC_1 = (0, 1)$,
- the 2-bit RGC sequence is obtained from RGC_1 by inserting a 0 and a 1 in front of RGC_1, resulting in the two sequences $(00, 01)$ and $(10, 11)$. Reversing the second sequence and concatenation yields $RGC_2 = (00, 01, 11, 10)$.
- For $k \geq 2$, the k-bit Gray code RGC_k is constructed from the $(k-1)$-bit Gray code $RGC_{k-1} = (b_1, \ldots, b_m)$ with $m = 2^{k-1}$ where each entry b_i for $1 \leq i \leq m$ is a binary string of length $k-1$. To construct RGC_k, RGC_{k-1} is duplicated; a 0 is inserted in front of each b_i of the original sequence, and a 1 is inserted in front of each b_i of the duplicated sequence. This results in sequences $(0b_1, \ldots, 0b_m)$ and $(1b_1, \ldots, 1b_m)$. RGC_k results by reversing the second sequence and concatenating the two sequences; thus $RGC_k = (0b_1, \ldots, 0b_m, 1b_m, \ldots, 1b_1)$.

The Gray code sequences RGC_k constructed in this way have the property that they contain all binary representations of a k-dimensional hypercube, since the construction corresponds to the construction of a k-dimensional cube from two $(k-1)$-dimensional cubes as described in the previous section. Two neighboring k-bit words of RGC_k differ in exactly one bit position, as can be shown by induction. The statement is surely true for RGC_1. Assuming that the statement is true for RGC_{k-1}, it is true for the first 2^{k-1} elements of RGC_k as well as for the last 2^{k-1} elements, since these differ only by a leading 0 or 1 from RGC_{k-1}. The statement is also true for the two middle elements $0b_m$ and $1b_m$ at which the two sequences of length 2^{k-1} are concatenated. Similarly, the first element $0b_1$ and the last element $1b_1$ of RGC_k differ only in the first bit. Thus, neighboring elements of RGC_k are connected by a

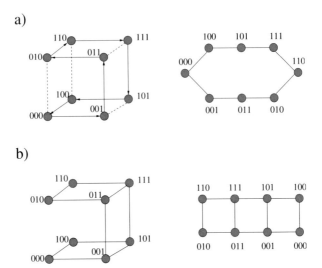

Fig. 2.12 Embeddings into a hypercube network: (a) embedding of a ring network with 8 nodes into a 3-dimensional hypercube and (b) embedding of a 2-dimensional 2×4 mesh into a 3-dimensional hypercube.

hypercube edge. An embedding of a ring into a k-dimensional cube can be defined by the mapping

$$\sigma : \{1,\ldots,n\} \rightarrow \{0,1\}^k \text{ with } \sigma(i) := RGC_k(i)$$

where $RGC_k(i)$ denotes the ith element of RGC_k. Figure 2.12 (a) shows an example for $k = 3$.

2.7.3.2 Embedding a 2-dimensional mesh into a hypercube network

The embedding of a two-dimensional mesh with $n = n_1 \cdot n_2$ nodes into a k-dimensional cube with $n = 2^k$ nodes can be obtained by a generalization of the embedding of a ring network. For k_1 and k_2 with $n_1 = 2^{k_1}$ and $n_2 = 2^{k_2}$, i.e., $k_1 + k_2 = k$, the Gray codes $RGC_{k_1} = (a_1,\ldots,a_{n_1})$ and $RGC_{k_2} = (b_1,\ldots,b_{n_2})$ are used to construct an $n_1 \times n_2$ matrix M whose entries are k-bit strings. In particular, it is

$$M = \begin{bmatrix} a_1b_1 & a_1b_2 & \cdots & a_1b_{n_2} \\ a_2b_1 & a_2b_2 & \cdots & a_2b_{n_2} \\ \vdots & \vdots & \vdots & \vdots \\ a_{n_1}b_1 & a_{n_1}b_2 & \cdots & a_{n_1}b_{n_2} \end{bmatrix}.$$

The matrix is constructed such that neighboring entries differ in exactly one bit position. This is true for neighboring elements in a row, since identical elements of RGC_{k_1} and neighboring elements of RGC_{k_2} are used. Similarly, this is true for neighboring elements in a column, since identical elements of RGC_{k_2} and neighboring elements of RGC_{k_1} are used. All elements of M are bit strings of length k and there are no identical bit strings according to the construction. Thus, the matrix M contains all bit representations of nodes in a k-dimensional cube and neighboring entries in M correspond to neighboring nodes in the k-dimensional cube, which are connected by an edge. Thus, the mapping

$$\sigma : \{1,\ldots,n_1\} \times \{1,\ldots,n_2\} \to \{0,1\}^k \text{ with } \sigma((i,j)) = M(i,j)$$

is an embedding of the two-dimensional mesh into the k-dimensional cube. Figure 2.12 (b) shows an example.

2.7.3.3 Embedding of a d-dimensional mesh into a hypercube network

In a d-dimensional mesh with $n_i = 2^{k_i}$ nodes in dimension i, $1 \le i \le d$, there are $n = n_1 \cdots n_d$ nodes in total. Each node can be represented by its mesh coordinates $(x_1,\ldots,x_d)$ with $1 \le x_i \le n_i$. The mapping

$$\sigma : \{(x_1,\ldots,x_d) \mid 1 \le x_i \le n_i, 1 \le i \le d\} \longrightarrow \{0,1\}^k$$

with $\sigma((x_1,\ldots,x_d)) = s_1 s_2 \ldots s_d$ and $s_i = RGC_{k_i}(x_i)$

(where s_i is the x_ith bit string in the Gray code sequence RGC_{k_i}) defines an embedding into the k-dimensional cube. For two mesh nodes $(x_1,\ldots,x_d)$ and $(y_1,\ldots,y_d)$ that are connected by an edge in the d-dimensional mesh, there exists exactly one dimension $i \in \{1,\ldots,d\}$ with $|x_i - y_i| = 1$ and for all other dimensions $j \ne i$, it is $x_j = y_j$. Thus, for the corresponding hypercube nodes $\sigma((x_1,\ldots,x_d)) = s_1 s_2 \ldots s_d$ and $\sigma((y_1,\ldots,y_d)) = t_1 t_2 \ldots t_d$, all components $s_j = RGC_{k_j}(x_j) = RGC_{k_j}(y_j) = t_j$ for $j \ne i$ are identical. Moreover, $RGC_{k_i}(x_i)$ and $RGC_{k_i}(y_i)$ differ in exactly one bit position. Thus, the hypercube nodes $s_1 s_2 \ldots s_d$ and $t_1 t_2 \ldots t_d$ also differ in exactly one bit position and are therefore connected by an edge in the hypercube network.

2.7.4 Dynamic Interconnection Networks

Dynamic interconnection networks are also called indirect interconnection networks. In these networks, nodes or processors are not connected directly with each other. Instead, switches are used and provide an *indirect* connection between the nodes, giving these networks their name. From the processors' point of view, such a network forms an interconnection unit into which data can be sent and from which data can be received. Internally, a dynamic network consists of switches that are connected by physical links. For a message transmission from one node to another node, the switches can be configured *dynamically* such that a connection is estab-

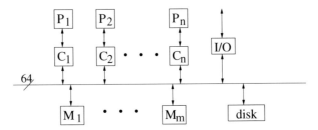

Fig. 2.13 Illustration of a bus network with 64 wires to connect processors $P_1,\ldots,P_n$ with caches $C_1,\ldots,C_n$ to memory modules $M_1,\ldots,M_m$.

lished. Dynamic interconnection networks can be characterized according to their topological structure. Popular forms are bus networks, multistage networks, and crossbar networks.

2.7.4.1 Bus networks

A bus essentially consists of a set of wires which can be used to transport data from a sender to a receiver, see Fig. 2.13 for an illustration. In some cases, several hundreds of wires are used to ensure a fast transport of large data sets. At each point in time, only one data transport can be performed via the bus, i.e., the bus must be used in a time-sharing way. When several processors attempt to use the bus simultaneously, a **bus arbiter** is used for the coordination. Because the likelihood for simultaneous requests of processors increases with the number of processors, bus networks are typically used for a small number of processors only.

2.7.4.2 Crossbar networks

An $n \times m$ crossbar network has n inputs and m outputs. The actual network consists of $n \cdot m$ switches as illustrated in Fig. 2.14 (left). For a system with a shared address space, the input nodes may be processors and the outputs may be memory modules. For a system with a distributed address space, both the input nodes and the output nodes may be processors. For each request from a specific input to a specific output, a connection in the switching network is established. Depending on the specific input and output nodes, the switches on the connection path can have different states (straight or direction change) as illustrated in Fig. 2.14 (right). Typically, crossbar networks are used only for a small number of processors because of the large hardware overhead required.

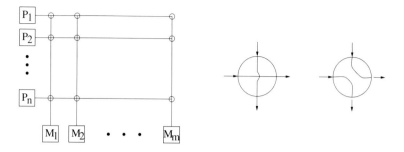

Fig. 2.14 Illustration of a $n \times m$ crossbar network for n processors and m memory modules (left). Each network switch can be in one of two states: straight or direction change (right).

2.7.4.3 Multistage switching networks

Multistage switching networks consist of several stages of switches with connecting wires between neighboring stages. The network is used to connect input devices to output devices. Input devices are typically the processors of a parallel system. Output devices can be processors (for distributed memory machines) or memory modules (for shared memory machines). The goal is to obtain a small distance for arbitrary pairs of input and output devices to ensure fast communication. The internal connections between the stages can be represented as a graph where switches are represented by nodes and wires between switches are represented by edges. Input and output devices can be represented as specialized nodes with edges going into the actual switching network graph. The construction of the switching graph and the degree of the switches used are important characteristics of multistage switching networks.

Regular multistage interconnection networks are characterized by a *regular* construction method using the same degree of incoming and outgoing wires for all switches. For the switches, $a \times b$ crossbars are often used where a is the input degree and b is the output degree. The switches are arranged in stages such that neighboring stages are connected by fixed interconnections, see Fig. 2.15 for an illustration. The input wires of the switches of the first stage are connected with the input devices. The output wires of the switches of the last stage are connected with the output devices. Connections from input devices to output devices are performed by selecting a path from a specific input device to the selected output device and setting the switches on the path such that the connection is established.

The actual graph representing a regular multistage interconnection network results from *gluing* neighboring stages of switches together. The connection between neighboring stages can be described by a directed acyclic graph of depth 1. Using w nodes for each stage, the degree of each node is $g = n/w$ where n is the number of edges between neighboring stages. The connection between neighboring stages can be represented by a permutation $\pi : \{1, \ldots, n\} \to \{1, \ldots, n\}$ which specifies which output link of one stage is connected to which input link of the next stage. This

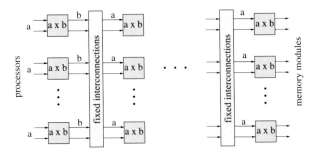

Fig. 2.15 Multistage interconnection networks with $a \times b$ crossbars as switches according to [116].

means that the output links $\{1,\ldots,n\}$ of one stage are connected to the input links $(\pi(1),\ldots,\pi(n))$ of the next stage. Partitioning the permutation $(\pi(1),\ldots,\pi(n))$ into w parts results in the ordered set of input links of nodes of the next stage. For regular multistage interconnection networks, the same permutation is used for all stages, and the stage number can be used as parameter.

Popular regular multistage networks are the omega network, the baseline network, and the butterfly network. These networks use 2×2 crossbar switches which are arranged in $\log n$ stages. Each switch can be in one of four states as illustrated in Fig. 2.16. In the following, we give a short overview of the omega, baseline, butterfly, Beneš, and fat-tree networks, see [144] for a detailed description.

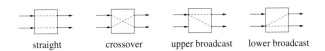

Fig. 2.16 Settings for switches in a omega, baseline, or butterfly network.

2.7.4.4 Omega network

An $n \times n$ omega network is based on 2×2 crossbar switches which are arranged in $\log n$ stages such that each stage contains $n/2$ switches where each switch has two input links and two output links. Thus, there are $(n/2) \cdot \log n$ switches in total, with $\log n \equiv \log_2 n$. Each switch can be in one of four states, see Fig. 2.16. In the omega network, the permutation function describing the connection between neighboring stages is the same for all stages, independently from the number of the stage. The switches in the network are represented by pairs (α, i) where $\alpha \in \{0,1\}^{\log n-1}$ is a bit string of length $\log n - 1$ representing the position of a switch within a stage and $i \in \{0,\ldots,\log n - 1\}$ is the stage number. There is an edge from node (α, i) in stage i to two nodes $(\beta, i+1)$ in stage $i+1$ where β is defined as follows:

1. β results from α by a cyclic left shift, or
2. β results from α by a cyclic left shift followed by an inversion of the last (right-most) bit.

An $n \times n$ omega network is also called $(\log n - 1)$-dimensional omega network. Figure 2.17 (a) shows a 16×16 (three-dimensional) omega network with four stages and eight switches per stage.

2.7.4.5 Butterfly network

Similar to the omega network, a k-dimensional butterfly network connects $n = 2^{k+1}$ inputs to $n = 2^{k+1}$ outputs using a network of 2×2 crossbar switches. Again, the switches are arranged in $k+1$ stages with 2^k nodes/switches per stage. This results in a total number $(k+1) \cdot 2^k$ of nodes. Again, the nodes are represented by pairs (α, i) where i for $0 \le i \le k$ denotes the stage number and $\alpha \in \{0, 1\}^k$ is the position of the node in the stage. The connection between neighboring stages i and $i+1$ for $0 \le i < k$ is defined as follows: Two nodes (α, i) and $(\alpha', i+1)$ are connected if and only if

1. α and α' are identical (straight edge), or
2. α and α' differ in precisely the $(i+1)$th bit from the left (cross edge).

Fig. 2.17 (b) shows a 16×16 butterfly network with four stages.

2.7.4.6 Baseline network

The k-dimensional baseline network has the same number of nodes, edges and stages as the butterfly network. Neighboring stages are connected as follows: node (α, i) is connected to node $(\alpha', i+1)$ for $0 \le i < k$ if and only if

1. α' results from α by a cyclic right shift on the last $k - i$ bits of α, or
2. α' results from α by first inverting the last (rightmost) bit of α and then performing a cyclic right shift on the last $k - i$ bits.

Fig. 2.17 (c) shows a 16×16 baseline network with four stages.

2.7.4.7 Beneš network

The k-dimensional Beneš network is constructed from two k-dimensional butterfly networks such that the first $k+1$ stages are a butterfly network and the last $k+1$ stages are a reversed butterfly network. The last stage $(k+1)$ of the first butterfly network and the first stage of the second (reverted) butterfly network are merged. In total, the k-dimensional Beneš network has $2k+1$ stages with 2^k switches in each stage. Fig. 2.18 (a) shows a three-dimensional Beneš network as an example.

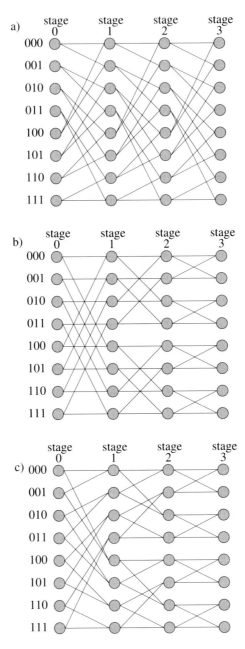

Fig. 2.17 Examples for dynamic interconnection networks: a) 16×16 omega network, b) $16 \times$ 16 butterfly network, c) 16×16 baseline network. All networks are 3-dimensional.

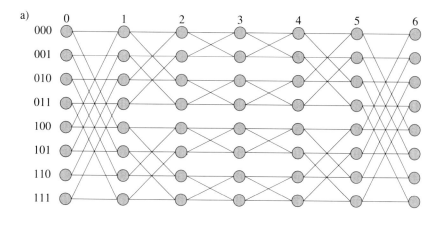

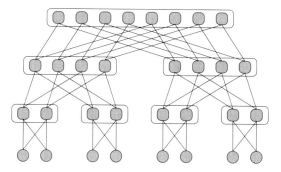

Fig. 2.18 Examples for dynamic interconnection networks: a) 3-dimensional Benes network and b) fat tree network for 16 processors.

2.7.4.8 Fat tree network

The basic structure of a *dynamic tree* or *fat tree* network is a complete binary tree. The difference to a normal tree is that the number of connections between the nodes increases towards the root to avoid bottlenecks. Inner tree nodes consist of switches whose structure depends on their position in the tree structure. The leaf level is level 0. For n processors, represented by the leaves of the tree, a switch on tree level i has 2^i input links and 2^i output links for $i = 1, \ldots, \log n$. This can be realized by assembling the switches on level i internally from 2^{i-1} switches with two input and two output links each. Thus, each level i consists of $n/2$ switches in total, grouped in $2^{\log n - i}$ nodes. This is shown in Fig. 2.18 (b) for a fat tree with four layers. Only the inner switching nodes are shown, not the leaf nodes representing the processors.

2.8 Routing and Switching

Direct and indirect interconnection networks provide the physical basis to send messages between processors. If two processors are not directly connected by a network link, a path in the network consisting of a sequence of nodes has to be used for message transmission. In the following, we give a short description on how to select a suitable path in the network (routing) and how messages are handled at intermediate nodes on the path (switching).

2.8.1 Routing Algorithms

A **routing algorithm** determines a path in a given network from a source node A to a destination node B. The path consists of a sequence of nodes such that neighboring nodes in the sequence are connected by a physical network link. The path starts with node A and ends at node B. A large variety of routing algorithms have been proposed in the literature, and we can only give a short overview in the following. For a more detailed description and discussion, we refer to [45, 56].

Typically, multiple message transmissions are being executed concurrently according to the requirements of one or several parallel programs. A routing algorithm tries to reach an even load on the physical network links as well as to avoid the occurrence of deadlocks. A set of messages is in a **deadlock situation** if each of the messages is supposed to be transmitted over a link that is currently used by another message of the set. A routing algorithm tries to select a path in the network connecting nodes A and B such that minimum costs result, thus leading to a fast message transmission between A and B. The resulting communication costs depend not only on the length of the path used, but also on the load of the links on the path. The following issues are important for the path selection:

- **network topology**: the topology of the network determines which paths are available in the network to establish a connection between nodes A and B;
- **network contention**: contention occurs when two or more messages should be transmitted at the same time over the same network link, thus leading to a delay in message transmission;
- **network congestion**: congestion occurs when too many messages are assigned to a restricted resource (like a network link or buffer) such that arriving messages have to be discarded since they cannot be stored anywhere. Thus, in contrast to contention, congestion leads to an overflow situation with message loss [173].

A large variety of routing algorithms have been proposed in the literature. Several classification schemes can be used for a characterization. Using the path length, **minimal** and **non-minimal** routing algorithms can be distinguished. Minimal routing algorithms always select the shortest message transmission, which means that when using a link of the path selected, a message always gets closer to the target node. But this may lead to congestion situations. Non-minimal routing algorithms do not always use paths with minimum length if this is necessary to avoid congestion at intermediate nodes.

A further classification can be made by distinguishing **deterministic** routing algorithms and **adaptive** routing algorithms. A routing algorithm is deterministic if the path selected for message transmission depends only on the source and destination node regardless of other transmissions in the network. Therefore, deterministic routing can lead to unbalanced network load. Path selection can be done *source oriented* at the sending node or *distributed* during message transmission at intermediate nodes. An example for deterministic routing is **dimension-order routing** which can be applied for network topologies that can be partitioned into several orthogonal dimensions as is the case for meshes, tori, and hypercube topologies. Using dimension-order routing, the routing path is determined based on the position of the source node and the target node by considering the dimensions in a fixed order and traversing a link in the dimension if necessary. This can lead to network contention because of the deterministic path selection.

Adaptive routing tries to avoid such contentions by dynamically selecting the routing path based on load information. Between any pair of nodes, multiple paths are available. The path to be used is dynamically selected such that network traffic is spread evenly over the available links, thus leading to an improvement of network utilization. Moreover, *fault tolerance* is provided, since an alternative path can be used in case of a link failure. Adaptive routing algorithms can be further categorized into minimal and non-minimal adaptive algorithms as described above. In the following, we give a short overview of important routing algorithms. For a more detailed treatment, we refer to [45, 116, 56, 144, 159].

2.8.1.1 Dimension-order routing

We give a short description of XY routing for two-dimensional meshes and E-cube routing for hypercubes as typical examples for dimension-order routing algorithms.

XY routing for two-dimensional meshes:

For a two-dimensional mesh, the position of the nodes can be described by an X-coordinate and an Y-coordinate where X corresponds to the horizontal and Y corresponds to the vertical direction. To send a message from a source node A with position (X_A, Y_A) to target node B with position (X_B, Y_B), the message is sent from the source node into (positive or negative) X-direction until the X-coordinate X_B of B is reached. Then, the message is sent into Y-direction until Y_B is reached. The length of the resulting path is $| X_A - X_B | + | Y_A - Y_B |$. This routing algorithm is deterministic and minimal.

E-cube routing for hypercubes:

In a k-dimensional hypercube, each of the $n = 2^k$ nodes has a direct interconnection link to each of its k neighbors. As introduced in Section 2.7.2, each of the nodes can be represented by a bit string of length k such that the bit string of one of the k neighbors is obtained by inverting one of the bits in the bit string. E-cube uses the bit representation of a sending node A and a receiving node B to select a routing path between them. Let $\alpha = \alpha_0 \ldots \alpha_{k-1}$ be the bit representation of A and $\beta = \beta_0 \ldots \beta_{k-1}$ be the bit representation of B. Starting with A, in each step a dimension is selected which determines the next node on the routing path. Let A_i with bit representation $\gamma = \gamma_0 \ldots \gamma_{k-1}$ be a node on the routing path $A = A_0, A_1, \ldots, A_l = B$ from which the message should be forwarded in the next step. For the forwarding from A_i to A_{i+1}, the following two substeps are made:

- The bit string $\gamma \oplus \beta$ is computed where $\oplus$ denotes the bitwise exclusive or computation (i.e., $0 \oplus 0 = 0, 0 \oplus 1 = 1, 1 \oplus 0 = 1, 1 \oplus 1 = 0$)
- The message is forwarded in dimension d where d is the rightmost bit position of $\gamma \oplus \beta$ with value 1. The next node A_{i+1} on the routing path is obtained by inverting the dth bit in γ, i.e., the bit representation of A_{i+1} is $\delta = \delta_0 \ldots \delta_{k-1}$ with $\delta_j = \gamma_j$ for $j \neq d$ and $\delta_d = \bar{\gamma}_d$. The target node B is reached when $\gamma \oplus \beta = 0$.

Example: For $k = 3$, let A with bit representation $\alpha = 010$ be the source node and B with bit representation $\beta = 111$ be the target node. First, the message is sent from A into direction $d = 2$ to A_1 with bit representation 011 (since $\alpha \oplus \beta = 101$). Then, the message is sent in dimension $d = 0$ to β since ($011 \oplus 111 = 100$). □

2.8.1.2 Deadlocks and routing algorithms

Usually, multiple messages are in transmission concurrently. A deadlock occurs if the transmission of a subset of the messages is blocked forever. This can happen in particular if network resources can be used only by one message at a time. If, for example, the links between two nodes can be used by only one message at a time and if a link can only be released when the following link on the path is free, then the mutual request for links can lead to a deadlock. Such deadlock situations can be avoided by using a suitable routing algorithm. Other deadlock situations that occur because of limited size of the input or output buffer of the interconnection links or because of an unsuited order of the send and receive operations are considered in Section 2.8.3 on switching strategies and Chapter 5 on message-passing programming.

To prove the deadlock freedom of routing algorithms, possible dependencies between interconnection channels are considered. A dependence from an interconnection channel l_1 to an interconnection channel l_2 exists, if it is possible that the routing algorithm selects a path which contains channel l_2 directly after channel l_1. These dependencies between interconnection channels can be represented by a **channel**

dependence graph which contains the interconnection channels as nodes; each dependence between two channels is represented by an edge. A routing algorithm is deadlock free for a given topology, if the channel dependence graph does not contain cycles. In this case, no communication pattern can ever lead to a deadlock.

For topologies that do not contain cycles, no channel dependence graph can contain cycles, and therefore each routing algorithm for such a topology must be deadlock free. For topologies with cycles, the channel dependence graph must be analyzed. In the following, we show that XY routing for 2D meshes with bidirectional links is deadlock free.

Deadlock freedom of XY routing:

The channel dependence graph for XY routing contains a node for each unidirectional link of the two-dimensional $n_x \times n_Y$ mesh, i.e., there are two nodes for each bidirectional link of the mesh. There is a dependence from link u to link v, if v can be directly reached from u in horizontal or vertical direction or by a 90° turn down or up. To show the deadlock freedom, all uni-directional links of the mesh are numbered as follows:

- Each horizontal edge from node (i, y) to node $(i + 1, y)$ gets number $i + 1$ for $i = 0, \ldots, n_x - 2$ for each valid value of y. The opposite edge from $(i + 1, y)$ to (i, y) gets number $n_x - 1 - (i + 1) = n_x - i - 2$ for $i = 0, \ldots, n_x - 2$. Thus, the edges in increasing x-direction are numbered from 1 to $n_x - 1$, the edges in decreasing x-direction are numbered from 0 to $n_x - 2$.
- Each vertical edge from (x, j) to $(x, j + 1)$ gets number $j + n_x$ for $j = 0, \ldots, n_y - 2$. The opposite edge from $(x, j + 1)$ to (x, j) gets number $n_x + n_y - (j + 1)$.

Figure 2.19 shows a 3×3 mesh and the resulting channel dependence graph for XY routing. The nodes of the graph are annotated with the numbers assigned to the corresponding network links. It can be seen that all edges in the channel dependence graph go from a link with a smaller number to a link with a larger number. Thus, a delay during message transmission along a routing path can occur only if the message has to wait after the transmission along a link with number i for the release of a successive link w with number $j > i$ currently used by another message transmission (delay condition). A deadlock can only occur if a set of messages $\{N_1, \ldots, N_k\}$ and network links $\{n_1, \ldots, n_k\}$ exists such that for $1 \leq i < k$ each message N_i uses a link n_i for transmission and waits for the release of link n_{i+1} which is currently used for the transmission of message N_{i+1}. Additionally, N_k is currently transmitted using link n_k and waits for the release of n_1 used by N_1. If $n()$ denotes the numbering of the network links introduced above, the delay condition implies that for the deadlock situation just described, it must be

$$n(n_1) < n(n_2) < \ldots < n(n_k) < n(n_1).$$

This is a contradiction, and thus it follows that no deadlock can occur. Each routing path selected by XY routing consists of a sequence of links with increasing numbers.

2D mesh with 3 x 3 nodes   channel dependence graph

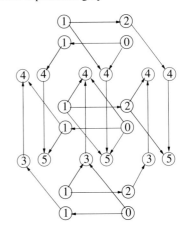

Fig. 2.19 3×3 mesh and corresponding channel dependence graph for XY routing.

Each edge in the channel dependence graph points to a link with a larger number than the source link. Thus, there can be no cycles in the channel dependence graph. A similar approach can be used to show deadlock freedom for E-cube routing [48].

2.8.1.3 Source-based routing

Source-based routing is a deterministic routing algorithm for which the source node determines the entire path for message transmission. For each node n_i on the path, the output link number a_i is determined, and the sequence of output link numbers $a_0, \ldots, a_{n-1}$ to be used is added as header to the message. When the message passes a node, the first link number is stripped from the front of the header and the message is forwarded through the specified link to the next node.

2.8.1.4 Table-driven routing

For table-driven routing, each node contains a routing table which contains for each destination node the output link to be used for the transmission. When a message arrives at a node, a lookup in the routing table is used to determine how the message is forwarded to the next node.

2.8.1.5 Turn model routing

The turn model [82, 159] tries to avoid deadlocks by a suitable selection of turns that are allowed for the routing. Deadlocks occur if the paths for message transmission

possible turns in a 2D mesh

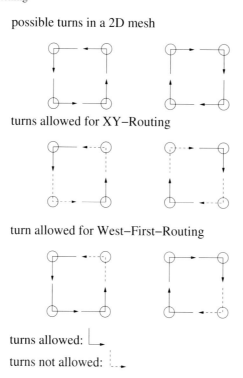

turns allowed for XY–Routing

turn allowed for West–First–Routing

turns allowed: ⌊→

turns not allowed: ⌊→

Fig. 2.20 Illustration of turns for a 2D mesh with all possible turns (top), allowed turns for *XY* routing (middle), and allowed turns for west-first routing (bottom).

contain turns that may lead to cyclic waiting in some situations. Deadlocks can be avoided by prohibiting some of the turns. An example is the *XY* routing on a 2D mesh. From the eight possible turns, see Fig. 2.20 (above), only four are allowed for *XY* routing, prohibiting turns from vertical into horizontal direction, see Fig. 2.20 (middle) for an illustration. The remaining four turns are not allowed in order to prevent cycles in the networks. This not only avoids the occurrence of deadlocks, but also prevents the use of adaptive routing. For *n*-dimensional meshes and, in the general case, *k*-ary *d*-cubes, the turn model tries to identify a minimum number of turns that must be prohibited for routing paths to avoid the occurrence of cycles. Examples are the west-first routing for 2D meshes and the *P*-cube routing for *n*-dimensional hypercubes.

The **west-first routing** algorithm for a two-dimensional mesh prohibits only two of the eight possible turns: turns to the west (left) are prohibited, and only the turns shown in Fig. 2.20 (bottom) are allowed. Routing paths are selected such that messages that must travel to the west must do so before making any turns. Such messages are sent to the west first until the requested *x*-coordinate is reached. Then the message can be adaptively forwarded to the south (bottom), east (right), or north (top). Figure 2.21 shows some examples for possible routing paths [159]. West-first routing is deadlock free, since cycles are avoided. For the selection of minimal routing paths, the algorithm is adaptive only if the target node lies to the east (right). Using non-minimal routing paths, the algorithm is always adaptive.

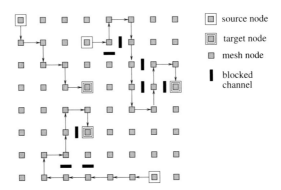

Fig. 2.21 Illustration of path selection for west-first routing in an 8×8 mesh. The links shown as blocked are used for other message transmissions and are not available for the current transmission. One of the paths shown is minimal, the other two are non-minimal, since some of the links are blocked.

Routing in the n-dimensional hypercube can be done with **P-cube routing**. To send a message from a sender A with bit representation $\alpha = \alpha_0 \ldots \alpha_{n-1}$ to a receiver B with bit representation $\beta = \beta_0 \ldots \beta_{n-1}$, the bit positions in which α and β differ are considered. The number of these bit positions is the Hamming distance between A and B which determines the minimum length of a routing path from A to B. The set $E = \{i \mid \alpha_i \neq \beta_i, i = 0, \ldots, n-1\}$ of different bit positions is partitioned into two sets $E_0 = \{i \in E \mid \alpha_i = 0 \text{ and } \beta_i = 1\}$ and $E_1 = \{i \in E \mid \alpha_i = 1 \text{ and } \beta_i = 0\}$. Message transmission from A to B is split into two phases accordingly: First, the message is sent into the dimensions in E_0 and then into the dimensions in E_1.

2.8.1.6 Virtual channels

The concept of *virtual channels* is often used for minimal adaptive routing algorithms. To provide multiple (virtual) channels between neighboring network nodes, each physical link is split into multiple virtual channels. Each virtual channel has its own separate buffer. The provision of virtual channels does not increase the number of physical links in the network, but can be used for a systematic avoidance of deadlocks.

Based on virtual channels, a network can be split into several virtual networks such that messages injected into a virtual network can only move into one direction for each dimension. This can be illustrated for a two-dimensional mesh which is split into two virtual networks, a $+X$ network and a $-X$ network, see Fig. 2.22 for an illustration. Each virtual network contains all nodes, but only a subset of the virtual channels. The $+X$ virtual network contains in the vertical direction all virtual channels between neighboring nodes, but in the horizontal direction only the virtual channels in positive direction. Similarly, the $-X$ virtual network contains in the horizontal direction only the virtual channels in negative direction, but all virtual channels in the vertical direction. The latter is possible by the definition of a suitable number of virtual channels in the vertical direction. Messages from a node A with x-coordinate x_A to a node B with x-coordinate x_B are sent in the $+X$ network, if $x_A < x_B$. Messages from A to B with $x_A > x_B$ are sent in the $-X$ network. For $x_A = x_B$, one of the two networks can be selected arbitrarily, possibly using load information for the selection. The resulting adaptive routing algorithm is deadlock free [159]. For other topologies like hypercubes or tori, more virtual channels might be needed to provide deadlock freedom [159].

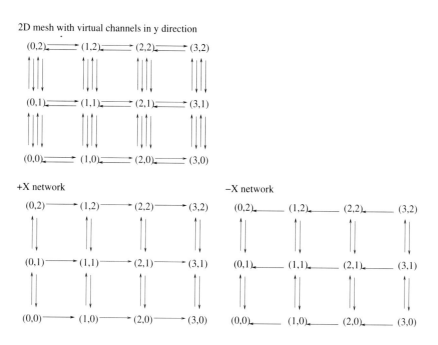

Fig. 2.22 Partitioning of a 2D mesh with virtual channels into a $+X$ network and a $-X$ network for applying a minimal adaptive routing algorithm.

A non-minimal adaptive routing algorithm can send messages over longer paths if no minimal path is available. **Dimension reversal routing** can be applied to arbitrary meshes and k-ary d-cubes. The algorithm uses r pairs of virtual channels between any pair of nodes that is connected by a physical link. Correspondingly, the network is split into r virtual networks where network i for $i = 0, \ldots, r-1$ uses all virtual channels i between the nodes. Each message to be transmitted is assigned a class c with initialization $c = 0$ which can be increased to $c = 1, \ldots, r-1$ during message transmission. A message with class $c = i$ can be forwarded in network i

in each dimension, but the dimensions must be traversed in increasing order. If a message must be transmitted in opposite order, its class is increased by 1 (reverse dimension order). The parameter r controls the number of dimension reversals that are allowed. If $c = r$ is reached, the message is forwarded according to dimension-ordered routing.

2.8.2 Routing in the Omega Network

The omega network introduced in Section 2.7.4 allows message forwarding using a distributed algorithm where each switch can forward the message without coordination with other switches. For the description of the algorithm, it is useful to represent each of the n input channels and output channels by a bit string of length $\log n$ [144]. To forward a message from an input channel with bit representation α to an output channel with bit representation β the receiving switch on stage k of the network, $k = 0, \ldots, \log n - 1$, considers the kth bit β_k (from the left) of β and selects the output link for forwarding the message according to the following rule:

- for $\beta_k = 0$, the message is forwarded over the upper link of the switch and
- for $\beta_k = 1$, the message is forwarded over the lower link of the switch.

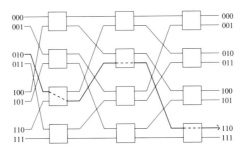

Fig. 2.23 8×8 omega network with path from 010 to 110 [14].

Figure 2.23 illustrates the path selected for message transmission from input channel $\alpha = 010$ to the output channel $\beta = 110$ according to the algorithm just described. In an $n \times n$ omega network, at most n messages from different input channels to different output channels can be sent concurrently without collision. An example of a concurrent transmission of $n = 8$ messages in a 8×8 omega network can be described by the permutation

$$\pi^8 = \begin{pmatrix} 0\ 1\ 2\ 3\ 4\ 5\ 6\ 7 \\ 7\ 3\ 0\ 1\ 2\ 5\ 4\ 6 \end{pmatrix}$$

which specifies that the messages are sent from input channel i ($i = 0,\ldots,7$) to output channel $\pi^8(i)$. The corresponding paths and switch positions for the eight paths are shown in Fig. 2.24.

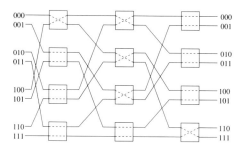

Fig. 2.24 8×8 Omega network with switch positions for the realization of π^8 from the text.

Many simultaneous message transmissions that can be described by permutations $\pi^8 : \{0,\ldots,n-1\} \to \{0,\ldots,n-1\}$ cannot be executed concurrently since **network conflicts** would occur. For example, the two message transmissions from $\alpha_1 = 010$ to $\beta_1 = 110$ and from $\alpha_2 = 000$ to $\beta_2 = 111$ in an 8×8 omega network would lead to a conflict. This type of conflict occurs, since there is exactly one path for any pair (α, β) of input and output channels, i.e., there is no alternative to avoid a critical switch. Networks with this characteristic are also called **blocking networks**. Conflicts in blocking networks can be resolved by multiple transmissions through the network.

There is a notable number of permutations that cannot be implemented in one switching of the network. This can be seen as follows. For the connection from the n input channels to the n output channels, there are in total $n!$ possible permutations, since each output channel must be connected to exactly one input channel. There are in total $n/2 \cdot \log n$ switches in the omega network, each of which can be in one of two positions. This leads to $2^{n/2 \cdot \log n} = n^{n/2}$ different switchings of the entire network, corresponding to n concurrent paths through the network. In conclusion, only $n^{n/2}$ of the $n!$ possible permutations can be performed without conflicts.

Other examples for blocking networks are the butterfly or Banyan network, the baseline network and the delta network [144]. In contrast, the Beneš network is a non-blocking network since there are different paths from an input channel to an output channel. For each permutation $\pi : \{0,\ldots,n-1\} \to \{0,\ldots,n-1\}$ there exists a switching of the Beneš network which realizes the connection from input i to output $\pi(i)$ for $i = 0,\ldots,n-1$ concurrently without collision, see [144] for more details. As example, the switching for the permutation

$$\pi^8 = \begin{pmatrix} 0\ 1\ 2\ 3\ 4\ 5\ 6\ 7 \\ 5\ 3\ 4\ 7\ 0\ 1\ 2\ 6 \end{pmatrix}$$

is shown in Fig. 2.25.

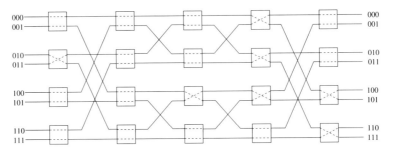

Fig. 2.25 8×8 Benes network with switch positions for the realization of π^8 from the text.

2.8.3 Switching

The switching strategy determines how a message is transmitted along a path that has been selected by the routing algorithm. In particular, the switching strategy determines

- whether and how a message is split into pieces, which are called packets or *flits* (for *flow control units*),
- how the transmission path from the source node to the destination node is allocated, and
- how messages or pieces of messages are forwarded from the input channel to the output channel of a switch or a router. The routing algorithm only determines *which* output channel should be used.

The switching strategy may have a large influence on the message transmission time from a source to a destination. Before considering specific switching strategies, we first consider the time for message transmission between two nodes that are directly connected by a physical link.

2.8.3.1 Message transmission between neighboring processors

Message transmission between two directly connected processors is implemented as a series of steps. These steps are also called *protocol*. In the following, we sketch a simple example protocol [103]. To send a message, the sending processor performs the following steps:

1. The message is copied into a system buffer.
2. A checksum is computed and a *header* is added to the message, containing the checksum as well as additional information related to the message transmission.
3. A timer is started and the message is sent out over the network interface.

To receive a message, the receiving processor performs the following steps:

1. The message is copied from the network interface into a system buffer.

2. The checksum is computed over the data contained. This checksum is compared with the checksum stored in the header. If both checksums are identical, an acknowledgment message is sent to the sender. In case of a mismatch of the checksums, the message is discarded. The message will be re-sent again after the sender timer has elapsed.
3. If the checksums are identical, the message is copied from the system buffer into the user buffer, provided by the application program. The application program gets a notification and can continue execution.

After having sent out the message, the sending processor performs the following steps:

1. If an acknowledgment message arrives for the message sent out, the system buffer containing a copy of the message can be released.
2. If the timer has elapsed, the message will be re-sent again. The timer is started again, possibly with a longer time.

In this protocol, it has been assumed that the message is kept in the system buffer of the sender to be re-sent if necessary. If message loss is tolerated, no re-send is necessary and the system buffer of the sender can be re-used as soon as the packet has been sent out. Message transmission protocols used in practice are typically much more complicated and may take additional aspects like network contention or possible overflows of the system buffer of the receiver into consideration. A detailed overview can be found in [139, 174].

The time for a message transmission consists of the actual transmission time over the physical link and the time needed for the software overhead of the protocol, both at the sender and the receiver side. Before considering the transmission time in more detail, we first review some performance measures that are often used in this context, see [104, 45] for more details.

- The **bandwidth** of a network link is defined as the maximum frequency at which data can be sent over the link. The bandwidth is measured in bits per second or bytes per second.
- The **byte transfer time** is the time which is required to transmit a single byte over a network link. If the bandwidth is measured in bytes per second, the byte transfer time is the reciprocal of the bandwidth.
- The **time of flight**, also referred to as *channel propagation delay*, is the time which the first bit of a message needs to arrive at the receiver. This time mainly depends on the physical distance between the sender and the receiver.
- The **transmission time** is the time needed to transmit the message over a network link. The transmission time is the message size in bytes divided by the bandwidth of the network link, measured in bytes per second. The transmission time does not take conflicts with other messages into consideration.
- The **transport latency** is the total time that is needed to transfer a message over a network link. This is the sum of the transmission time and the time of flight, capturing the entire time interval from putting the first bit of the message onto the network link at the sender and receiving the last bit at the receiver.

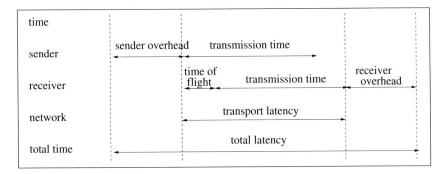

Fig. 2.26 Illustration of performance measures for the point-to-point transfer between neighboring nodes, see [103].

- The **sender overhead**, also referred to as *startup time*, is the time that the sender needs for the preparation of message transmission. This includes the time for computing the checksum, appending the header, and executing the routing algorithm.
- The **receiver overhead** is the time that the receiver needs to process an incoming message, including checksum comparison and generation of an acknowledgment if required by the specific protocol.
- The **throughput** of a network link is the effective bandwidth experienced by an application program.

Using these performance measures, the total latency $T(m)$ of a message of size m can be expressed as

$$T(m) = O_{send} + T_{delay} + m/B + O_{recv} \qquad (2.3)$$

where O_{send} and O_{recv} are the sender and receiver overhead, respectively, T_{delay} is the time of flight, and B is the bandwidth of the network link. This expression does not take into consideration that a message may need to be transmitted multiple times because of checksum errors, network contention, or congestion.

The performance parameters introduced are illustrated in Fig. 2.26. Equation (2.3) can be reformulated by combining constant terms, yielding

$$T(m) = T_{overhead} + m/B \qquad (2.4)$$

with $T_{overhead} = T_{send} + T_{recv}$. Thus, the latency consists of an overhead which does not depend on the message size and a term which linearly increases with the message size. Using the byte transfer time $t_B = 1/B$, Eq. (2.4) can also be expressed as

$$T(m) = T_{overhead} + t_B \cdot m \qquad (2.5)$$

This equation is often used to describe the message transmission time over a network link. When transmitting a message between two nodes that are not directly connected in the network, the message must be transmitted along a path between the two nodes. For the transmission along the path, several switching techniques can be used, including circuit switching, packet switching with store-and-forward routing, virtual cut-through routing, and wormhole routing. We give a short overview in the following.

2.8.3.2 Circuit switching

The two basic switching strategies are circuit switching and packet switching, see [45, 103] for a detailed treatment. In **circuit switching**, the entire path from the source node to the destination node is established and reserved until the end of the transmission of this message. This means that the path is established exclusively for this message by setting the switches or routers on the path in a suitable way. Internally, the message can be split into pieces for the transmission. These pieces can be so-called *physical units* (*phits*) denoting the amount of data that can be transmitted over a network link in one cycle. The size of the phits is determined by the number of bits that can be transmitted over a physical channel in parallel. Typical phit sizes lie between 1 bit and 256 bits. The transmission path for a message can be established by using short *probe messages* along the path. After the path is established, all phits of the message are transmitted over this path. The path can be released again by a message trailer or by an acknowledgment message from the receiver of the sender.

Sending a control message along a path of length l takes time $l \cdot t_c$ where t_c is the time to transmit the control message over a single network link. If m_c is the size of the control message, it is $t_c = t_B \cdot m_c$. After the path has been established, the transmission of the actual message of size m takes time $m \cdot t_B$. Thus, the total time of message transmission along a path of length l with circuit switching is

$$T_{cs}(m,l) = T_{overhead} + t_c \cdot l + t_B \cdot m \tag{2.6}$$

If m_c is small compared to m, this can be reduced to $T_{overhead} + t_B \cdot m$ which is linear in m, but independent of l. Message transfer with circuit switching is illustrated in Fig. 2.28 a).

2.8.3.3 Packet switching

For **packet switching** the message to be transmitted is partitioned into a sequence of packets which are transferred independently from each other through the network from the sender to the receiver. Using an adaptive routing algorithm, the packets can be transmitted over different paths. Each packet consists of three parts: (i) a header, containing routing and control information, (ii) the data part, containing a part of the original message, and (iii) a trailer which may contain an error control code. Each packet is sent separately to the destination according to the routing information con-

tained in the packet. Fig. 2.27 illustrates the partitioning of a message into packets.
The network links and buffers are used by one packet at a time.

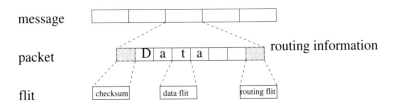

Fig. 2.27 Illustration of the partitioning of a message into packets and of packets into *flits*
(*flow control units*).

Packet switching can be implemented in several ways. Packet switching with
store-and-forward routing sends a packet along a path such that the entire packet
is received by each switch on the path (*store*) before it is sent to the next switch
on the path (*forward*). The connection between two switches A and B on the path
is released for reuse by another packet as soon as the packet has been stored at B.
This strategy is useful if the links connecting the switches on a path have different
bandwidth as this is typically the case in *wide area networks* (WANs). In this case,
store-and-forward routing allows the utilization of the full bandwidth for every link
on the path. Another advantage is that a link on the path can be quickly released
as soon as the packet has passed the links, thus reducing the danger of deadlocks.
The drawback of this strategy is that the packet transmission time increases with the
number of switches that must be traversed from source to destination. Moreover, the
entire packet must be stored at each switch on the path, thus increasing the memory
demands of the switches.

The time for sending a packet of size m over a single link takes time $t_h + t_B \cdot m$
where t_h is the constant time needed at each switch to store the packet in a receive
buffer and to select the output channel to be used by inspecting the header information of the packet. Thus, for a path of length l, the entire time for packet transmission
with store-and-forward routing is

$$T_{sf}(m,l) = t_S + l(t_h + t_B \cdot m). \tag{2.7}$$

Since t_h is typically small compared to the other terms, this can be reduced to
$T_{sf}(m,l) \approx t_S + l \cdot t_B \cdot m$. Thus, the time for packet transmission depends linearly
on the packet size and the length l of the path. Packet transmission with store-and-
forward routing is illustrated in Fig. 2.28 (b). The time for the transmission of an
entire message, consisting of several packets, depends on the specific routing algorithm used. When using a deterministic routing algorithm, the message transmission
time is the sum of the transmission time of all packets of the message, if no network
delays occur. For adaptive routing algorithms, the transmission of the individual
packets can be overlapped, thus potentially leading to a smaller message transmission time.

a)

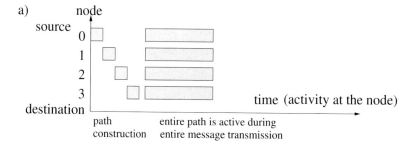

path   entire path is active during
construction entire message transmission

b)

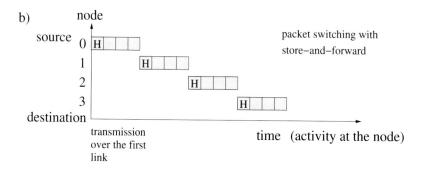

c)

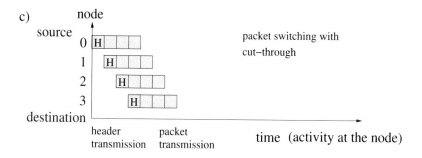

Fig. 2.28 Illustration of the latency of a point-to-point transmission along a path of length $l = 4$ for a) circuit-switching, b) packet-switching with store-and-forward and c) packet-switching with cut-through.

If all packets of a message are transmitted along the same path, **pipelining** can be used to reduce the transmission time of messages: using pipelining, the packets of a message are sent along a path such that the links on the path are used by successive packets in an overlapping way. Using pipelining for a message of size m and packet size m_p, the time of message transmission along a path of length l is described by

$$t_S + (m - m_p)t_B + l(t_h + t_B \cdot m_p) \approx t_S + m \cdot t_B + (l - 1)t_B \cdot m_p \qquad (2.8)$$

where $l(t_h + t_B \cdot m_p)$ is the time that elapses before the first packet arrives at the destination node. After this time, a new packet arrives at the destination in each time step of size $m_p \cdot t_B$, assuming the same bandwidth for each link on the path.

2.8.3.4 Cut-through routing

The idea of the pipelining of message packets can be extended by applying pipelining to the individual packets. This approach is taken by **cut-through routing**. Using this approach, a message is again split into packets as required by the packet-switching approach. The different packets of a message can take different paths through the network to reach the destination. Each individual packet is sent through the network in a pipelined way. To do so, each switch on the path inspects the first few *phits* (*physical units*) of the packet header, containing the routing information, and then determines over which output channel the packet is forwarded. Thus, the transmission path of a packet is established by the packet header and the rest of the packet is transmitted along this path in a pipelined way. A link on this path can be released as soon as all *phits* of the packet, including a possible trailer, have been transmitted over this link.

The time for transmitting a header of size m_H along a single link is given by $t_H = t_B \cdot m_H$. The time for transmitting the header along a path of length l is then given by $t_H \cdot l$. After the header has arrived at the destination node, the additional time for the arrival of the rest of the packet of size m is given by $t_B(m - m_H)$. Thus, the time for transmitting a packet of size m along a path of length l using packet-switching with cut-through routing can be expressed as

$$T_{ct}(m,l) = t_S + l \cdot t_H + t_B \cdot (m - m_H) . \tag{2.9}$$

If m_H is small compared to the packet size m, this can be reduced to $T_{ct}(m,l) \approx t_S + t_B \cdot m$. If all packets of a message use the same transmission path, and if packet transmission is also pipelined, this formula can also be used to describe the transmission time of the entire message. Message transmission time using packet-switching with cut-through routing is illustrated in Fig. 2.28 (c).

Until now, we have considered the transmission of a single message or packet through the network. If multiple transmissions are performed concurrently, network contention may occur because of conflicting requests to the same links. This increases the communication time observed for the transmission. The switching strategy must react appropriately if contention happens on one of the links of a transmission path. Using store-and-forward routing, the packet can simply be buffered until the output channel is free again.

With cut-through routing, two popular options are available: *virtual cut-through routing* and *wormhole routing*. Using **virtual cut-through routing**, in case of a blocked output channel at a switch, all phits of the packet in transmission are collected in a buffer at the switch until the output channel is free again. If this happens at every switch on the path, cut-through routing degrades to store-and-forward routing. Using *partial cut-through routing*, the transmission of the buffered phits of a packet can continue as soon as the output channel is free again, i.e., not all phits of a packet need to be buffered.

The **wormhole routing** approach is based on the definition of *flow control units (flits)* which are usually at least as large as the packet header. The header *flit* establishes the path through the network. The rest of the flits of the packet follow in a pipelined way on the same path. In case of a blocked output channel at a switch, only a few flits are stored at this switch, the rest is kept on the preceding switches of the path. Therefore, a blocked packet may occupy buffer space along an entire path or at least a part of the path. Thus, this approach has some similarities to circuit switching at packet level. Storing the *flits* of a blocked message along the switches of a path may cause other packets to block, leading to network saturation. Moreover, deadlocks may occur because of cyclic waiting, see Fig. 2.29 [159, 199]. An advantage of the wormhole routing approach is that the buffers at the switches can be kept small, since they need to store only a small portion of a packet.

Since buffers at the switches can be implemented large enough with today's technology, virtual cut-through routing is the more commonly used switching technique [103]. The danger of deadlocks can be avoided by using suitable routing algorithms like dimension-ordered routing or by using virtual channels, see Section 2.8.1.

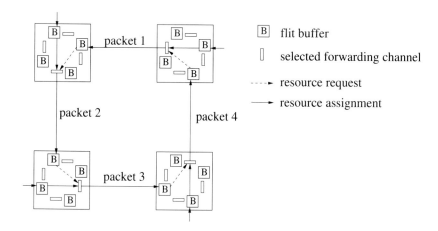

Fig. 2.29 Illustration of a deadlock situation with wormhole routing for the transmission of four packets over four switches. Each of the packets occupies a flit buffer and requests another flit buffer at the next switch; but this flit buffer is already occupied by another packet. A deadlock occurs, since none of the packets can be transmitted to the next switch.

2.8.4 Flow control mechanisms

A general problem in networks may arise from the fact that multiple messages can be in transmission at the same time and may attempt to use the same network links at the same time. If this happens, some of the message transmissions must be blocked while others are allowed to proceed. Techniques to coordinate concurrent message transmissions in networks are called *flow control mechanisms*. Such techniques are important in all kinds of networks, including local and wide area networks, and popular protocols such as TCP contain sophisticated mechanisms for flow control to obtain a high effective network bandwidth, see [139, 174] for more details. Flow control is especially important for networks of parallel computers, since these must be able to transmit a large number of messages fast and reliably. A loss of messages cannot be tolerated, since this would lead to errors in the parallel program currently executed.

Flow control mechanisms typically try to avoid congestion in the network to guarantee fast message transmission. An important aspect are the flow control mechanisms at the link level which considers message or packet transmission over a single link of the network. The link connects two switches A and B. We assume that a packet should be transmitted from A to B. If the link between A and B is free, the packet can be transferred from the output port of A to the input port of B from which it is forwarded to the suitable output port of B. But if B is busy, there might be the situation that B does not have enough buffer space in the input port available to store the packet from A. In this case, the packet must be retained in the output buffer of A until there is enough space in the input buffer of B. But this may cause back pressure to switches preceding A, leading to the danger of network congestion. The idea of link-level flow control mechanisms is that the receiving switch provides a feedback to the sending switch, if not enough input buffer space is available to prevent the transmission of additional packets. This feedback rapidly propagates backwards in the network until the original sending node is reached. This sender can then reduce its transmission rate to avoid further packet delays.

Link-level flow control can help to reduce congestion, but the feedback propagation might be too slow and the network might already be congested when the original sender is reached. An *end-to-end flow control* with a direct feedback to the original sender may lead to a faster reaction. A windowing mechanism as it is used by the TCP protocol is one possibility for an implementation. Using this mechanism, the sender is provided with the available buffer space at the receiver and can adapt the number of packets sent such that no buffer overflow occurs. More information can be found in [139, 174, 104, 45].

2.9 Caches and Memory Hierarchy

A significant characteristic of the hardware development during the last decades has been the increasing gap between processor cycle time and main memory access time as described in Section 2.1. One of the reasons for this development has been the large increase of the clock frequency of the processors, whereas the memory access time could not be reduced significantly. We now have a closer look at this development: The main memory is constructed based on **DRAM** (dynamic random access memory). For a typical processor with a clock frequency of 3 GHz, corresponding to a cycle time of 0.33 ns, a memory access takes between 60 and 210 machine cycles, depending on the DRAM chip used. The use of caches can help to reduce this memory access time significantly. Caches are built with **SRAM** (static random access memory) chips, and SRAM chips are significantly faster than DRAM, but have a smaller capacity per unit area and are more costly.

In 2022, typical cache access times are about 1 ns for an L1 cache, between 3 and 10 ns for an L2 cache, between 10 and 20 ns for an L3 cache, and between 50 and 70 ns for DRAM main memory [171]. For an Intel Core i7 processor (Cascade Lake), the access time to the L1 cache is 4-5 machine cycles, the access time to the L2 cache is 14 machine cycles, and the access time to the L3 cache is 50-70 machine cycles, see the detailed description in Section 2.10.1. On the other hand, the overall main memory latency is 81 ns, taking the memory control into consideration. If the working space of an application is larger than the capacity of the main memory, a part of the working space must be kept on disk (magnetic disk or SSD (Solid-state drive)). However, an access to SSD or magnetic disk takes considerably longer than an access to main memory. For magnetic disks, the large access time is mainly caused by the mechanical movement of the read/write head. Correspondingly, the typical access time for magnetic disks lies between 10 and 20 ms, which corresponds to 30 million to 100 million machine cycles for a typical CPU frequency. Therefore, storing application data on disk can slow down the execution time of programs considerably. A disk access loads a complete page block, which typically has the size 4 Kbytes or 8 Kbytes to amortize the time for disk accesses in situations when the main memory is too small for the application data.

The simplest form of a memory hierarchy is the use of a single cache between the processor and main memory (one-level cache, L1 cache). The cache contains a subset of the data stored in the main memory, and a replacement strategy is used to bring new data from the main memory into the cache, replacing data elements that are no longer accessed. The goal is to keep those data elements in the cache which are currently used most. A detailed description of cache replacement strategies can be found in [120]. Today, two or three levels of caches are used for each processor, using a small and fast L1 cache and larger, but slower L2 and L3 caches.

For multiprocessor systems where each processor uses a separate local cache, there is the additional problem of keeping a consistent view to the shared address space for all processors. It must be ensured that a processor accessing a data element always accesses the most recently written data value, also in the case that another processor has written this value. This is also referred to as **cache coherence problem**, and will be considered in more detail in Section 2.9.3.

For multiprocessors with a shared address space, the top level of the memory hierarchy is the shared address space that can be accessed by each of the processors. The design of a memory hierarchy may have a large influence on the execution time of parallel programs, and memory accesses should be ordered such that a given memory hierarchy is used as efficiently as possible. Moreover, techniques to keep a memory hierarchy consistent may also have an important influence. In this section, we therefore give an overview of memory hierarchy designs and discuss issues of cache coherence and memory consistency. Since caches are the building blocks of memory hierarchies and have a significant influence on the memory consistency, we give a short overview of caches in the following subsection. For a more detailed treatment, we refer to [45, 105, 100, 205, 171].

2.9.1 Characteristics of Caches

A cache is a small, but fast memory between the processor and the main memory. Caches are built with SRAM. Typical access times are 0.5 - 2.5 ns (ns = nanoseconds $= 10^{-9}$ seconds) compared to 50-70 ns for DRAM [105]). In the following, we first consider a one-level cache. A cache contains a copy of a subset of the data in main memory. Data is moved in blocks, containing a small number of words, between the cache and main memory, see Fig. 2.30. These blocks of data are called **cache blocks** or **cache lines**. The size of the cache lines is fixed for a given architecture and cannot be changed during program execution.

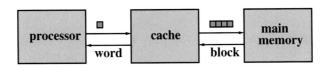

Fig. 2.30 Data transport between cache and main memory is done by the transfer of memory blocks comprising several words whereas the processor accesses single words in the cache.

Cache control is decoupled from the processor and is performed by a separate cache controller. During program execution, the processor specifies memory addresses to be read or to be written as given by the load and store operations of the machine program. The processor forwards the memory addresses to the memory system and waits until the corresponding values are returned or written. The processor specifies memory addresses independently of the organization of the memory system, i.e., the processor does not need to know the architecture of the memory system. After having received a memory access request from the processor, the cache controller checks whether the memory address specified belongs to a cache line which is currently stored in the cache. If this is the case, a **cache hit** occurs, and the requested word is delivered to the processor from the cache. If the corresponding cache line is not stored in the cache, a **cache miss** occurs, and the cache line is first copied from main memory into the cache before the requested word is delivered to the processor. The corresponding delay time is also called **miss penalty**. Since the access time to main memory is significantly larger than the access time to the cache, a cache miss leads to a delay of operand delivery to the processor. Therefore, it is desirable to reduce the number of cache misses as much as possible.

The exact behavior of the cache controller is hidden from the processor. The processor observes that some memory accesses take longer than others, leading to a delay in operand delivery. During such a delay, the processor can perform other operations that are independent of the delayed operand. This is possible, since the processor is not directly occupied for the operand access from the memory system. Techniques like *operand prefetch* can be used to support an anticipated loading of operands so that other independent operations can be executed, see [105].

The number of cache misses may have a significant influence on the resulting execution time of a program. If many memory accesses lead to cache misses, the processor may often have to wait for operands, and program execution may be quite slow. Since the cache management is implemented in hardware, the programmer cannot directly specify which data should reside in the cache at which point in program execution. However, the order of memory accesses in a program can have a large influence on the resulting program execution time, and a smart reordering of the memory accesses may lead to a significant reduction of the execution time. In this context, the **locality of memory accesses** is often used as a characterization of the memory access behavior of a program. Spatial and temporal locality can be distinguished as follows:

- The memory accesses of a program have a high **spatial locality**, if the program often accesses memory locations with neighboring or nearby addresses at successive points in time during program execution. Thus, for programs with high spatial locality there is often the situation that after an access to a memory location, one or more memory locations of the same cache line are also accessed shortly afterwards. In such situations, after loading a cache block, several of the following memory locations can be loaded from this cache block, thus avoiding expensive cache misses. The use of cache blocks comprising several memory words is based on the assumption that most programs exhibit spatial locality, i.e., when loading a cache block not only one but several memory words of the cache block are accessed before the cache block is replaced again.

- The memory accesses of a program have a high **temporal locality**, if it often happens that the *same* memory location is accessed multiple times at successive points in time during program execution. Thus, for programs with a high temporal locality there is often the situation that after loading a cache block in the cache, the memory words of the cache block are accessed multiple times before the cache block is replaced again.

For programs with small spatial locality there is often the situation that after loading a cache block, only one of the memory words contained is accessed before the cache block is replaced again by another cache block. For programs with small temporal locality, there is often the situation that after loading a cache block because of a memory access, the corresponding memory location is accessed only once before the cache block is replaced again. Many program transformations have been proposed to increase temporal or spatial locality of programs, see [12, 223] for more details.

In the following, we give a short overview of important characteristics of caches. In particular, we consider cache size, mapping of memory blocks to cache blocks, replacement algorithms, and write-back policies. We also consider the use of multi-level caches.

2.9.1.1 Cache size

Using the same hardware technology, the access time of a cache increases (slightly) with the size of the cache because of an increased complexity of the addressing. But using a larger cache leads to a smaller number of replacements as a smaller cache, since more cache blocks can be kept in the cache. The size of the caches is limited by the available chip area. Typical sizes for L1 caches lie between 8K and 128K memory words where a memory word is four or eight bytes long, depending on the architecture. During recent years, the typical size of L1 caches has not been increased significantly.

If a cache miss occurs when accessing a memory location, an entire cache block is brought into the cache. For designing a memory hierarchy, the following points have to be taken into consideration when fixing the size of the cache blocks:

- Using larger blocks reduces the number of blocks that fit in the cache when using the same cache size. Therefore, cache blocks tend to be replaced earlier when using larger blocks compared to smaller blocks. This suggests to set the cache block size as small as possible.
- On the other hand, it is useful to use blocks with more than one memory word, since the transfer of a block with x memory words from main memory into the cache takes less time than x transfers of a single memory word. This suggests to use larger cache blocks.

As a compromise, a medium block size is used. Typical sizes for L1 cache blocks are four or eight memory words. For example, the Intel Cascade Lake architecture uses cache blocks of size 64 bytes, see Section 2.10.1.

2.9.1.2 Mapping of memory blocks to cache blocks

Data is transferred between main memory and cache in blocks of a fixed length. Because the cache is significantly smaller than the main memory, not all memory blocks can be stored in the cache at the same time. Therefore, a mapping algorithm must be used to define at which position in the cache a memory block can be stored. The mapping algorithm used has a significant influence on the cache behavior and determines how a stored block is localized and retrieved from the cache. For the mapping, the notion of **associativity** plays an important role. Associativity determines at how many positions in the cache a memory block can be stored. The following methods are distinguished:

- for a **direct mapped** cache, each memory block can be stored at exactly one position in the cache;
- for a **fully associative** cache, each memory block can be stored at an arbitrary position in the cache;
- for a **set associative** cache, each memory block can be stored at a fixed number of positions.

In the following, we consider these three mapping methods in more detail for a memory system which consists of a main memory and a cache. We assume that the main memory comprises $n = 2^s$ blocks which we denote as B_j for $j = 0, \ldots n-1$. Furthermore, we assume that there are $m = 2^r$ cache positions available; we denote the corresponding cache blocks as $\bar{B}_i$ for $i = 0, \ldots, m-1$. The memory blocks and the cache blocks have the same size of $l = 2^w$ memory words. At different points of program execution, a cache block may contain different memory blocks. Therefore, for each cache block a **tag** must be stored, which identifies the memory block that is currently stored. The use of this tag information depends on the specific mapping algorithm and will be described in the following. As running example, we consider a memory system with a cache of size 64 Kbytes which uses cache blocks of 4 bytes. A memory word is assumed to be one byte, i.e., four words fit into one cache block. In total, $16K = 2^{14}$ blocks of four bytes each fit into the cache. With the notation from above, it is $r = 14$ and $w = 2$. The main memory is 4 Gbytes $= 2^{32}$ bytes large, i.e., it is $s = 30$ if we assume that a memory word is one byte. We now consider the three mapping methods in turn.

2.9.1.3 Direct mapped caches

The simplest form to map memory blocks to cache blocks is implemented by **direct mapped** caches. Each memory block B_j can be stored at only one specific cache location. The mapping of a memory block B_j to a cache block $\bar{B}_i$ is defined as follows:

B_j is mapped to $\bar{B}_i$ if $i = j \bmod m$.

Thus, there are $n/m = 2^{s-r}$ different memory blocks that can be stored in one specific cache block $\bar{B}_i$. Based on the mapping, memory blocks are assigned to cache positions as follows:

cache block	may contain memory block
0	$0, m, 2m, \ldots, 2^s - m$
1	$1, m+1, 2m+1, \ldots, 2^s - m + 1$
$\vdots$	$\vdots$
$m-1$	$m-1, 2m-1, 3m-1, \ldots, 2^s - 1$

Since the cache size m is a power of 2, the modulo operation specified by the mapping function can be computed by using the low-order bits of the memory address specified by the processor. Since a cache block contains $l = 2^w$ memory words, the memory address can be partitioned into a word address and a block address. The block address specifies the position of the corresponding memory block in main memory. It consists of the s most significant (leftmost) bits of the memory address. The word address specifies the position of the memory location in the memory block, relative to the first location of the memory block. It consists of the w least significant (rightmost) bits of the memory address.

For a direct-mapped cache, the r rightmost bits of the block address of a memory location define at which of the $m = 2^r$ cache positions the corresponding memory block must be stored if the block is loaded into the cache. The remaining $s - r$ bits can be interpreted as tag which specifies which of the 2^{s-r} possible memory blocks is currently stored at a specific cache position. This tag must be stored with the cache block. Thus each memory address is partitioned as follows:

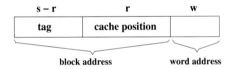

For the running example, the tags consist of $s - r = 16$ bits for a direct mapped cache.

Memory access is illustrated in Fig. 2.31 (a) for an example memory system with block size 2 ($w = 1$), cache size 4 ($r = 2$) and main memory size 16 ($s = 4$). For each memory access specified by the processor, the cache position at which the requested memory block must be stored is identified by considering the r rightmost bits of the block address. Then the tag stored for this cache position is compared with the $s - r$ leftmost bits of the block address. If both tags are identical, the referenced memory block is currently stored in the cache and the memory access can be done via the cache. A *cache hit* occurs. If the two tags are different, the requested memory block must first be loaded into the cache at the given cache position before the memory location specified can be accessed.

Direct mapped caches can be implemented in hardware without great effort, but they have the disadvantage that each memory block can be stored at only one cache position. Thus, it can happen that a program repeatedly specifies memory addresses in different memory blocks that are mapped to the same cache position. In this situation, the memory blocks will be continually loaded and replaced in the cache, leading to a large number of cache misses and therefore a large execution time. This phenomenon is also called *thrashing*.

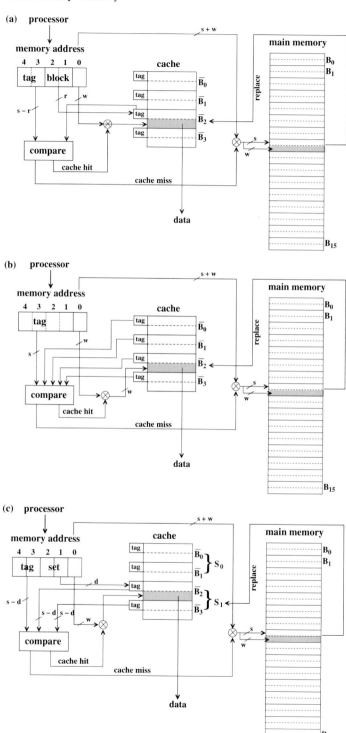

Fig. 2.31 Illustration of the mapping of memory blocks to cache blocks for a cache with $m = 4$ cache blocks ($r = 2$) and a main memory with $n = 16$ memory blocks ($s = 4$). Each block contains two memory words ($w = 1$). (a) direct mapped cache; (b) fully associative cache; (c) set associative cache with $k = 2$ blocks per set, using $v = 2$ sets ($d = 1$).

2.9.1.4 Fully associative caches

In a fully associative cache, each memory block can be placed at *any* cache position, thus overcoming the disadvantage of direct mapped caches. As for direct mapped caches, a memory address can again be partitioned into a block address (*s* leftmost bits) and a word address (*w* rightmost bits). Since each cache block can contain any memory block, the entire block address must be used as tag and must be stored with the cache block to allow the identification of the memory block stored. Thus, each memory address is partitioned as follows:

To check whether a given memory block is stored in the cache, all entries in the cache must be searched, since the memory block can be stored at any cache position. This is illustrated in Fig. 2.31 (b).

The advantage of fully associative caches lies in the increased flexibility when loading memory blocks into the cache. The main disadvantage is that for each memory access all cache positions must be considered to check whether the corresponding memory block is currently held in the cache. To make this search practical, it must be done in parallel using a separate comparator for each cache position, thus increasing the required hardware effort significantly. Another disadvantage is that the tags to be stored for each cache block are significantly larger as for direct mapped caches. For the example cache introduced above, the tags must be 30 bits long for a fully associated cache, i.e., for each 32 bit memory block, a 30 bit tag must be stored. Because of the large search effort, a fully associative mapping is useful only for caches with a small number of positions.

2.9.1.5 Set associative caches

Set associative caches are a compromise between direct mapped and fully associative caches. In a set associative cache, the cache is partitioned into v sets $S_0, \ldots, S_{v-1}$ where each set consists of $k = m/v$ blocks. A memory block B_j is not mapped to an individual cache block, but to a unique set in the cache. Within the set, the memory block can be placed in any cache block of that set, i.e., there are k different cache blocks in which a memory block can be stored. The set of a memory block B_j is defined as follows:

$$B_j \text{ is mapped to set } S_i, \text{ if } i = j \bmod v.$$

for $j = 0, \ldots, n-1$. A memory access is illustrated in Fig. 2.31 (c). Again, a memory address consists of a block address (*s* bits) and a word address (*w* bits). The $d = \log v$ rightmost bits of the block address determine the set S_i to which the corresponding memory block is mapped. The leftmost $s - d$ bits of the block address are the tag that is used for the identification of the memory blocks stored in the individual cache blocks of a set. Thus, each memory address is partitioned as follows:

When a memory access occurs, the hardware first determines the set to which the memory block is assigned. Then, the tag of the memory block is compared with the tags of all cache blocks in the set. If there is a match, the memory access can be performed via the cache. Otherwise, the corresponding memory block must first be loaded into one of the cache blocks of the set.

For $v = m$ and $k = 1$, a set associative cache reduces to a direct mapped cache. For $v = 1$ and $k = m$, a fully associative cache results. Typical cases are $v = m/4$ and $k = 4$, leading to a *4-way set associative cache*, and $v = m/8$ and $k = 8$, leading to an *8-way set associative cache*. For the example cache, using $k = 4$ leads to 4K sets; $d = 12$ bits of the block address determine the set to which a memory block is mapped. The tags used for the identification of memory blocks within a set are 18 bits long.

2.9.1.6 Block replacement methods

When a cache miss occurs, a new memory block must be loaded into the cache. To do this for a fully occupied cache, one of the memory blocks in the cache must be replaced. For a direct mapped cache, there is only one position at which the new memory block can be stored, and the memory block occupying that position must be replaced. For a fully associative or set associative cache, there are several positions at which the new memory block can be stored. The block to be replaced is selected using a *replacement method*. Many replacement policies have been proposed, including recency-based and frequency-based policies, see [120] for a detailed overview.

A popular recency-based replacement method is **least recently used (LRU)** which replaces the block in a set that has not been used for the longest time. For the implementation of the LRU method, the hardware must keep track for each block of a set when the block has been used last. The corresponding time entry must be updated at each usage time of the block. This implementation requires additional space to store the time entries for each block and additional control logic to update the time entries. For a 2-way set associative cache the LRU method can be implemented more easily by keeping a USE bit for each of the two blocks in a set. When a cache block of a set is accessed, its USE bit is set to 1 and the USE bit of the other block in the set is set to 0. This is performed for each memory access. Thus, the block whose USE bit is 1 has been accessed last, and the other block should be replaced if a new block has to be loaded into the set. Many variants of LRU have been proposed, see [120] for an overview.

An alternative to LRU is the frequency-based **least frequently used (LFU)** method which replaces the block of a set that has experienced the fewest references. However, the LFU method also requires additional control logic, since for each block a frequency counter has to be maintained which must be updated for each memory access. Hybrid policies that combine recency-based and frequency-based information have also been proposed. An example is the LRFU (Least Recently/Frequently Used) policy that is based on a combined metric that uses a weighting function between LRU and LFU to determine the block to be replaced, see [120] for a detailed treatment.

For a larger associativity, an exact implementation of LRU or LFU as described above is often considered as too costly [104, 171], and approximations or other schemes are used. Often, **random replacement** is used in practice, since this policy can be implemented easily in hardware. Moreover, simulations have shown that random replacement leads to only slightly inferior performance as more sophisticated methods such as LRU or LFU and in some cases random replacement can even lead to a better cache utilization than approximations of LRU. [171, 207].

2.9.2 Write Policy

A cache contains a subset of the memory blocks. When the processor issues a *write access* to a memory block that is currently stored in the cache, the referenced block is definitely updated in the cache, since the next read access must return the most recent value. There remains the question: When is the corresponding memory block in the main memory updated? The earliest possible update time for the main memory is immediately after the update in the cache; the latest possible update time for the main memory is when the cache block is replaced by another block. The exact replacement time and update method is captured by the write policy. The most popular policies are **write-through** and **write-back**.

2.9.2.1 Write-through policy

Using write-through, a modification of a block in the cache using a write access is immediately transferred to main memory, thus keeping the cache and the main memory consistent. An advantage of this approach is that other devices such as I/O modules that have direct access to main memory always get the newest value of a memory block. This is also important for multicore systems, since after a write by one core, all other cores always get the most recently written value when accessing the same memory block. A drawback of the write through policy is that every write in the cache causes also a write to main memory which typically takes at least 100 processor cycles to complete. This could slow down the processor if it had to wait for the completion. To avoid processor waiting, a *write buffer* can be used to store pending write operations into the main memory [171, 104]. After writing the data into the cache and into the write buffer, the processor can continue its execution without waiting for the completion of the write into the main memory. A write buffer entry can be freed after the write into main memory completes. When the processor performs a write and the write buffer is full, a write stall occurs, and the processor must wait until there is a free entry in the write buffer.

2.9.2.2 Write-back policy

Using write-back, a write operation to a memory block that is currently held in the cache is performed only in the cache; the corresponding memory entry is not updated immediately. Thus, the cache may contain newer values than the main memory. The modified memory block is written to the main memory when the cache block is replaced by another memory block. To check whether a write to main memory is necessary when a cache block is replaced, a separate bit (*dirty bit*) is held for each cache block which indicates whether the cache block has been modified. The dirty bit is initialized with 0 when a block is loaded into the cache. A write access to a cache block sets the dirty bit to 1, indicating that a write to main memory must be performed when the cache block is replaced.

Using write-back policy usually leads to fewer write operations to main memory than write-through, since cache blocks can be written multiple times before they are written back to main memory. The drawback of write-back is that the main memory may contain invalid entries, and hence I/O modules can access main memory only through the cache.

If a write to a memory location goes to a memory block that is currently not in the cache, most caches use the *write-allocate* method: the corresponding memory block is first brought into the cache and then the modification is performed as described above. An alternative approach is *write no allocate*, which modifies in main memory without loading it into the cache. However, this approach is used less often.

2.9.2.3 Number of caches

So far, we have considered the behavior of a single cache which is placed between the processor and main memory and which stores data blocks of a program in execution. Such caches are also called **data caches**.

Besides the program data, a processor also accesses instructions of the program in execution before they are decoded and executed. Because of loops in the program, an instruction can be accessed multiple times. To avoid multiple loading operations from main memory, instructions are also held in cache. To store instructions and data, a single cache can be used (*unified cache*). But often, two separate caches are used on the first level, an **instruction cache** to store instructions and a separate data cache to store data. This approach is also called *split caches*. This enables a greater flexibility for the cache design, since the data and instruction caches can work independently of each other, and may have different size and associativity depending on the specific needs.

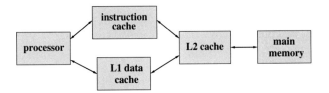

Fig. 2.32 Illustration of a two-level cache hierarchy.

In practice, multiple levels of caches are typically used as illustrated in Fig. 2.32. Today, most desktop processors have a three-level cache hierarchy, consisting of a first level (L1) cache, a second level (L2) cache, and a third level (L3) cache. All these caches are integrated onto the chip area. For the L1 cache, split caches are typically used, for the remaining levels, unified caches are standard. In 2021, typical cache sizes lie between 8 Kbytes and 64 Kbytes for the L1 cache, between 128 Kbytes and 1 Mbytes for the L2 cache, and between 2 Mbytes and 48 Mbytes for the L3 cache. Typical sizes of the main memory lie between 8 Gbytes and 64 Gbytes. The caches are hierarchically organized, and for three levels, the L1 caches contain a subset of the L2 cache, which contains a subset of the L3 cache, which contains a subset of the main memory. Typical access times are one or a few processor cycles for the L1 cache, between 15 and 25 cycles for the L2 cache, between 100 and 300 cycles for main memory, and between 10 million to 100 million cycles for the hard disk [104]

2.9.3 Cache coherency

Using a memory hierarchy with multiple levels of caches can help to bridge large access times to main memory. But the use of caches introduces the effect that memory blocks can be held in multiple copies in caches and main memory, and after an update in the L1 cache, other copies might become invalid, in particular if a write-back policy is used. This does not cause a problem as long as a single processor is the only accessing device. But if there are multiple accessing devices, as is the case for multicore processors, inconsistent copies can occur and should be avoided, and each execution core should always access the most recent value of a memory location. The problem of keeping the different copies of a memory location consistent is also referred to as *cache coherency problem*. A detailed description can be found in [205].

In a multiprocessor system with different cores or processors, in which each processor has a separate local cache, the same memory block can be held as copy in the local cache of multiple processors. If one or more of the processors update a copy of a memory block in their local cache, the other copies become invalid and contain inconsistent values. The problem can be illustrated for a bus-based system with three processors [45] as shown in the following example.

Example: We consider a bus-based SMP system with three processors P_1, P_2, P_3 where each processor P_i has a local cache C_i for $i = 1, 2, 3$. The processors are connected to a shared memory M via a central bus. The caches C_i use a write-through strategy. We consider a variable u with initial value 5 which is held in main memory before the following operations are performed at times t_1, t_2, t_3, t_4:

t_1: Processor P_1 reads variable u. The memory block containing u is loaded into cache C_1 of P_1.

t_2: Processor P_3 reads variable u. The memory block containing u is also loaded into cache C_3 of P_3.

t_3: Processor P_3 writes the value 7 into u. This new value is also written into main memory because write-through is used.

t_4: Processor P_1 reads u by accessing the copy in its local cache.

At time t_4, processor P_1 reads the old value 5 instead of the new value 7, i.e., a cache coherency problem occurs. This is the case for both write-through and write-back caches: For write-through caches, at time t_3 the new value 7 is directly written into main memory by processor P_3, but the cache of P_1 will not be updated. For write-back caches, the new value of 7 is not even updated in main memory, i.e., if another processor P_2 reads the value of u after time t_3, it will obtain the old value, even when the variable u is not held in the local cache of P_2.   □

For a correct execution of a parallel program on a shared address space, it must be ensured that for each possible order of read and write accesses performed by the participating processors according to their program statements, each processor obtains the right value, no matter whether the corresponding variable is held in cache or not.

The behavior of a memory system for read and write accesses performed by *different* processors to the *same* memory location is captured by the **coherency of the memory system**. Informally, a memory system is coherent if for each memory location any read access returns the most recently written value of that memory location. Since multiple processors may perform write operations to the same memory location at the same time, we first must define more precisely what the most recently written value is. For this definition, the order of the memory accesses in the parallel program executed is used as time measure, not the physical point in time at which the memory accesses are executed by the processors. This makes the definition independent from the specific execution environment and situation.

Using the program order of memory accesses, a memory system is coherent, if the following conditions are fulfilled [103].

1. If a processor P writes into a memory location x at time t_1 and reads from the same memory location x at time $t_2 > t_1$ and if between t_1 and t_2 no other processor performs a write into x, then P obtains at time t_2 the value written by itself at time t_1. Thus, for each processor the order of the memory accesses in its program is preserved despite a parallel execution.

2. If a processor P_1 writes into a memory location x at time t_1 and if another processor P_2 reads x at time $t_2 > t_1$, then P_2 obtains the value written by P_1, if between t_1 and t_2 no other processors write into x and if the period of time $t_2 - t_1$ is sufficiently large. Thus, a value written by one of the processors must become visible to the other processors after a certain amount of time.

3. If two processors write into the same memory location x, these write operations are *serialized* so that all processors see the write operations in the *same order*. Thus, a global **write serialization** is performed.

To be coherent, a memory system must fulfill these three properties. In particular, for a memory system with caches which can store multiple copies of memory blocks, it must be ensured that each processor has a coherent view to the memory system through its local caches. To ensure this, hardware-based *cache coherence protocols* are used. These protocols are based on the approach to maintain information about the current modification state for each memory block while it is stored in a local cache. Depending on the coupling of the different components of a parallel system, *snooping protocols* and *directory-based protocols* are used. Snooping protocols are based on the existence of a shared medium over which the memory accesses of the processor cores are transferred. The shared medium can be a shared bus or a shared cache. For directory-based protocols, it is not necessary that such a shared medium exists. In the following, we give a short overview of these protocols and refer to [45, 103, 150] for a more detailed treatment.

2.9.3.1 Snooping protocols

The technique of bus snooping has first been used for bus-based SMP systems for which the local caches of the processors use a write-through policy. Snooping protocols can be used if all memory accesses are performed via a shared medium which can be observed by all processor cores. For SMP systems, the central bus has been used as shared medium. For current multicore processors, such as the Intel Core i7 processors, the shared medium is the connection between the private L1 and L2 caches of the cores and the shared L3 cache.

In the following, we assume the existence of a shared medium and consider a memory system with a write-through policy of the caches. Especially, we consider the use of a central bus as shared medium. In this case, the snooping technique relies on the property that on such systems all memory accesses are performed via the central bus, i.e., the bus is used as broadcast medium. Thus, all memory accesses can be observed by the cache controller of each processors. When the cache controller observes a write into a memory location that is currently held in the local cache, it updates the value in the cache by copying the new value from the bus into the cache.

Thus, the local caches always contain the most recently written values of memory locations. These protocols are also called *update-based protocols*, since the cache controllers directly perform an update. There are also *invalidation-based protocols* in which the cache block corresponding to a memory block is invalidated so that the next read access must first perform an update from main memory. Using an update-based protocol in the example from page 87, processor P_1 can observe the write operation of P_3 at time t_3 and can update the value of u in its local cache C_1 accordingly. Thus, at time t_4, P_1 reads the correct value 7.

The technique of bus snooping has first been used for a write-through policy and relies on the existence of a broadcast medium so that each cache controller can observe all write accesses to perform updates or invalidations. For newer architectures interconnection networks like crossbars or point-to-point networks are used. This makes updates or invalidations more complicated, since the interprocessor links are not shared, and the coherency protocol must use broadcasts to find potentially shared copies of memory blocks, see [103] for more details. Due to the coherence protocol, additional traffic occurs in the interconnection network, which may limit the effective memory access time of the processors. Snooping protocols are not restricted to write-through caches. The technique can also be applied to write-back caches as described in the following.

2.9.3.2 Write-back invalidation protocol

In the following, we describe a basic write-back invalidation protocol, see [45, 103] for more details. In the protocol, each cache block can be in one of three states [45]:

M (modified) means that the cache block contains the current value of the memory block and that all other copies of this memory block in other caches or in the main memory are invalid, i.e., the block has been updated in the cache;

S (shared) means that the cache block has not been updated in this cache and that this cache contains the current value, as does the main memory and zero or more other caches;

I (invalid) means that the cache block does not contain the most recent value of the memory block.

According to these three states, the protocol is also called **MSI protocol**. The same memory block can be in different states in different caches. Before a processor modifies a memory block in its local cache, all other copies of the memory block in other caches and the main memory are marked as invalid (I). This is performed by an operation on the broadcast medium. After that, the processor can perform one or several write operations to this memory block without performing other invalidations. The memory block is marked as modified (M) in the cache of the writing processor. The protocol provides three operations on the broadcast medium, which is a shared bus in the simplest case:

- **Bus Read** (BusRd): This operation is generated by a read operation (PrRd) of a processor to a memory block that is currently not stored in the cache of this processor. The cache controller requests a copy of the memory block by specifying the corresponding memory address. The requesting processor does not intend to modify the memory block. The most recent value of the memory block is provided from main memory or from another cache.

- **Bus Read Exclusive** (BusRdEx): This operation is generated by a write operation (PrWr) of a processor to a memory block that is currently not stored in the cache of this processor or that is currently not in the M state in this cache. The cache controller requests an exclusive copy of the memory block that it intends to modify; the request specifies the corresponding memory address. The memory system provides the most recent value of the memory block. All other copies of this memory block in other caches are marked invalid (I).

- **Write Back** (BusWr): The cache controller writes a cache block that is marked as modified (M) back to main memory. This operation is generated if the cache block is replaced by another memory block. After the operation, the main memory contains the latest value of the memory block.

The processor performs the usual read and write operations (PrRd, PrWr) to memory locations, see Fig. 2.33 (right). The cache controller provides the requested memory words to the processor by loading them from the local cache. In case of a cache miss, this includes the loading of the corresponding memory block using a bus operation. The exact behavior of the cache controller depends on the state of the cache block addressed and can be described by a state transition diagram that is shown in Fig. 2.33 (left).

A read and write operation to a cache block marked with M can be performed in the local cache without a bus operation. The same is true for a read operation to a cache block that is marked with S. To perform a write operation to a cache block marked with S, the cache controller first must execute a BusRdEx operation to become the exclusive owner of the cache block. The local state of the cache block is transformed from S to M. The cache controllers of other processors that have a local copy of the same cache block with state S observe the BusRdEx operation and perform a local state transition from S to I for this cache block.

When a processor tries to read a memory block that is not stored in its local cache or that is marked with I in its local cache, the corresponding cache controller performs a BusRd operation. This causes a valid copy to be stored in the local cache marked with S. If another processor observes a BusRd operation for a memory block, for which it has the only valid copy (state M), it puts the value of the memory block on the bus and marks its local copy with state S (shared).

When a processor tries to write into a memory block that is not stored in its local cache or that is marked with I, the cache controller performs a BusRdEx operation. This provides a valid copy of the memory block in the local cache, which is marked with M, i.e., the processor is the exclusive owner of this memory block. If another processor observes a BusRdEx operation for a memory block which is marked with M in its local cache, it puts the value of the memory block on the bus and performs a local state transition from M to I.

A drawback of the MSI protocol is that a processor which first reads a memory location and then writes into a memory location must perform two bus operations

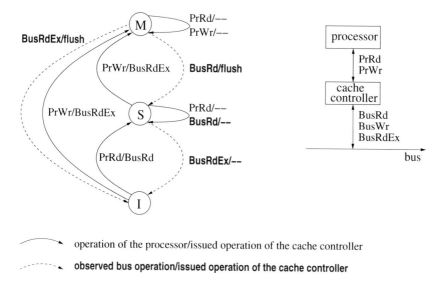

operation of the processor/issued operation of the cache controller

observed bus operation/issued operation of the cache controller

Fig. 2.33 Illustration of the MSI protocol: Each cache block can be in one of the states M (modified), S (shared), or I (invalid). State transitions are shown by arcs that are annotated with operations. A state transition can be caused by
(a) Operations of the processor (PrRd, PrWr) (solid arcs); the bus operations initiated by the cache controller are annotated behind the slash sign. If no bus operation is shown, the cache controller only accesses the local cache.
(b) Operations on the bus observed by the cache controller and issued by the cache controller of other processors (dashed arcs). Again, the corresponding operations of the local cache controller are shown behind the slash sign. The operation *flush* means that the cache controller puts the value of the requested memory block on the bus, thus making it available to other processors. If no arc is shown for a specific bus operation observed for a specific state, no state transition occurs and the cache controller does not need to perform an operation.

BusRd and BusRdEx, even if no other processor is involved. The BusRd provides the memory block in S state, the BusRdEx causes a state transition from S to M. This drawback can be eliminated by adding a new state E (exclusive):

E (exclusive) means that the cache contains the only (exclusive) copy of the memory block and that this copy has not been modified. The main memory contains a valid copy of the block, but no other processor is caching this block.

If a processor requests a memory block by issuing a PrRd and if no other processor has a copy of this memory block in its local cache, then the block is marked with E (instead of S in the MSI protocol) in the local cache after being loaded from main memory with a BusRd operation. If at a later time, this processor performs a write into this memory block, a state transition from E to M is performed before the write. In this case, no additional bus operation is necessary. If between the local read and write operation, another processor performs a read to the same memory block, the local state is changed from E to S. The local write would then cause the same actions as in the MSI protocol. The resulting protocol is called **MESI protocol** according to the abbreviation of the four states. A more detailed discussion and a detailed description of several variants can be found in [45]. Variants of the MESI protocol are supported by many processors and the protocols play an important role for multicore processors to ensure the coherence of the local caches of the cores.

The MSI and MESI protocols are invalidation protocols. An alternative are **write-back update protocols** for write-back caches. In these protocols, after an update of a cache block with state M, all other caches which also contain a copy of the corresponding memory block are also updated. Therefore, the local caches always contain the most recent values of the cache blocks. In practice, these protocols are rarely used because they cause more traffic on the bus.

2.9.3.3 Directory-based cache coherence protocols

Snooping protocols rely on the existence of a shared broadcast medium like a bus or a switch through which all memory accesses are transferred. This is typically the case for multicore processors or small SMP systems. But for larger systems, such a shared medium often does not exist and other mechanisms have to be used.

A simple solution would be not to support cache coherence at hardware level. Using this approach, the local caches would only store memory blocks of the local main memory. There would be no hardware support to store memory blocks from the memory of other processors in the local cache. Instead, software support could be provided, but this requires more support from the programmer and is typically not as fast as a hardware solution.

An alternative to snooping protocols are **directory-based protocols**. These do not rely on a shared broadcast medium. Instead, a central directory is used to store the state of every memory block that may be held in cache. Instead of observing a shared broadcast medium, a cache controller can get the state of a memory block by a lookup in the directory. The directory can be held shared, but it could also be distributed among different processors to avoid bottlenecks when the directory is accessed by many processors. In the following, we give a short overview of directory-based protocols. For a more detailed description, we refer again to [45, 103].

As example, we consider a parallel machine with a distributed memory. We assume that for each local memory a directory is maintained that specifies for each memory block of the local memory which caches of other processors currently store a copy of this memory block. For a parallel machine with p processors the directory can be implemented by maintaining a bit vector with p *presence bits* and a number of state bits for each memory block. Each presence bit indicates whether a specific processor has a valid copy of this memory block in its local cache (value 1) or not (value 0). An additional *dirty bit* is used to indicate whether the local memory contains a valid copy of the memory block (value 0) or not (value 1). Each directory is maintained by a *directory controller* which updates the directory entries according to the requests observed on the network.

Fig. 2.34 illustrates the organization. In the local caches, the memory blocks are marked with M (modified), S (shared), or I (invalid), depending on their state, similar to the snooping protocols described above. The processors access the memory system via their local cache controllers. We assume a global address space, i.e., each memory block has a memory address which is unique in the entire parallel system.

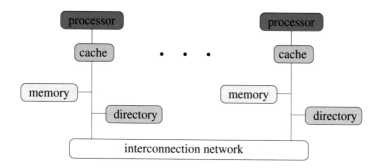

Fig. 2.34 Directory-based cache coherency.

When a read miss or write miss occurs at a processor i, the associated cache controller contacts the local directory controller to obtain information about the accessed memory block. If this memory block belongs to the local memory and the local memory contains a valid copy (dirty bit 0), the memory block can be loaded into the cache with a local memory access. Otherwise, a non-local (remote) access must be performed. A request is sent via the network to the directory controller at the processor owning the memory block (home node). For a read miss, the receiving directory controller reacts as follows:

- If the dirty bit of the requested memory block is 0, the directory controller retrieves the memory block from local memory and sends it to the requesting node via the network. The presence bit of the receiving processor i is set to 1 to indicate that i has a valid copy of the memory block.
- If the dirty bit of the requested memory block is 1, there is exactly one processor j which has a valid copy of the memory block; the presence bit of this processor is 1. The directory controller sends a corresponding request to this processor j. The cache controller of j sets the local state of the memory block from M to S and sends the memory block both to the home node of the memory block and the processor i from which the original request came. The directory controller of the home node stores the current value in the local memory, sets the dirty bit of the memory block to 0, and sets the presence bit of processor i to 1. The presence bit of j remains 1.

For a write miss, the receiving directory controller performs the following steps:

- If the dirty bit of the requested memory block is 0, the local memory of the home node contains a valid copy. The directory controller sends an invalidation request to all processors j for which the presence bit is 1. The cache controllers of these processors set the state of the memory block to I. The directory controller waits for an acknowledgment from these cache controllers, sets the presence bit for these processors to 0, and sends the memory block to the requesting processor i. The presence bit of i is set to 1, the dirty bit is also set to 1. After having received the memory block, the cache controller of i stores the block in its cache and sets its state to M.
- If the dirty bit of the requested memory block is 1, the memory block is requested from the processor j whose presence bit is 1. Upon arrival, the memory block is forwarded to processor i, the presence bit of i is set to 1, and the presence bit of j is set to 0. The dirty bit remains at 1. The cache controller of j sets the state of the memory block to I.

When a memory block with state M should be replaced by another memory block in the cache of processor i, it must be written back into its home memory, since this is the only valid copy of this memory block. To do so, the cache controller of i sends the memory block to the directory controller of the home node. This one writes the memory block back to the local memory and sets the dirty bit of the block and the presence bit of processor i to 0.

A cache block with state S can be replaced in a local cache without sending a notification to the responsible directory controller. Sending a notification avoids the responsible directory controller sending an unnecessary invalidation message to the replacing processor in case of a write miss as described above.

The directory protocol just described is kept quite simple. Directory protocols used in practice are typically more complex and contain additional optimizations to reduce the overhead as far as possible. Directory protocols are typically used for distributed memory machines as described. But they can also be used for shared memory machines. An example are the Sun T1 and T2 processors, see [103] for more details.

2.9.4 Memory consistency

Cache coherence ensures that each processor of a parallel system has the same consistent view of the memory through its local cache. Thus, at each point in time, each processor gets the same value for each variable if it performs a read access. But cache coherence does not specify in which order write accesses become visible to the other processors. This issue is addressed by memory consistency models. These models provide a formal specification of how the memory system will appear to the programmer. The consistency model sets some restrictions on the values that can be returned by a read operation in a shared address space. Intuitively, a read operation should always return the value that has been written last. In uniprocessors, the program order uniquely defines which value this is. In multiprocessors, different processors execute their programs concurrently and the memory accesses may take place in different order depending on the relative progress of the processors.

The following example illustrates the variety of different results of a parallel program if different execution orders of the program statements by the different processors are considered, see also [116].

Example: We consider three processors P_1, P_2, P_3 which execute a parallel program with shared variables x_1, x_2, x_3. The three variables x_1, x_2, x_3 are assumed to be initialized with 0. The processors execute the following programs:

processor	P_1	P_2	P_3
program	(1) $x_1 = 1$;	(3) $x_2 = 1$;	(5) $x_3 = 1$;
	(2) print x_2, x_3;	(4) print x_1, x_3;	(6) print x_1, x_2;

Processor P_i sets the value of $x_i, i = 1, 2, 3$ to 1 and prints the values of the other variables x_j for $j \neq i$. In total, six values are printed which may be 0 or 1. Since there are no dependencies between the two statements executed by P_1, P_2, P_3, their order can be arbitrarily reversed. If we allow such a reordering and if the statements of the different processors can be mixed arbitrarily, there are in total $2^6 = 64$ possible output combinations consisting of 0 and 1. Different global orders may lead to the same output. If the processors are restricted to execute their statements in program order (e.g. P_1 must execute (1) before (2)), then output 000000 is *not* possible, since at least one of the variables x_1, x_2, x_3 must be set to 1 before a print operation occurs. A possible sequentialization of the statements is (1), (2), (3), (4), (5), (6). The corresponding output is 001011. □

To clearly describe the behavior of the memory system in multiprocessor environments, the concept of consistency models has been introduced. Using a consistency model, there is a clear definition of the allowable behavior of the memory system which can be used by the programmer for the design of parallel programs. The situation can be described as follows [208]: The input to the memory system is a set of memory accesses (read or write) which are partially ordered by the program order of the executing processors. The output of the memory system is a collection of values returned by the read accesses executed. A consistency model can be seen as a function that maps each input to a set of allowable outputs. The memory system using a specific consistency model guarantees that for any input, only outputs from the set of allowable outputs are produced. The programmer must write parallel programs such that they work correctly for any output allowed by the consistency model. The use of a consistency model also has the advantage that it abstracts from the specific physical implementation of a memory system and provides a clear abstract interface for the programmer.

In the following, we give a short overview of popular consistency models. For a more detailed description, we refer to [3, 45, 103, 140, 208].

Memory consistency models can be classified according to the following two criteria:

- Are the memory access operations of each processor executed in program order?
- Do all processors observe the memory access operations performed in the same order?

Depending on the answer to these questions, different consistency models can be identified.

2.9.4.1 Sequential consistency

A popular model for memory consistency is the *sequential consistency model* (SC model) [140]. This model is an intuitive extension of the uniprocessor model and places strong restrictions on the execution order of the memory accesses. A memory system is sequentially consistent, if the memory accesses of each single processor are performed in the program order described by that processor's program and if the global result of all memory accesses of all processors appears to all processors in the *same* sequential order which results from an arbitrary interleaving of the memory accesses of the different processors. Memory accesses must be performed as *atomic* operations, i.e., the effect of each memory operation must become globally visible to all processors before the next memory operation of any processor is started.

The notion of program order leaves some room for interpretation. Program order could be the order of the statements performing memory accesses in the *source program*, but it could also be the order of the memory access operations in a machine program generated by an optimizing compiler which could perform statement reordering to obtain a better performance. In the following, we assume that the order in the source program is used.

Using sequential consistency, the memory operations are treated as atomic operations that are executed in the order given by the source program of each processor and that are centrally sequentialized. This leads to a *total order* of the memory operations of a parallel program which is the same for all processors of the system. In the example given above, not only output 001011, but also 111111 conforms to the SC model. The output 011001 is not possible for sequential consistency.

The requirement of a total order of the memory operations is a stronger restriction as it has been used for the coherence of a memory system in the last section (page 87). For a memory system to be coherent it is required that the write operations to the *same* memory location are sequentialized such that they appear to all processors in the same order. But there is no restriction on the order of write operations to different memory locations. On the other hand, sequential consistency requires that all write operations (to arbitrary memory locations) appear to all processors in the same order.

The following example illustrates that the atomicity of the write operations is important for the definition of sequential consistency and that the requirement of a sequentialization of the write operations alone is not sufficient.

Example: Three processors P_1, P_2, P_3 execute the following statements:

processor	P_1	P_2	P_3
program	(1) $x_1 = 1$;	(2) while($x_1 == 0$);	(4) while($x_2 == 0$);
		(3) $x_2 = 1$;	(5) print(x_1);

The variables x_1 and x_2 are initialized with 0. Processor P_2 waits until x_1 has value 1 and then sets x_2 to 1. Processor P_3 waits until x_2 has value 1 and then prints the value of x_1. Assuming atomicity of write operations, the statements are executed in the order (1), (2), (3), (4), (5), and processor P_3 prints the value 1 for x_1, since write operation (1) of P_1 must become visible to P_3 before P_2 executes write operation (3). Using a sequentialization of the write operations of a variable without requiring atomicity and global sequentialization as required for sequential consistency would allow the execution of statement (3) before the effect of (1) becomes visible for P_3. Thus, (5) could print the value 0 for x_1.

To further illustrate this behavior, we consider a directory-based protocol and assume that the processors are connected via a network. In particular, we consider a directory-based invalidation protocol to keep the caches of the processors coherent. We assume that the variables x_1 and x_2 have been initialized with 0 and that they are both stored in the local caches of P_2 and P_3. The cache blocks are marked as shared (S).

The operations of each processor are executed in program order and a memory operation is not started before the preceding operations of the same processor have been completed. Since no assumptions on the transfer of the invalidation messages in the network are made, the following execution order is possible:

1. P_1 executes the write operation (1) to x_1. Since x_1 is not stored in the cache of P_1, a write miss occurs. The directory entry of x_1 is accessed and invalidation messages are sent to P_2 and P_3.
2. P_2 executes the read operation (2) to x_1. We assume that the invalidation message of P_1 has already reached P_2 and that the memory block of x_1 has been marked invalid (I) in the cache of P_2. Thus, a read miss occurs, and P_2 obtains the current value 1 of x_1 over the network from P_1. The copy of x_1 in the main memory is also updated.

   After having received the current value of x_1, P_1 leaves the while loop and executes the write operation (3) to x_2. Because the corresponding cache block is marked as shared (S) in the cache of P_2, a write miss occurs. The directory entry of x_2 is accessed and invalidation messages are sent to P_1 and P_3.
3. P_3 executes the read operation (4) to x_2. We assume that the invalidation message of P_2 has already reached P_3. Thus, P_3 obtains the current value 1 of x_2 over the network. After that, P_3 leaves the while loop and executes the print operation (5). Assuming that the invalidation message of P_1 for x_1 has not yet reached P_3, P_3 accesses the old value 0 for x_1 from its local cache, since the corresponding cache block is still marked with S. This behavior is possible if the invalidation messages may have different transfer times over the network.

In this example, sequential consistency is violated, since the processors observe different orders of the write operation: Processor P_2 observes the order $x_1 = 1, x_2 = 1$ whereas P_3 observes the order $x_2 = 1, x_1 = 1$ (since P_3 gets the *new* value of x_2, but the *old* value of x_1 for its read accesses).   □

In a parallel system, sequential consistency can be guaranteed by the following *sufficient conditions* [45, 57, 198]:

- Every processor issues its memory operations in program order. In particular, the compiler is not allowed to change the order of memory operations, and no out-of-order executions of memory operations are allowed.
- After a processor has issued a write operation, it waits until the write operation has been completed before it issues the next operation. This includes that for a write miss all cache blocks which contain the memory location written must be marked invalid (I) before the next memory operation starts.
- After a processor has issued a read operation, it waits until this read operation and the write operation whose value is returned by the read operation has been entirely completed. This includes that the value returned to the issuing processor becomes visible to all other processors before the issuing processor submits the next memory operation.

These conditions do not contain specific requirements concerning the interconnection network, the memory organization or the cooperation of the processors in the parallel system. In the example from above, condition (3) ensures that after reading x_1, P_2 waits until the write operation (1) has been completed before it issues the next memory operation (3). Thus, P_3 always reads the new value of x_1 when it reaches statement (5). Therefore, sequential consistency is ensured.

For the programmer, sequential consistency provides an easy and intuitive model. But the model has a performance disadvantage, since all memory accesses must be atomic and since memory accesses must be performed one after another. Therefore, processors may have to wait for quite a long time before memory accesses that they have issued have been completed. To improve performance, consistency models with fewer restrictions have been proposed. We give a short overview in the following and refer to [45, 103] for a more detailed description. The goal of the less restricted models is to enable a more efficient implementation while keeping the model simple and intuitive.

2.9.4.2 Relaxed consistency models

Sequential consistency requires that the read and write operations issued by a processor maintain the following orderings where $X \rightarrow Y$ means that the operation X must be completed before operation Y is executed:

- $R \rightarrow R$: The read accesses are performed in program order.
- $R \rightarrow W$: A read operation followed by a write operation is executed in program order. If both operations access the same memory location, an *anti-dependence*

occurs. In this case, the given order must be preserved to ensure that the read operation accesses the correct value.

- $W \rightarrow W$: The write accesses are performed in program order. If both operations access the same memory location, an *output dependence* occurs. In this case, the given order must be preserved to ensure that the correct value is written last.
- $W \rightarrow R$: A write operation followed by a read operation is executed in program order. If both operations access the same memory location, a *flow dependence* (also called *true dependence*) occurs.

If there is a dependence between the read and write operation, the given order must be preserved to ensure the correctness of the program. If there is no such dependence, the given order must be kept to ensure sequential consistency. *Relaxed consistency models* abandon one or several of the orderings required for sequential consistency, if the data dependencies allow this.

Processor consistency models relax the $W \rightarrow R$ ordering to be able to partially hide the latency of write operations. Using this relaxation, a processor can execute a read operation even if a preceding write operation has not yet been completed, if there are no dependencies. Thus, a read operation can be performed even if the effect of a preceding write operation is not visible yet to all processors. Processor consistency models include *total store ordering* (**TSO model**) and *processor consistency* (**PC model**). In contrast to the TSO model, the PC model does not guarantee atomicity of the write operations. The differences between sequential consistency and the TSO or PC model are illustrated in the following example.

Example: Two processors P_1 and P_2 execute the following statements:

processor	P_1	P_2
program	(1) $x_1 = 1$;	(3) $x_2 = 1$;
	(2) print(x_2);	(4) print(x_1);

Both variables x_1 and x_2 are initialized with 0. Using sequential consistency, statement (1) must be executed before statement (2), and statement (3) must be executed before statement (4). Thus, it is not possible that the value 0 is printed for both x_1 and x_2. But using TSO or PC, this output is possible, since, for example, the write operation (3) does not need to be completed before P_2 reads the value of x_1 in (4). Thus, both P_1 and P_2 may print the old value for x_1 and x_2, respectively. ☐

Partial store ordering (PSO) models relax both the $W \rightarrow W$ and the $W \rightarrow R$ ordering required for sequential consistency. Thus in PSO models, write operations can be completed in a different order as given in the program if there is no output dependence between the write operations. Successive write operations can be overlapped which may lead to a faster execution, in particular when write misses occur.

The following example illustrates the differences between the four models sequential consistency, processor consistency (PC), total store ordering (TSO), and partial store ordering (PSO).

Example: We assume that the variables x_1 and *flag* are initialized with 0. Two processors P_1 and P_2 execute the following statements:

processor P_1 P_2
program   (1) $x_1 = 1$;   (3) while(flag == 0);
             (2) flag = 1;   (4) print(x_1);

Using sequential consistency, PC or TSO, it is *not* possible that the value 0 is printed for x_1. But using the PSO model, the write operation (2) can be completed *before* $x_1 = 1$. Thus, it is possible that the value 0 is printed for x_1 in statement (4). This output does not conform with intuitive understanding of the program behavior in the example, making this model less attractive for the programmer.   □

Weak ordering models additionally relax the $R \rightarrow R$ and $R \rightarrow W$ orderings. Thus, no completion order of the memory operations is guaranteed. To support programming, these models provide additional synchronization operations to ensure the following properties:

- All read and write operations which lie in the program *before* the synchronization operation are completed before the synchronization operation.
- The synchronization operation is completed before read or write operations are started which lie in the program *after* the synchronization operation.

The advent of multicore processors has led to an increased availability of parallel systems and most processors provide hardware support for a memory consistency model. Often, relaxed consistency models are supported, as it is the case for the PowerPC architecture of IBM or the different Intel architectures. But different hardware manufacturers favor different models, and there is no standardization as yet.

2.10 Examples for hardware parallelism

Hardware for executing calculations is parallel is available at several levels, including instruction level parallelism (ILP) using multiple functional units, multithreading parallelism using multiple cores of a processor, and message-passing parallelism using cluster systems. This section considers examples for the different levels. In particular, Subsection 2.10.1 describes the architecture of a typical desktop or server processor with internal instruction level parallelism that is controlled by hardware without support by the programmer. Subsection 2.10.2 briefly describes the Top500 list capturing the architectural details of current supercomputers, which exhibit parallelism at large scale connecting multiple processors with a fast interconnection network.

2.10.1 Intel Cascade Lake and Ice Lake Architectures

An example for the design of a typical multicore processor is the Intel Cascade Lake architecture, which is available in many configurations with up to 56 cores. The Cascade Lake micro-architecture has been introduced in 2019 as server architecture and it is also used for the Core i7 and Core i9 high-end desktop systems. A more detailed description is given in [105, 17, 34].

The Intel Cascade Lake micro-architecture is the successor of the Skylake micro-architecture, which has been used by Intel since 2017. The high-level design of the individual cores of the Cascade Lake micro-architecture and the Skylake micro-architecture are identical [34]. The architecture supports the Intel x86-64 architecture, which is a 64-bit extension of the x86-architecture that has already been used for the Pentium processors. The Cascade Lake processors are available in many different configurations with up to 56 cores and up to 12 DDR4 memory channels. Most of the Cascade Lake processors support hyper-threading with two threads, i.e., the resources of each physical core are shared between two independent threads. The cores are internally connected by a two-dimensional mesh network. The parallelism provided by the architecture with multiple cores requires the provision of a parallel program that controls the execution of the different cores. Suitable parallel programming models and environments are described in the following chapters.

Two versions of the Cascade Lake architecture are available: a server configuration and a client configuration. The core of the server configuration is considerably larger than the core of the client configuration, featuring **Advanced Vector Extensions 512** (AVX-512). Mainstream desktop and laptop processors are based on the client configuration and high-end server processors are based on the server configuration. The following overview of the architecture describes the internal parallelism at instruction level within each single physical core using the information in [34]. This parallelism is completely controlled by hardware and its exploitation does not require the support by a parallel program. Instead, the sequential control flow provided for a core is inspected dynamically during program execution by the hardware to find instructions that can be executed concurrently to each other due to the absence of dependencies.

The architecture of each physical core consists of the **front-end**, the **back–end** (also called **execution engine**), and the **memory subsystem**, see Fig. 2.35 for a high-level overview by a block diagram [34]. The **front-end** is responsible for fetching x86 instructions from the memory system, decoding them, and delivering them to the execution units. The **back-end** is responsible for the actual execution of machine instructions. The **memory subsystem** is responsible for the provision of data that the execution engine needs for its computations. All three components are now described in more detail.

Front-End: Since the execution engine can execute multiple instructions in parallel in one machine cycle due to the existence of multiple functional units, it is important that the front-end is also able to decode multiple instructions in one cycle. Therefore, multiple decoders are employed. The **x86 instructions** to be decoded by the front-end may be complex CISC (complex instruction set computer) instruc-

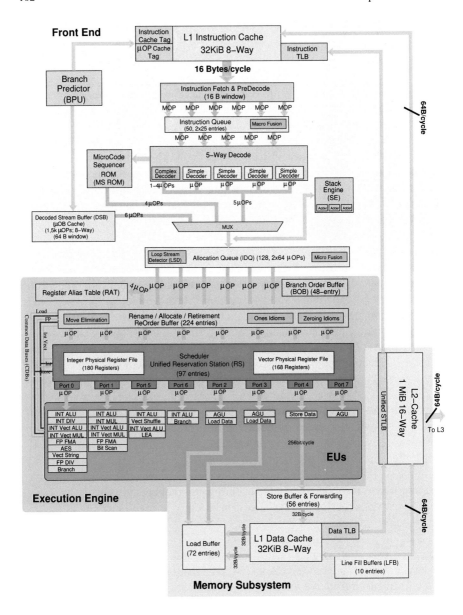

Fig. 2.35 Block diagram illustrating the internal organization of one core of the Intel Cascade Lake micro-architecture [34].

tions, may have variable length and may represent multiple operations. Therefore, decoders of different type are available. For the decoding, the un-decoded instructions are loaded from the L1 instruction cache into the fetch window (16 Bytes). In the first step of the decoding, the current content of the fetch window is inspected and the instruction boundaries are detected and marked. The resulting pre-decoded instructions are stored in the Instruction Queue (IQ), which has 25 entries for each thread. From the IQ, up to five pre-decoded instructions can be sent to the decoders in each cycle. The decoders read in pre-decoded instructions and emit regular, so-called μ**OPs** of fixed length. For the decoding, four simple decoders and one complex decoder are available. The simple decoders are capable of translating instructions that emit a single fused-μOP. The complex decoder is used to decode complex pre-decoded instructions that may result in up to four fused-μOPs. Even more complex instructions that require more than four μOPs are decoded by the microcode sequencer (MS) ROM. Stack instructions (such as PUSH or POP) are handled by the separate **Stack Engine**. All μOPs emitted by the decoders are sent to the **Instruction Decode Queue (IDQ)**, which can store up to 64 entries per thread. The IDQ can be considered as an interface buffer between the front-end and the execution engine, providing a steady flow of executable instructions as μOPs to keep the execution engine busy. The IDQ contains a Loop Stream Detector (LSD) to detect loops that fit entirely into the IDQ. Such loops can be executed without reloading instructions into the IDQ, i.e., the front-end need not provide new decoded instructions for the execution of the loop and, thus, the execution engine can work at full speed.

Execution Engine: The execution engine is responsible for the actual execution of the μOP instructions. To do so, μOPs are loaded from the IDQ into the reorder buffer. In each machine cycle, up to six μOPs can be loaded into the reorder buffer, which can store up to 224 μOPs in total. Before the execution of the instructions, register renaming is performed, i.e., the architectural registers are mapped onto the physical registers of the core. An integer register file with 180 registers and a vector register file with 168 registers are available. Register renaming is supported by the Register Alias Table (RAT). **Eight ports** are available to dispatch μOPs to functional units for execution, i.e., up to eight instructions can be executed in parallel, thus providing Instruction Level Parallelism (ILP). Different ports are used for different types of instructions as described in the following subsection. In general, the degree of ILP is limited by dependencies between instructions that have to be determined by hardware from a sequential flow of control provided to the hardware core. Therefore, increasing the number of ports would not necessarily lead to an increased degree of ILP.

The actual assignment of μOPs to functional units is performed by the scheduler that is shared by two threads (hyperthreading). This scheduler is referred to as **unified Reservation Station** (RS) in Fig. 2.35 and can store up to 97 entries. The scheduler uses speculative execution, i.e., μOPs are executed even if it is not clear whether they are on the correct execution path (e.g. when using branch prediction). Speculative execution is supported by the Branch Order Buffer (BOB): if speculative execution has selected the wrong branch, the architectural state is rolled back to the last valid architectural state. Only μOPs that are on the correct execution path are finally retired, i.e., their results are written into the target registers and any resources used are released. The assignment of μOPs to functional units (also called execution units, EUs) is performed via **eight port**, where each port leads to a subset of the functional units as depicted in Fig. 2.35. For example, Port 0 can be used to issue integer arithmetic operations (INT ALU), integer divisions (INT DIV), integer vector arithmetic operations (INT Vect ALU), integer vector multiplications (INT MUL), floating-point fused multiply-add operations (FP FMA), AES encryption (AES), string operations (Vect String), floating-point divisions (FP DIV), or branch instructions (Branch). Two ports (Port 2 and Port 3) can be used to simultaneously load two data elements into registers and one port (Port 4) can be used to store a data element from a register into the memory system. Thus, along with some arithmetic operations, two operands can be loaded and one results can be stored simultaneously in one cycle.

Memory Subsystem: The memory subsystem is responsible for the data transfer between the memory system and the internal registers. The memory subsystem is activated by the provision of memory addresses by load and store instructions executed by the execution engine. The memory subsystem consists of several levels:

- Each physical core has its own **L1 split cache** (32 KB data, 32 KB instructions), shared by the two threads when hyperthreading is supported. The L1 data cache can be accessed in 4–5 cycles, depending on the complexity of the address calculation.
- Each physical core also has its own **unified L2 cache** containing instructions and data (1 MB per core). The L2 cache is also shared by the two threads and can be accessed in 14 cycles.
- There is one **unified L3 cache** with up to 1.375 MB per core that is shared among all physical cores of the processor. The L3 cache can be accessed in 50-70 cycles.
- The **off-chip main memory** has a configurable size. Six channels are available for accessing the main memory, each providing a bandwidth of about 23.5 GB/s.

Table 2.4 contains a summary of the cache characteristics of the Cascade Lake architecture. Each core has a separate L1 and L2 cache, the L3 cache is shared among all cores. The cache line size is 64 bytes for all caches. All caches use a write-back replacement policy and have a block size of 64 bytes. The memory subsystem is able to sustain two memory reads (on Ports 2 and 3) and one memory write (on Port 4) in each cycle. The target of each memory operation can be any register of size up to 256 bits. The bandwidth between the L1 and the L2 cache is 64 Bytes in each cycle in both directions. The bandwidth between the L2 and the shared L3 cache is 32 Bytes in each cycle. A more detailed description of the different aspects of the Cascade Lake architecture can be found in [34].

characteristic	L1	L2	L3
size	32 KB instructions 32 KB data	1 MB/core	1.375 MB/core
associativity	8-way instructions 11-way data	16-way	16-way
access latency	4-5 cycles	14 cycles	50-70 cycles
replacement policy	write-back	write-back	write-back

Table 2.4 Description of the cache architecture of the Intel Xeon server processor (Cascade Lake) 2019.

2.10.2 Top500 list

Parallelism for high-performance codes in scientific computing is usually provided by large parallel computers, also called supercomputers or high-performance computers. The characteristics and architecture of these parallel computers have been changed over the years. The most recent parallel computers are given in the Top500 list, which has been introduced in 1993 to collect statistics on high-performance computers, see www.top500.org. The effort is based on an initiative by Hans Meuer from 1986. The list is updated twice a year in June and November in connection with two international supercomputing conferences (June: ISC High Performance in Germany, November: SC: International Conference for High Performance, Networking, Storage, and Analysis in the US). The Top500 list provides information on the 500 most powerful computer systems. The systems are ranked according to their performance on the LINPACK Benchmark, which solves a dense system of linear equations using a decompositional approach, see [54] for a detailed description. In particular, the HPL (High-Performance Linpack) benchmark is used for solving random dense linear system in double precision (64 bits) arithmetic on distributed-memory computers using MPI (Message Passing Interface), see chapter 5.1. Most of the computation time of the benchmark is used to multiply a scalar α with a vector x and add the result to another vector y.

The Top500 list provides detailed information on the computer systems in the list, including the LINPACK performance, the peak performance, the number and type of cores, the memory system, the power consumption, the interconnection network, and the operating system. In many cases, a system URL with more detailed information about the system is also provided. Statistics about the vendor shares, country shares, interconnection shares, or operating system shares are also provided.

To broaden the perspective, two alternative lists are maintained on the Top500 web page: To use an alternative benchmark, the HPCG list is maintained, using the High-Performance Conjugate Gradient (HPCG) benchmark for ranking the systems. To draw the attention to energy efficiency, the Green500 list is maintained, collecting information on the 500 most energy-efficient system. The ranking is performed according to the power efficiency of the systems when executing the HPL benchmark, using GFlops per Watt as performance measure.

2.11 Exercises for Chapter 2

Exercise 2.1. Consider a two-dimensional mesh network with m columns and n rows. Determine the bisection bandwidth of this network?

Exercise 2.2. Consider a shuffle-exchange network with $n = 2^k$ nodes, $k > 1$. How many of the $3 \cdot 2^{k-1}$ edges are shuffle edges and how many are exchange edges? Draw a shuffle-exchange network for $k = 4$.

Exercise 2.3. In Section 2.7.2, page 44, we have shown that there exist k independent paths between any two nodes of a k-dimensional hypercube network. For $k = 5$, determine all paths between the following pairs of nodes: (i) nodes 01001 and 00011; (ii) nodes 00001 and 10000.

Exercise 2.4. Write a (sequential) program that determines all paths between any two nodes for hypercube networks of arbitrary dimension.

Exercise 2.5. The RGC sequences RGC_k can be used to compute embeddings of different networks into a hypercube network of dimension k. Determine RGC_3, RGC_4, and RGC_5. Determine an embedding of a 3-dimensional mesh with $4 \times 2 \times 4$ nodes into a 5-dimensional hypercube network.

Exercise 2.6. Show how a complete binary tree with n leaves can be embedded into a butterfly network of dimension $\log n$. The leaves of the trees correspond to the butterfly nodes at level $\log n$.

Exercise 2.7. Construct an embedding of an three-dimensional torus network with $8 \times 8 \times 8$ nodes into a nine-dimensional hypercube network according to the construction in Section 2.7.3, page 49.

Exercise 2.8. A k-dimensional Beneš network consists of two connected k-dimensional butterfly networks, leading to $2k + 1$ stages, see page 53. A Beneš network is *non-blocking*, i.e., any permutation between input nodes and output nodes can be realized without blocking. Consider an 8×8 Benes network and determine the switch positions for the following two permutations:

$$\pi_1 = \begin{pmatrix} 0 & 1 & 2 & 3 & 4 & 5 & 6 & 7 \\ 0 & 1 & 2 & 4 & 3 & 5 & 7 & 6 \end{pmatrix} \quad \pi_2 = \begin{pmatrix} 0 & 1 & 2 & 3 & 4 & 5 & 6 & 7 \\ 2 & 7 & 4 & 6 & 0 & 5 & 3 & 1 \end{pmatrix}$$

Exercise 2.9. The cross-product $G_3 = (V_3, E_3) = G_1 \otimes G_2$ of two graphs $G_1 = (V_1, E_1)$ and $G_2 = (V_2, E_2)$ can be defined as follows: $V_3 = V_1 \times V_2$ and $E_3 = \{((u_1, u_2), (v_1, v_2)) \mid ((u_1 = v_1) \text{ and } (u_2, v_2) \in E_2) \text{. or } ((u_2 = v_2) \text{ and } (u_1, v_1) \in E_1)\}$. The symbol $\otimes$ can be used as abbreviation with the following meaning

$$\bigotimes_{i=a}^{b} G_i = ((\cdots (G_a \otimes G_{a+1}) \otimes \cdots) \otimes G_b)$$

Draw the following graphs and determine their network characteristics (degree, node connectivity, edge connectivity, bisection bandwidth, and diameter)

(a) linear array of size 4 $\otimes$ linear array of size 2
(b) 2-dimensional mesh with 2×4 nodes $\otimes$ linear array of size 3
(c) linear array of size 3 $\otimes$ complete graph with 4 nodes
(d) $\overset{4}{\underset{i=2}{\otimes}}$ linear array of size i
(e) $\overset{k}{\underset{i=1}{\otimes}}$ linear array of size 23. Draw the graph for $k = 4$, but determine the characteristics for general values of k.

Exercise 2.10. Consider a three-dimensional hypercube network and prove that E-cube routing is deadlock-free for this network, see Section 2.8.1, page 58.

Exercise 2.11. In the directory-based cache coherence protocol described in Section 2.9.3, page 93, in case of a read miss with dirty bit 1, the processor which has the requested cache block sends it to both the directory controller and the requesting processor. Instead, the owning processor could send the cache block to the directory controller and this one could forward the cache block to the requesting processor. Specify the details of this protocol.

Exercise 2.12. Consider the following sequence of memory accesses

$$2, 3, 11, 16, 21, 13, 64, 48, 19, 11, 3, 22, 4, 27, 6, 11$$

Consider a cache of size 16 Bytes. For the following configurations of the cache determine for each of the memory accesses in the sequence whether it leads to a cache hit or a cache miss. Show the resulting cache state that results after each access with the memory locations currently held in cache. Determine the resulting miss rate.

a) direct-mapped cache with block-size 1
b) direct-mapped cache with block-size 4
c) two-way set-associative cache with block-size 1, LRU replacement strategy
d) two-way set-associative cache with block-size 4, LRU replacement strategy
e) fully-associative cache with block-size 1, LRU replacement
f) fully-associative cache with block-size 4, LRU replacement

Exercise 2.13. Consider the MSI protocol from Fig. 2.33, page 91, for a bus-based system with three processors P_1, P_2, P_3. Each processor has a direct-mapped cache. The following sequence of memory operations access two memory locations A and B which are mapped to the same cache line:

processor	action
P_1	write $A, 4$
P_3	write $B, 8$
P_2	read A
P_3	read A
P_3	write A, B
P_2	read A
P_1	read B
P_1	write $B, 10$

Assume that the variables are initialized with $A = 3$ and $B = 3$ and that the caches are initially empty. For each memory access determine

- the cache state of each processor after the memory operations,
- the content of the cache and the memory location for A and B,
- the processor actions (PrWr, PrRd) caused by the access, and
- the bus operations (BusRd, BusRdEx, flush) caused by the MSI protocol.

Exercise 2.14. Consider three processes P_1, P_2, P_3 performing the following memory accesses:

P_1	P_2	P_3
(1) $A = 1;$	(1) $B = A;$	(1) $D = C;$
(2) $C = 1;$		

The variables A, B, C, D are initialized with 0. Using the sequential consistency model, which values can the variables B and D have?

Exercise 2.15. Visit the Top500 webpage at www.top500.org and determine important characteristics of the five fastest parallel computers, including number of processors or core, interconnection network, processors used, and memory hierarchy.

Exercise 2.16. Consider the following two implementations of a matrix computation with a column-wise traversal of the matrix (left) and a row-wise traversal of the matrix (right):

```
for (j=0; j<1500; j++)              for (i=0; i<1500; i++)
   for (i=0; i<1500; i++)              for (j=0; j<1500; j++)
      x[i][j] = 2 · x[i][j];             x[i][j] = 2 · x[i][j];
```

Assume a cache of size 8 Kbytes with a large enough associativity so that no conflict misses occur. The cache line size is 32 bytes. Each entry of the matrix x occupies 8 bytes. The implementations of the loops are given in C which uses a row-major storage order for matrices. Compute the number of cache lines that must be loaded for each of the two loop nests. Which of the two loop nests leads to a better spatial locality?

Chapter 3
Parallel Programming Models

About this chapter

Programming models are a suitable way to abstract away from details of the architecture of parallel systems and to provide a high-level view which is appropriate for the application programmer. This chapter introduces models for parallel programming and provides mechanisms and definitions that are important for the organization of parallel programs for shared or distributed address spaces. Models for parallel systems, the parallelization of programs, SIMD computations, data distributions for arrays, information exchange techniques as well as processes and threads are covered.

The sections of this chapter are organized as follows: Section 3.1 explains the difference between architectural models, computational models, and programming models. Section 3.2 describes the steps that are required for the parallelization of programs. Section 3.3 considers different levels of parallelism such as data parallelism, loop parallelism, or functional parallelism. Patterns for the structuring of parallel programs are also addressed. Section 3.4 addresses SIMD computations. Section 3.5 describes data distributions for arrays. Section 3.6 examines the information exchange in parallel programs and introduces common communication operations for distributed address spaces. Section 3.7 studies the parallel computation of a matrix-vector product as an example. Section 3.8 investigates the organization of parallel programs using processes and threads.

3.1 Models for parallel systems

The coding of a parallel program for a given algorithm is strongly influenced by the parallel computing system to be used. The term computing system comprises all hardware and software components which are provided to the programmer and which form the programmer's view of the machine. The hardware architectural aspects have been presented in Chapter 2. The software aspects include the specific operating system, the programming language and the compiler, or the runtime libraries. The same parallel hardware can result in different views for the program-

© The Author(s), under exclusive license to Springer Nature Switzerland AG 2023
T. Rauber, G. Rünger, *Parallel Programming*, https://doi.org/10.1007/978-3-031-28924-8_3

mer, i.e. in different parallel computing systems when used with different software installations. A very efficient coding can usually be achieved when the specific hardware and software installation is taken into account. But in contrast to sequential programming there are many more details and diversities in parallel programming and a machine dependent programming can result in a large variety of different programs for the same algorithm. In order to study more general principles in parallel programming, parallel computing systems are considered in a more abstract way with respect to some properties, such as the organization of memory as shared or private. A systematic way to do this is to consider models which step back from details of single systems and provide an abstract view for the design and analysis of parallel programs.

In the following, the types of models used for parallel processing according to [108] are presented. Models for parallel processing can differ in their level of abstraction. The four basic types are machine models, architectural models, computational models, and programming models. The **machine model** is at the lowest level of abstraction and consists of a description of hardware and operating system, e.g. the registers or the input and output buffers. Assembly languages are based on this level of models. **Architectural models** are at the next level of abstraction. Properties described at this level include the interconnection network of parallel platforms, the memory organization, the synchronous or asynchronous processing, or the execution mode of single instructions by SIMD or MIMD.

The **computational model** (or model of computation) is at the next higher level of abstraction and offers an abstract or more formal model of a corresponding architectural model. It provides cost functions reflecting the time needed for the execution of an algorithm on the resources of a computer given by an architectural model. Thus, a computational model provides an analytical method for designing and evaluating algorithms. The complexity of an algorithm should reflect the performance on a real computer. For sequential computing, the RAM model (random access machine) is a computational model for the von Neumann architectural model. The RAM model describes a sequential computer by a memory and one processor accessing the memory. The memory consists of an unbounded number of memory locations each of which can contain an arbitrary value. The processor executes a sequential algorithm consisting of a sequence of instructions step by step. Each instruction comprises the load of data from memory into registers, the execution of an arithmetic or logical operation, and storing the result into memory. The RAM model is suitable for theoretical performance prediction although real computers have a much more diverse and complex architecture. A computational model for parallel processing is the PRAM model (parallel random access machine), which is a generalization of the RAM model and is described in Chapter 4.

The **programming model** is at the next higher level of abstraction and describes a parallel computing system in terms of the semantics of the programming language or programming environment. A parallel programming model specifies the programmer's view on the parallel computer by defining how the programmer can code an algorithm. This view is influenced by the architectural design and the language, the compiler, or the runtime libraries, and, thus, there exist many different parallel programming models even for the same architecture. There are several criteria concerning which the parallel programming models can differ.

- The level of parallelism which is exploited in the parallel execution (instruction level, statement level, procedural level or parallel loops).
- The implicit or user defined explicit specification of parallelism.
- The way how parallel program parts are specified.
- The execution mode of parallel units (SIMD or SPMD, synchronous or asynchronous).
- The modes and pattern of communication among computing units for the exchange of information (explicit communication or shared variables).
- Synchronization mechanisms to organize computation and communication between parallel units.

Each parallel programming language or environment implements the criteria given above and there is a large number of different possibilities for combination. Parallel programming models provide methods to support parallel programming.

The goal of a programming model is to provide a mechanism with which the programmer can specify parallel programs. To do so, a set of basic tasks must be supported. A parallel program specifies computations which can be executed in parallel. Depending on the programming model, the computations can be defined at different levels: a computation can be (i) a sequence of *instructions* performing arithmetic or logical operations, (ii) a sequence of *statements* where each statement may capture several instructions, or (iii) a function or method invocation which typically consists of several statements. Many parallel programming models provide the concept of *parallel loops*; the iterations of a parallel loop are independent of each other and can therefore be executed in parallel, see Section 3.3.3 for an overview. Another concept is the definition of independent **tasks** (or *modules*) which can be executed in parallel and which are mapped to the processors of a parallel platform such that an efficient execution results. The mapping may be specified explicitly by the programmer or performed implicitly by a runtime library.

A parallel program is executed by the processors of a parallel execution environment such that on each processor one or multiple control flows are executed. Depending on the specific coordination, these control flows are referred to as **processes** or **threads**. The thread concept is a generalization of the process concept: a process can consist of several threads which share a common address space whereas each process works on a different address space. Which of these two concepts is more suitable for a given situation depends on the physical memory organization of the execution environment. The process concept is usually suitable for distributed memory organizations whereas the thread concept is typically used for shared memory machines, including multicore processors. In the following chapters, programming models based on the process or thread concept are discussed in more detail.

The processes or threads executing a parallel program may be created statically at program start. They may also be created during program execution according to the specific execution needs. Depending on the execution and synchronization modi supported by a specific programming model, there may or may not exist a hierarchical relation between the threads or processes. A fixed mapping from the threads or processes to the execution cores or processors of a parallel system may be used. In this case, a process or thread cannot be migrated to another processor or core during program execution. The partitioning into tasks and parallel execution modes for parallel programs are considered in more detail in Sections 3.2 - 3.3.6. Data distributions for structured data types like vectors or matrices are considered in Section 3.5.

An important classification for parallel programming models is the *organization of the address space*. There are models with a shared or distributed address space, but there are also hybrid models which combine features of both memory organizations. The address space has a significant influence on the information exchange between the processes or threads. For a shared address space, shared variables are often used. Information exchange can be performed by write or read accesses of the processors or threads involved. For a distributed address space, each process has a local memory, but there is no shared memory via which information or data could be exchanged. Therefore, information exchange must be performed by additional message-passing operations to send or receive messages containing data or information. More details will be described in Section 3.6.

3.2 Parallelization of programs

The parallelization of a given algorithm or program is typically performed on the basis of the programming model used. Independent of the specific programming model, typical steps can be identified to perform the parallelization. In this section, we will describe these steps. We assume that the computations to be parallelized are given in form of a sequential program or algorithm. To transform the sequential computations into a parallel program, their control and data dependencies have to be taken into consideration to ensure that the parallel program produces the same results as the sequential program for all possible input values. The main goal is usually to reduce the program execution time as much as possible by using multiple processors or cores. The transformation into a parallel program is also referred to as **parallelization**. To perform this transformation in a systematic way, it can be partitioned into several steps:

1. **Decomposition of the computations**: The computations of the sequential algorithm are decomposed into tasks and dependencies between the tasks are determined. The tasks are the smallest units of parallelism. Depending on the target system, they can be identified at different execution levels: instruction level, data parallelism, or functional parallelism, see Section 3.3. In principle, a task is a sequence of computations which is executed by a single processor or core. Depending on the memory model, a task may involve accesses to the shared address space or may execute message-passing operations. Depending on the specific application, the decomposition into tasks may be done in an initialization phase at program start (static decomposition), but tasks can also be created dynamically

during program execution. In this case, the number of tasks available for execution can vary significantly during the execution of a program. At any point in program execution, the number of executable tasks is an upper bound on the available degree of parallelism and, thus, the number of cores that can be usefully employed. The goal of task decomposition is therefore to generate enough tasks to keep all cores busy at all times during program execution. But on the other hand, the tasks should contain enough computations such that the task execution time is large compared to the scheduling and mapping time required to bring the task to execution. The computation time of a task is also referred to as **granularity**: Tasks with many computations have a coarse-grained granularity, tasks with only a few computations are fine-grained. If task granularity is too fine-grained, the scheduling and mapping overhead is large and constitutes a significant amount of the total execution time. Thus, the decomposition step must find a good compromise between the number of tasks and their granularity.

2. **Assignment of tasks to processes or threads**: A process or a thread represents a flow of control executed by a physical processor or core. A process or thread can execute different tasks one after another. The number of processes or threads does not necessarily need to be the same as the number of physical processors or cores, but often the same number is used. The main goal of the assignment step is to assign the tasks such that a good **load balancing** results, i.e. each process or thread should have about the same number of computations to perform. But the number of memory accesses (for shared address space) or communication operations for data exchange (for distributed address space) should also be taken into consideration. For example, when using a shared address space, it is useful to assign two tasks which work on the same data set to the same thread, since this leads to a good cache usage. The assignment of tasks to processes or threads is also called **scheduling**. For a static decomposition, the assignment can be done in the initialization phase at program start (static scheduling). But scheduling can also be done during program execution (dynamic scheduling).

3. **Mapping of processes or threads to physical processes or cores**: In the simplest case, each process or thread is mapped to a separate processor or core, also called execution unit in the following. If less cores than threads are available, multiple threads must be mapped to a single core. This mapping can be done by the operating system, but it could also be supported by program statements. The main goal of the mapping step it to get an equal utilization of the processors or cores while keeping communication between the processors as small as possible.

The parallelization steps are illustrated in Fig. 3.1.

In general, a **scheduling algorithm** is a method to determine an efficient execution order for a set of tasks of a given duration on a given set of execution units. Typically, the number of tasks is much larger than the number of execution units. There may be dependencies between the tasks, leading to *precedence constraints*. Since the number of execution units is fixed, there are also *capacity constraints*. Both types of constraints restrict the schedules that can be used. Usually, the scheduling algorithm considers the situation that each task is executed sequentially by one processor or core (single-processor tasks). But in some models, a more general case is

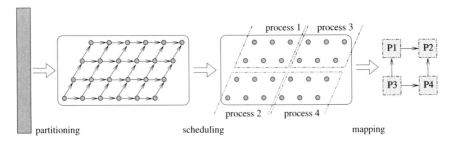

Fig. 3.1 Illustration of typical parallelization steps for a given sequential application algorithm. The algorithm is first split into tasks and dependencies between the tasks are identified. These tasks are then assigned to processes by the scheduler. Finally, the processes are mapped to the physical processors P1, P2, P3 and P4.

also considered which assumes that several execution units can be employed for a single task (parallel tasks), thus leading to a smaller task execution time. The overall goal of a scheduling algorithm is to find a schedule for the tasks which defines for each task a starting time and an execution unit such that the precedence and capacity constraints are fulfilled and such that a given objective function is optimized. Often, the overall completion time (also called *makespan*) should be minimized. This is the time elapsed between the start of the first task and the completion of the last task of the program. For realistic situations, the problem of finding an optimal schedule is NP-complete or NP-hard [76]. A good overview of scheduling algorithms is given in [26].

Often, the number of processes or threads is adapted to the number of execution units such that each execution unit performs exactly one process or thread, and there is no migration of a process or thread from one execution unit to another during execution. In these cases, the terms "process" and "processor" or "thread" and "core" are used interchangeably.

3.3 Levels of parallelism

The computations performed by a given program provide opportunities for a parallel execution at different levels: instruction level, statement level, loop level, and function level. Depending on the level considered, tasks of different **granularity** result. Considering the instruction or statement level, fine-grained tasks result when a small number of instructions or statements are grouped to form a task. On the other hand, considering the function level, tasks are coarse-grained when the functions used to form a task comprise a significant amount of computations. On the loop level medium-grained tasks are typical, since one loop iteration usually consists of several statements. Tasks of different granularity require different scheduling methods to use the available potential of parallelism. In this section, we give a short overview of the available degree of parallelism at different levels and how it can be exploited in different programming models.

3.3.1 Parallelism at instruction level

Multiple instructions of a program can be executed in parallel at the same time, if they are independent of each other. In particular, the existence of one of the following **data dependencies** between instructions I_1 and I_2 inhibits their parallel execution:

- **Flow dependency** (also called *true dependency*): There is a flow dependency from instruction I_1 to I_2, if I_1 computes a result value in a register or variable which is then used by I_2 as operand.
- **Anti-dependency**: There is an anti-dependency from I_1 to I_2, if I_1 uses a register or variable as operand which is later used by I_2 to store the result of a computation.
- **Output dependency**: There is an output dependency from I_1 to I_2, if I_1 and I_2 use the same register or variable to store the result of a computation.

Figure 3.2 shows examples for the different dependency types [227]. In all three cases, instructions I_1 and I_2 cannot be executed in opposite order or in parallel, since this would result in an erroneous computation: For the flow dependency, I_2 would use an old value as operand if the order is reversed. For the anti-dependency, I_1 would use the wrong value computed by I_2 as operand, if the order is reversed. For the output dependency, the subsequent instructions would use a wrong value for R_1, if the order is reversed. The dependencies between instructions can be illustrated by a data dependency graph. Figure 3.3 shows the data dependency graph for a sequence of instructions.

$$I_1: \underline{R_1} \leftarrow R_2 + R_3 \qquad I_1: R_1 \leftarrow \underline{R_2} + R_3 \qquad I_1: \underline{R_1} \leftarrow R_2 + R_3$$
$$I_2: R_5 \leftarrow \underline{R_1} + R_4 \qquad I_2: \underline{R_2} \leftarrow R_4 + R_5 \qquad I_2: \underline{R_1} \leftarrow R_4 + R_5$$

flow dependency   anti dependency   output dependency

Fig. 3.2 Different types of data dependencies between instructions using registers $R_1, \ldots, R_5$. For each type, two instructions are shown which assign a new value to the registers on the left hand side (represented by an arrow). The new value results by applying the operation on the right hand side to the register operands. The register causing the dependency is underlined.

Superscalar processors with multiple functional units can execute several instructions in parallel. They employ a dynamic instruction scheduling realized in hardware, which extracts independent instructions from a sequential machine program by checking whether one of the dependency types discussed above exists. These

I_1: $R_1 \leftarrow A$

I_2: $R_2 \leftarrow R_2 + R_1$

I_3: $R_1 \leftarrow R_3$

I_4: $B \leftarrow R_1$

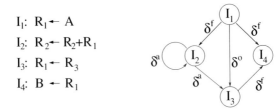

Fig. 3.3 Data dependency graph for a sequence I_1, I_2, I_3, I_4 of instructions using registers R_1, R_2, R_3 and memory addresses A, B. The edges representing a flow dependency are annotated with δ^f. Edges for anti-dependencies and output dependencies are annotated with δ^a and δ^o, respectively. There is a flow dependency from I_1 to I_2 and to I_4, since these two instructions use register R_1 as operand. There is an output dependency from I_1 to I_3, since both instructions use the same output register. Instruction I_2 has an anti-dependency to itself caused by R_2. The flow dependency form I_3 to I_4 is caused by R_1. Finally, there is an anti-dependency from I_2 to I_3 because of R_1.

independent instructions are then assigned to the functional units for execution. For VLIW processors, static scheduling by the compiler is used to identify independent instructions and to arrange a sequential flow of instructions in appropriate long instruction words such that the functional units are explicitly addressed. For both cases, a sequential program is used as input, i.e. no explicit specification of parallelism is used. Appropriate compiler techniques like software pipelining and trace scheduling can help to rearrange the instructions such that more parallelism can be extracted, see [60, 12, 6] for more details.

3.3.2 Data parallelism

In many programs, the same operation must be applied to different elements of a larger data structure. In the simplest case, this could be an array structure. If the operations to be applied are independent of each other, this could be used for a parallel execution: The elements of the data structure are distributed evenly among the processors and each processor performs the operation on its assigned elements. This form of parallelism is called **data parallelism** and can be used in many programs, especially from the area of scientific computing. To use data parallelism, sequential programming languages have been extended to **data parallel programming languages**. Similar to sequential programming languages, *one single* control flow is used, but there are special constructs to express data parallel operations on data structures like arrays. The resulting execution scheme is also referred to as SIMD model, see Section 2.4.

Often, data parallel operations are only provided for arrays. A typical example is the *array assignment* of Fortran 90/95, see [61, 223, 154]. Other examples for data parallel programming languages are C* and data-parallel C [101], PC++ [24],

DINO [190] and High Performance Fortran (HPF) [66, 70]. An example for an array assignment in Fortran 90 is

$$a(1:n) = b(0:n-1) + c(1:n).$$

The computations performed by this assignment are identical to those computed by the following loop:

```
for (i=1:n)
    a(i) = b(i-1) + c(i)
endfor
```

Similar to other data-parallel languages, the semantics of an array assignment in Fortran 90 is defined as follows: first, all array accesses and operations on the right hand side of the assignment are performed. After the complete right hand side is computed, the actual assignment to the array elements on the left hand side is performed. Thus, the following array assignment

$$a(1:n) = a(0:n-1) + a(2:n+1)$$

is not identical to the loop

```
for (i=1:n)
    a(i) = a(i-1) + a(i+1)
endfor
```

The array assignment uses the old values of $a(0:n-1)$ and $a(2:n+1)$ whereas the for loop uses the old value only for $a(i+1)$; for $a(i-1)$ the new value is used, which has been computed in the preceding iteration.

Data parallelism can also be exploited for MIMD models. Often, the SPMD model (**S**ingle **P**rogram **M**ultiple **D**ata) is used which means that *one* parallel program is executed by all processors in parallel. Program execution is performed *asynchronously* by the participating processors. Using the SPMD model, data parallelism results if each processor gets a part of a data structure for which it is responsible. For example, each processor could get a part of an array identified by a lower and an upper bound stored in private variables of the processor. For each processor, the processor ID can be used to compute its part assigned. Different data distributions can be used for arrays, see Section 3.5 for more details. Figure 3.4 shows a part of an SPMD program to compute the scalar product of two vectors.

In practice, most parallel programs are SPMD programs, since SPMD programs are usually easier to understand than general MIMD programs, but provide enough expressiveness to formulate typical parallel computation patterns. In principle, each processor can execute a different program part, depending on its processor ID. Most parallel programs shown in the rest of the book are SPMD programs.

Data parallelism can be exploited for both shared and distributed address space. For a distributed address space, the program data must be distributed among the processors such that each processor can access the data that it needs for its computations directly from its local memory. The processor is then called *owner* of its local data. Often, the distribution of data and computation is done in the same way such that each processor performs the computations specified in the program on the data that it stores in its local memory. This is called **owner-computes rule**, since the owner of the data performs the computations on this data.

```
local_size = size/p;
local_lower = me * local_size;
local_upper = (me+1) * local_size - 1;
local_sum = 0.0;

for (i=local_lower; i<=local_upper; i++)
  local_sum += x[i] * y[i];

Reduce(&local_sum, &global_sum, 0, SUM);
```

Fig. 3.4 SPMD program to compute the scalar product of two vectors x and y. All variables are assumed to be private, i.e., each processor can store a different value in its local instance of a variable. The variable p is assumed to be the number of participating processors, me is the rank of the processor, starting from rank 0. The two arrays x and y of size size and the corresponding computations are distributed blockwise among the processors. The size of a data block of each processor is computed in local_size, the lower and upper bounds of the local data block are stored in local_lower and local_upper, respectively. For simplicity, we assume that size is a multiple of p. Each processor computes in local_sum the partial scalar product for its local data block of x and y. These partial scalar products are accumulated with the reduction function Reduce() at processor 0. Assuming a distribution address space, this reduction can be obtained by calling the MPI function MPI_Reduce(&local_sum, &global_sum, 1, MPI_FLOAT, MPI_SUM, 0, MPI_COMM_WORLD), see Section 5.2.

3.3.3 Loop parallelism

Many algorithms perform computations by iteratively traversing a large data structure. The iterative traversal is usually expressed by a loop provided by imperative programming languages. A loop is usually executed *sequentially* which means that the computations of the ith iteration are started not before all computations of the $(i-1)$th iteration are completed. This execution scheme is called *sequential loop* in the following. If there are no dependencies between the iterations of a loop, the iterations can be executed in arbitrary order, and they can also be executed in parallel by different processors. Such a loop is then called a *parallel loop*. Depending on their exact execution behavior, different types of parallel loops can be distinguished as will be described in the following [223, 12].

3.3.3.1 `forall` loop

The body of a `forall` loop can contain one or several assignments to array elements. If a `forall` loop contains a single assignment, it is equivalent to an array assignment, see Section 3.3.2, i.e., the computations specified by the right hand side of the assignment are first performed in any order, and then the results are assigned to their corresponding array elements, again in any order. Thus, the loop

```
forall (i = 1:n)
  a(i) = a(i-1) + a(i+1)
endforall
```

is equivalent to the array assignment

$$a(1:n) = a(0:n-1)+a(2:n+1)$$

in Fortran 90/95. If the `forall` loop contains multiple assignments, these are executed *one after another* as array assignments, such that the next array assignment is started not before the previous array assignment has been completed. A `forall` loop is provided in Fortran 95, but not in Fortran 90, see [154] for details.

3.3.3.2 `dopar` loop

The body of a `dopar` loop may contain one or several assignments to array elements, but also other statements and even other loops. The iterations of a `dopar` loop are executed by multiple processors in parallel. Each processor executes its iterations in any order one after another. The instructions of each iteration are executed sequentially in program order, using the variable values of the initial state before the `dopar` loop is started. Thus, variable updates performed in one iteration are not visible to the other iterations. After all iterations have been executed, the updates of the single iterations are combined and a new global state is computed. If two different iterations update the same variable, one of the two updates becomes visible in the new global state, resulting in a *nondeterministic* behavior.

The overall effect of `forall` and `dopar` loops with the same loop body may differ if the loop body contains more than one statement. This is illustrated by the following example [223].

Example: We consider the following three loops:

```
for (i=1:4)            forall (i=1:4)          dopar (i=1:4)
  a(i)=a(i)+1            a(i)=a(i)+1             a(i)=a(i)+1
  b(i)=a(i-1)+a(i+1)     b(i)=a(i-1)+a(i+1)      b(i)=a(i-1)+a(i+1)
endfor                 endforall               enddopar
```

In the sequential `for` loop, the computation of `b(i)` uses the value of `a(i-1)` that has been computed in the preceding iteration and the value of `a(i+1)` valid before the loop. The two statements in the forall loop are treated as separate array assignments. Thus, the computation of b(i) uses for both a(i-1) and a(i+1) the new value computed by the first statement. In the dopar loop, updates in one iteration are not visible to the other iterations. Since the computation of b(i) does not use the value of a(i) that is computed in the same iteration, the old values are used for a(i-1) and a(i+1). The following table shows an example for the values computed:

	start values		after for-loop	after forall-loop	after dopar-loop
a(0)	1				
a(1)	2	b(1)	4	5	4
a(2)	3	b(2)	7	8	6
a(3)	4	b(3)	9	10	8
a(4)	5	b(4)	11	11	10
a(5)	6				

□

A dopar loop in which an array element computed in an iteration is only used in that iteration is sometimes called doall loop. The iterations of such a doall loop are independent of each other and can be executed sequentially, or in parallel in any order without changing the overall result. Thus, a doall loop is a *parallel loop* whose iterations can be distributed arbitrarily among the processors and can be executed without synchronization. On the other hand, for a general dopar loop, it has to be made sure that the different iterations are separated, if a processor executes multiple iterations of the same loop. A processor is not allowed to use array values that it has computed in another iteration. This can be ensured by introducing temporary variables to store those array operands of the right-hand side that might cause conflicts and using these temporary variables on the right-hand side. On the left-hand side, the original array variables are used. This is illustrated by the following example:

Example: The following dopar loop

```
dopar (i=2:n-1)
  a(i) = a(i-1) + a(i+1)
enddopar
```

is equivalent to the following program fragment

```
doall (i=2:n-1)
  t1(i) = a(i-1)
  t2(i) = a(i+1)
enddoall
doall (i=2:n-1)
  a(i) = t1(i) + t2(i)
enddoall
```

where t1 and t2 are temporary array variables.   □

More information on parallel loops and their execution as well as on transformations to improve parallel execution can be found in [177, 223]. Parallel loops play an important role in programming environments like OpenMP, see Section 6.5 for more details.

3.3.4 Functional parallelism

Many sequential programs contain program parts that are independent of each other and can be executed in parallel. The independent program parts can be single statements, basic blocks, loops, or function calls. Considering the independent program parts as tasks, this form of parallelism is called **task parallelism** or **functional parallelism**. To use task parallelism, the tasks and their dependencies can be represented as a **task graph** where the nodes are the tasks and the edges represent the dependencies between the tasks. A dependency graph is used for the conjugate gradient method discussed in Section 8.4. Depending on the programming model used, a single task can be executed sequentially by *one* processor, or in parallel by *multiple* processors. In the latter case, each task can be executed in a data-parallel way, leading to mixed task and data parallelism.

To determine an execution plan (schedule) for a given task graph on a set of processors, a starting time has to be assigned to each task such that the dependencies are fulfilled. Typically, a task cannot be started before all tasks which it depends on are finished. The goal of a scheduling algorithm is to find a schedule that minimizes the overall execution time, see also Section 4.4. Static and dynamic scheduling algorithms can be used. A *static* scheduling algorithm determines the assignment of tasks to processors deterministically at program start or at compile time. The assignment may be based on an estimation of the execution time of the tasks, which might be obtained by runtime measurements or an analysis of the computational structure of the tasks, see Section 4.4. A detailed overview of static scheduling algorithms for different kinds of dependencies can be found in [26]. If the tasks of a task graph are *parallel tasks*, the scheduling problem is sometimes called *multiprocessor task scheduling*.

A *dynamic scheduling algorithm* determines the assignment of tasks to processors during program execution. Therefore, the schedule generated can be adapted to the observed execution times of the tasks. A popular technique for dynamic scheduling is the use of a **task pool** in which tasks that are ready for execution are stored and from which processors can retrieve tasks if they have finished the execution of their current task. After the completion of the task, all depending tasks in the task graph whose predecessors have been terminated can be stored in the task pool for execution. The task pool concept is particularly useful for shared address space machines since the task pool can be held in the global memory. The task pool concept is discussed further in Section 6.1 in the context of pattern programming. The implementation of task pools with Pthreads and their provision in Java is considered in more detail in Chapter 6. A detailed treatment of task pools is considered in [145, 200, 135, 114]. Information on the construction and scheduling of task graphs can be found in [20, 81, 177, 179]. The use of task pools for irregular applications is considered in [193]. Programming with multiprocessor tasks is supported by library-based approaches like Tlib [182].

Task parallelism can also be provided at language level for appropriate language constructs which specify the available degree of task parallelism. The management and mapping can then be organized by the compiler and the runtime system. This approach has the advantage that the programmer is only responsible for the specification of the degree of task parallelism. The actual mapping and adaptation to specific details of the execution platform is done by the compiler and runtime system, thus providing a clear separation of concerns. Some language approaches are based on *coordination languages* to specify the degree of task parallelism and dependencies between the tasks. Some approaches in this direction are TwoL (*Two Level parallelism*) [180], P3L (*Pisa Parallel Programming Language*) [172], and PCN (*Program Composition Notation*) [71]. A more detailed treatment can be found in [58, 99]. Many thread-parallel programs are based on the exploitation of functional parallelism, since each thread executes independent function calls. The implementation of thread parallelism will be considered in detail in Chapter 6.

3.3.5 Explicit and implicit representation of parallelism

Parallel programming models can also be distinguished depending on whether the available parallelism, including the partitioning into tasks and specification of communication and synchronization, is represented explicitly in the program or not. The development of parallel programs is facilitated if no explicit representation must be included, but in this case an advanced compiler must be available to produce efficient parallel programs. On the other hand, an explicit representation is more effort for program development, but the compiler can be much simpler. In the following, we briefly discuss both approaches. A more detailed treatment can be found in [201].

3.3.5.1 Implicit parallelism

For the programmer, the most simple model results, when no explicit representation of parallelism is required. In this case, the program is mainly a specification of the computations to be performed, but no parallel execution order is given. In such a model, the programmer can concentrate on the details of the (sequential) algorithm to be implemented and does not need to care about the organization of the parallel execution. We give a short description of two approaches in this direction: parallelizing compilers and functional programming languages.

The idea of **parallelizing compilers** is to transform a sequential program into an efficient parallel program by using appropriate compiler techniques. This approach is also called *automatic parallelization*. To generate the parallel program, the compiler must first analyze the dependencies between the computations to be performed. Based on this analysis, the computation can then be assigned to processors for execution such that a good load balancing results. Moreover, for a distributed address space, the amount of communication should be reduced as much as possible, see [177, 223, 12, 6]. In practice, automatic parallelization is difficult to perform because dependency analysis is difficult for pointer-based computations or indirect addressing and because the execution time of function calls or loops with unknown bounds is difficult to predict at compile time. Therefore, automatic parallelization often produces parallel programs with unsatisfactory runtime behavior and, hence, this approach is not often used in practice.

Functional programming languages describe the computations of a program as the evaluation of mathematical functions without side effects; this means the evaluation of a function has the only effect that the output value of the function is computed. Thus, calling a function twice with the same input argument values always produces the same output value. Higher-order functions can be used; these are functions which use other functions as arguments and yield functions as arguments. Iterative computations are usually expressed by recursion. The most popular functional programming language is Haskell, see [115, 215, 22]. Function evaluation in functional programming languages provides potential for a parallel execution, since the arguments of the function can always be evaluated in parallel. This is possible because of the lack of side effects. The problem of an efficient execution is to extract the parallelism at the right level of recursion: On the upper level of recursion, a parallel evaluation of the arguments may not provide enough potential of parallelism. On a lower level of recursion, the available parallelism may be too fine-grained, thus making an efficient assignment to processors difficult. In the context of multicore processors, the degree of parallelism provided at the upper level of recursion may be enough to efficiently supply a few cores with computations. The advantage of using functional languages would be that new language constructs are not necessary to enable a parallel execution as this is the case for non-functional programming languages.

3.3.5.2 Explicit parallelism with implicit distribution

Another class of parallel programming models comprises models which require an explicit representation of parallelism in the program, but which do not demand an explicit distribution and assignment to processes or threads. Correspondingly, no explicit communication or synchronization is required. For the compiler, this approach has the advantage that the available degree of parallelism is specified in the program and does not need to be retrieved by a complicated data dependency analysis. This class of programming models includes parallel programming languages which extend sequential programming languages by *parallel loops* with independent iterations, see Section 3.3.3.

The parallel loops specify the available parallelism, but the exact assignments of loop iterations to processors is not fixed. This approach has been taken by the library OpenMP where parallel loops can be specified by compiler directives, see Section 6.5 for more details on OpenMP. High Performance Fortran (HPF) [66] has been another approach in this direction which adds constructs for the specification of array distributions to support the compiler in the selection of an efficient data distribution, see [128] on the history of HPF.

3.3.5.3 Explicit distribution

A third class of parallel programming models not only requires an explicit representation of parallelism, but also an explicit partitioning into tasks or an explicit assignment of work units to threads. The mapping to processors or cores as well as communication between processors is implicit and does not need to be specified. An example for this class is the BSP (bulk synchronous parallel) programming model which is based on the BSP computation model described in more detail in Section 4.6.2 [109, 110]. An implementation of the BSP model is BSPLib. A BSP program is explicitly partitioned into threads, but the assignment of threads to processors is done by the BSPLib library.

3.3.5.4 Explicit assignment to processors

The next class captures parallel programming models which require an explicit partitioning into tasks or threads and also need an explicit assignment to processors. But the communication between the processors does not need to be specified. An example for this class is the coordination language Linda [29, 28] which replaces the usual point-to-point communication between processors by a **tuple space** concept. A tuple space provides a global pool of data in which data can be stored and from which data can be retrieved. The following three operations are provided to access the tuple space:

- **in**: read and remove a tuple from the tuple space;
- **read**: read a tuple from the tuple space without removing it;
- **out**: write a tuple in the tuple space.

A tuple to be retrieved from the tuple space is identified by specifying required values for a part of the data fields which are interpreted as a key. For distributed address spaces, the access operations to the tuple space must be implemented by communication operations between the processes involved: If in a Linda program, a process A writes a tuple into the tuple space which is later retrieved by a process B, a communication operation from process A (send) to process B (recv) must be generated. Depending on the execution platform, this communication may produce a significant amount of overhead. Other approaches based on a tuple space are TSpaces from IBM and JavaSpaces [23] which is part of the Java Jini technology.

3.3.5.5 Explicit communication and synchronization

The last class comprises programming models in which the programmer must specify all details of a parallel execution, including the required communication and synchronization operations. This has the advantage that a standard compiler can be used and that the programmer can control the parallel execution explicitly with all details. This usually provides efficient parallel programs, but it also requires a significant amount of work for program development. Programming models belonging to this class are message-passing models like MPI, see Chapter 5, as well as thread-based models, such as Pthreads, see Chapter 6.

3.3.6 Parallel programming patterns

Parallel programs consist of a collection of tasks that are executed by processes or threads on multiple processors. To structure a parallel program, several forms of organizations can be used which can be captured by specific programming patterns. These patterns provide specific coordination structures for processes or threads, which have turned out to be effective for a large range of applications. We give a short overview of useful programming patterns in the following. More information and details on the implementation in specific environments can be found in [151]. Some of the patterns are presented as programs in Chapter 6.

3.3.6.1 Creation of processes or threads

The creation of processes or threads can be carried out statically or dynamically, see Fig. 3.5 for an illustration. In the static case, a fixed number of processes or threads is created at program start. These processes or threads exist during the entire execution of the parallel program and are terminated when program execution is finished. An alternative approach is to allow creation and termination of processes or threads dynamically at arbitrary points during program execution. At program start, a single process or thread is active and executes the main program. In the following, we describe well-known parallel programming patterns. For simplicity, we restrict our attention to the use of threads, but the patterns can as well be applied for the coordination of processes.

3.3.6.2 Fork-join

The fork-join construct is a simple concept for the creation of processes or threads [35] which was originally developed for process creation, but the pattern can also be used for threads. Using the concept, an existing thread T creates a number of child threads $T_1, \ldots, T_m$ with a fork statement, see Fig. 3.5 for an illustration. The child threads work in parallel and execute a given program part or function. The creating parent thread T can execute the same or a different program part or function and can then wait for the termination of $T_1, \ldots, T_m$ by using a join call.

The fork-join concept can be provided as a language construct or as a library function. It is usually provided for shared address space, but it can also be used for distributed address space. The fork-join concept is, for example, used in OpenMP for the creation of threads executing a parallel loop, see Section 6.5 for more details. The **spawn** and **exit** operations provided by message-passing systems such as MPI, see Sect. 5, provide a similar action pattern as fork-join. The concept of fork-join is simple, yet flexible, since by a nested use, arbitrary structures of parallel activities can be built. Specific programming languages and environments provide specific variants of the pattern, see Chapter 6 for details on Pthreads and Java threads.

3.3.6.3 Parbegin-parend

A similar pattern as fork-join for thread creation and termination is provided by the **parbegin-parend** construct which is sometimes also called **cobegin-coend**. The construct allows the specification of a sequence of statements, including function calls, to be executed by a set of processors in parallel. When an executing thread reaches a parbegin-parend construct, a set of threads is created and the statements of the construct are assigned to these threads for execution, see Fig. 3.5 for an illustration.. The statement following the parbegin-parend construct are executed not before all these threads have finished their work and have been terminated. The parbegin-parend construct can be provided as a language construct or by compiler directives. An example are parallel sections in OpenMP, see Sect. 6.5.

3.3.6.4 SPMD and SIMD

The SIMD (**s**ingle-**i**nstruction, **m**ultiple-**d**ata) and SPMD (**s**ingle-**p**rogram, **m**ultiple-**d**ata) programming models use a (fixed) number of threads which apply the same program to different data. In the SIMD approach, the single instructions are executed synchronously by the different threads to different data. This is sometimes called *data parallelism in the strong sense*. SIMD is useful if the same instruction

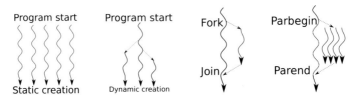

Fig. 3.5 Illustration of a static thread creation (left), dynamic thread creation (second from left) fork-join (second from right), and parbegin-parend (right).

must be applied to a large set of data, as it is often the case for graphics applications. Therefore, graphics processors often provide SIMD instructions, and some standard processors also provide SIMD extensions.

In the SPMD approach, the different threads work asynchronously with each other and different threads may execute different parts of the parallel program. This effect can be caused by different speeds of the executing processors or by delays of the computations because of slower access to global data. But the program could also contain control statements to assign different program parts to different threads. There is no implicit synchronization of the executing threads, but synchronization can be achieved by explicit synchronization operations. The SPMD approach is one of the most popular models for parallel programming. MPI is based on this approach, see Section 5, but thread-parallel programs are usually also SPMD programs.

3.3.6.5 Master-Slave or Master-Worker

In the SIMD and SPMD models, all threads have equal rights. In the master-slave model, also called master-worker model, there is one master which controls the execution of the program. The master thread often executes the main function of a parallel program and creates worker threads at appropriate program points to perform the actual computations, see Fig. 3.6(left) for an illustration. Depending on the specific system, the worker threads may be created statically or dynamically. The assignment of work to the worker threads is usually done by the master thread, but worker threads could also generate new work for computation. In this case, the master thread would only be responsible for coordination and could, e.g. perform initializations, timings, and output operations.

3.3.6.6 Client-Server

The coordination of parallel programs according to the client-server model is similar to the general MPMD (multiple-program, multiple-data) model. The client-server model originally comes from distributed computing where multiple client computers have been connected to a mainframe which acts as a server and provides re-

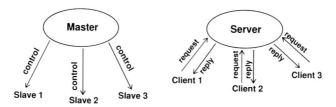

Fig. 3.6 Illustration of the master-slave model (left) and the client-server model (right).

sponses to access requests to a database. On the server side, parallelism can be used by computing requests from different clients concurrently or even by using multiple threads to compute a single request if this includes enough work.

Employing the client-server model for the structuring of parallel programs, multiple client threads are used which generate requests to a server and then perform some computations on the result, see Fig. 3.6(right) for an illustration. After having processed a request of a client, the server delivers the result back to the client. The client-server model can be applied in many variations: there may be several server threads, or the threads of a parallel program may play the role of both clients and servers, generating requests to other threads and processing requests from other threads. Section 6.2.3 shows an example for a Pthreads program using the client-server model. The client-server model is important for parallel programming in heterogeneous systems and is also often used in grid computing and cloud computing.

3.3.6.7 Pipelining

The pipelining model describes a special form of coordination of different threads in which data elements are forwarded from thread to thread to perform different processing steps. The threads are logically arranged in a predefined order, $T_1, \ldots, T_p$, such that thread T_i receives the output of thread T_{i-1} as input and produces an output which is submitted to the next thread T_{i+1} as input, $i = 2, \ldots, p - 1$. Thread T_1 receives its input from another program part and thread T_p provides its output to another program part. Thus, each of the pipeline threads processes a stream of input data in sequential order and produces a stream of output data. Despite the dependencies of the processing steps, the pipeline threads can work in parallel by applying their processing step to different data.

The pipelining model can be considered as a special form of functional decomposition where the pipeline threads process the computations of an application algorithm one after another. A parallel execution is obtained by partitioning the data into a stream of data elements which flow through the pipeline stages one after another. At each point in time, different processing steps are applied to different elements of the data stream. The pipelining model can be applied for both shared and distributed address space. In Section 6.2.2, the pipelining pattern is implemented as Pthreads program.

3.3.6.8 Task pools

In general, a task pool is a data structure in which tasks to be performed are stored and from which they can be retrieved for execution. A task comprises computations to be executed and a specification of the data to which the computations should be applied. The computations are often specified as a function call. A *fixed* number of threads is used for the processing of the tasks. The threads are created at program start by the main thread and they are terminated not before all tasks have been pro-

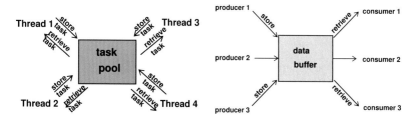

Fig. 3.7 Illustration of a task pool (left) and a producer-consumer model (right).

cessed. For the threads, the task pool is a common data structure which they can access to retrieve tasks for execution, see Fig. 3.7(left) for an illustration. During the processing of a task, a thread can generate new tasks and insert them into the task pool. Access to the task pool must be synchronized to avoid race conditions. Using a task-based execution, the execution of a parallel program is finished, when the task pool is empty and when each thread has terminated the processing of its last task. Task pools provide a flexible execution scheme which is especially useful for adaptive and irregular applications for which the computations to be performed are not fixed at program start. Since a fixed number of threads is used, the overhead for thread creation is independent of the problem size and the number of tasks to be processed.

Flexibility is ensured, since tasks can be generated dynamically at any point during program execution. The actual task pool data structure could be provided by the programming environment used or it could be included in the parallel program. An example for the first case is the Executor interface of Java, see Sect. 6.4 for more details. A simple task pool implementation based on a shared data structure is described in Sect. 6.2.1 using Pthreads. For fine-grained tasks, the overhead of retrieval and insertion of tasks from or into the task pool becomes important, and sophisticated data structures should be used for the implementation, see [114].

3.3.6.9 Producer-Consumer

The producer-consumer model distinguishes between producer threads and consumer threads. Producer threads produce data which are used as input by consumer threads. For the transfer of data from producer threads to consumer threads, a common data structure is used, which is typically a data buffer of fixed length and which can be accessed by both types of threads. Producer threads store the data elements generated into the buffer, consumer threads retrieve data elements from the buffer for further processing, see Fig. 3.7(right) for an illustration. A producer thread can only store data elements into the buffer, if this is not full. A consumer thread can only retrieve data elements from the buffer, if this is not empty. Therefore, synchronization has to be used to ensure a correct coordination between producer and consumer threads. The producer-consumer model is considered in more detail in Sect. 6.3.1 for Pthreads and Sect. 6.4.3 for Java threads.

3.4 SIMD Computations

In this section, we give a short overview of different methods of SIMD computations of data. Such computations are supported by several processor architectures in the form of SIMD multimedia extensions as well as by GPUs. The basic idea of SIMD computations is to apply the same (arithmetic) operation (successively or concurrently) to multiple data elements. To do so, appropriate vector instructions are provided. This kind of computations has the advantage that a single vector instruction issues the computation of several pairs of source operands, in contrast to a scalar instruction where one instruction issues exactly one computation.

To use vector instructions, the application program must provide suitable operations, e.g., by applying the same operation to a set of data elements controlled by a loop. This means that the program should provide a potential of data parallelism. This requirement is typically fulfilled for application programs from scientific computing, where the same arithmetic operation is often applied to all elements of a vector or a matrix.

This section first starts with a short view at the execution of vector operations as they are supported by vector architectures, and then considers SIMD instruction set extensions such as the Streaming SIMD Extensions (SSE) and the Advanced Vector Extensions (AVX) that are supported by X86 architectures. For a more detailed description of SIMD computations, we refer to [103].

3.4.1 Execution of vector operations

Specialized vector computers have been successfully used in the area of high performance computing for many years, e.g., for programs from scientific computing. Many of these vector computers have been manufactured by Cray (Cray Research or Cray Inc.) or the Japanese company NEC. NEC offers vector computing in form of a Vector Engine (VE) that is used in the NEC SX series supercomputers. In 2022, the newest version of a VE processor (named SX-Aurora TSUBASA) integrates eight vector-cores and 48 GB of high bandwidth memory (HBM2). Each VE core can execute 32 double precision floating point operations per cycle using vector registers that can store 256 floating point values. With three fused-multiply-add units (FMA) each core has a peak performance of 192 FLOP per cycle or up to 307 GigaFLOPS (double precision).

In addition to the normal scalar general-purpose registers, vector processors provide a set of vector registers of fixed length. A typical length is 32 or 64, thus allowing to store 32 or 64 floating-point values. The vector registers can be loaded with data from main memory using specialized vector load instructions. These can be used to load a set of memory elements with consecutive addresses, and often the load of elements with a fixed distance between their memory addresses (called *stride*) as well as the load of memory elements whose addresses are collected in a special vector address register is also supported. These advanced load capabilities are important when dealing with sparse vectors or matrices.

Floating-point values in vector registers can be used as operands of vector instructions. For example, a vector addition instruction (VADD) can be used to add floating-point values in two operand vector registers element-wise and to store the resulting sums in a destination vector register, see Fig. 3.8 for an illustration with source operands VOpnd1 and VOpnd2 and destination operand VOpnd3.

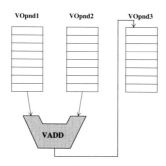

Fig. 3.8 Illustration of a vector addition operation c = a + b.

Vector instructions can be applied to all values stored in the operand registers, but it is also possible to use only a part of the values in the operand registers by specifying the length of the vector operands in a specialized vector length register (VLR). Vector instructions can also use scalar registers as operands: when using a vector register and a scalar register as operands of a VADD instruction, the result vector is obtained by adding the value in the scalar register to each element of the vector register. The use of vector instructions has the advantage that for the internal processing by the hardware, an efficient pipelining can be used because there are no dependencies between the values to be processed.

There are two popular possibilities for programming vector computers: the use of vectorizing compilers or the use of a programming language that provides vector operations at the language level. Fortran 90/95 is an example for such a language. Vector operations can be used to apply an arithmetic operation to arrays or array section specified by an index range. A typical example for a vector operation is the following vector assignment

$$a[1 : 100] = b[1 : 100] + a[1 : 100] \cdot c;$$

with vector operands $a[1 : 100]$ and $b[1 : 100]$ as well as the scalar operand c. The array section $a[1 : 100]$ specifies all 100 elements of array a between $a[1]$ and $a[100]$ with the borders included. A vector operation is executed by first loading all elements of the vector sections used as operands into a vector register and then applying the arithmetic operation specified. If the array sections specified do not fit into the vector registers provided by the hardware, multiple slices must be used. After all computations are performed, the result vector is stored into the array section on the left hand side of the vector assignment. The use of vectorizing compilers is based on the approach to identify loops in the source program for which there is no dependency from one loop iteration to a following loop iteration. Such loops are called *vectorizable* and can be transformed into an equivalent vector statement. For example, the following loop

$$\text{for } (i = 1; i \leq 100; i + +)$$
$$a[i] = b[i] + a[i] \cdot c;$$

is equivalent to the vector assignment from above. Program transformations can be used to transform some sort of non-vectorizable loops into vectorizable loops, see [12] for a detailed treatment of such loop transformations.

3.4.2 SIMD instructions

Vector operations, as they are described in the last subsection, can also be useful for standard microprocessors. Therefore, specialized SIMD instructions have been included in addition to the standard instructions that use 32 or 64 bit operands. These SIMD instructions were originally intended for the fast processing of multimedia data, but they can be used for other data as well. For the x86 architecture, SIMD instructions have been added in 1996 in the form of MMX instructions (MultiMedia eXtensions), which were working on 64-bit floating-point registers. In 1999 the Streaming SIMD Extensions (SSE) were introduced. The SSE instructions are working on separate 128-bit floating-point registers, which can store eight 16-bit values, four 32-bit values, or two 64-bit values. The SSE instructions perform operations simultaneously on these values. In 2010, the Advanced Vector Extensions (AVX) have been added. These instructions are working on 256-bit registers and are supported by the Intel Core i7 processors starting with the Sandy Bridge architecture as well as by the AMD processors starting with the Bulldozer architecture. In 2017, AVX-512 has been added, supporting 512-bit registers.

For the current AVX-256 instructions, the special 256-bit registers can be considered as vector registers which can, e.g., be used to store four 64-bit floating-point values to which an arithmetic operation can be applied simultaneously in one machine cycle. Vector computers typically provide much larger vector registers. However, it can be expected that the support of SIMD instructions will be continued in future processor architectures, along with a further increase of the width of the special registers to 1024 bits.

The addressing modes and load operations provided for the AVX instructions are quite simple compared to vector computers, and support for a convenient loading of non-contiguous memory locations into the 256-bit register is missing, see [103]. However, the basic approach of computation is similar to the computations of vector computers and it can be expected that future extensions to AVX will provide additional addressing modes for loading the special registers.

For the execution of AVX-256 instructions, 16 256-bit registers are provided, denoted as YMM0 to YMM15. These registers can be loaded from memory with 256-bit blocks of data using special load operations. Examples are the MOVAPD instruction to load a 256-bit block of double values that is aligned to 256-bit borders or the MOVUPD instruction to load blocks that are not aligned. The data stored in the YMM registers can be interpreted as single-precision (32 bits) or double-precision (64 bits) floating-point values, and special instructions are provided to apply arithmetic operations (such as addition, subtraction, multiplication, division, square root, minimum, maximum, rounding) to all values stored in a YMM register. For compatibility reasons, AVX also supports the SSE data types and the non-arithmetic operations of SSE, see [148, 37] for a detailed description of AVX instructions. A detailed performance analysis of the AVX extensions is given in [103].

Extensions to AVX have already been proposed, and they contain support for *fused multiply-add* (FMA) instructions, which can perform operations of the form

$$a = a + b \cdot c$$

in one cycle. The operands a, b, c are then stored in YMM registers, and the FMA instruction is applied to all floating-point values stored in these YMM registers. An example is the VFMADDPD instruction for double-precision floating-point values. FMA operations can be used for many numerical computations, such as the inner product of vectors or matrix multiplication, and they may reduce the resulting computation time significantly. SIMD computations are also important for GPUs. In chapter 7, the architecture of GPUs is described and the programming models of CUDA and OpenCL are discussed. In particular, programming techniques to write efficient progress for GPU architectures will be considered.

3.5 Data distributions for arrays

Many algorithms, especially from numerical analysis and scientific computing, are based on vectors and matrices. The corresponding programs use one-, two- or higher dimensional arrays as basic data structures. For those programs, a straightforward parallelization strategy decomposes the array-based data into subarrays and assigns the subarrays to different processors. The decomposition of data and the mapping to different processors is called **data distribution**, **data decomposition** or **data partitioning**. In a parallel program, the processors perform computations only on their part of the data.

Data distributions can be used for parallel programs for distributed as well as for shared memory machines. For distributed memory machines, the data assigned to a processor reside in its local memory and can only be accessed by this processor. Communication has to be used to provide data to other processors. For shared memory machines, all data reside in the same shared memory. Still a data decomposition is useful for designing a parallel program since processors access different parts of the data and conflicts such as race conditions or critical regions are avoided. This simplifies the parallel programming and supports a good performance. In this section, we present regular data distributions for arrays, which can be described by a mapping from array indices to processor numbers. The set of processors is denoted as $P = \{P_1, \ldots, P_p\}$.

3.5.1 Data distribution for one-dimensional arrays

For one-dimensional arrays the blockwise and the cyclic distribution of array elements are typical data distributions. For the formulation of the mapping, we assume that the enumeration of array elements starts with 1; for an enumeration starting with 0 the mappings have to be modified correspondingly.

The **blockwise data distribution** of an array $v = (v_1, \ldots, v_n)$ of length n cuts the array into p blocks with $\lceil n/p \rceil$ consecutive elements each. Block j, $1 \leq j \leq p$, contains the consecutive elements with indices $(j-1) \cdot \lceil n/p \rceil + 1, \ldots, j \cdot \lceil n/p \rceil$ and is assigned to processor P_j. When n is not a multiple of p, the last block contains less than $\lceil n/p \rceil$ elements. For $n = 14$ and $p = 4$ the following blockwise distribution results:

P_1:   owns $v_1, v_2, v_3, v_4,$
P_2:   owns $v_5, v_6, v_7, v_8,$
P_3:   owns $v_9, v_{10}, v_{11}, v_{12},$
P_4:   owns $v_{13}, v_{14}.$

Alternatively, the first $n \bmod p$ processors get $\lceil n/p \rceil$ elements and all other processors get $\lfloor n/p \rfloor$ elements.

The **cyclic data distribution** of a one-dimensional array assigns the array elements in a round robin way to the processors so that array element v_i is assigned to processor $P_{(i-1) \bmod p + 1}$, $i = 1, \ldots, n$. Thus, processor P_j owns the array elements $j, j+p, \ldots, j+p \cdot (\lceil n/p \rceil - 1)$ for $j \leq n \bmod p$ and $j, j+p, \ldots, j+p \cdot (\lceil n/p \rceil - 2)$ for $n \bmod p < j \leq p$. For the example $n = 14$ and $p = 4$ the cyclic data distribution

P_1:   owns $v_1, v_5, v_9, v_{13},$
P_2:   owns $v_2, v_6, v_{10}, v_{14},$
P_3:   owns $v_3, v_7, v_{11},$
P_4:   owns $v_4, v_8, v_{12}.$

results, where P_j for $1 \leq j \leq 2 = 14 \bmod 4$ owns the elements $j, j+4, j+4*2, j+4*(4-1)$ and P_j for $2 < j \leq 4$ owns the elements $j, j+4, j+4*(4-2)$.

The **block-cyclic data distribution** is a combination of the blockwise and the cyclic distribution. Consecutive array elements are structured into blocks of size b, where $b \ll n/p$ in most cases. When n is not a multiple of b, the last block contains less than b elements. The blocks of array elements are assigned to processors in a round robin way. Figure 3.9(a) shows an illustration of the array decompositions for one-dimensional arrays.

3.5.2 Data distribution for two-dimensional arrays

For two-dimensional arrays, combinations of blockwise and cyclic distributions in only one or in both dimensions are used.

For the distribution in one dimension, columns or rows are distributed in a block-wise, cyclic or block-cyclic way. The blockwise columnwise (or rowwise) distribution builds p blocks of contiguous columns (or rows) of equal size and assigns block i to processor P_i, $i = 1, \ldots, p$. When n is not a multiple of p, the same adjustment as for one-dimensional arrays is used. The cyclic columnwise (or rowwise) distribution assigns columns (or rows) in a round robin way to processors and uses the adjustments of the last blocks as described for the one-dimensional case, when n is not a multiple of p. The block-cyclic columnwise (or rowwise) distribution forms blocks of contiguous columns (or rows) of size b and assigns these blocks in a round robin way to processors. Figure 3.9(b) illustrates the distribution in one dimension for two-dimensional arrays.

A distribution of array elements of a two-dimensional array of size $n_1 \times n_2$ in both dimensions uses **checkerboard distributions** which distinguish between blockwise cyclic and block-cyclic checkerboard patterns. The processors are arranged in a virtual mesh of size $p_1 \cdot p_2 = p$ where p_1 is the number of rows and p_2 is the number of columns in the mesh. Array elements (k,l) are mapped to processors $P_{i,j}$, $i = 1, \ldots, p_1, j = 1, \ldots, p_2$.

In the **blockwise checkerboard distribution**, the array is decomposed into $p_1 \cdot p_2$ blocks of elements where the row dimension (first index) is divided into p_1 blocks and the column dimension (second index) is divided into p_2 blocks. Block (i, j), $1 \leq i \leq p_1$, $1 \leq j \leq p_2$, is assigned to the processor with position (i, j) in the processor mesh. The block sizes depend on the number of rows and columns of the array. Block (i, j) contains the array elements (k, l) with $k = (i-1) \cdot \lceil n_1/p_1 \rceil + 1, \ldots, i \cdot \lceil n_1/p_1 \rceil$ and $l = (j-1) \cdot \lceil n_2/p_2 \rceil + 1, \ldots, j \cdot \lceil n_2/p_2 \rceil$. Figure 3.9(c) shows an example for $n_1 = 4, n_2 = 8$ and $p_1 \cdot p_2 = 2 \cdot 2 = 4$.

The **cyclic checkerboard distribution** assigns the array elements in a round robin way in both dimensions to the processors in the processor mesh so that a cyclic assignment of row indices $k = 1, \ldots, n_1$ to mesh rows $i = 1, \ldots, p_1$ and a cyclic assignment of column indices $l = 1, \ldots, n_2$ to mesh columns $j = 1, \ldots, p_2$ results. Array element (k, l) is thus assigned to the processor with mesh position $((k-1) \bmod p_1 + 1, (l-1) \bmod p_2 + 1)$. When n_1 and n_2 are multiples of p_1 and p_2, respectively, the processor at position (i, j) owns all array elements (k, l) with $k = i + s \cdot p_1$ and $l = j + t \cdot p_2$ for $0 \leq s < n_1/p_1$ and $0 \leq t < n_2/p_2$. An alternative way to describe the cyclic checkerboard distribution is to build blocks of size $p_1 \times p_2$ and to map element (i, j) of each block to the processor at position (i, j) in the mesh. Figure 3.9(c) shows a cyclic checkerboard distribution with $n_1 = 4, n_2 = 8$, $p_1 = 2$ and $p_2 = 2$. When n_1 or n_2 are not a multiple of p_1 or p_2, respectively, the cyclic distribution is handled as in the one-dimensional case.

The **block-cyclic checkerboard distribution** assigns blocks of size $b_1 \times b_2$ cyclically in both dimensions to the processors in the following way: Array element (m, n) belongs to the block (k, l), with $k = \lceil m/b_1 \rceil$ and $l = \lceil n/b_2 \rceil$. Block (k, l) is assigned to the processor at mesh position $((k-1) \bmod p_1 + 1, (l-1) \bmod p_2 + 1)$. The cyclic checkerboard distribution can be considered as special case of the block-cyclic distribution with $b_1 = b_2 = 1$ and the blockwise checkerboard distribution can be considered as special case with $b_1 = n_1/p_1$ and $b_2 = n_2/p_2$. Figure 3.9(c) illustrates the block-cyclic distribution for $n_1 = 4, n_2 = 12, p_1 = 2$ and $p_2 = 2$.

3.5.3 Parameterized data distribution

A data distribution is defined for a d-dimensional array A with index set $I_A \subset \mathbb{N}^d$. The size of the array is $n_1 \times \ldots \times n_d$ and the array elements are denoted as $A[i_1, \ldots, i_d]$ with an index $\mathbf{i} = (i_1, \ldots, i_d) \in I_A$. Array elements are assigned to p processors which are arranged in a d-dimensional mesh of size $p_1 \times \ldots \times p_d$ with $p = \prod_{i=1}^{d} p_i$. The data distribution of A is given by a **distribution function** $\gamma_A : I_A \subset \mathbb{N}^d \to 2^P$, where 2^P denotes the power set of the set of processors P. The meaning of γ_A is that the array element $A[i_1, \ldots, i_d]$ with $\mathbf{i} = (i_1, \ldots, i_d)$ is assigned to all processors in $\gamma_A(\mathbf{i}) \subseteq P$, i.e. array element $A[\mathbf{i}]$ can be assigned to more than one processor. A data distribution is called **replicated**, if $\gamma_A(\mathbf{i}) = P$ for all $\mathbf{i} \in I_A$. When each array element is uniquely assigned to a processor, then $|\gamma_A(\mathbf{i})| = 1$ for all $\mathbf{i} \in I_A$; examples are the block-cyclic data distribution described above. The function $L(\gamma_A) : P \to 2^{I_A}$ delivers all elements assigned to a specific processor, i.e.

$$\mathbf{i} \in L(\gamma_A)(q) \quad \text{if and only if} \quad q \in \gamma_A(\mathbf{i}).$$

Generalizations of the block-cyclic distributions in the one- or two-dimensional case can be described by a distribution vector in the following way. The array elements are structured into blocks of size $b_1, \ldots, b_d$ where b_i is the blocksize in dimension $i, i = 1, \ldots, d$. The array element $A[i_1, \ldots, i_d]$ is contained in block $(k_1, \ldots, k_d)$ with $k_j = \lceil i_j/b_j \rceil$ for $1 \leq j \leq d$. The block $(k_1, \ldots, k_d)$ is then assigned to the processor at mesh position $((k_1 - 1) \bmod p_1 + 1, \ldots, (k_d - 1) \bmod p_d + 1)$. This block-cyclic distribution is called **parameterized data distribution** with distribution vector

$$((p_1, b_1), \ldots, (p_d, b_d)). \tag{3.1}$$

This vector uniquely determines a block-cyclic data distribution for a d-dimensional array of arbitrary size. The blockwise and the cyclic distributions of a d-dimensional array are special cases of this distribution. Parameterized data distributions are used in the applications of later sections, e.g. the Gaussian elimination in Section 8.1.

3.6 Information exchange

To control the coordination of the different parts of a parallel program, information must be exchanged between the executing processors. The implementation of such

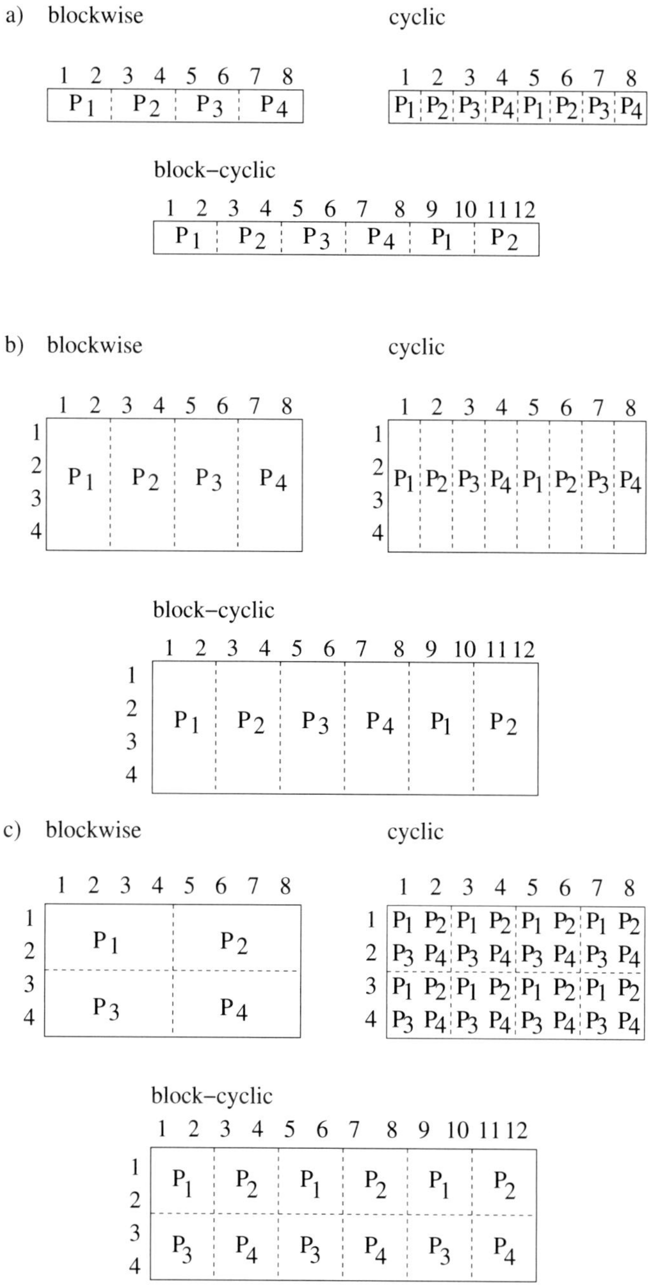

a) blockwise   cyclic

block–cyclic

b) blockwise   cyclic

block–cyclic

c) blockwise   cyclic

block–cyclic

Fig. 3.9 Illustration of the data distributions for arrays: a) for one-dimensional arrays, b) for two-dimensional arrays within one of the dimensions and c) for two-dimensional arrays with checkerboard distribution.

an information exchange strongly depends on the memory organization of the parallel platform used. In the following, we give a first overview on techniques for information exchange for shared address space in Subsection 3.6.1 and for distributed address space in Subsection 3.6.2. More details will be discussed in the following chapters. As example, parallel matrix-vector multiplication is considered for both memory organizations in Section 3.7.

3.6.1 Shared variables

Programming models with a shared address space are based on the existence of a global memory which can be accessed by all processors. Depending on the model, the executing control flows may be referred to as *processes* or *threads*, see Section 3.8 for more details. In the following, we will use the notion *threads*, since this is more common for shared address space models. Each thread will be executed by one processor or by one core for multicore processors. Each thread can access shared data in the global memory. Such shared data can be stored in **shared variables** which can be accessed as normal variables. A thread may also have private data stored in **private variables**, which cannot be accessed by other threads. There are different ways how parallel program environments define shared or private variables. The distinction between shared and private variables can be made by using annotations like `shared` or `private` when declaring the variables. Depending on the programming model, there can also be declaration rules which can, for example, define that global variables are always shared and local variables of functions are always private. To allow a coordinated access to a shared variable by multiple threads, synchronization operations are provided to ensure that concurrent accesses to the same variable are synchronized. Usually, a **sequentialization** is performed such that concurrent accesses are done one after another. Chapter 6 considers programming models and techniques for shared address spaces in more detail and describes different systems, like Pthreads, Java threads, and OpenMP. In the current section, a few basic concepts are given for a first overview.

A central concept for information exchange in shared address space are shared variables. When a thread T_1 wants to transfer data to another thread T_2, it stores the data in a shared variable such that T_2 obtains the data by reading this shared variable. To ensure that T_2 reads the variable not before T_1 has written the appropriate data, a **synchronization operation** is used. T_1 stores the data into the shared variable before the corresponding synchronization point and T_2 reads the data after the synchronization point.

When using shared variables, multiple threads accessing the same shared variable by a read or write at the same time must be avoided, since this may lead to race conditions. The term **race condition** describes the effect that the result of a parallel execution of a program part by multiple execution units depends on the order in which the statements of the program part are executed by the different units. In the presence of a race condition it may happen that the computation of a program part leads to different results, depending on whether thread T_1 executes the program part before T_2 or vice versa. Usually, race conditions are undesirable, since the relative execution speed of the threads may depend on many factors (like execution speed of the executing cores or processors, the occurrence of interrupts, or specific values of the input data) which cannot be influenced by the programmer. This may lead to **nondeterministic behavior**, since, depending on the execution order, different results are possible, and the exact outcome cannot be predicted.

Program parts in which concurrent accesses to shared variables by multiple threads may occur, thus holding the danger of the occurrence of inconsistent values, are called **critical sections**. An error-free execution can be ensured by letting only one thread at a time execute a critical section. This is called **mutual exclusion**. Programming models for shared address space provide mechanisms to ensure mutual exclusion. The techniques used have originally been developed for multi-tasking operating systems and have later been adapted to the needs of parallel programming environments. For a concurrent access of shared variables, race conditions can be avoided by a **lock mechanism**, which will be discussed in more detail in Section 3.8.3.

3.6.2 Communication operations

In programming models with a distributed address space, exchange of data and information between the processors is performed by *communication operations* which are *explicitly* called by the participating processors. The execution of such a communication operation causes that one processor receives data which is stored in the local memory of another processor. The actual data exchange is realized by the transfer of messages between the participating processors. The corresponding programming models are therefore called **message passing programming models**.

To send a message from one processor to another, one send and one receive operation have to be used as a pair. A send operation sends a data block from the local address space of the executing processor to another processor as specified by the operation. A receive operation receives a data block from another processor and stores it in the local address space of the executing processor. This kind of data exchange is also called *point-to-point communication*, since there is exactly one send point and one receive point. Additionally, **global communication operations** are often provided in which a larger set of processors is involved. These global communication operations typically capture a set of regular communication patterns often used in parallel programs [21, 124].

3.6.2.1 A set of communication operations

In the following, we consider a typical set of global communication operations which will be used in the following chapters to describe parallel implementations for platforms with a distributed address space [21]. We consider p identical processors $P_1, \ldots, P_p$ and use the index i, $i \in \{1, \ldots, p\}$, as processor rank to identify the processor P_i.

- **Single transfer**: For a single transfer operation, a processor P_i (sender) sends a message to processor P_j (receiver) with $j \neq i$. Only these two processors participate in this operation. To perform a single transfer operation, P_i executes a send operation specifying a send buffer in which the message is provided as well as the processor rank of the receiving processor. The receiving processor P_j executes a corresponding receive operation which specifies a receive buffer to store the received message as well as the processor rank of the processor from which the message should be received. For each send operation, there must be a corresponding receive operation, and vice versa. Otherwise, deadlocks may occur, see Sections 3.8.4.2 and 5.1.1 for more details. Single transfer operations are the basis of each communication library. In principle, any communication pattern can be assembled with single transfer operations. For regular communication patterns, it is often beneficial to use global communication operations, since they are typically easier to use and more efficient.

- **Single-broadcast**: For a single broadcast operation, a specific processor P_i sends the *same* data block to all other processors. P_i is also called *root* in this context. The effect of a single broadcast operation with processor P_1 as root and message x can be illustrated as follows:

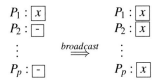

Before the execution of the broadcast, the message x is only stored in the local address space of P_1. After the execution of the operation, x is also stored in the local address space of all other processors. To perform the operation, each processor explicitly calls a broadcast operation which specifies the root processor of the broadcast. Additionally, the root processor specifies a send buffer in which the broadcast message is provided. All other processors specify a receive buffer in which the message should be stored upon receipt.

- **Single-accumulation**: For a single accumulation operation, each processor provides a block of data with the same type and size. By performing the operation, a given reduction operation is applied element by element to the data blocks provided by the processors, and the resulting accumulated data block of the same length is collected at a specific root processor P_i. The reduction operation is a binary operation which is associative and commutative. The effect of a single accumulation operation with root processor P_1 to which each processor P_i provides a data block x_i for $i = 1, \ldots, p$ can be illustrated as follows:

$$P_1 : \boxed{x_1}$$
$$P_2 : \boxed{x_2}$$
$$\vdots \quad \overset{accumulation}{\Longrightarrow}$$
$$P_p : \boxed{x_p}$$

$$P_1 : \boxed{x_1 + x_2 + \ldots + x_p}$$
$$P_2 : \boxed{x_2}$$
$$\vdots$$
$$P_p : \boxed{x_p}$$

The addition is used as reduction operation. To perform a single accumulation, each processor explicitly calls the operation and specifies the rank of the root processor, the reduction operation to be applied, and the local data block provided. The root processor additionally specifies the buffer in which the accumulated result should be stored.

- **Gather**: For a gather operation, each processor provides a data block, and the data blocks of all processors are collected at a specific root processor P_i. No reduction operation is applied, i.e. processor P_i gets p messages. For root processor P_1, the effect of the operation can be illustrated as follows:

$$P_1 : \boxed{x_1}$$
$$P_2 : \boxed{x_2}$$
$$\vdots \quad \overset{gather}{\Longrightarrow}$$
$$P_p : \boxed{x_p}$$

$$P_1 : \boxed{x_1 \parallel x_2 \parallel \ldots \parallel x_p}$$
$$P_2 : \boxed{x_2}$$
$$\vdots$$
$$P_p : \boxed{x_p}$$

Here, the symbol $\parallel$ denotes the concatenation of the received data blocks. To perform the gather, each processor explicitly calls a gather operation and specifies the local data block provided as well as the rank of the root processor. The root processor additionally specifies a receive buffer in which all data blocks are collected. This buffer must be large enough to store all blocks. After the operation is completed, the receive buffer of the root processor contains the data blocks of all processors in rank order.

- **Scatter**: For a scatter operation, a specific root processor P_i provides a separate data block for each other processor. For root processor P_1, the effect of the operation can be illustrated as follows:

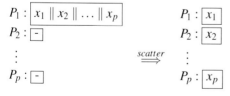

To perform the scatter, each processor explicitly calls a scatter operation and specifies the root processor as well as a receive buffer. The root processor additionally specifies a send buffer in which the data blocks to be sent are provided in rank order of the rank $i = 1, \ldots, p$.

- **Multi-broadcast**: The effect of a multi-broadcast operation is the same as of the execution of several single-broadcast operations, one for each processor, i.e. each processor sends the *same* data block to every other processor. From the receiver's point of view, each processor receives a data block from every other processor. Different receivers get the same data block from the same sender. The operation can be illustrated as follows:

$$
\begin{array}{ll}
P_1 : \boxed{x_1} & P_1 : \boxed{x_1 \| x_2 \| \ldots \| x_p} \\
P_2 : \boxed{x_2} & P_2 : \boxed{x_1 \| x_2 \| \ldots \| x_p} \\
\;\vdots \quad \xRightarrow{\;multi-broadcast\;} & \;\vdots \\
P_p : \boxed{x_p} & P_p : \boxed{x_1 \| x_2 \| \ldots \| x_p}
\end{array}
$$

In contrast to the global operations considered so far, there is *no* root processor. To perform the multi-broadcast, each processor explicitly calls a multi-broadcast operation and specifies a send buffer which contains the data block as well as a receive buffer. After the completion of the operation, the receive buffer of every processor contains the data blocks provided by all processors in rank order, including its own data block. Multi-broadcast operations are useful to collect blocks of an array that have been computed in a distributed way and to make the entire array available to all processors.

- **Multi-accumulation**: The effect of a multi-accumulation operation is that each processor executes a single-accumulation operation, i.e. each processor provides for every other processor a potentially different data block. The data blocks for the same receiver are combined with a given reduction operation such that *one* (reduced) data block arrives at the receiver. There is no root processor, since each processor acts as a receiver for one accumulation operation. The effect of the operation with the addition as reduction operation can be illustrated as follows:

$$
\begin{array}{ll}
P_1 : \boxed{x_{11} \| x_{12} \| \ldots \| x_{1p}} & P_1 : \boxed{x_{11} + x_{21} + \ldots + x_{p1}} \\
P_2 : \boxed{x_{21} \| x_{22} \| \ldots \| x_{2p}} & P_2 : \boxed{x_{12} + x_{22} + \ldots + x_{p2}} \\
\;\vdots \quad \xRightarrow{\;multi-accumulation\;} & \;\vdots \\
P_p : \boxed{x_{p1} \| x_{p2} \| \ldots \| x_{pp}} & P_p : \boxed{x_{1p} + x_{2p} + \ldots + x_{pp}}
\end{array}
$$

The data block provided by processor P_i for processor P_j is denoted as x_{ij}, $i, j = 1, \ldots, p$. To perform the multi-accumulation, each processor explicitly calls a multi-accumulation operation and specifies a send buffer, a receive buffer and a reduction operation. In the send buffer, each processor provides a separate data block for each other processor, stored in rank order. After the completion of the operation, the receive buffer of each processor contains the accumulated result for this processor.

- **Total exchange**: For a total exchange operation, each processor provides for each other processor a potentially different data block. These data blocks are sent to their intended receivers, i.e. each processor executes a scatter operation. From a receiver's point of view, each processor receives a data block from every other processor. In contrast to a multi-broadcast, different receivers get different data blocks from the same sender. There is no root processor. The effect of the operation can be illustrated as follows:

$P_1 :$ | $x_{11} \parallel x_{12} \parallel \ldots \parallel x_{1p}$

$P_2 :$ | $x_{21} \parallel x_{22} \parallel \ldots \parallel x_{2p}$

$\vdots$

$P_p :$ | $x_{p1} \parallel x_{p2} \parallel \ldots \parallel x_{pp}$

$\overset{total\ exchange}{\Longrightarrow}$

$P_1 :$ | $x_{11} \parallel x_{21} \parallel \ldots \parallel x_{p1}$

$P_2 :$ | $x_{12} \parallel x_{22} \parallel \ldots \parallel x_{p2}$

$\vdots$

$P_p :$ | $x_{1p} \parallel x_{2p} \parallel \ldots \parallel x_{pp}$

To perform the total exchange, each processor specifies a send buffer and a receive buffer. The send buffer contains the data blocks provided for the other processors in rank order. After the completion of the operation, the receive buffer of each processor contains the data blocks gathered from the other processors in rank order.

Section 4.4.1 considers the implementation of these global communication operations for different networks and derives running times. Chapter 5 describes how these communication operations are provided by the MPI library.

3.6.2.2 Duality of communication operations

A single-broadcast operation can be implemented by using a *spanning tree* with the sending processor as root. Edges in the tree correspond to physical connections in the underlying interconnection network. Using a graph representation $G = (V, E)$ of the network, see Section 2.7.2, a spanning tree can be defined as a subgraph $G' = (V, E')$ which contains all nodes of V and a subset $E' \subseteq E$ of the edges such that E' represents a tree. The construction of a spanning tree for different networks is considered in Section 4.4.1.

Given a spanning tree, a single-broadcast operation can be performed by a top-down traversal of the tree such that starting from the root each node forwards the message to be sent to its children as soon as the message arrives. The message can be forwarded over different links at the same time. For the forwarding, the tree edges can be partitioned into stages such that the message can be forwarded concurrently over all edges of a stage. Fig. 3.10 (left) shows a spanning tree with root P_1 and three stages 0, 1, 2.

Similar to a single-broadcast, a single-accumulation operation can also be implemented by using a spanning tree with the accumulating processor as root. The reduction is performed at the inner nodes according to the given reduction operation. The accumulation results from a bottom-up traversal of the tree, see Fig. 3.10 (right). Each node of the spanning tree receives a data block from each of its children (if present), combines these blocks according to the given reduction operation, including its own data block, and forwards the results to its parent node. Thus, one data block is sent over each edge of the spanning tree, but in the opposite direction as it has been done for a single-broadcast. Since the same spanning trees can be used, single-broadcast and single-accumulation are *dual* operations.

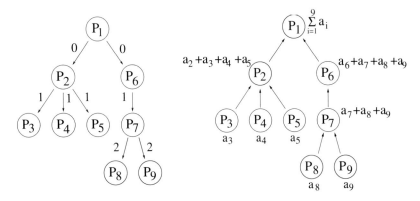

Fig. 3.10 Implementation of a single-broadcast operation using a spanning tree (left). The edges of the tree are annotated with the stage number. The right tree illustrates the implementation of a single-accumulation with the same spanning tree. Processor P_i provides a value a_i for $i = 1, \ldots, 9$. The result is accumulated at the root processor P_1 [21].

A duality relation also exists between a gather and a scatter operation as well as between a multi-broadcast and a multi-accumulation operation.

A scatter operation can be implemented by a top-down traversal of a spanning tree where each node (except the root) receives a set of data blocks from its parent node and forwards those data blocks that are meant for a node in a subtree to its corresponding child node being the root of that subtree. Thus, the number of data blocks forwarded over the tree edges decreases on the way from the root to the leaves. Similarly, a gather operation can be implemented by a bottom-up traversal of the spanning tree where each node receives a set of data blocks from each of its child nodes (if present) and forwards all data blocks received, including its own data block, to its parent node. Thus, the number of data blocks forwarded over the tree edges increases on the way from the leaves to the root. On each path to the root, over each tree edge the same number of data blocks is sent as for a scatter operation, but in opposite direction. Therefore, gather and scatter are dual operations. A multi-broadcast operation can be implemented by using p spanning trees where each spanning tree has a different root processor. Depending on the underlying network, there may or may not be physical network links that are used multiple times in different spanning trees. If no links are shared, a transfer can be performed concurrently over all spanning trees without waiting, see Section 4.4.1 for the construction of such sets of spanning trees for different networks. Similarly, a multi-accumulation can also be performed by using p spanning trees, but compared to a multi-broadcast, the transfer direction is reversed. Thus, multi-broadcast and multi-accumulation are also dual operations.

3.6.2.3 Hierarchy of communication operations

The communication operations described form a hierarchy in the following way: starting from the most general communication operation (total exchange), the other communication operations result by a step-wise *specialization*. A total exchange is the most general communication operation, since each processor sends a potentially *different* message to each other processor. A multi-broadcast is a special case of a total exchange in which each processor sends the *same* message to each of the other processors, i.e. instead of p different messages, each processor provides only one message. A multi-accumulation is also a special case of a total exchange for which the messages arriving at an intermediate node are combined according to the given reduction operation before they are forwarded. A gather operation with root P_i is a special case of a multi-broadcast which results from considering only one of the receiving processors, P_i, which receives a message from every other processor. A scatter operation with root P_i is a special case of multi-accumulation which results by using a special reduction operation which forwards the messages of P_i and ignores all other messages. A single-broadcast is a special case of a scatter operation in which the root processor sends the *same* message to every other processor, i.e. instead of p different messages the root processor provides only one message. A single-accumulation is a special case of a gather operation in which a reduction is performed at intermediate nodes of the spanning tree such that only *one* (combined) message results at the root processor. A single transfer between processors P_i and P_j is a special case of a single-broadcast with root P_i for which only the path from P_i to P_j is relevant. A single transfer is also a special case of a single-accumulation with root P_j using a special reduction operation which forwards only the message from P_i. In summary, the hierarchy in Fig. 3.11 results.

3.7 Parallel matrix-vector product

The matrix-vector multiplication is a frequently used component in scientific computing. It computes the product $\mathbf{A}\mathbf{b} = \mathbf{c}$ where $\mathbf{A} \in \mathbb{R}^{n \times m}$ is an $n \times m$-matrix and $\mathbf{b} \in \mathbb{R}^m$ is a vector of size m. (In this section, we use boldfaced type for the notation of matrices or vectors and normal type for scalar values.) The sequential computation of the matrix-vector product

$$c_i = \sum_{j=1}^{m} a_{ij} b_j, \quad i = 1, ..., n,$$

with $\mathbf{c} = (c_1, ..., c_n) \in \mathbb{R}^n$, $\mathbf{A} = (a_{ij})_{i=1,...,n, j=1,...,m}$, and $\mathbf{b} = (b_1, ..., b_m)$ can be implemented in two ways, differing in the loop order of the loops over i and j. First, the matrix-vector product is considered as the computation of n scalar products between rows $\mathbf{a}_1, ..., \mathbf{a}_n$ of $\mathbf{A}$ and vector $\mathbf{b}$, i.e.

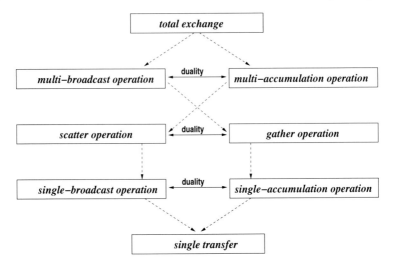

Fig. 3.11 Hierarchy of global communication operations. The horizontal arrows denote duality relations. The dashed arrows show specializations relations [21].

$$\mathbf{A} \cdot \mathbf{b} = \begin{pmatrix} (\mathbf{a}_1, \mathbf{b}) \\ \vdots \\ (\mathbf{a}_n, \mathbf{b}) \end{pmatrix},$$

where $(\mathbf{x}, \mathbf{y}) = \sum_{j=1}^{m} x_j y_j$ for $\mathbf{x}, \mathbf{y} \in \mathbb{R}^m$ with $\mathbf{x} = (x_1, \ldots, x_m)$ and $\mathbf{y} = (y_1, \ldots, y_m)$ denotes the scalar product (or inner product) of two vectors. The corresponding algorithm (in C-notation) is:

```
for (i=0; i<n; i++) c[i] = 0;
for (i=0; i<n; i++)
    for (j=0; j<m; j++)
        c[i] = c[i] + A[i][j] * b[j];
```

The matrix $\mathbf{A} \in \mathbb{R}^{n \times m}$ is implemented as two-dimensional array A and the vectors $\mathbf{b} \in \mathbb{R}^m$, and $\mathbf{c} \in \mathbb{R}^n$ are implemented as one-dimensional arrays b and c. (The indices start with 0 as usual in C.) For each $i = 0, \ldots, n - 1$, the inner loop body consists of a loop over j computing one of the scalar products. Second, the matrix-vector product can be written as a linear combination of columns $\tilde{\mathbf{a}}_1, \ldots, \tilde{\mathbf{a}}_m$ of $\mathbf{A}$ with coefficients $b_1, \ldots, b_m$, i.e.

$$\mathbf{A} \cdot \mathbf{b} = \sum_{j=1}^{m} b_j \tilde{\mathbf{a}}_j .$$

The corresponding algorithm (in C-notation) is:

```
for (i=0; i<n; i++) c[i] = 0;
for (j=0; j<m; j++)
    for (i=0; i<n; i++)
        c[i] = c[i] + A[i][j] * b[j] ;
```

For each $j = 0,\dots,m-1$, a column $\tilde{a}_j$ is added to the linear combination. Both sequential programs are equivalent since there are no dependencies and the loops over i and j can be exchanged. For a parallel implementation, the row- and column-oriented representations of matrix A give rise to different parallel implementation strategies.

(a) The row-oriented representation of matrix A in the computation of n scalar products (a_i, b), $i = 1,\dots,n$, of rows of A with vector b leads to a parallel implementation in which each processor of a set of p processors computes approximately n/p scalar products.

(b) The column-oriented representation of matrix A in the computation of the linear combination $\sum_{j=1}^{m} b_j \tilde{a}_j$ of columns of A leads to a parallel implementation in which each processor computes a part of this linear combination with approximately m/p column vectors.

In the following, we consider these parallel implementation strategies for the case that n and m are multiples of the number of processors p.

3.7.1 Parallel computation of scalar products

For a parallel implementation of a matrix-vector product on a distributed memory machine, the data distributions of A and b are chosen such that the processor computing the scalar product (a_i, b), $i \in \{1,\dots,n\}$ accesses only data elements stored in its private memory, i.e. row a_i of A and vector b are stored in the private memory of the processor computing the corresponding scalar product. Since vector $b \in \mathbb{R}^m$ is needed for all scalar products, b is stored in a replicated way. For matrix A, a row-oriented data distribution is chosen such that a processor computes the scalar product for which the matrix row can be accessed locally. Row-oriented blockwise as well as cyclic or block-cyclic data distributions can be used.

For the row-oriented blockwise data distribution of matrix A, processor P_k, $k = 1,\dots,p$, stores the rows a_i, $i = n/p \cdot (k-1)+1,\dots,n/p \cdot k$, in its private memory and computes the scalar products (a_i, b). The computation of (a_i, b) needs no data from other processors and, thus, no communication is required. According to the row-oriented blockwise computation the result vector $c = (c_1,\dots,c_n)$ has a blockwise distribution.

When the matrix-vector product is used within a larger algorithm like iteration methods, there are usually certain requirements for the distribution of c. In iteration methods, there is often the requirement that the result vector c has the same data distribution as the vector **b**. To achieve a replicated distribution for **c**, each processor P_k, $k = 1,\ldots,p$, sends its block $(c_{n/p \cdot (k-1)+1},\ldots,c_{n/p \cdot k})$ to all other processors. This can be done by a multi-broadcast operation. A parallel implementation of the matrix-vector product including this communication is given in Figure 3.12. The program is executed by all processors P_k, $k = 1,\ldots,p$, in the SPMD style. The communication operation includes an implicit barrier synchronization. Each processor P_k stores a different part of the $n \times m$-array A in its local array local_A of dimension local_n $\times$ m. The block of rows stored by P_k in local_A contains the global elements

$$\texttt{local\_A}[\texttt{i}][\texttt{j}] = \texttt{A}[\texttt{i} + (\texttt{k} - 1) * \texttt{n}/\texttt{p}][\texttt{j}]$$

with $i = 0,\ldots,n/p-1$, $j = 0,\ldots,m-1$ and $k = 1,\ldots,p$. Each processor computes a local matrix-vector product of array local_A with array b and stores the result in array local_c of size local_n. The communication operation

multi_broadcast(local_c,local_n,c)

performs a multi-broadcast operation with the local arrays local_c of all processors as input. After this communication operation, the global array c contains the values

$$\texttt{c}[\texttt{i} + (\texttt{k} - 1) * \texttt{n}/\texttt{p}] = \texttt{local\_c}[\texttt{i}]$$

for $i = 0,\ldots,n/p-1$ and $k = 1,\ldots,p$, i.e. the array c contains the values of the local vectors in the order of the processors and has a replicated data distribution. See Figure 3.15(1) for an illustration of the data distribution of A, b and c for the program given in Figure 3.12.

```
/* Matrix-vector product Ab = c with parallel inner products*/
/* Row-oriented blockwise distribution of A */
/* Replicated distribution of vectors b and c */
local_n = n/p;
for (i=0; i<local_n; i++) local_c[i] = 0;
for (i=0; i<local_n; i++)
    for (j=0; j<m; j++)
        local_c[i] = local_c[i] + local_A[i][j] * b[j];
multi_broadcast(local_c,local_n,c);
/* Multi-broadcast operation of (c[0],...,c[localₙ]) to globalᶜ*/
```

Fig. 3.12 Program fragment in C notation for a parallel program of the matrix-vector product with row-oriented blockwise distribution of the matrix **A** and a final redistribution of the result vector **c**.

For a row-oriented cyclic distribution, each processor P_k, $k = 1,\ldots,p$, stores the rows $\mathbf{a}_i$ of matrix **A** with $i = k + p \cdot (l-1)$ for $l = 1,\ldots,n/p$ and computes the corresponding scalar products. The rows in the private memory of processor P_k are stored within one local array local_A of dimension local_n $\times$ m. After the parallel computation of the result array local_c, the entries have to be reordered correspondingly to get the global result vector in the original order.

For the implementation of the matrix-vector product on a **shared memory** machine, the row-oriented distribution of the matrix A and the corresponding distribution of the computation can be used. Each processor of the shared memory machine computes a set of scalar products as described above. A processor P_k computes n/p elements of the result vector c and uses n/p corresponding rows of matrix A in a blockwise or cyclic way, $k = 1, \ldots, p$. The difference to the implementation on a distributed memory machine is that an explicit distribution of the data is not necessary since the entire matrix A and vector b reside in the common memory accessible by all processors.

The distribution of the computation to processors according to a row-oriented distribution, however, causes the processors to access different elements of A and compute different elements of c. Thus, the write accesses to c cause no conflict. Since the accesses to matrix A and vector b are read accesses they also cause no conflict. Synchronization and locking are not required for this shared memory implementation. Figure 3.13 shows an SPMD program for a parallel matrix-vector multiplication accessing the global array A, b, c. The variable k denotes the processor id of the processor P_k, $k = 1, \ldots, p$. Because of this processor number k, each processor P_k computes different elements of the result array c. The program fragment ends with a barrier synchronization synch() to guarantee that all processors reach this program point and the entire array c is computed before any processor executes subsequent program parts. (The same program can be used for a distributed memory machine when the entire arrays A, b, c are allocated in each private memory; this approach needs much more memory since the arrays are allocated p times.)

```
/* Matrix-vector product Ab=c with parallel inner products*/
/* Row-oriented distribution of the computation */
local_n = n/p;
for (i=0; i<local_n; i++) c[i+(k-1)*local_n] = 0;
for (i=0; i<local_n; i++)
    for (j=0; j<m; j++)
        c[i+(k-1)*local_n] =
        c[i+(k-1)*local_n] + A[i+(k-1)*local_n][j] * b[j];
synch();
```

Fig. 3.13 Program fragment in C notation for a parallel program of the matrix-vector product with row-oriented blockwise distribution of the computation. In contrast to the program in Fig. 3.12, the program uses the global arrays A, b, and c for a shared memory system.

3.7.2 Parallel computation of the linear combinations

For a distributed memory machine, the parallel implementation of the matrix-vector product in the form of the linear combination uses a **column-oriented** distribution of the matrix **A**. Each processor computes the part of the linear combination for which it owns the corresponding columns $\tilde{\mathbf{a}}_i$, $i \in \{1,\ldots,m\}$. For a blockwise distribution of the columns of **A**, processor P_k owns the columns $\tilde{\mathbf{a}}_i$, $i = m/p \cdot (k-1) + 1, \ldots, m/p \cdot k$, and computes the n-dimensional vector

$$\mathbf{d}_k = \sum_{j=m/p \cdot (k-1)+1}^{m/p \cdot k} b_j \tilde{\mathbf{a}}_j,$$

which is a partial linear combination and a part of the total result, $k = 1, \ldots, p$. For this computation only a block of elements of vector **b** is accessed and only this block needs to be stored in the private memory. After the parallel computation of the vectors $\mathbf{d}_k$, $k = 1, \ldots, p$, these vectors are added to give the final result $\mathbf{c} = \sum_{k=1}^{p} \mathbf{d}_k$. Since the vectors $\mathbf{d}_k$ are stored in different local memories, this addition requires communication, which can be performed by an accumulation operation with the addition as reduction operation. Each of the processors P_k provides its vector $\mathbf{d}_k$ for the accumulation operation. The result of the accumulation is available on one of the processors. When the vector is needed in a replicated distribution, a broadcast operation is performed. The data distribution before and after the communication is illustrated in Figure 3.15(2a). A parallel program in the SPMD style is given in Figure 3.14. The local arrays `local_b` and `local_A` store blocks of **b** and blocks of columns of **A** so that each processor P_k owns the elements

$$\text{local\_A}[i][j] = A[i][j + (k-1) * m/p]$$

and

$$\text{local\_b}[j] = b[j + (k-1) * m/p],$$

where $j = 0, \ldots, m/p - 1$, $i = 0, \ldots, n-1$ and $k = 1, \ldots, p$. The array d is a private vector allocated by each of the processors in its private memory containing different data after the computation. The operation

`single_accumulation(d,local_m,c,ADD,1)`

denotes an accumulation operation, for which each processor provides its array d of size n, and ADD denotes the reduction operation. The last parameter is 1 and means that processor P_1 is the root processor of the operation, which stores the result of the addition into the array c of length n. The final `single_broadcast(c,1)` sends the array c from processor P_1 to all other processors and a replicated distribution of c results.

Alternatively to this final communication, a multi-accumulation operation can be applied which leads to a block-wise distribution of array c. This program version may be advantageous if c is required to have the same distribution as array b. Each

```
/* Matrix-vector product Ab=c with parallel linear combination*/
/* Column-oriented distribution of A */
/* Replicated distribution of vectors b and c */
local_m=m/p;
for (i=0;  i<n;  i++) d[i] = 0;
for (j=0;  j<local_m;  j++)
    for (i=0 ;i<n; i++)
        d[i] = d[i] + local_b[j] * local_A[i][j];
single_accumulation(d,n,c,ADD,1);
single_broadcast(c,1);
```

Fig. 3.14 Program fragment in C notation for a parallel program of the matrix-vector product with column-oriented blockwise distribution of the matrix **A** and reduction operation to compute the result vector **c**. The program uses local array d for the parallel computation of partial linear combinations.

processor accumulates the n/p elements of the local arrays d, i.e. each processor computes a block of the result vector c and stores it in its local memory. This communication is illustrated in Figure 3.15(2b).

For shared memory machines, the parallel computation of the linear combinations can also be used but special care is needed to avoid access conflicts for the write accesses when computing the partial linear combinations. To avoid write conflicts, a separate array d_k of length n should be allocated for each of the processors P_k to compute the partial result in parallel without conflicts. The final accumulation needs no communication, since the data d_k are stored in the common memory, and can be performed in a blocked way.

The computation and communication time for the matrix-vector product is analyzed in Section 4.5.2.

3.8 Processes and Threads

Parallel programming models are often based on processors or threads. Both are abstractions for a flow of control, but there are some differences which we will consider in this section in more detail. As described in Section 3.2, the principal idea is to decompose the computation of an application into tasks and to employ multiple control flows running on different processors or cores for their execution, thus obtaining a smaller overall execution time by parallel processing.

3.8.1 Processes

In general, a process is defined as a program in execution. The process comprises the executable program along with all information that is necessary for the execution

1) Parallel computation of inner products

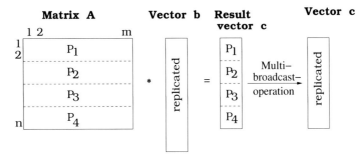

2) Parallel computation of linear combination

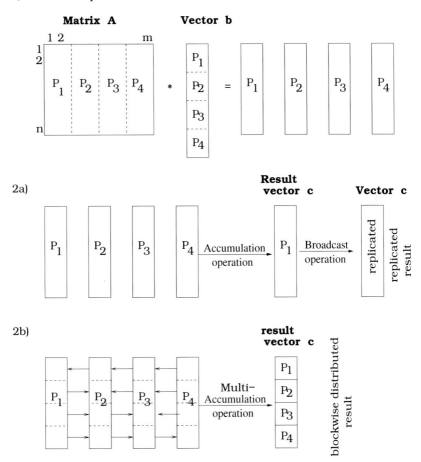

Fig. 3.15 Parallel matrix-vector multiplication with 1) parallel computation of scalar products and replicated result and 2) parallel computation of linear combinations with a) replicated result and b) blockwise distribution of the result.

of the program. This includes the program data on the runtime stack or the heap, the current values of the registers, as well as the content of the program counter which specifies the next instruction to be executed. All this information changes dynamically during the execution of the process. Each process has its own address space, i.e. the process has exclusive access to its data. When two processes want to exchange data, this has to be done by explicit communication.

A process is assigned to execution resources (processors or cores) for execution. There may be more processes than execution resources. If this is the case, an execution resource typically executes several processes in rotation at different points in time, e.g. in a round-robin fashion. If the execution resource is assigned to another process by the scheduler of the operating system, the state of the suspended process must be saved to allow a continuation of the execution at a later time with the process state before suspension. This switching between processes is called **context switch**, and it may cause a significant overhead, depending on the hardware support [104]. Often time slicing is used to switch between the processes. If there is a single execution resource only, the active processes are executed concurrently in a time-sliced way, but there is no real parallelism. If several execution resources are available, different processes can be executed by different execution resources, thus indeed leading to a parallel execution.

When a process is generated, it must obtain the data required for its execution. In Unix systems, a process P_1 can create a new process P_2 with the `fork` system call. The new child process P_2 is an identical copy of the parent process P_1 at the time of the `fork` call. This means, that the child process P_2 works on a *copy* of the address space of the parent process P_1 and executes the same program as P_1, starting with the instruction following the `fork` call. The child process gets its own process number and, depending on this process number, it can execute different statements as the parent process. Since each process has its own address space and since process creation includes the generation of a copy of the address space of the parent process, process creation and management may be quite time-consuming. Data exchange between processes is often done via socket communication which is based on TCP/IP or UDP/IP communication. This may lead to a significant overhead, depending on the socket implementation and the speed of the interconnection between the execution resources assigned to the communicating processes.

3.8.2 Threads

The thread model is an extension of the process model. In the thread model, each process may consist of *multiple* independent control flows which are called **threads**. The word *thread* is used to indicate that a potentially long continuous sequence of instructions is executed. During the execution of a process, the different threads of this process are assigned to execution resources by a scheduling method.

3.8.2.1 Basic concepts of threads

A significant feature of threads is that the threads of *one* process share the address space of the process, i.e. they have a common address space. When a thread stores a value in the shared address space, another thread of the same process can access this value afterwards. Threads are typically used if the execution resources used have access to a physically shared memory, as is the case for the cores of a multicore processor. In this case, information exchange is fast compared to socket communication. Thread generation is usually much faster than process generation: no copy of the address space is necessary, since the threads of a process share a common address space. Therefore, the use of threads is often more flexible than the use of processes, yet providing the same advantages concerning a parallel execution. In particular, the different threads of a process can be assigned to different cores of a multicore processor, thus providing parallelism within the processes.

Threads can be provided by the runtime system as **user-level threads** or by the operating system as **kernel threads**. User-level threads are managed by a *thread library* without specific support of the operating system. This has the advantage that a switch from one thread to another can be done without interaction of the operating system and is therefore quite fast. Disadvantages of the management of threads at user-level result from the fact that the operating system has no knowledge about the existence of threads and manages entire processes only. Therefore, the operating system cannot map different threads of the same process to different execution resources and all threads of one process are executed on the same execution resource. Moreover, the operating system cannot switch to another thread if one thread executes a blocking I/O operation. Instead, the CPU scheduler of the operating system suspends the entire process and assigns the execution resource to another process.

These disadvantages can be avoided by using kernel threads, since the operating system is aware of the existence of threads and can react correspondingly. This is especially important for an efficient use of the cores of a multicore system. Most operating systems support threads at the kernel level.

3.8.2.2 Execution models for threads

If there is no support for thread management by the operating system, the thread library is responsible for the entire thread scheduling. In this case, *all* user-level threads of a user process are mapped to *one* process of the operating system. This is called **N:1 mapping**, or *many-to-one mapping*, see Fig. 3.16 for an illustration. At each point in time, the library scheduler determines which of the different threads comes to execution. The mapping of the processes to the execution resources is done by the operating system. If several execution resources are available, the operating system can bring several processes to execution concurrently, thus exploiting parallelism. But with this organization the execution of different threads of one process on different execution resources is not possible.

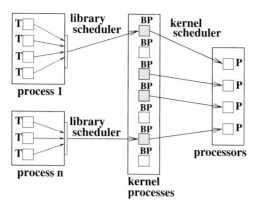

Fig. 3.16 Illustration of a N:1 mapping for thread management without kernel threads. The scheduler of the thread library selects the next thread T of the user process for execution. Each user process is assigned to exactly one process BP of the operating system. The scheduler of the operating system selects the processes to be executed at a certain time and maps them to the execution resources P.

If the operating system supports thread management, there are two possibilities for the mapping of user-level threads to kernel threads. The first possibility is to generate a kernel thread for each user-level thread. This is called **1:1 mapping**, or *one-to-one mapping*, see Fig. 3.17 for an illustration. The scheduler of the operating system selects which kernel threads are executed at which point in time. If multiple execution resources are available, it also determines the mapping of the kernel threads to the execution resources. Since each user-level thread is assigned to exactly one kernel thread, there is no need for a library scheduler. Using a 1:1 mapping, different threads of a user process can be mapped to different execution resources, if enough resources are available, thus leading to a parallel execution within a single process.

The second possibility is to use a two-level scheduling where the scheduler of the thread library assigns the user-level threads to a given set of kernel threads. The scheduler of the operating system maps the kernel threads to the available exccution resources. This is called **N:M mapping**, or *many-to-many mapping*, see Fig. 3.18 for an illustration. At different points in time, a user thread may be mapped to a different kernel thread, i.e. no fixed mapping is used. Correspondingly, at different points in time, a kernel thread may execute different user threads. Depending on the thread library, the programmer can influence the scheduler of the library, e.g. by selecting a scheduling method as this is the case for the Pthreads library, see Section 6.3.2 for more details. The scheduler of the operating system on the other hand is tuned for an efficient use of the hardware resources, and there is typically no possibility for the programmer to directly influence the behavior of this scheduler. This second mapping possibility usually provides more flexibility than a 1:1 mapping, since the programmer can adapt the number of user-level threads to the specific algorithm or application. The operating system can select the number

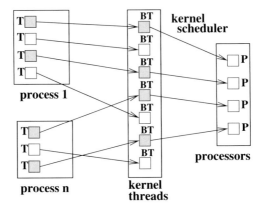

Fig. 3.17 Illustration of a 1:1 mapping for thread management with kernel threads. Each user-level thread T is assigned to one kernel thread BT. The kernel threads BT are mapped to execution resources P by the scheduler of the operating system.

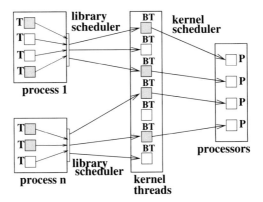

Fig. 3.18 Illustration of a N:M mapping for thread management with kernel threads using a two-level scheduling. User-level threads T of different processes are assigned to a set of kernel threads BT (N:M mapping) which are then mapped by the scheduler of the operating system to execution resources P.

of kernel threads such that an efficient management and mapping of the execution resources is facilitated.

3.8.2.3 Thread states

A thread can be in one of the following states:

- **newly generated**, i.e. the thread has just been generated, but has not yet performed any operation;

- **executable**, i.e. the thread is ready for execution, but is currently not assigned to any execution resources;
- **running**, i.e. the thread is currently being executed by an execution resource;
- **waiting**, i.e. the thread is waiting for an external event to occur; the thread cannot be executed before the external event happens;
- **finished**, i.e. the thread has terminated all its operations.

Fig. 3.19 illustrates the transition between these states. The transitions between the states *executable* and *running* are determined by the scheduler. A thread may enter the state *waiting* because of a blocking I/O operation or because of the execution of a synchronization operation which causes it to be blocked. The transition from state *waiting* to *executable* may be caused by a termination of a previously issued I/O operation or because another thread releases the resource which this thread is waiting for.

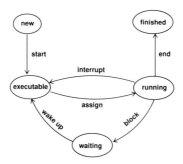

Fig. 3.19 States of a thread. The nodes of the diagram show the possible states of a thread and the arrows show possible transitions between them.

3.8.2.4 Visibility of data

The different threads of a process share a common address space. This means that the global variables of a program and all dynamically allocated data objects can be accessed by any thread of this process, no matter which of the threads has allocated the object. But for each thread, there is a private runtime stack for controlling function calls of this thread and to store the local variables of these functions, see Fig. 3.20 for an illustration. The data kept on the runtime stack is local data of the corresponding thread and the other threads have no direct access to this data. It is in principle possible to give them access by passing an address, but this is dangerous, since it cannot be predicted how long the data remains accessible. The stack frame of a function call is freed as soon as the function call is terminated. The runtime stack of a thread exists only as long as the thread is *active*; it is freed as soon as the thread is terminated. Therefore, a return value of a thread should not be passed via its runtime stack. Instead, a global variable or a dynamically allocated data object should be used, see Chapter 6 for more details.

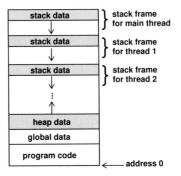

Fig. 3.20 Runtime stack for the management of a program with multiple threads.

3.8.3 Synchronization mechanisms

When multiple threads execute a parallel program in parallel, their execution has to be coordinated to avoid race conditions. Synchronization mechanisms are provided to enable a coordination, e.g. to ensure a certain execution order of the threads or to control access to shared data structures. Synchronization for shared variables is mainly used to avoid a concurrent manipulation of the same variable by different threads, which may lead to non-deterministic behavior. This is important for multi-threaded programs, no matter whether a single execution resource is used in a time-slicing way, or whether several execution resources execute multiple threads in parallel. Different synchronization mechanisms are provided for different situations. In the following, we give a short overview.

3.8.3.1 Lock synchronization

For a concurrent access of shared variables, race conditions can be avoided by a **lock mechanism** based on predefined **lock variables**, which are also called **mutex variables** as they help to ensure mutual exclusion. A lock variable l can be in one of two states: *locked* or *unlocked*. Two operations are provided to influence this state: lock(l) and unlock(l). The execution of lock(l) locks l such that it cannot be locked by another thread; after the execution, l is in the *locked* state and the thread that has executed lock(l) is the *owner* of l. The execution of unlock(l) unlocks a previously locked lock variable l; after the execution, l is in the *unlocked* state and has no owner. To avoid race conditions for the execution of a program part, a lock variable 1 is assigned to this program part and each thread executes lock(1) before entering the program part and unlock(1) after leaving the program part. To avoid race conditions, *each* of the threads must obey this programming rule.

A call of lock(1) for a lock variable 1 has the effect that the executing thread T_1 becomes the owner of 1, if 1 has been in the *unlocked* state before. But if there is already another owner T_2 of 1 before T_1 calls lock(1), T_1 is blocked until T_2 has called unlock(1) to release 1. If there are blocked threads waiting for 1 when unlock(1) is called, one of the waiting threads is woken up and becomes the new owner of 1. Thus, using a lock mechanism in the described way leads to a *sequentialization* of the execution of a program part which ensures that at each point in time, only one thread executes the program part. The provision of lock mechanisms in libraries like Pthreads, OpenMP, or Java threads is described in Chapter 6.

It is important to see that mutual exclusion for accessing a shared variable can only be guaranteed if all threads use a lock synchronization to access the shared variable. If this is not the case, a race condition may occur, leading to an incorrect program behavior. This can be illustrated by the following example where two threads T_1 and T_2 access a shared integer variable s which is protected by a lock variable 1 [141]

Thread T_1	Thread T_2
lock(1);	
s = 1;	s = 2;
if (s!=1) fire_missile();	
unlock(1);	

In this example, thread T_1 may get interrupted by the scheduler and thread T_2 can set the value of s to 2; if T_1 resumes execution, s has value 2 and fire_missile() is called. For other execution orders, fire_missile() will not be called. This non-deterministic behavior can be avoided if T_2 also uses a lock mechanism with 1 to access s.

Another mechanism to ensure mutual exclusion is provided by **semaphores** [51]. A semaphore is a data structure which contains an integer counter s and to which two atomic operations $P(s)$ and $V(s)$ can be applied. A *binary semaphore* s can only have values 0 or 1. For a *counting semaphore*, s can have any positive integer value. The operation $P(s)$, also denoted as wait(s), waits until the value of s is larger than 0. When this is the case, the value of s is decreased by 1, and execution can continue with the subsequent instructions. The operation $V(s)$, also denoted as signal(s), increments the value of s by 1. To ensure mutual exclusion for a critical section, the section is protected by a semaphore s in the following form:

```
wait(s)
critical section
signal(s)
```

Different threads may execute operations $P(s)$ or $V(s)$ for a semaphore s to access the critical section. After a thread T_1 has successfully executed the operation wait(s) with waiting it can enter the critical section. Every other thread T_2 is blocked when it executes `wait(s)` and can therefore not enter the critical section. When T_1 executes `signal(s)` after leaving the critical section, one of the waiting threads will be woken up and can enter the critical section.

Another concept to ensure mutual exclusion is the concept of **monitors** [111]. A monitor is a language construct which allows the definition of data structures and access operations. These operations are the *only* means with which the data of a monitor can be accessed. The monitor ensures that the access operations are executed with mutual exclusion, i.e., at each point in time, only one thread is allowed to execute any of the access methods provided.

3.8.3.2 Thread execution control

To control the execution of multiple threads, barrier synchronization and condition synchronization can be used. A **barrier synchronization** defines a synchronization point where each thread must wait until all other threads have also reached this synchronization point. Thus, none of the threads executes any statement after the synchronization point before all other threads have also arrived at this point. A barrier synchronization also has the effect that it defines a global state of the shared address space in which all operations specified before the synchronization point have been executed. Statements after the synchronization point can be sure that this global state has been established.

Using a **condition synchronization**, a thread T_1 is blocked until a given condition has been established. The condition could, for example, be that a shared variable contains a specific value or has a specific state like a shared buffer containing at least one entry. The blocked thread T_1 can only be woken up by another thread T_2, e.g. after T_2 has established the condition which T_1 waits for. When T_1 is woken up, it enters the state *executable*, see Section 3.8.2.2, and will later be assigned to an execution resource, then entering the state *running*. Thus, after being woken up, T_1 may not be immediately executed, e.g., if not enough execution resources are available. Therefore, although T_2 may have established the condition which T_1 waits for, it is important that T_1 checks the condition again as soon as it is running. The reason for this additional check is that in the meantime another thread T_3 may have performed some computations which might have led to the fact that the condition is not fulfilled any more. Condition synchronization can be supported by condition variables. These are for example provided by Pthreads and must be used together with a lock variable to avoid race conditions when evaluating the condition, see Section 6.1 for more details. A similar mechanism is provided in Java by `wait()` and `notify()`, see Section 6.4.3.

3.8.4 Developing efficient and correct thread programs

Depending on the requirements of an application and the specific implementation by the programmer, synchronization leads to a complicated interaction between the executing threads. This may cause problems like performance degradation by sequentializations, or even deadlocks. This section contains a short discussion of this topic and gives some suggestions how efficient thread-based programs can be developed.

3.8.4.1 Number of threads and sequentialization

Depending on the design and implementation, the runtime of a parallel program based on threads can be quite different. For the design of a parallel program it is important

- to use a suitable number of threads which should be selected according to the degree of parallelism provided by the application and the number of execution resources available and
- to avoid sequentialization by synchronization operations whenever possible.

When synchronization is necessary, e.g. to avoid race conditions, it is important that the resulting critical section which is executed sequentially, is made as small as possible to reduce the resulting waiting times.

The creation of threads is necessary to exploit parallel execution. A parallel program should create a sufficiently large number of threads to provide enough work for all cores of an execution platform, thus using the available resources efficiently. But the number of threads created should not be too large to keep the overhead for thread creation, management and termination small. For a large number of threads, the work per thread may become quite small, giving the thread overhead a signification portion of the overall execution time. Moreover, many hardware resources, in particular caches, may be shared by the cores, and performance degradations may result, if too many threads share the resources; in the case of caches, a degradation of the read/write bandwidth might result.

The threads of a parallel program must be coordinated to ensure a correct behavior. An example is the use of synchronization operations to avoid race conditions. But too many synchronizations may lead to situations where only one or a small number of threads are active while the other threads are waiting because of a synchronization operation. In effect, this may result in a **sequentialization** of the thread execution, and the available parallelism cannot be used. In such situations, increasing the number of threads does not lead to faster program execution, since the new threads are waiting most of the time.

3.8.4.2 Deadlock

Non-deterministic behavior and race conditions can be avoided by synchronization mechanisms like lock synchronization. But the use of locks can lead to **deadlocks**, when program execution comes into a state where each thread waits for an event that can only be caused by another thread, but this thread is also waiting.

Generally, a deadlock occurs for a set of activities, if each of the activities waits for an event that can only be caused by one of the other activities, such that a cycle of mutual waiting occurs. A deadlock may occur in the following example where two threads T_1 and T_2 both use two locks s1 and s2:

Thread T_1	Thread T_2
lock(s1);	lock(s2);
lock(s2);	lock(s1):
do_work();	do_work();
unlock(s2)	unlock(s1)
unlock(s1)	unlock(s2)

A deadlock occurs for the following execution order:

- a thread T_1 first tries to set a lock s_1, and then s_2; after having locked s_1 successfully, T_1 is interrupted by the scheduler;
- a thread T_2 first tries to set lock s_2 and then s_1; after having locked s_2 successfully, T_2 waits for the release of s_1.

In this situation, s_1 is locked by T_1 and s_2 by T_2. Both threads T_1 and T_2 wait for the release of the missing lock by the other thread. But this cannot occur, since the other thread is waiting.

It is important to avoid such mutual or cyclic waiting situations, since the program cannot be terminated in such situations. Specific techniques are available to avoid deadlocks in cases where a thread must set multiple locks to proceed. Such techniques are described in Section 6.1.2.

3.8.4.3 Memory access times and cache effects

Memory access times may constitute a significant portion of the execution time of a parallel program. A memory access issued by a program causes a data transfer from the main memory into the cache hierarchy of that core which has issued the memory access. This data transfer is caused by the read and write operations of the cores. Depending on the specific pattern of read and write operations, there is not only a transfer from main memory to the local caches of the cores, but there may also be a transfer between the local caches of the cores. The exact behavior is controlled by hardware, and the programmer has no direct influence on this behavior.

The transfer within the memory hierarchy can be captured by dependencies between the memory accesses issued by different cores. These dependencies can be categorized as read-read dependency, read-write dependency and write-write dependency. A read-read dependency occurs if two threads running on different cores access the same memory location. If this memory location is stored in the local caches of both cores, both can read the stored values from their cache, and no access to main memory needs to be done. A read-write dependency occurs, if one thread T_1 executes a write into a memory location which is later read by another thread T_2 running on a different core. If the two cores involved do not share a common cache, the memory location that is written by T_1 must be transferred into main memory after the write before T_2 executes its read which then causes a transfer from main memory into the local cache of the core executing T_2. Thus, a read-write dependency consumes memory bandwidth.

A write-write dependency occurs, if two threads T_1 and T_2, running on different cores perform a write into the same memory location in a given order. Assuming that T_1 writes before T_2, a cache coherency protocol, see Section 2.9.3, must ensure that the caches of the participating cores are notified when the memory accesses occur. The exact behavior depends on the protocol and the cache implementation as write-through or write-back, see Section 2.9.1. In any case, the protocol causes a certain amount of overhead to handle the write-write dependency.

False sharing occurs if two threads T_1 and T_2, running on different cores, access different memory locations that are held in the same cache line. In this case, the same memory operations must be performed as for an access to the same memory locations, since a cache line is the smallest transfer unit in the memory hierarchy. False sharing can lead to a significant amount of memory transfers and to notable performance degradations. It can be avoided by an alignment of variables to cache line boundaries; this is supported by some compilers.

3.9 Further parallel programming approaches

For the programming of parallel architectures, a large number of approaches have been developed during the last years. A first classification of these approaches can be made according to the memory view provided, shared address space or distributed address space, as discussed earlier. In the following, we give a detailed description of the most popular approaches for both classes. For a distributed address space, MPI is by far the most often used environment, see Chapter 5 for a detailed description. The use of MPI is not restricted to parallel machines with a physically distributed memory organization. It can also be used for parallel architectures with a physically shared address space like multicore architectures. Popular programming approaches for shared address space include Pthreads, Java threads, and OpenMP, see Chapter 6 for a detailed treatment. But besides these popular environments, there are many other interesting approaches aiming at making parallel programming easier by providing the right abstraction. We give a short overview in this section.

The advent of multicore architectures and their use in normal desktop computers has led to an intensifying of the research efforts to develop a simple, yet efficient parallel language. An important argument for the need of such a language is that parallel programming with processes or threads is difficult and is a big step for programmers used to sequential programming [143]. It is often mentioned that, for example, thread programming with lock mechanisms and other forms of synchronization are too low-level and too error-prone, since problems like race conditions or deadlocks can easily occur. Current techniques for parallel software development are therefore sometimes compared to assembly programming [212].

In the following, we give a short description of language approaches which attempt to provide suitable mechanisms at the right level of abstraction.

We now give a short overview of interesting approaches for new parallel languages that are already in use but are not yet popular enough to be described in great detail in an introductory textbook on parallel computing. Some of the approaches described have been developed in the area of high performance computing, but they can also be used for small parallel systems, including multicore systems.

Unified Parallel C (UPC) has been proposed as an extension to C for the use of parallel machines and cluster systems [59]. UPC is based on the model of a *partitioned global address space* (PGAS) [42], in which shared variables can be stored. Each such variable is associated with a certain thread, but the variable can also be read or manipulated by other threads. But typically, the access time for the variable is smaller for the associated thread than for another thread. Additionally, each thread can define private data to which it has exclusive access.

In UPC programs, parallel execution is obtained by creating a number of threads at program start. The UPC language extensions to C define a parallel execution model, memory consistency models for accessing shared variables, synchronization operations, and parallel loops. A detailed description is given in [59], see also upc.lbl.gov. UPC compilers are available for several platforms, see upc.lbl.gov for more information. Other languages based on the PGAS model are the Co-Array Fortran Language (CAF) which is based on Fortran, and Titanium, which is similar to UPC, but is based on Java instead of C.

In the context of the DARPA HPCS (*High Productivity Computing Systems*) program, several programming languages have been proposed and implemented, which support programming with a shared address space. These languages include Fortress, X10, and Chapel.

Fortress has been developed by Sun Microsystems. Fortress is an object-oriented language based on Fortran which facilitates program development for parallel systems by providing a mathematical notation [11]. The language Fortress supports the parallel execution of programs by parallel loops and by the parallel evaluation of function arguments with multiple threads. Many constructs provided are implicitly parallel, meaning that the threads needed are created without an explicit control in the program.

A separate thread is, for example, implicitly created for each argument of a function call without any explicit thread creation in the program. Additionally, explicit threads can be created for the execution of program parts. Thread synchronization is performed with `atomic` expressions which guarantee that the effect on the memory becomes atomically visible immediately after the expression has been completely evaluated; see also the next section on transactional memory.

X10 has been developed by IBM as an extension to Java targeting at high-performance computing. Similar to UPC, X10 is based on the PGAS memory model and extends this model to the GALS model (*globally asynchronous, locally synchronous*) by introducing logical *places* [32]. The threads of a place have a locally synchronous view to their shared address space, but threads of different places work asynchronously with each other. X10 provides a variety of operations to access array variables and parts of array variables. Using array distributions, a partitioning of an array to different places can be specified. For the synchronization of threads, atomic blocks are provided which support an atomic execution of statements. By using atomic blocks, the low-level details of synchronization are performed by the runtime system, and no low-level lock synchronization must be performed.

Chapel has been developed by Cray Inc. as a new parallel language for high-performance computing [47]. Some of the language constructs provided are similar to High-Performance Fortran (HPF). Like Fortress and X10, Chapel also uses the model of a global address space in which data structures can be stored and accessed. The parallel execution model supported is based on threads. At program start, there is a single main thread; using language constructs like parallel loops, more threads can be created. The threads are managed by the runtime system and the programmer does not need to start or terminate threads explicitly. For the synchronization of computations on shared data, synchronization variables and **atomic** blocks are provided.

The global array (GA) approach has been developed to support program design for applications from scientific computing which mainly use array-based data structures, like vectors or matrices [161].

The GA approach is provided as a library with interfaces for C, C++, and Fortran for different parallel platforms. The GA approach is based on a global address space in which global array can be stored such that each process is associated with a logical block of the global array; access to this block is faster than access to the other blocks. The GA library provides basic operations (like put, get, scatter, gather) for the shared address space, as well as atomic operations and lock mechanisms for accessing global arrays. Data exchange between processes can be performed via global arrays. But a message-passing library like MPI can also be used. An important application area for the GA approach is the area of chemical simulations.

3.10 Exercices for Chapter 3

Exercise 3.1. Consider the following sequence of instructions I_1, I_2, I_3, I_4, I_5:

$$I_1: \quad R_1 \leftarrow R_1 + R_2$$
$$I_2: \quad R_3 \leftarrow R_1 + R_2$$
$$I_3: \quad R_5 \leftarrow R_3 + R_4$$
$$I_4: \quad R_4 \leftarrow R_3 + R_1$$
$$I_5: \quad R_2 \leftarrow R_2 + R_4$$

Determine all flow, anti, and output dependences and draw the resulting data dependence graph. Is it possible to execute some of these instructions parallel to each other?

Exercise 3.2. Consider the following two loops

```
for (i=0 : n-1)              forall (i=0 : n-1)
    a(i) = b(i) +1;              a(i) = b(i) + 1;
    c(i) = a(i) +2;              c(i) = a(i) + 2;
    d(i) = c(i+1)+1;             d(i) = c(i+1) + 1;
endfor                       endforall
```

Do these loops perform the same computations? Explain your answer.

Exercise 3.3. Consider the following sequential loop:

```
for (i=0 : n-1)
    a(i+1) = b(i) + c;
    d(i) = a(i) + e;
endfor
```

Can this loop be transformed into an equivalent `forall` loop? Explain your answer.

Exercise 3.4. Consider a 3×3 mesh network and the global communication operation scatter. Give a spanning tree which can be used to implement a scatter operation as defined in Section 3.6.2. Explain how the scatter operation is implemented on this tree. Also explain why the scatter operation is the dual operation to the gather operation and how the gather operation can be implemented.

Exercise 3.5. Consider a matrix of dimension 100×100. Specify the distribution vector $((p_1, b_1), (p_2, b_2))$ to describe the following data distributions for p processors:

- column-cyclic distribution
- row-cyclic distribution
- blockwise column-cyclic distribution with blocksize 5
- blockwise row-cyclic distribution with blocksize 5

Exercise 3.6. Consider a matrix of size 7×11. Describe the data distribution which results for the distribution vector $((2,2),(3,2))$ by specifying which matrix element is stored by which of the six processors.

Exercise 3.7. Consider the matrix-vector multiplication programs in Section 3.7. Based on the notation used in this section, develop an SPMD program for computing a matrix-matrix multiplication $C = A \cdot B$ for a distributed address space. Use the notation from Section 3.7 for the communication operations. Assume the following distributions for the input matrices A and B:

(a) A is distributed in row-cyclic, B is distributed in column-cyclic order;
(b) A is distributed in column-blockwise, B in row-blockwise order;
(c) A and B are distributed in checkerboard order as it has been defined on page 135;

In which distribution is the result matrix C computed?

Exercise 3.8. The transposition of an $n \times n$ matrix A can be computed sequentially as follows

```
for (i=0; i<n; i++)
    for (j=0; j<n; j++)
        B[i][j] = A[j][i];
```

where the result is stored in B. Develop an SPMD program for performing a matrix transposition for a distributed address space using the notation from Section 3.7. Consider both a row-blockwise and a checkerboard order distribution of A.

Exercise 3.9. The statement fork(m) creates m child threads $T_1, \ldots, T_m$ of the calling thread T, see Section 3.3.6, page 125. Assume a semantics that a child thread executes the same program code as its parent thread starting at the program statement directly after the fork() statement and that a join() statement matches the last unmatched fork() statement. Consider a shared memory program fragment

```
fork(3);
fork(2);
join();
join();
```

Give the tree of threads created by this program fragment.

Exercise 3.10. Two threads T_0 and T_1 access a shared variable in a critical section. Let int flag[2] be an array with flag[i] = 1, if thread i wants to enter the critical section. Consider the following approach for coordinating the access to the critical section:

thread T_0 thread T_1

```
repeat {                                repeat {
   while (flag[1]) do no_op();             while (flag[0]) do no_op();
   flag[0] = 1;                            flag[1] = 1;
   - - - critical section - - -;          - - - critical section - - -;
   flag[0] = 0;                            flag[1] = 0;
   - - - uncritical section - - -;        - - - uncritical section - - -;
until 0;                                until 0;
```

Does this approach guarantee mutual exclusion, if both threads are executed on
the same execution core? Explain your answer.

Exercise 3.11. Consider the following implementation of a lock mechanism:

```
int me;
int flag[2];
int lock() {
   int other = 1 - me;
   flag[me] = 1;
   while (flag[other]) ; // wait
}
int unlock() {
   flag[me] = 0;
}
```

Assume that two threads with ID 0 and 1 execute this piece of program to access a data structure concurrently and that each thread has stored its ID in its
local vaiable me. Does this implementation guarantee mutual exclusion when
the functions lock() and unlock() are used to protect critical sections, see
Section 3.8.3? Can this implementation lead to a deadlock? Explain your answer.

Exercise 3.12. Consider the following example for the use of an atomic block
[141]:
bool flag_A = false; bool flag_B = false;

Thread 1   Thread 2
```
atomic {                                atomic {
    while (!flag_A) ;                        flag_A = true ;
    flag_B = true;                           while (!flag_B);
}                                       }
```
Why is this code incorrect?

Chapter 4
Performance Analysis of Parallel Programs

About this Chapter

The most important motivation for using a parallel system is the reduction of the execution time of computation-intensive application programs. In this chapter, performance measures for an analysis and comparison of different versions of a parallel program are considered in more detail. The topics covered include the quantitative performance evaluation of computer systems with performance metrics and benchmarks, performance metrics for evaluating parallel programs, execution times for global communication operations, the analysis of parallel execution times, parallel computation models, as well as loop scheduling and loop tiling methods.

The execution time of a parallel program depends on many factors, including the architecture of the execution platform, the compiler and operating system used, the parallel programming environment and the parallel programming model on which the environment is based, as well as properties of the application program such as locality of memory references or dependencies between the computations to be performed. In principle, all these factors have to be taken into consideration when developing a parallel program. However, there may be complex interactions between these factors, and it is therefore difficult to consider them all when designing a parallel program.

To facilitate the development and analysis of parallel programs, *performance measures* are often used to abstract from some of the influencing factors. Such performance measures can be based on theoretical cost models but also on measured execution times for a specific parallel system.

Section 4.1 starts with a discussion of different methods for a performance analysis of (sequential and parallel) execution platforms, which are mainly directed toward a performance evaluation of the architecture of the execution platform, without considering a specific user-written application program. Section 4.2 gives an overview of popular performance measures for parallel programs, such as speedup or efficiency. These performance measures mainly aim at a comparison of the execution time of a parallel program with the execution time of a corresponding sequential program. Section 4.4 analyses the running time of global communication operations, such as broadcast or scatter operations, in the distributed memory model with different interconnection networks. Optimal algorithms and asymptotic running times are derived. Section 4.5 shows how runtime functions (in closed form) can be used for a runtime analysis of application programs. This is demonstrated for parallel computations of a scalar product and of a matrix-vector multiplication. Section 4.6 contains a short overview of popular theoretical cost models like BSP and LogP. Section 4.7 considers approaches for the scheduling of independent loop iterations and introduces the loop tiling transformation, which is typically used to improve the locality of the memory references in loop-based computations.

4.1 Performance Evaluation of Computer Systems

The performance of a computer system is one of the most important aspects of its evaluation. Depending on the point of view, different criteria are important to evaluate performance. The user of a computer system is interested in small **response times**, where the response time of a program is defined as the time between the start and the termination of the program. On the other hand, a large computing center is mainly interested in high **throughputs**, where the throughput is the average number of work units that can be executed per time unit.

4.1.1 Evaluation of CPU Performance

In the following, we first consider a *sequential* computer system and use the response times as performance criteria. The performance of a computer system increases, if the response times for a given set of application programs decreases. The response time of a program A can be split into

- the **user CPU time** of A, capturing the time that the CPU spends for executing A;
- the **system CPU time** of A, capturing the time that the CPU spends for the execution of routines of the operating system issued by A;
- the **waiting time** of A, caused by waiting for the completion of I/O operations and by the execution of other programs because of time sharing.

So the response time of a program includes the waiting times, but these waiting times are not included in the CPU time. For Unix systems, the `time` command can be used to get information on the fraction of the CPU and waiting times of the overall response time. In the following, we ignore the waiting times, since these strongly depend on the load of the computer system. We also neglect the system CPU time, since this time mainly depends on the implementation of the operating system, and concentrate on the execution times that are directly caused by instructions for statements of the application program [170].

The user CPU time depends both on the translation of the statements of the program into equivalent sequences of instructions by the compiler and the execution time for the single instructions. The latter time is strongly influenced by the **cycle time** of the CPU (also called *clock cycle time*) which is the reciprocal of the **clock rate**. For example, a processor with a clock rate of 2 GHz $= 2 \cdot 10^9 \cdot 1/s$ has cycle time of $1/(2 \cdot 10^9)s = 0.5 \cdot 10^{-9}s = 0.5ns$ (s denotes seconds and ns denotes nanoseconds). In the following, the cycle time is denoted as t_{cycle} and the user CPU time of a program A is denoted as $T_{U\_CPU}(A)$. This time is given by the product of t_{cycle} and the total number $n_{cycle}(A)$ of CPU cycles needed for all instructions of A:

$$T_{U\_CPU}(A) = n_{cycle}(A) \cdot t_{cycle} . \tag{4.1}$$

Different instructions may have different execution times. To get a relation between the number of cycles and the number of instructions executed for program A, the *average number of CPU cycles* used for instructions of program A is considered. This number is called **CPI** (Clock cycles **P**er **I**nstruction). The CPI value depends on the program A to be executed, since the specific selection of instructions has an influence on CPI. Thus, for the same computer system, different programs may lead to different CPI values. Using CPI, the user CPU time of a program A can be expressed as

$$T_{U\_CPU}(A) = n_{instr}(A) \cdot CPI(A) \cdot t_{cycle}, \tag{4.2}$$

where $n_{instr}(A)$ denotes the total number of instructions executed for A. This number depends on many factors. The architecture of the computer system has a large influence on $n_{instr}(A)$, since the behavior of the instruction provided by the architecture determines how efficient constructs of the programming language can be translated into sequences of instructions. Another important influence comes from the *compiler*, since the compiler selects the instructions to be used in the machine program. An efficient compiler can make the selection such that a small number $n_{instr}(A)$ results.

For a given program, the CPI value strongly depends on the implementation of the instructions, which depends on the internal organization of the CPU and the memory system. The CPI value also depends on the compiler, since different instructions may have different execution times and since the compiler can select instructions such that a smaller or a larger CPI value results.

We consider a processor which provides n types of instructions, $I_1, \ldots, I_n$. The average number of CPU cycles needed for instructions of type I_i is denoted by CPI_i and $n_i(A)$ be the number of instructions of type I_i executed for a program A, $i = 1, \ldots, n$. Then the total number of CPU cycles used for the execution of A can be expressed as

$$n_{cycle}(A) = \sum_{i=1}^{n} n_i(A) \cdot CPI_i, \tag{4.3}$$

The total number of machine instructions executed for a program A is an exact measure for the number of CPU cycles and the resulting execution time of A only if all instructions require the same number of CPU cycles, i.e., have the same values for CPI_i. This is illustrated by the following example, see [170].

Example: We consider a processor with three instruction classes $\mathscr{I}_1, \mathscr{I}_2, \mathscr{I}_3$ containing instructions which require 1, 2, or 3 cycles for their execution, respectively. We assume that there are two different possibilities for the translation of a programming language construct using different instructions according to the following table:

| | Instruction classes | | | Sum of the | |
Translation	$\mathscr{I}_1$	$\mathscr{I}_2$	$\mathscr{I}_3$	instructions	n_{cycle}
1	2	1	2	5	10
2	4	1	1	6	9

Translation 2 needs less cycles than translation 1, although translation 2 uses a larger number of instructions. Thus, translation 1 leads to a CPI value of $10/5 = 2$, whereas translation 2 leads to a CPI value of $9/6 = 1.5$. ☐

4.1.2 MIPS and MFLOPS

A performance measure that is sometimes used in practice to evaluate the performance of a computer system is the **MIPS rate** (**M**illion **I**nstructions **P**er **S**econd). Using the notation from the previous subsection for the number of instructions $n_{instr}(A)$ of a program A and for the user CPU time $T_{U\_CPU}(A)$ of A, the MIPS rate of A is defined as

$$MIPS(A) = \frac{n_{instr}(A)}{T_{U\_CPU}(A) \cdot 10^6} \ . \tag{4.4}$$

Using Equation (4.2), this can be transformed into

$$MIPS(A) = \frac{r_{cycle}}{CPI(A) \cdot 10^6} \ ,$$

where $r_{cycle} = 1/t_{cycle}$ is the clock rate of the processor. Therefore, faster processors lead to larger MIPS rates than slower processors. Because the CPI value depends on the program A to be executed, the resulting MIPS rate also depends on A.

Using MIPS rates as performance measure has some drawbacks. First, the MIPS rate only considers the *number* of instructions. But more powerful instructions usually have a longer execution time, but fewer of such powerful instructions are needed for a program. This favors processors with simple instructions over processors with more complex instructions. Second, the MIPS rate of a program does not necessarily correspond to its execution time: comparing two programs A and B on a processor X, it can happen that B has a higher MIPS rate than A, but A has a smaller execution time. This can be illustrated by the following example.

Example: Again, we consider a processor X with three instruction classes $\mathscr{I}_1, \mathscr{I}_2, \mathscr{I}_3$ containing instructions which require 1, 2, or 3 cycles for their execution, respectively.

We assume that processor X has a clock rate of 2 GHz and, thus, the cycle time is 0.5 ns. Using two different compilers for the translation of a program may lead to two different machine programs A_1 and A_2 for which we assume the following numbers of instructions from the different classes:

Program	$\mathscr{I}_1$	$\mathscr{I}_2$	$\mathscr{I}_3$
A_1	$5 \cdot 10^9$	$1 \cdot 10^9$	$1 \cdot 10^9$
A_2	$10 \cdot 10^9$	$1 \cdot 10^9$	$1 \cdot 10^9$

For the CPU time of $A_j, j = 1, 2$, we get from Equations (4.2) and (4.3):

$$T_{U\_CPU}(A_j) = \sum_{i=1}^{3} n_i(A_j) \cdot CPI_i(A_j) \cdot t_{cycle},$$

where $n_i(A_j)$ is the number of instruction executions from the table and $CPI_i(A_j)$ is the number of cycles needed for instructions of class $\mathscr{I}_i$ for $i = 1, 2, 3$. Thus, machine program A_1 leads to an execution time of 5 sec, whereas A_2 leads to an execution time of 7.5 sec. The MIPS rates of A_1 and A_2 can be computed with Equation (4.4). For A_1, in total $7 \cdot 10^9$ instructions are executed, leading to a MIPS rate of 1400 (1/s). For A_2, a MIPS rate of 1600 (1/s) results. This shows that A_2 has a higher MIPS rate than A_1, but A_1 has a smaller execution time.   □

For programs from scientific computating, the MFLOPS rate (**M**illion **F**loating-point **O**perations **P**er **S**econd) is sometimes used. The MFLOPS rate of a program A is defined by

$$MFLOPS(A) = \frac{n_{flp\_op}(A)}{T_{U\_CPU}(A) \cdot 10^6} \, [1/s], \tag{4.5}$$

where $n_{flp\_op}(A)$ is the number of floating-point operations executed by A. The MFLOPS rate is not based on the number of instructions executed, as is the case for the MIPS rate, but on the number of arithmetic operations on floating point values performed by the execution of their instructions. Instructions that do not perform floating-point operations have no effect on the MFLOPS rate. Since the effective number of operations performed is used, the MFLOPS rate provides a fair comparison of different program versions performing the same operations, and larger MFLOPS rates correspond to faster execution times.

A drawback of using the MFLOPS rate as performance measure is that there is no differentiation between different types of floating-point operations performed.

In particular, operations like division and square root that typically take quite long to perform are counted in the same way as operations like addition and multiplication that can be performed much faster. Thus, programs with simpler floating-point operations are favored over programs with more complex operations. However, the MFLOPS rate is well suited to compare program versions that perform the same floating-point operations.

4.1.3 Performance of Processors with a Memory Hierarchy

According to Equation (4.1), the user CPU time of a program A can be represented as the product of the number of CPU cycles $n_{cycles}(A)$ for A and the cycle time t_{cycle} of the processor. By taking the access time to the memory system into consideration, this can be refined to

$$T_{U\_CPU}(A) = \left(n_{cycles}(A) + n_{mm\_cycles}(A)\right) \cdot t_{cycle}, \qquad (4.6)$$

where $n_{mm\_cycles}(A)$ is the number of additional machine cycles caused by memory accesses of A. In particular, this includes those memory accesses that lead to the loading of a new cache line because of a cache miss, see Section 2.9. We first consider a one-level cache. If we assume that cache hits do not cause additional machine cycles, they are captured by $n_{cycles}(A)$. Cache misses can be caused by read misses or write misses:

$$n_{mm\_cycles}(A) = n_{read\_cycles}(A) + n_{write\_cycles}(A).$$

The number of cycles needed for read accesses can be expressed as

$$n_{read\_cycles}(A) = n_{read\_op}(A) \cdot r_{read\_miss}(A) \cdot n_{miss\_cycle},$$

where $n_{read\_op}(A)$ is the total number of read operations of A, $r_{read\_miss}(A)$ is the read miss rate for A, n_{miss_cycle} is the number of machine cycles needed to load a cache line into the cache in case of a read miss; this number is also called *read miss penalty*. A similar expression can be given for the number of cycles n_{write_cycles} needed for write accesses. The effect of read and write misses can be combined for simplicity which results in the following expression for the user CPU time

$$T_{U\_CPU}(A) = n_{instr}(A) \cdot \left(CPI(A) + n_{rw\_op}(A) \cdot r_{miss}(A) \cdot n_{miss\_cycle}\right) \cdot t_{cycle} \qquad (4.7)$$

where $n_{rw\_op}(A)$ is the total number of read or write operations of A, $r_{miss}(A)$ is the (read and write) miss rate of A, and n_{miss_cycles} is the number of additional cycles needed for loading a new cache line. Equation (4.7) is derived from Equations (4.2) and (4.6).

Example: We consider a processor for which each instruction takes two cycles to execute, i.e., it is CPI = 2 ,see [170]. The processor uses a cache for which the loading of a cache block takes 100 cycles. We consider a program A for which the (read and write) miss rate is 2% and in which 33% of the instructions executed are load and store operations, i.e., it is $n_{rw\_op}(A) = n_{instr}(A) \cdot 0.33$. According to Equation (4.7) it is

$$T_{U\_CPU}(A) = n_{instr}(A) \cdot (2 + 0.33 \cdot 0.02 \cdot 100) \cdot t_{cycle}$$
$$= n_{instr}(A) \cdot 2.66 \cdot t_{cycle}.$$

This can be interpreted such that the ideal CPI value of 2 is increased to the real CPI value of 2.66 if the data cache misses are taken into consideration. This does not take instruction cache misses into consideration. The equation for $T_{U\_CPU}(A)$ can also be used to compute the benefit of using a data cache: without a data cache, each memory access would take 100 cycles, leading to a real CPI value of $2 + 100 \cdot 0.33 = 35$.

Doubling the clock rate of the processor without changing the memory system leads to an increase of the cache loading time to 200 cycles, resulting in a real CPI value of $2 + 0.33 \cdot 0.02 \cdot 200 = 3.32$. Using t_{cycle} for the original cycle time, the CPU time on the new processor with half of the cycle time yields

$$\tilde{T}_{U\_CPU}(A) = n_{instr}(A) \cdot 3.32 \cdot t_{cycle}/2.$$

Thus, the new processor needs 1.66 instead of 2.66 original cycle time units. Therefore, doubling the clock rate of the processor leads to a decrease of the execution time of the program to 1.66/2.66, which is about 62.4% of the original execution time, but not 50% as one might expect. This shows that the memory system has an important influence on program execution time. □

The influence of memory access times using a memory hierarchy can be captured by defining an *average memory access time* [170]. The average read access time $t_{read\_access}(A)$ of a program A can be defined as

$$t_{read\_access}(A) = t_{read\_hit} + r_{read\_miss}(A) \cdot t_{read\_miss}, \tag{4.8}$$

where t_{read_hit} is the time for a read access to the cache. The additional time needed for memory access in the presence of cache misses can be captured by multiplying the cache read miss rate $r_{read\_miss}(A)$ with the read miss penalty time t_{read_miss} needed for loading a cache line. In Equation (4.7), t_{read_miss} has been calculated from n_{miss_cycle} and t_{cycle}. The time t_{read_hit} for a read hit in the cache was assumed to be included in the time for the execution of an instruction.

It is beneficial if the access time to the cache is adapted to the cycle time of the processor, since this avoids delays for memory accesses in case of cache hits. To do this, the first-level (L1) cache must be kept small and simple and an additional second-level (L2) cache is used, which is large enough such that most memory accesses go to the L2 cache and not to main memory. For performance analysis, the modeling of the average read access time is based on the performance values of the L1 cache. In particular, for Equation (4.8), we have

$$t_{read\_access}(A) = t_{read\_hit}^{(L1)} + r_{read\_miss}^{(L1)}(A) \cdot t_{read\_miss}^{(L1)},$$

where $r^{(L1)}_{read\_miss}(A)$ is the cache read miss rate of A for the L1 cache, calculated by dividing the total number of read accesses causing an L1 cache miss by the total number of read accesses. To model the reload time $t^{(L1)}_{read\_miss}$ of the L1 cache, the access time and miss rate of the L2 cache can be used. More precisely, we get

$$t^{L1}_{read\_miss} = t^{(L2)}_{read\_hit} + r^{(L2)}_{read\_miss}(A) \cdot t^{(L2)}_{read\_miss}$$

where $r^{(L2)}_{read\_miss}(A)$ is the read miss rate of A for the L2 cache, calculated by dividing the total number of read misses of the L2 cache by the total number of read misses of the L1 cache. Thus, the global read miss rate of program A can be calculated by $r^{(L1)}_{read\_miss}(A) \cdot r^{(L2)}_{read\_miss}(A)$.

4.1.4 Benchmark Programs

The performance of a computer system may vary significantly, depending on the program considered. For two programs A and B, the situation can occur that program A has a smaller execution time on a computer system X than on a computer system Y, whereas program B has a smaller execution time on Y than on X.

For the user, it is important to base the selection of a computer system on a set of programs that are often executed by the user. These programs may be different for different users. Ideally, the programs would be weighted with their execution time and their execution frequency. But often, the programs to be executed on a computer system are not known in advance. Therefore, **benchmark programs** have been developed which allow a standardized performance evaluation of computer systems based on specific characteristics that can be measured on a given computer system. Different benchmark programs have been proposed and used, including the following approaches, listed in increasing order of their usefulness:

- **Synthetic benchmarks**, which are typically small artificial programs containing a mixture of statements which are selected such that they are representative for a large class of real applications. Synthetic benchmarks usually do not execute meaningful operations on a large set of data. This bears the risk that some program parts may be removed by an optimizing compiler. Examples for synthetic benchmarks are *Whetstone* [46, 49], which has originally been formulated in Fortran to measure floating-point performance, and *Dhrystone* [222] to measure integer performance in C. The performance measured by Whetstone or Dhrystone is measured in specific units as `KWhetstone/s` or `KDhrystone/s`. The largest drawback of synthetic benchmarks is that they are not able to match the profile and behavior of large application programs with their complex interactions between computations of the processor and accesses to the memory system. Such interactions have a large influence on the resulting performance, yet they can not be captured by synthetic benchmarks. Another drawback is that a compiler or system can be tuned towards simple benchmark programs to let the computer system appear faster than it is for real applications.

- **Kernel benchmarks** with small but relevant parts of real applications which typically capture a large portion of the execution time of real applications. Compared to real programs, kernels have the advantage that they are much shorter and easier to analyze. Examples for kernel collections are the *Livermore Loops* (Livermore Fortran Kernels, LFK) [153, 62] consisting of 24 loops extracted from scientific simulations, and *Linpack* [52] capturing a piece of a Fortran library with linear algebra computations. Both kernels compute the performance in MFLOPS. The drawback of kernels is that the performance values they produce are often too large for applications that come from other areas than scientific computing. A variant of kernels are toy programs, which are small, but complete programs performing useful computations. Examples are quicksort for sorting or the sieve of Erathostenes for prime test.
- **Real application benchmarks** comprise several entire programs which reflect a workload of a standard user. Such collections are often called *benchmark suites*. They have the advantage that all aspects of the selected programs are captured. The performance results produced are meaningful for users for which the benchmark suite is representative for typical workloads. Examples for benchmark suites are the SPEC benchmarks, described in the following for desktop computers, the TPC benchmarks (Transaction Processing Performance Council) for applications with a focus on on-line transaction processing for database systems (see www.tpc.org for more information), and the EEMBC benchmarks (EDV Embedded Microprocessor Benchmark Consortium) for embedded systems (see www.eembc.org for more information). EEMBC provides benchmarks for different application areas, including AutoBench for automotive applications such as signal processing, DENBench for multimedia applications for digital entertainment products such as TV or digital cameras, or Networking 2.0 for network algorithms for routers and switches.

The most popular benchmark suites for desktop and server systems are the **SPEC benchmarks** (System Performance Evaluation Cooperation), see www.spec.org for detailed information. The SPEC cooperation has been founded in 1988 as a non-profit corporation with the goal to define a standardized performance evaluation method for computer systems and to facilitate a performance comparison. The SPEC corporation maintains benchmark suites for different application areas including cloud computing, CPU performance, High Performance Computing, and Java client/server computations. There are also benchmark suites for file servers (SPECSFC), web servers (SPECWeb), or parallel systems such as SPECOpenMP.

For the CPU performance of desktop and server systems, SPEC has published six generations of benchmark suites: SPEC CPU89, SPEC CPU92, SPEC CPU95, SPEC CPU2000, SPEC CPU2006, and SPEC CPU2017. The suites are changed after some years to adapt the mix of benchmarks to the current needs of the users of the systems, see [106, 136, 104] for a detailed description of the development of these benchmark suites.

Integer performance

SPECrate Integer	SPECspeed Integer	Language	KLOC	Application Area
perlbench_r	perlbench_s	C	362	Perl interpreter
gcc_r	gcc_s	C	1,304	GNU C compiler
mcf_r	mcf_s	C	3	Route planning
omnetpp_r	omnetpp_s	C++	134	Discrete Event simulation - network
xalancbmk_r	xalancbmk_s	C++	520	XML to HTML conversion via XSLT
x264_r	x264_s	C	96	Video compression
deepsjeng_r	deepsjeng_s	C++	10	AI: alpha-beta tree search (Chess)
leela_r	leela_s	C++	21	AI: Monte Carlo tree search (Go)
exchange2_r	exchange2_s	Fortran	1	AI: solution generator (Sudoku)
xz_r	xz_s	C	33	General data compression

Floating Point performance

SPECrate FlPoint	SPECspeed Floating Point	Language	KLOC	Application Area
bwaves_r	bwaves_s	Fortran	1	Explosion modeling
cactuBSSN_r	cactuBSSN_s	C++, C, Fortran	257	Physics: relativity
namd_r		C++	8	Molecular dynamics
parest_r		C++	427	Biomedical imaging: tomography
povray_r		C++, C	170	Ray tracing
lbm_r	lbm_s	C	1	Fluid dynamics
wrf_r	wrf_s	Fortran, C	991	Weather forecasting
blender_r		C++, C	1,577	3D rendering and animation
cam4_r	cam4_s	Fortran, C	407	Atmosphere modeling
	pop2_s	Fortran, C	338	Wide-scale ocean modeling (climate)
imagick_r	638.imagick_s	C	259	Image manipulation
nab_r	nab_s	C	24	Molecular dynamics
fotonik3d_r	fotonik3d_s	Fortran	14	Computational Electromagnetics
roms_r	roms_s	Fortran	210	Regional ocean modeling

Table 4.1 Overview of the SPEC CPU 2017 benchmark suite according to www.spec.org. KLOC stands for Kilo line count and specifies the number of code lines in thousands.

The focus of the newest benchmark suite SPEC CPU2017 lies on compute intensive performance, capturing the performance of the processor and the memory system as well as the quality of the compiler used for the translation of the source code. The SPEC CPU2017 comprises 43 benchmarks, organized into four suites that are used to compute four different metrics: SPECspeed int, SPECspeed fp, SPECrate int, and SPECrate fp. The four suites focus on different types of compute intensive performance: the two SPEC int suites measure integer performance, the two SPEC fp suites measure floating-point performance. The SPECspeed suites focus on the execution time of the benchmarks in the suite, i.e., the result is a time-based metric. The evaluation is based on one execution of the benchmarks on the computer system to be evaluated. The SPECrate suites focus on the throughput obtained for the benchmarks. Therefore, the evaluation is based on multiple runs of each benchmark.

Table 4.1 gives an overview of the benchmarks in the different suites along with a short description of the application area. For the integer performance, the same benchmarks are used for the SPECrate and SPECspeed integer metrics. However, different workloads, compiler flags and execution rules may apply to compute the SPECrate and SPECspeed metrics. Therefore, the benchmarks are denoted with different suffix r for rate and s for speed. The integer benchmarks include, for example, three benchmarks from AI (Artificial Intelligence) using different search strategies, the GNU C compiler (gcc), and a video compression program (x264). For the floating point performance, the benchmark sets that are used for the SPECrate and SPECspeed floating point metrics are slightly different. The floating point benchmarks include, for example, an explosion modeling (bwaves), a molecular dynamics simulation (namd), an application for biomedical imaging for optical tomography with finite elements (parest), a fluid dynamics simulation (lbm), and a ray tracing application (povray).

The performance metrics for the SPEC CPU2017 benchmarks are given as the relative performance with respect to a fixed reference computer, which is specified by the specific SPEC suite. Larger values of the performance metrics correspond to a higher performance of the computer system tested. For SPEC CPU2017 suite, the reference computer is a Sun Fire V490 with 2100 MHz UltraSPARC-IV+ chips. This reference computer gets a SPECspeed int and SPECspeed fp score of 1.0. The SPECspeed int and SPECspeed fp values are determined separately by using the SPEC integer and floating-point benchmarks, respectively. To perform the benchmark evaluation and to compute the performance metrics SPECspeed int and SPECspeed fp, the following three steps are executed:

1. Each of the programs is executed three times on the computer system U to be tested. For each of the programs A_i an average execution time $T_U(A_i)$ in seconds is determined by taking the median of the three execution times measured, i.e., the middle value.
2. For each program, the execution time $T_U(A_i)$ determined in step (1) is normalized with respect to the reference computer R by dividing the execution time $T_R(A_i)$ on R by the execution time $T_U(A_i)$ on U. This yields an execution factor $F_U(A_i) = T_R(A_i)/T_U(A_i)$ for each of the programs A_i which expresses how much faster machine U is compared to R for program A_i.
3. The SPECspeed int value is computed as the geometric mean of the execution factors of the 10 SPEC integer programs, i.e., a global factor G_U^{int} is computed as

$$G_U^{int} = \sqrt[10]{\prod_{i=1}^{10} F_U(A_i)}$$

The SPECspeed int value G_U^{int} expresses how much faster U is in comparison to R. The SPECspeed fp value is defined similarly, using the geometric mean of the 10 floating-point programs.

The SPECrate metrics are computed in a similar way using the SPECrate benchmarks in Table 4.1. Since each of the benchmarks is executed multiple times, the computation of the execution factors F_U takes the number of copies into consideration. Again, the geometric mean of the execution factors is taken to compute the final SPECrate metric. The SPEC CPU2017 suite also supports corresponding energy metrics. The calculations of the energy metrics are performed in the same way but use the energy consumption instead of the execution time.

An alternative to the geometric means would be the *arithmetic means* to compute the global execution factors, by calculation, for example, $A_U^{int} = 1/10 \sum_{i=1}^{10} F_U(A_i)$. But using the geometric means has some advantages. The most important advantage is that the comparison between two machines is independent from the choice of the reference computer. This is not necessarily the case when the arithmetic means is used instead; this is illustrated by the following example calculation.

Example: Two programs A_1 and A_2 and two machines X and Y are considered, see also [103]. Assuming the following execution times

$$T_X(A_1) = 1 \text{ s} \quad T_Y(A_1) = 10 \text{ s}$$
$$T_X(A_2) = 500 \text{ s} \quad T_Y(A_2) = 50 \text{ s}$$

results in the following execution factors if Y is used as reference computer

$$F_X(A_1) = 10, F_X(A_2) = 0.1, F_Y(A_1) = F_Y(A_2) = 1 .$$

This yields the following performance score for the arithmetic means A and the geometric means G:

$$G_X = \sqrt{10 \cdot 0.1} = 1, A_X = \frac{1}{2}(10 + 0.1) = 5.05, G_Y = A_Y = 1 .$$

Using X as reference computer yields the following execution factors:

$$F_X(P_1) = F_X(P_2) = 1, F_Y(P_1) = 0.1, F_Y(P_2) = 10$$

resulting in the following performance scores

$$G_X = 1, A_X = 1, G_Y = \sqrt{10 \cdot 0.1} = 1, A_Y = \frac{1}{2}(0.1 + 10) = 5.05$$

Thus, considering the arithmetic means, using Y as reference computer yields the statement that $X = 5.05$ times faster than Y. Using X as reference computer yields the opposite result. Such contradictory statements are avoided by using the geometric means, which states that X and Y have the same performance, independently from the reference computer.

A drawback of the geometric means is that it does not provide information about the actual execution time of the programs. This can be seen at the example just given. Executing A_1 and A_2 only once requires 501 sec on X and 60 sec on Y, i.e. Y is more than eight times faster than X. □

A detailed discussion of benchmark programs and program optimization issues can be found in [53, 113, 83, 136], which also contain references to further literature.

4.2 Performance Metrics for Parallel Programs

As considered in the last subsection, benchmark programs provide a set of application programs with the goal to evaluate the performance of a computer system for a wide range of applications. However, it is also important to evaluate the performance of a specific application, which is usually done with performance metrics. Well-established performance metrics for evaluating parallel applications are, for example, parallel runtime, cost, speedup and efficiency, which are defined in the following.

4.2.1 Parallel Runtime and Cost

An important criterion for the quality of the parallelization of a parallel program is its runtime (also called execution time) on a specific execution platform.

The **parallel runtime** of a program is the time between the start of the program and the end of the execution on all participating processors. The end of the execution is the point in time when the last processor finishes its execution for this program. The parallel runtime $T_p(n)$ is usually expressed for a specific number p of participating processors as a function of the problem size n. The problem size is given by the size of the input data, which can for example be the size of a matrix or the number of equations of an equation system to be solved. Depending on the architecture of the execution platform, the parallel runtime comprises the following times:

- the runtime for the execution of *local computations* of each participating processor; these are the computations that each processor performs using data in its local memory;
- the runtime for the *exchange of data* between processors, e.g. by performing explicit communication operations in the case of a distributed address space;
- the runtime for the *synchronization* of the participating processors when accessing shared data structures in the case of a shared address space;
- *waiting times* occurring because of an unequal load distribution of the processors; waiting times can also occur when a processor has to wait in order to ensure mutual exclusion before it can access a shared data structure.

The time spent for data exchange and synchronization as well as waiting times can be considered as **overhead** of the parallel computation, since they do not contribute directly to the computations to be performed.

The **cost of a parallel program** accumulates the runtime of all participating processors that are executing the program. More precisely, the cost $C_p(n)$ of a parallel program with input size n executed on p processors or processor cores is defined by

$$C_p(n) = p \cdot T_p(n).$$

Thus, $C_p(n)$ is a measure for the total amount of work performed by all processors that are participating in the program execution. Therefore, the cost of a parallel program is also called *work* or processor-runtime product.

A parallel program is called **cost-optimal** if $C_p(n) = T^\star(n)$, i.e., if it executes the same total number of operations as the fastest sequential program which has runtime $T^\star(n)$. Using asymptotic execution times, this means that a parallel program is cost-optimal if $T^\star(n)/C_p(n) \in \Theta(1)$ (see Section 4.4.1 for the Θ definition).

4.2.2 Speedup and Efficiency

For the analysis of parallel programs, a comparison with the execution time of a sequential implementation is especially important to see the benefit of parallelism. Such a comparison is often based on the relative saving in execution time as expressed by the notion of **speedup**.

4.2.2.1 Speedup

The speedup $S_p(n)$ of a parallel program with parallel execution time $T_p(n)$ is defined as

$$S_p(n) = \frac{T^*(n)}{T_p(n)}, \tag{4.9}$$

where p is the number of processors used to solve a problem of size n. Again, $T^\star(n)$ is the execution time of the best sequential implementation solving the same problem. The speedup of a parallel implementation expresses the relative saving of execution time compared to the best sequential implementation when using p processors for the parallel execution. The concept of speedup is used both for a theoretical analysis of algorithms based on the asymptotic notation, see Section 4.4.1, as well as for the practical evaluation of parallel programs based on measurements.

Theoretically, $S_p(n) \leq p$ always holds. This can be seen by the following argument: If $S_p(n) > p$ would hold, then a new sequential algorithm could be constructed which is faster than the sequential algorithm that has been used for the computation of the speedup. The new sequential algorithm is derived from the parallel algorithm by a round robin simulation of the computation steps of the participating p processors, i.e. the new sequential algorithm uses its first p steps to simulate the first step of all p processors in a fixed order. Similarly, the next p steps are used to simulate the second step of all p processors, and so on. Thus, the new sequential algorithm performs p times more steps than the parallel algorithm. Because of $S_p(n) > p$, the new sequential algorithm would have execution time

$$p \cdot T_p(n) = p \cdot \frac{T^*(n)}{S_p(n)} < T^*(n).$$

This is a contradiction to the assumption that the best sequential algorithm with execution time $T^\star(n)$ has been used for the speedup computation, since the new algorithm is faster.

As described, the speedup definition given above requires a comparison with the fastest sequential algorithm. However, this algorithm may be difficult to determine or construct. Possible reasons are:

- The best sequential algorithm may not be known. There might be the situation that a lower bound for the execution time of a solution method for a given problem can be determined, but until now, no algorithm with this asymptotic execution time has yet been constructed.
- There exists an algorithm with the optimum asymptotic execution time, but depending on the size and the characteristics of a specific input set, other algorithms lead to lower execution times in practice. For example, the use of balanced trees for the dynamic management of data sets should be preferred only if the data set is large enough and if enough access operations are performed.
- The sequential algorithm which leads to the smallest execution times requires a large effort to be implemented.

Because of these reasons, the speedup is often computed by using a sequential version of the parallel implementation instead of the best sequential algorithm.

In practice, superlinear speedup can sometimes be observed, i.e., $S_p(n) > p$ can occur. The reason for this behavior often lies in cache effects: A typical parallel program assigns only a fraction of the entire data set to each processor. The fraction is selected such that the processor performs its computations on its assigned local data set. In this situation, it can occur that the entire data set does not fit into the cache of a single processor executing the program sequentially, thus leading to cache misses during the computation. But when several processors execute the program with the same amount of data in parallel, it may well be that the fraction of the data set assigned to each processor fits into its local cache, thus avoiding cache misses. However, superlinear speedup does not occur often. A more typical situation is that a parallel implementation does not even reach **linear speedup** $(S_p(n) = p)$, since the parallel implementation requires additional overhead for the management of parallelism. This overhead might be caused by the necessity to exchange data between processors, by synchronization between processors, or by waiting times caused by an unequal load balancing between the processors as already mentioned above. Also, a parallel program might have to perform more computations than the sequential program version because replicated computations are performed to avoid data exchanges. This is called **computational redundancy**. The parallel program might also contain computations that must be executed sequentially by only one of the processors because of data dependencies. During such sequential computations, the other processors must wait. Input and output operations are a typical example for sequential program parts.

4.2.2.2 Efficiency

An alternative measure for the performance of a parallel program is the **efficiency**. The efficiency captures the fraction of time in which a processor is usefully employed by computations that also have to be performed by a sequential program. The definition of efficiency is based on the cost of a parallel program and can be expressed as

$$E_p(n) = \frac{T^*(n)}{C_p(n)} = \frac{S_p(n)}{p} = \frac{T^*(n)}{p \cdot T_p(n)} \qquad (4.10)$$

where $T^*(n)$ is the sequential execution time of the best sequential algorithm and $T_p(n)$ is the parallel execution time on p processors. If no superlinear speedup occurs, then $E_p(n) \leq 1$. An ideal speedup $S_p(n) = p$ corresponds to an efficiency of $E_p(n) = 1$.

4.2.2.3 Amdahl's law

The parallel execution time of programs cannot be arbitrarily reduced by employing parallel resources. As shown, the number of processors is an upper bound for the speedup that can be obtained. Other restrictions may be caused by data dependencies within the algorithm to be implemented, which may limit the degree of parallelism. An important restriction comes from program parts that have to be executed sequentially. The effect on the obtainable speedup can be captured quantitatively by **Amdahl's law**[15], which is presented in the following.

Amdahl's law assumes that the program contains a program part which cannot be parallelized, i.e., has to be executed sequentially, and a program part that is perfectly parallelizable. The sequential part is given quantitatively as a (constant) fraction f, $0 \leq f \leq 1$, of the total runtime of the sequential algorithm. When a fraction f of a parallel program must be executed sequentially, then the parallel execution time of the program is composed of a fraction of the sequential execution time $f \cdot T^*(n)$ and the execution time of the fraction $(1 - f) \cdot T^*(n)$, which is fully parallelized for p processors, i.e. $(1 - f)/p \cdot T^*(n)$. The attainable speedup is therefore:

$$S_p(n) = \frac{T^*(n)}{f \cdot T^*(n) + \frac{1-f}{p} T^*(n)} = \frac{1}{f + \frac{1-f}{p}} \leq \frac{1}{f} \qquad (4.11)$$

since $\frac{1-f}{p} > 0$. The estimation (4.11) assumes that the best sequential algorithm is used and that the parallel part of the program can be perfectly parallelized. The effect of the sequential computations on the attainable speedup can be demonstrated by considering the following example: if 20% of a program have to be executed sequentially, then the attainable speedup is limited to $1/f = 5$ according to Amdahl's law, no matter how many processors are used. Program parts that must be executed sequentially must be taken into account in particular when a large number of processors is employed.

4.2.3 Weak and Strong Scalability

The term **scalability** is a general term which applies in many areas to systems which may have growing resources. Examples are business, administration or compute systems. Broadly speaking scalability describes the ability of a system to increase its resources while keeping its functionality and to make effective use of the additional resources. However, the usages of the term scalability are quite diverse.

In parallel computing there are two aspects of scalability, which are **hardware scalability** and **software scalability**. **Hardware scalability** captures the ability of a computer system to be expandable, i.e., to work correctly and to deliver more performance if more hardware resources are added to the system. An important aspect of hardware scalability for parallel systems is the property of the interconnection network to be expandable so that more compute nodes can be added to the system, see Sect. 2.7. **Software scalability** is related to the terms speedup and efficiency and describes the ability to increase the number of compute resources (processors or cores) for executing the software and to make full use of them.

4.2.3.1 Software scalability

Scalability definitions and investigations in theoretical as well as in practical ways have accompanied parallel programming from the very beginning and are still important in today's program development, especially on HPC platforms for real world applications. The term scalability is often associated with a growing speedup when more resources are added for the computation. The specific meaning and applicability may however differ in the specific context. An overview of the terms and definitions used and developed over time and their applicability to modern systems is for example collected in [158]. In the following, software scalability and scalability analysis are considered from a practical point of view.

In parallel scientific computing, scalability is an important property of scientific parallel programs and is investigated in a **scalability analysis**. The scalability analysis considers the performance behaviour of the parallel program with respect to a growing number p of compute units (processors or cores). Is is distinguished between **strong and weak scaling**, where **strong scaling** considers a constant problem size which means that the overall computation load remains unchanged with growing p, while **weak scaling** allows an increasing work load with growing p.

Strong scaling is related to **Amdahls's law** described earlier in Sect. 4.2.2.3, which also assumes a constant workload (the constant part plus the parallelizable part), but a growing number of processors p: Only the parallelizable program part can be distributed among the additional resources added. The non-parallelizable part remains constant and requires a larger portion of the overall execution time, since the execution time of the parallelizable program part is reduced due to the usage of a larger number of resources. The limitations of strong scalability for parallel software is obvious, since the constant part of the computation gains prevalence with an increasing number of processors or cores, which may cause the speedup to shrink. Thus, a **saturation** of the speedup curve can be observed when the number p of processors is increased and a fixed problem size n or a fixed workload is considered. In practice, usually an enormous programming effort has to be invested to avoid the speedup curve to reach the saturation too early, i.e. to keep strong scalability as long as possible in the sense that a constant efficiency is reached even if more compute resources are added while the workload remains constant.

Another observation is that increasing the problem size but keeping the number of processors fixed usually leads to an increase of the attained speedup. This observation leads to the analysis of weak scaling which considers both, an increasing problem size and an increasing number of processors. **Weak scaling** is the property of a parallel implementation that the efficiency can be kept constant if both the number p of processors and the problem size n are increased together. This property is important for parallel programs and expresses that larger problems can be solved in the same time as smaller problems if a sufficiently large number of processors is employed. However, also for weak scalabilty there is usually a point where the speedup cannot be increased further even if the problem size increases. This is often due to the overhead in parallel computations. For message passing programs on distributed memory platforms, the overhead of communication can be essential for the resulting performance, which is examined in the next subsection.

4.2.3.2 Performance behavior in practice

The performance behavior of a parallel program in terms of parallel runtime, speedup or efficiency is a very complex metric, which is influenced by many factors of the parallel platform and the parallel program itself. The influence of all these factors on the performance is difficult to assess. Thus, the investigation of the influence of the number p of compute units (processors or cores) on the performance is one way to reduce the complexity of the analysis and the varying performance and speedup is evaluated depending only on p, which is done in a **scalability analysis**. This means that the performance, speedup or efficiency is considered as a function depending on p. Several observations can be summarized for the performance, speedup and efficiency behavior in practice.

For parallel applications that are often used and that are executed on a variety of different parallel systems, it is important to investigate the exact scalability behavior to evaluate their suitability for a specific parallel system. The factors influencing the performance of a parallel program and its speedup include the distribution of the computational work between the hardware resources of the parallel system, the memory access behavior of the program and, for distributed address spaces, the data distribution and the resulting communication operations, see also Sect. 3.2. These factors also influence the **scalability properties** of a parallel program. For example, when using more hardware resources, the communication overhead might increase due to the use of expensive collective communication operations, and correspondingly the overall performance might be decreased, see also Sect. 4.4. Depending on the outcome of the analysis, it may be advantageous to modify the communication operations of the implementation.

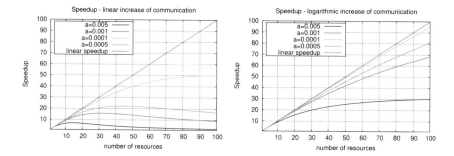

Fig. 4.1 Illustration of the communication overhead on the resulting speedup for a fixed problem size $n = 10000$ and different amounts $a \cdot n$ of data to be communicated: linear communication overhead (left) vs. logarithmic communication overhead (right).

As already mentioned, a saturation of the speedup can be observed when more and more resources are added to the execution while keeping the input size constant. The saturation starts earlier if the communication overhead $T_{comm}(n, p)$ has a larger increase depending on the number p of resources. This effect is illustrated in the following example and in Fig. 4.1, where the speedup is shown for an increasing number of resources.

Example: In this example, $T_{seq}(n)$, $T_{par}(n, p)$, and $T_{comm}(n, p)$, denote the sequential, the parallel and the communication time, respectively. For the parallel execution time $T_{par}(n, p)$, it is assumed that $T_{par}(n, p) = T_{seq}(n)/p + T_{comm}(n, p)$, i.e., the computations of the sequential program can be evenly distributed among the p resources. In Fig. 4.1, it is assumed that the computational effort is linear in n. In the left diagram, it is assumed that the communication depends linearly on the problem size n as well as the number p of resources and that the fraction a of the data is sent by communication operations, i.e., $T_{comm}(n, p) = a \cdot n \cdot p$. As speedup, the following function is depicted:

$$S_p(n) = \frac{T_{seq}(n)}{T_{seq}(n)/p + a \cdot n \cdot p}$$

for fixed $n = 10000$. In the right diagram, it is assumed that the communication depends logarithmically on the number p of resources, i.e., $T_{comm}(n, p) = a \cdot n \cdot \log p$. As speedup, the following function is depicted:

$$S_p(n) = \frac{T_{seq}(n)}{T_{seq}(n)/p + a \cdot n \cdot \log p}$$

for fixed $n = 10000$. The left diagram shows that the saturation depends on the fraction a and that the saturation starts quite early, if a is large, i.e. the communication overhead is larger. In the right diagram, it can be seen, that the saturation starts much later due to the logarithmic dependence of the communication time on the number of processors. The two diagrams demonstrate that the communication times may have a large influence on the speedup saturation and that the speedup behavior can be significantly influenced by replacing expensive communication operations (linear dependence on p) by less expensive communication operations (logarithmic dependence on p).   □

   This example considers a fixed problem size for each of the curves, i.e., strong scaling has been investigated. In contrast, weak scaling captures the performance behavior of a parallel program for the case that the problem size is increased proportionally to the number of resources used. Thus, the amount of work for each of the resources remains constant, if more resources are used for program execution. Therefore, the computation time for each resource does not change and in the ideal case, the parallel execution time remains the same, although the problem size has been increased. Since the sequential execution time is also increased due to the increased problem size, the speedup could increase linearly with the number of resources used. However, depending on the organization of the communication, increasing the number of resources could increase the communication overhead. In this case, the speedup would not remain linear. Instead, the speedup would saturate, similar to the situation described for strong scaling. However, the saturation is typically not as significant as for strong scaling, i.e. for fixed problem size or fixed workload.

4.2.3.3 Gustafson's law

The speedup for an increasing problem size n or an increasing work load cannot be captured by Amdahl's law, since the sequential fraction is assumed to remain constant, even if the problem size is increased. An extension to Amdahl's law to express the speedup for increasing work load is Gustafson's law[95, 158], which is based on the assumption that the parallel execution time consists of a sequential fraction f_{seq} and a parallel fraction f_{par} with $f_{seq} + f_{par} = 1$. Assuming a computational work $W(n)$ for problem size n and a time t_{op} for a unit of computational work according to [158], the parallel execution time is

$$T_{par}(n) = (f_{seq} + f_{par}) \cdot W(n) \cdot t_{op}$$

and the sequential execution time is

$$T_{seq}(n) = (f_{seq} + p \cdot f_{par}) \cdot W(n) \cdot t_{op}$$

This results in the following speedup:

$$S_p(n) = \frac{T_{seq}(n)}{T_{par}(n)} = \frac{f_{seq} + p \cdot f_{par}}{f_{seq} + f_{par}}.$$

Because of $f_{seq} + f_{par} = 1$ and $f_{par} = 1 - f_{seq}$, it results

$$S_p(n) = f_{seq} + p \cdot f_{par} = f_{seq} + p \cdot (1 - f_{seq}) = p + (1 - p) \cdot f_{seq}.$$

The estimations of the speedup according to Amdahl's law and to Gustafson's law can be quite different in a specific situation, which is shown by the following example.

Example: For both scenarios, the sequential fraction is set to 0.2, i.e., $f = 0.2$ for Amdahl's law and $f_{seq} = 0.2$ for Gustafson's law. For Amdahl's law, a speedup $S_p(n) \leq 1/f = 5$ results for any number p of processors, while for Gustafson's law the speedup $S_p(n) = p + (1 - p) \cdot f_{seq} = 80.2$ results for $p = 100$, and the speedup $S_p(n) = 800,2$ results for $p = 1000$. □

There exist more complex scalability analysis methods which try to capture how the problem size n must be increased relative to the number p of processors to obtain a constant efficiency. An example are isoefficiency functions as introduced in [89] which express the required change of the problem size n as a function of the number of processors p.

4.2.3.4 Scaled speedup and convergence behavior

The next approach is also a variant of Amdahl's law, which assumes that the sequential program part is not a constant fraction f of the total amount of computations, but that it decreases with the input size. In this case, for an arbitrary number p of processors, the intended speedup $S_p(n) \leq p$ can be obtained by setting the problem size to a large enough value. The convergence behavior of the scaled speedup for increasing n can be considered for specific runtime functions as shown in the following.

It is assumed that τ_f is the constant execution time of the sequential program part and that $\tau_v(n, p)$ is the execution time of the parallelizable program part for problem size n and p processors. Since $\tau_v(n, 1)$ increases with increasing n, the fraction of the parallelizable part is increasing with n and therefore, the fraction of the sequential part is decreasing in relation to the parallelizable program part. Then the **scaled speedup** of the program on p processors is expressed by

$$S_p(n) = \frac{\tau_f + \tau_v(n, 1)}{\tau_f + \tau_v(n, p)} \tag{4.12}$$

To derive an estimation for the scaled speedup, the following calculations are performed: Using the notation $T^*(n)$ for the execution time of the best sequential implementation, the sequential execution of the parallelizable part can be expressed as $\tau_v(n,1) = T^*(n) - \tau_f$. Assuming that the parallelizable part of the sequential program is perfectly parallelized, leads to $\tau_v(n,p) = (T^*(n) - \tau_f)/p$. Inserting both formulas into the speedup formula (4.12) results in:

$$S_p(n) = \frac{\tau_f + T^*(n) - \tau_f}{\tau_f + (T^*(n) - \tau_f)/p} = \frac{\frac{\tau_f}{T^*(n) - \tau_f} + 1}{\frac{\tau_f}{T^*(n) - \tau_f} + \frac{1}{p}} \tag{4.13}$$

If $T^\star(n)$ increases strongly monotonically with n, then

$$\lim_{n \to \infty} S_p(n) = p$$

results. That means, the scaled speedup converges to the ideal speedup.

Example: An example with a strongly monotonically execution time is $\tau_v(n,p) = n^2/p$, which describes the amount of parallel computations for many iteration methods on two-dimensional meshes:

$$\lim_{n \to \infty} S_p(n) = \lim_{n \to \infty} \frac{\tau_f + n^2}{\tau_f + n^2/p} = \lim_{n \to \infty} \frac{\tau_f/n^2 + 1}{\tau_f/n^2 + 1/p} = p \tag{4.14}$$

Figure 4.2 shows the speedup and the efficiency for the runtime function in this example for $\tau_f = 100$, i.e., $T_p(n) = \tau_f + n^2/p$. Four curves are shown for different system sizes $n = 298, 423, 671, 948$. The speedup figure left shows a better speedup for higher values of n. The efficiency figure shows that an efficiency of 0.9 is achieved for all these input sizes, if the number of processors is chosen appropriately, i.e., $p = 100, 200, 500, 1000$.   □

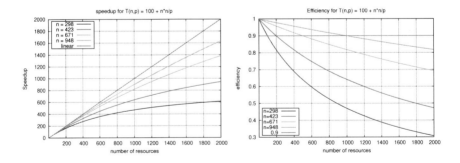

Fig. 4.2 Speedup (left) and efficiency (right) for the parallel execution time $\tau_v(n,p) = n^2/p$ and $\tau_f = 100$ for $n = 298, 423, 671, 948$.

4.3 Energy Measurement and Energy Metrics

The energy consumption of computer systems is getting more and more important due to environmental concerns. The energy consumption for the execution of a program is influenced by the hardware of the parallel system on which the program is executed. But the implementation of the program also has an influence on the energy consumption, since the implementation determines how efficiently the computational resources of the hardware are used. An efficient use of these resources typically leads to a smaller execution time and therefore a smaller energy consumption. Due to the growing importance of energy efficiency of parallel programs, measurement techniques for the power consumption and the energy consumption have been developed. The following subsection gives a short overview.

4.3.1 Performance and Energy Measurement Techniques

Several technologies have been developed to improve the energy efficiency of computer systems at hardware level. These technologies include power capping DVFS (Dynamic voltage and frequency scaling), see Section 2.2 for DVFS and [98] for a detailed description of power capping. The technologies are typically controlled by hardware, but there are also possibilities for the user or the programmer to influence or control the technologies. Therefore, measurement techniques should be available that allow a comparison of the energy efficiency of different implementation variants. An important aspect of performance and energy measurements is the fact that each measurement causes a **measurement overhead** that can lead to a distortion of the real behavior of the program without measurement, see [136] for a detailed treatment. Several measurement techniques as well as tools to control DVFS are discussed in the following.

Because of the importance of energy efficiency, hardware manufacturers have developed energy measurement methods and techniques. An important example is the running average power limit (RAPL) model that has been introduced by Intel starting with the Sandy Bridge architecture in 2013 [191, 50]. RAPL is based on a power-management architecture and a Package Control Unit (PCU) to collect information on the power consumption of the cores and to control the state transitions of the cores. The PCU constantly collects power and thermal information from the cores and uses this information to control the power consumption of the cores [191]. Internally, RAPL is based on a power model and the PCU estimates these power consumption values based on performance counters [50]. The results of the power model are stored in so-called model-specific registers (MSRs) that can be accessed by special instructions (rdmsr and wrmsr) in kernel mode. The accuracy of the power model has been verified by several investigations, see e.g., [186, 50]. Technology with a similar functionality has also been incorporated in AMD architectures [97, 98] and the resulting accuracy of the measurement also has been investigated [97].

```
#include <papi.h>
int main()
{
    long_long cy_start, cy_end, time_start, time_end;
    if (PAPI_library_init(PAPI_VER_CURRENT) != PAPI_VER_CURRENT)
        exit(1);
    /* request start time in clock cycles */
    cy_start = PAPI_get_real_cyc();
    /* request start time in microseconds */
    time_start = PAPI_get_real_usec();
    Do_Work(); /* program part to be measured */
    /* request start time in clock cycles */
    cy_end = PAPI_get_real_cyc();
    /* request end time in microseconds */
    time_end = PAPI_get_real_usec();
    printf("number of clock cycles: %lld\n",cy_end - cy_start);
    printf("runtime in microseconds: %lld\n", time_end - time_start);
}
```

Fig. 4.3 Program fragment illustrating the use of PAPI for measuring the number of cycles and the runtime of a program segment, see [214, 213] for more details.

The direct use of the MSR registers by the programmer is quite tedious and several peculiarities of the counter such as overflows have to be taken into consideration. It is much more convenient to use performance and energy measurement tools that provide a more convenient interface for the programmer. Popular tools are the Performance Application Programming Interface (PAPI) [98], see icl.utk.edu/papi for a detailed introduction and description, and likwid [217], see hpc.fau.de/research/tools/likwid for more information.

In the following, a short introduction to the **PAPI toolset** is given according to [214, 213]. In 2022, the newest PAPI version is 7.0. Figure 4.3 shows an example for using PAPI for measuring the number of cycles and the runtime of a program segment Do_Work(). The usage of the PAPI library requires to include the header file <papi.h> into the program code. The PAPI library is initialized by calling PAPI_library_init(int version) where PAPI_VER_CURRENT represents the PAPI version installed. This value is returned after a successful initialization. The function calls PAPI_get_real_cyc() and PAPI_get_real_usec() return the number clock cycles or the number of microseconds passed since some starting point. Calling these functions twice (before and after Do_Work()) and subtracting the values returned provides the number of cycles or microseconds, respectively, required for the execution of Do_Work().

PAPI supports the creation of so-called event sets into which specific events can be added according to the needs of the programmer. An event can, for example, be related to a counter capturing the execution time or the power consumption of the program to be evaluated. An event set is created by calling the PAPI function

```
PAPI_create_eventset (int *eventset)
```

```
#include <papi.h>
int main()
{
    int EventSet = PAPI_NULL;
    long_long values[1];
    if (PAPI_library_init(PAPI_VER_CURRENT) != PAPI_VER_CURRENT)
        exit(1);
    if (PAPI_create_eventset(&EventSet) != PAPI_OK)
        exit(1);
    if (PAPI_add_event(EventSet, PAPI_TOT_INS) != PAPI_OK)
        exit(1);
    if (PAPI_start(EventSet) != PAPI_OK)
        exit(1);
    Do_Work(); /* first program part to be measured */
    if (PAPI_read(EventSet, values) != PAPI_OK)
        exit(1);
    printf("number of instructions after first part: %lld\n", values[0]);
    Do_More_Work(); /* second program part to be measured */
    if (PAPI_stop(EventSet, values) != PAPI_OK)
        exit(1);
    printf("number of instructions after second part: %lld\n", values[0]);
}
```

Fig. 4.4 Program fragment illustrating the use of PAPI events for measuring the total number of instructions of a program segment, see [214, 213] for more details.

where the argument must be initialized to PAPI_NULL before the call. The event set generated by the call is represented by an integer handle that is returned via the argument. Upon successful creation, the return value of the funtion call is PAPI_OK. After the successful creation of an event set, events can be added to the set by calling

PAPI_add_event(int eventset, int eventcode).

where the first argument specifies the event set to which an event is added and the second argument specifies a specific event such as PAPI_TOT_INS for counting the total number of instructions executed or PAPI_L1_TCM for the total L1 cache misses. The return value PAPI_OK of the function call signals a successful addition, which is usually the case if the architecture supports the requested event. Multiple events can be added by calling PAPI_add_event() several times. After the definition of an event set, the following functions can be used to start, read, and stop the counters in the event set:

PAPI_start(eventset),
PAPI_read(int eventset, long_long * values),
PAPI_stop(int eventset, long_long * values).

The current values of the events in the event set are returned in the array values for PAPI_read() and PAPI_stop(). Using the concept of event sets, all counters that are supported by the hardware on which the program is executed can be added, including power and energy related counters. Figure 4.4 illustrates the use of event sets in PAPI. A single counter PAPI_TOT_INS is added to the event set. The PAPI function PAPI_read() is used to measure the total number of instructions after the first program part Do_Work(), PAPI_stop() is used to measure the total number of instructions after the second program part Do_More_Work().

Another popular method for performance and energy measurements is provided by the **likwid toolset** [217]. The toolset comprises several tools for different purposes related to performance and energy measurement, including:

- likwid-topology provides information about the execution platform such as number of cores and threads as well as the cache topology.
- likwid-perfctr supports the collection of performance counters using predefined or custom event sets; examples for predefined event sets are L2CACHE for collecting counters related to the miss rate of the L2 cache or ENERGY to collect counters related to the power and energy consumption. The tool can be used in wrapper mode to capture an entire application or in marker mode where the parts of the application to be measured are marked in the source code. The wrapper mode does not require a change in the source code of the program to be analyzed.
- likwid-pin offers support for the pinning of threads to hardware cores to support a consistent measurement of the power and energy consumption.
- likwid-powermeter provides information about the power and energy consumption of an application program based on the Intel RAPL interface.
- likwid-setFrequencies allows to set the operational frequency of the processor to a specific value (using DVFS) or to select a specific governor. Only values that are provided by the processor can be used.

More information about these and further tools of likwid can be found in [92].

Example: Figure 4.5 shows (a part of) the result of calling likwid-topology for an Intel Core i7-9700 desktop processor. The entry CPU stepping lists the number of operational frequencies that are available via DVFS and that can be set by likwid-setFrequencies. The result provides further information about the cache topology, showing that the processor has eight cores where each core has a private L1 cache of size 32 KB and a private L2 cache of size 256 MB. The L3 cache of size 12 MB is shared between the cores. There is only one thread per core, i.e., the processor does not support hyperthreading.   □

Example: Figure 4.6 shows (a part of) the result of calling likwid-perfctr on the same processor for executing a multithreaded program test_program using four threads. A detailed listing of the different performance counters is also provided by the call, but is omitted in the figure. likwid-perfctr is started for four threads with the performance group ENERGY focusing on performance counters that are related to the power consumption (measured in Watt) and the energy consumption (measured in Joule) during the execution of the program test_program. The result of likwid-perfctr shown in the figure gives information on the power and energy consumption during program execution, including runtime (in seconds), clock

```
>likwid-topology
---------------------------------------------------------------------------
CPU name: Intel(R) Core(TM) i7-9700 CPU @ 3.00GHz
CPU type: Intel Coffeelake processor
CPU stepping: 13
***************************************************************************
Hardware Thread Topology
***************************************************************************
Sockets: 1
Cores per socket: 8
Threads per core: 1
---------------------------------------------------------------------------
HWThread Thread Core Socket Available
0 0 0 0 *
1 0 1 0 *
2 0 2 0 *
3 0 3 0 *
4 0 4 0 *
5 0 5 0 *
6 0 6 0 *
7 0 7 0 *
---------------------------------------------------------------------------
Socket 0: ( 0 1 2 3 4 5 6 7 )
---------------------------------------------------------------------------
***************************************************************************
Cache Topology
***************************************************************************
Level: 1
Size: 32 kB
Cache groups: ( 0 ) ( 1 ) ( 2 ) ( 3 ) ( 4 ) ( 5 ) ( 6 ) ( 7 )
---------------------------------------------------------------------------
Level: 2
Size: 256 kB
Cache groups: ( 0 ) ( 1 ) ( 2 ) ( 3 ) ( 4 ) ( 5 ) ( 6 ) ( 7 )
---------------------------------------------------------------------------
Level: 3
Size: 12 MB
Cache groups: ( 0 1 2 3 4 5 6 7 )
---------------------------------------------------------------------------
```

Fig. 4.5 Result of calling `likwid-topology` for an Intel Core i7-9700 desktop processor. Only a part of the output is shown in the figure.

frequency (in MHz), temperature (in Celsius), power consumption (in Watt) and energy consumption (in Joule) of power plane PP0 and of the DRAM. PP0 contains the processor cores, DRAM refers to the memory system.   □

Another approach to capture the power and energy consumption is the modeling with functions. This is considered in the next subsection for DVFS.

```
>likwid-perfctr -C N:0-3 -g ENERGY test_program
-----------------------------------------------------------------------------
CPU name:        Intel(R) Core(TM) i7-9700 CPU @ 3.00GHz
CPU type:        Intel Coffeelake processor
CPU clock:       3.00 GHz
-----------------------------------------------------------------------------
+-------------------------+------------+------------+------------+------------+
|          Metric         |    Sum     |    Min     |    Max     |    Avg     |
+-------------------------+------------+------------+------------+------------+
| Runtime (RDTSC) [s] STAT|    62.9440 |    15.7360 |    15.7360 |    15.7360 |
| Runtime unhalted [s] STAT|   59.9797 |    14.7299 |    15.3324 |    14.9949 |
|         Clock [MHz] STAT | 11972.0653 |  2993.0153 |  2993.0175 |  2993.0163 |
|                 CPI STAT |     1.0384 |     0.2541 |     0.2655 |     0.2596 |
|     Temperature [C] STAT |        247 |         60 |         64 |    61.7500 |
|          Energy [J] STAT |   362.0966 |          0 |   362.0966 |    90.5242 |
|           Power [W] STAT |    23.0107 |          0 |    23.0107 |     5.7527 |
|      Energy PP0 [J] STAT |   347.4393 |          0 |   347.4393 |    86.8598 |
|       Power PP0 [W] STAT |    22.0793 |          0 |    22.0793 |     5.5198 |
|      Energy PP1 [J] STAT |          0 |          0 |          0 |          0 |
|       Power PP1 [W] STAT |          0 |          0 |          0 |          0 |
|     Energy DRAM [J] STAT |     8.7662 |          0 |     8.7662 |     2.1915 |
|      Power DRAM [W] STAT |     0.5571 |          0 |     0.5571 |     0.1393 |
+-------------------------+------------+------------+------------+------------+
```

Fig. 4.6 Result of calling `likwid-perfctr` for an Intel Core i7-9700 desktop processor for executing a multi-threaded program `test_program` using four threads. Only a part of the output is given in the figure.

4.3.2 Modeling of Power and Energy Consumption for DVFS

This section addresses the power and energy modeling using DVFS according to [183] and shows how optimal scaling factors can be computed. The energy model from Sect. 2.2 considers the power consumption (measured in Watt) and the energy consumption (measured in Joule). The power consumption consists of the dynamic power consumption and the static power consumption. The dynamic power consumption P_{dyn} increases cubically with the operational frequency f. For DVFS processors, the frequency f can be scaled down within a predefined interval $[f_{min}, f_{max}]$ where f_{max} is the normal operational frequency that can be scaled down. The scaling can be expressed by a dimensionless scaling factor $s \geq 1$ which describes a smaller frequency $\tilde{f}$ as $\tilde{f} = f/s$.

When using $P_{dyn} = \alpha \cdot C_L \cdot V^2 \cdot f$, see Sect. 2.2, and the linear dependency of the operational frequency f on the supply voltage V, i.e., $V = \beta \cdot f$ with some constant β, this leads to a cubic dependency of P_{dyn} from f. Thus, the dynamic power consumption decreases by a factor s^{-3} when reducing the frequency by a scaling factor $s \geq 1$. In the following, the scaling factor s is used as a parameter of the dynamic power consumption, denoted as $P_{dyn}(s)$, and $P_{dyn}(s) = s^{-3} \cdot P_{dyn}(1)$ holds where $P_{dyn}(1)$ is the power consumption for frequency f_{max}. According to [226], the static power consumption P_{static} can be assumed to be constant and independent from the scaling factor s.

The execution time of a program increases linearly with the scaling factor s due to the increased cycle time. When $T(s)$ denotes the execution time for scaling factor s and $T(1)$ the execution time for $s = 1$, then $T(s) = s \cdot T(1)$ holds. For the corresponding dynamic energy consumption $E_{dyn}(s)$ results $E_{dyn}(s) = P_{dyn}(s) \cdot (T(1) \cdot s)$. The static energy consumption can be modeled by $E_{static}(s) = P_{static} \cdot (T(1) \cdot s)$ Therefore, the total energy consumption can be modeled by the sum of these two components:

$$
\begin{aligned}
E(s) &= E_{dyn}(s) + E_{static}(s) \\
&= P_{dyn}(s) \cdot (T(1) \cdot s) + P_{static} \cdot (T(1) \cdot s) \\
&= (s^{-2} \cdot P_{dyn}(1) + s \cdot P_{static}) \cdot T(1) .
\end{aligned}
\tag{4.15}
$$

Thus, the energy consumption $E(s)$ is a function of s. The equation shows how increasing s has two effects: it increases the static energy consumption (linearly) because it slows down the computation, but it decreases the dynamic energy (quadratically). The optimal scaling factor can be computed analytically by minimizing this function, i.e., by determining the root of the first derivative. The resulting optimal scaling factor is

$$
s_{opt} = \left(\frac{2 \cdot P_{dyn}(1)}{P_{static}} \right)^{1/3} .
\tag{4.16}
$$

Figure 4.7 illustrates the dependency of $E(s)$ from s. The figure shows that a larger energy consumption results for

- $s < s_{opt}$ because the dynamic energy $E_{dyn}(s) = s^{-2} \cdot P_{dyn}(1) \cdot T(1)$ dominates and
- for $s > s_{opt}$ because the static energy $E_{static}(s) = s \cdot P_{static} \cdot T(1)$ dominates.

For the total energy consumption, a typical u-shape of the energy results, which is often observed in practice, when frequency scaling is applied.

4.3.3 Energy Metrics for Parallel Programs

A computer system is considered to be energy-efficient if a large number of computations can be performed per power unit. This is often captured by the number of Floating-Point operations (Flops) per Watt or variations of this metric such as GFlops/Watt (GigaFlops per Watt). This metric is also used for the Green500 list that has been introduced in 2013 as addition to the Top500 list, see Section 2.10.2 and www.top500.org, to set an additional focus on the energy efficiency for the computers on the Top500 list. As for the Top500 list, the evaluation is based on the Linpack benchmark for solving a dense system of linear equations. Only systems on the Top500 list are qualified for the Green500 list. In 2022, a high energy-efficiency value for systems on the list is 65 GFlops/Watt.

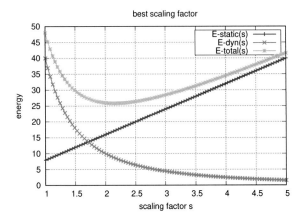

Fig. 4.7 Optimal scaling factor for a processor with $P_{static} = 8W$ and $P_{dyn}(1) = 40W$ and program execution time $T(1) = 1sec$. According to Equ. (4.16), the optimal scaling factor $s_{opt} = 2.15$ results. For a processor with $f_{max} = 3$ GHz, this corresponds to an optimal frequency of f_{opt} = 1.40 GHz.

Due to the importance of energy efficiency, there are also other proposals for capturing the energy efficiency of parallel programs as well as the development of the energy efficiency when increasing the number of resources used or varying the operational frequency of the resources (processors or cores). In the following, several such energy metrics for DVFS systems with p_{max} cores and frequencies f between the minimum frequency f_{min} and maximum frequency f_{max} are presented according to [187]. As in the previous subsection, s denotes the frequency scaling factor. The traditional runtime speedup is defined by

$$S(p,s) = \frac{T_{seq}(s)}{T_{par}(p,s)}, \qquad \text{with } s \text{ fixed,} \tag{4.17}$$

expressing the relative saving of execution time that can be obtained by using p processors instead of one processor for a fixed frequency. For different operational frequencies, different speedups may result. Also the execution time varies for varying frequencies and increases when the frequency is reduced. This can be captured by the *runtime reduction factor*:

$$R(p,s) = \frac{T_{par}(p,s)}{T_{par}(p,1)}, \qquad \text{with } p \text{ fixed.} \tag{4.18}$$

The runtime reduction factor describes the relative increase of the execution time that arises when the operational frequency $f = f_{max}/s$ is used instead of f_{max} for a fixed number p of processors. Equ. (4.18) can also be used for the sequential case, using $T_{seq}(s) = T_{par}(1,s)$.

The energy consumption of the execution of an application program also changes with varying frequencies and/or numbers of processors, which is caused by changes of the execution time as well as changes of the power consumption. Thus, it is reasonable to quantify the changes in the energy consumption. Analogously to the runtime speedup $S(p,s)$, the *energy speedup* can be defined as follows:

$$ES(p,s) = \frac{E(1,s)}{E(p,s)}, \qquad \text{with } s \text{ fixed.} \tag{4.19}$$

where $E(p,s)$ is the energy consumption using p processors and scaling factor s. The energy speedup captures the relative difference of the energy consumption that occurs by using p processors compared to one processor for a fixed frequency. A corresponding *energy reduction factor* for varying frequency can be defined as:

$$ER(p,s) = \frac{E(p,s)}{E(p,1)}, \qquad \text{with } p \text{ fixed.} \tag{4.20}$$

The energy reduction factor expresses the relative difference of the energy consumption when using frequency $f = f_{max}/s$ instead of f_{max} for a fixed number of processors.

The **energy-delay product EDP** is a metric that combines effects of execution time and energy consumption for different configurations of an application and captures the translation of energy into useful work. The EDP is defined as the energy consumed by an application program multiplied by its execution time [189]. When considering the scaling factor s explicitly, the EDP is expressed as

$$EDP(p,s) = E(p,s) \cdot (T(p,1) \cdot s) \qquad [Watt \cdot sec^2] \tag{4.21}$$
$$= (s^{-1} \cdot P_{dyn}(p,1) + s^2 \cdot P_{static}(p)) \cdot T(p,1)^2$$

using Equ. (4.15) with p explicitly shown. Determining the root of the first derivative of the function $EDP(s)$ yields the optimum scaling factor

$$s_{opt}^{EDP}(p) = \left(\frac{P_{dyn}(p,1)}{2 \cdot P_{static}(p,1)} \right)^{1/3} \tag{4.22}$$

for the EDP. It can be seen that $s_{opt}^{EDP}(p) < s_{opt}(p)$ from Equ. (4.16), i.e., to minimize the EDP, a smaller scaling factor must be used than for the minimization of the energy consumption.

The EDP is related to the **energy efficiency** in the following sense: The energy efficiency of an application is defined as performance (flop/s) per energy unit ($Watt \cdot s$) and is measured as $flop/(Watt \cdot sec^2)$. Given two EDP values EDP_1 and EDP_2 with $EDP_1 < EDP_2$ yields $1/EDP_1 > 1/EDP_2$, both sides measured in $1/(Watt \cdot sec^2)$, i.e., smaller EDP values indicate a better energy efficiency, and thus, a larger performance per energy unit. Therefore, the EDP is a good measure for the energy efficiency of a parallel program, and s_{opt}^{EDP} optimizes the energy efficiency of that program.

4.4 Asymptotic Times for Global Communication

In this section, we consider the analytical modeling of the execution time of parallel programs. For the implementation of parallel programs, many design decisions have to be made concerning, for example, the distribution of program data and the mapping of computations to resources of the execution platform. Depending on these decisions, different communication or synchronization operations must be performed, and different load balancing may result, leading to different parallel execution times for different program versions. Analytical modeling can help to perform a pre-selection by determining which program versions are promising and which program versions lead to significantly larger execution times, e.g., because of a potentially large communication overhead. In many situations, analytical modeling can help to favor one program version over many others. For distributed memory organizations, the main difference of the parallel program versions often is the data distribution and the resulting communication requirements.

For different programming models, different challenges arise for the analytical modeling. For programming models with a distributed address space, communication and synchronization operations are called explicitly in the parallel program, which facilitates the performance modeling. The modeling can capture the actual communication times quite accurately, if the runtime of the single communication operations can be modeled quite accurately. This is typically the case for many execution platforms. For programming models with a shared address space, accesses to different memory locations may result in different access times, depending on the memory organization of the execution platform. Therefore, it is typically much more difficult to analytically capture the access time caused by a memory access. In the following, we consider programming models with a distributed address space.

The time for the execution of local computations can often be estimated by the number of (arithmetical or logical) operations to be performed. But there are several sources of inaccuracy that must be taken into consideration:

- It may not be possible to determine the number of arithmetical operations exactly, since loop bounds may not be known at compile time or since adaptive features are included to adapt the operations to a specific input situation. Therefore, for some operations or statements, the frequency of execution may not be known. Different approaches can be used to support analytical modeling in such situations. One approach is that the programmer can give hints in the program about the estimated number of iterations of a loop or the likelihood of a condition to be true or false. These hints can be included by pragma statements and could then be processed by a modeling tool.

  Another possibility is the use of profiling tools with which typical numbers of loop iterations can be determined for similar or smaller input sets. This information can then be used for the modeling of the execution time for larger input sets, e.g., using extrapolation.

- For different execution platforms, arithmetical operations may have distinct execution times, depending on their internal implementation. Larger differences may occur for more complex operations like division, square root, or trigonometric functions. However, these operations are not used very often. If larger differences occur, a differentiation between the operations can help for a more precise performance modeling.

- Each processor typically has a local memory hierarchy with several levels of caches. This results in varying memory access times for different memory locations. For the modeling, average access times can be used, computed from cache miss and cache hit rates, see Section 4.1.3. These rates can be obtained by profiling.

The time for data exchange between processors can be modeled by considering the communication operations executed during program execution in isolation. For a theoretical analysis of communication operations, asymptotic running times can be used. We consider these for different interconnection networks in the following.

4.4.1 Implementing Global Communication Operations

In this section, we study the implementation and asymptotic running times of various global communication operations introduced in 3.6.2 on static interconnection networks according to [21]. Specifically, we consider the linear array, the ring, a symmetric mesh and the hypercube, as defined in Section 2.7.2. The parallel execution of global communication operations depends on the number of processors and the message size. The parallel execution time also depends on the topology of the network and the properties of the hardware realization. For the analysis, we make the following assumptions about the links and input and output ports of the network.

1. The links of the network are bidirectional, i.e., messages can be sent simultaneously in both directions. For real parallel systems, this property is usually fulfilled.
2. Each node can simultaneously send out messages on all its outgoing links; this is also called **all-port communication**. For parallel computers this can be organized by separate output buffers for each outgoing link of a node with corresponding controllers responsible for the transmission along that link. The simultaneous sending results from controllers working in parallel.
3. Each node can simultaneously receive messages on all its incoming links. In practice, there is a separate input buffer with controllers for each incoming link responsible for the receipt of messages.
4. Each message consists of several bytes, which are transmitted along a link without any interruption.
5. The time for transmitting a message consists of the startup time t_S, which is independent from the message size, and the byte transfer time $m \cdot t_B$, which is proportional to the size of the message m. The time for transmitting a single byte is denoted by t_B. Thus, the time for sending a message of size m from a node to a

directly connected neighbor node takes time $T(m) = t_S + m \cdot t_B$, see also Formula (2.5) from Section 2.8.3.

6. Packet switching with store-and-forward is used as switching strategy, see also Section 2.8.3. The message is transmitted along a path in the network from the source node to a target node and the length of the path determines the number of **time steps** of the transmission. Thus, the time for a communication also depends on the path length and the number of processors involved.

Given an interconnection network with these properties and parameters t_S and t_B, the time for a communication is mainly determined by the message size m and the path length p. For an implementation of global communication operations, several messages have to be transmitted and several paths are involved. For an efficient implementation, these paths should be planned carefully such that no conflicts occur. A conflict can occur when two messages are to be sent along the same link in the same time step; this usually leads to a delay of one of the messages, since the messages have to be sent one after another. The careful planning of the communication paths is a crucial point in the following implementation of global communication operations and the estimations of their running times. The execution times are given as asymptotic running time, which we briefly summarize now.

4.4.1.1 Asymptotic notation

Asymptotic running times describe how the execution time of an algorithm increases with the size of the input, see, e.g., [36]. The notation for the asymptotic running time uses functions whose domains are the natural numbers $\mathbb{N}$. The function describes the essential terms for the asymptotic behavior and ignores less important terms such as constants and terms of lower increase. The asymptotic notation comprises the O-notation, the Ω-notation and the Θ-notation, which describe boundaries of the increase of the running time. The asymptotic upper bound is given by the O-notation:

$$O(g(n)) = \{f(n) \mid \text{ there exists a positive constant } c \text{ and } n_0 \in \mathbb{N},$$
$$\text{such that for all } n \geq n_0 : 0 \leq f(n) \leq cg(n)\}$$

The asymptotic lower bound is given by the Ω-notation:

$$\Omega(g(n)) = \{f(n) \mid \text{ there exists a positive constant } c \text{ and } n_0 \in \mathbb{N},$$
$$\text{such that for all } n \geq n_0 : 0 \leq cg(n) \leq f(n)\}$$

The Θ-notation bounds the function from above and below

$$\Theta(g(n)) = \{f(n) \mid \text{ there exist positive constants } c_1, c_2 \text{ and } n_0 \in \mathbb{N},$$
$$\text{such that for all } n \geq n_0 : 0 \leq c_1 g(n) \leq f(n) \leq c_2 g(n)\}$$

Figure 4.8 illustrates the boundaries for the O-notation, the Ω-notation and the Θ-notation according to [36].

Table 4.2 Asymptotic running times of the implementation of global communication operations depending on the number p of processors in the static network. The linear array has the same asymptotic times as the ring.

Operation	ring	mesh	hypercube
Single-broadcast	$\Theta(p)$	$\Theta(\sqrt[d]{p})$	$\Theta(\log p)$
Scatter	$\Theta(p)$	$\Theta(p)$	$\Theta(p/\log p)$
Multi-broadcast	$\Theta(p)$	$\Theta(p)$	$\Theta(p/\log p)$
Total exchange	$\Theta(p^2)$	$\Theta(p^{(d+1)/d})$	$\Theta(p)$

The asymptotic running times of global communication operations with respect to the number of processors in the static interconnection network are given in Table 4.2. Running times for global communication operations are presented often in the literature, see e.g. [124, 89]. The analysis of running times mainly differs in the assumptions made about the interconnection network. In [89], one-port communication is considered, i.e., a node can send out only one message at a specific time step along one of its output ports; the communication times are given as functions in closed form depending on the number of processors p and the message size m for store-and-forward as well as cut-through switching. Here we use the assumptions given above according to [21].

The analysis uses the duality and hierarchy properties of global communication operations given in Fig. 3.11 of Section 3.6.2. Thus, from the asymptotic running times of one of the global communication operations it follows that a global communication operation which is less complex can be solved in no additional time and that a global communication operations which is more complex cannot be solved faster. For example, the scatter operation is less expensive than a multi-broadcast on the same network, but more expensive than a single-broadcast operation. Also a global communication operation has the same asymptotic time as its dual operation in the hierarchy. For example, the asymptotic time derived for a scatter operation can be used as asymptotic time of the gather operation.

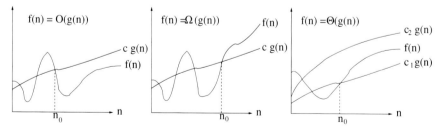

Fig. 4.8 Graphic examples of the O-, Ω- and Θ-notation. As value for n_0 the minimal value which can be used in the definition is shown.

4.4.1.2 Complete graph

A complete graph has a direct link between every pair of nodes. With the assumption of bidirectional links and a simultaneous sending and receiving of each output port, a total exchange can be implemented in one time step. Thus, all other communication operations such as broadcast, scatter and gather operations can also be implemented in one time step and the asymptotic time is $\Theta(1)$.

4.4.1.3 Linear array

A linear array with p nodes is represented by a graph $G = (V, E)$ with a set of nodes $V = \{1, \ldots, p\}$ and a set of edges $E = \{(i, i+1) | 1 \leq i < p\}$, i.e., each node except the first and the final is connected with its left and right neighbors. For an implementation of a **single-broadcast operation**, the root processor sends the message to its left and its right neighbor in the first step; in the next steps each processor sends the message received from a neighbor in the previous step to its other neighbor. The number of steps depends on the position of the root processor. For a root processor at the end of the linear array, the number of steps is $p - 1$. For a root processor in the middle of the array, the time is $\lfloor p/2 \rfloor$. Since the diameter of a linear array is $p - 1$, the implementation cannot be faster and the asymptotic time $\Theta(p)$ results.

A **multi-broadcast operation** can also be implemented in $p - 1$ time steps using the following algorithm. In the first step, each node sends its message to both neighbors. In the step $k = 2, \ldots, p - 1$, each node i with $k \leq i < p$ sends the message received in the previous step from its left neighbor to the right neighbor $i + 1$; this is the message originating from node $i - k + 1$. Simultaneously, each node i with $2 \leq i \leq p - k + 1$ sends the message received in the previous step from its right neighbor to the left neighbor $i - 1$; this is the message originally coming from node $i + k - 1$. Thus, the messages sent to the right make one hop to the right per time step and the messages sent to the left make one hop to the left in one time step. After $p - 1$ steps, all messages are received by all nodes. Figure 4.9 shows a linear array with 4 nodes as example; a multi-broadcast operation on this linear array can be performed in 3 time steps.

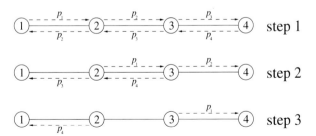

Fig. 4.9 Implementation of a multi-broadcast operation in three steps on a linear array with four nodes.

For the **scatter operation** on a linear array with p nodes, the asymptotic time $\Theta(p)$ results. Since the scatter operation is a specialization of the multi-broadcast operation it needs at most $p-1$ steps, and since the scatter operation is more general than a single-broadcast operation, it needs at least $p-1$ steps, see also the hierarchy of global communication operations in Figure 3.11. When the root node of the scatter operation is not one of the end nodes of the array, a scatter operation can be faster. The messages for more distant nodes are sent out earlier from the root node, i.e., the messages are sent in the reverse order of their distance from the root node. All other nodes send the messages received in one step from one neighbor to the other neighbor in the next step.

The number of time steps for a **total exchange** can be determined by considering an edge $(k, k+1)$, $1 \le k < p$, which separates the linear array into two subsets with k and $p-k$ nodes. Each node of the subset $\{1, \ldots, k\}$ sends $p-k$ messages along this edge to the other subset and each node of the subset $\{k+1, \ldots, p\}$ sends k messages in the other direction along this link. Thus, a total exchange needs at least $k \cdot (p-k)$ time steps or $p^2/4$ for $k = \lfloor p/2 \rfloor$. On the other hand, a total exchange can be implemented by p consecutive scatter operations, which lead to p^2 steps. Altogether, an asymptotic time $\Theta(p^2)$ results.

4.4.1.4 Ring

A ring topology has the nodes and edges of a linear array and an additional edge between node 1 and node p. All implementations of global communication operations are similar to the implementations on the linear array, but take only half of the time due to this additional link.

A **single-broadcast operation** is implemented by sending the message from the root node in both directions in the first step; in the following steps each node sends the message received into the opposite direction. This results in $\lfloor p/2 \rfloor$ time steps. Since the diameter of the ring is $\lceil p/2 \rceil$, the broadcast operation cannot be implemented faster and the time $\Theta(p)$ results.

A **multi-broadcast operation** is also implemented as for the array but in $\lfloor p/2 \rfloor$ steps. In the first step, each processor sends its message in both directions. In the following steps k, $2 \le k \le \lfloor p/2 \rfloor$, each processor sends the messages received in the opposite directions. Since the diameter is $\lceil p/2 \rceil$, the time $\Theta(p)$ results. Figure 4.10 illustrates a multi-broadcast operation for $p = 6$ processors.

The **scatter operation** also needs time $\Theta(p)$ since it cannot be faster than a single-broadcast and it is not slower than a multi-broadcast operation. To compute a lower bound for a **total exchange**, the ring is divided into two sets of $p/2$ nodes each (for p even). Each node of one of the subsets sends $p/2$ messages into the other subset across two links. This results in $p^2/8$ time steps, since one message needs one time step to be sent along one link. Thus, $\Theta(p^2)$ is a lower bound for a total exchange. An implementation in p^2 steps is obtained by executing p scatter operations from the different ring nodes one after another. Each of these scatter operations is done in p steps.

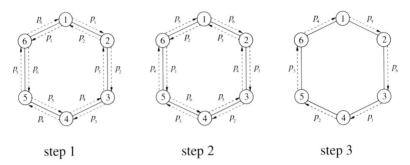

Fig. 4.10 Implementation of a multi-broadcast operation on a ring with 6 nodes. The message sent out by node i is denoted by p_i, $i = 1, \ldots, 6$.

4.4.1.5 Mesh

For a d-dimensional mesh with p nodes and $\sqrt[d]{p}$ nodes in each dimension, the diameter is $d(p^{1/d} - 1)$ and, thus, a **single-broadcast operation** can be executed in time $\Theta(p^{1/d})$, using a spanning tree with the root process of the broadcast operation as root of the spanning tree. For the **scatter operation**, an upper bound is $O(p)$ since a linear array with p nodes can be embedded into the mesh and a scatter operation needs time p on a linear array. A scatter operation needs at least time $p - 1$, since $p - 1$ messages have to be sent along the d outgoing links of the root node, which takes $\lceil \frac{p-1}{d} \rceil$ time steps. Together, the time $\Theta(p)$ results. The time $\Theta(p)$ for the **multi-broadcast operation** results in a similar way.

For the **total exchange**, we consider a mesh with an even number p of nodes and subdivide the mesh into two submeshes of equal size $p/2$, using a hyperplane of dimension $d - 1$ for the subdivision. For a total exchange, each node of a submesh sends $p/2$ messages into the other submesh. The messages have to be sent over the links of the hyperplane that has been used for the subdivision. This $(d - 1)$-dimensional hyperplane corresponds to $(\sqrt[d]{p})^{d-1}$ links. Thus, at least $p^{\frac{d+1}{d}}$ time steps are needed for the total exchange (because of $p^2/(4p^{\frac{d-1}{d}}) = 1/(4p^{\frac{d-1-2d}{d}}) = \frac{1}{4}p^{\frac{d+1}{d}}$).

To show that a total exchange can be performed in time $O(p^{\frac{d+1}{d}})$, we consider an algorithm implementing the total exchange in time $p^{\frac{d+1}{d}}$. Such an algorithm can be defined inductively from total exchange operations on meshes with lower dimension. For $d = 1$, the mesh is identical to a linear array for which the total exchange has a time complexity $O(p^2)$. Now we assume that an implementation on a $(d - 1)$-dimensional symmetric mesh with time $O(p^{\frac{d}{d-1}})$ is given. The total exchange operation on the d-dimensional symmetric mesh can be executed in two phases. The d-dimensional symmetric mesh is subdivided into disjoint meshes of dimension $d - 1$ which results in $\sqrt[d]{p}$ meshes. This can be done by fixing the value for the component in the last dimension x_d of the nodes $(x_1, \ldots, x_d)$ to one of the values $x_d = 1, \ldots, \sqrt[d]{p}$. In the first phase, total exchange operations are performed on the

$(d-1)$-dimensional meshes in parallel. Since each $(d-1)$-dimensional mesh has $p^{\frac{d-1}{d}}$ nodes, in one of the total exchange operations $p^{\frac{d-1}{d}}$ messages are exchanged. Since p messages have to be exchanged in each $d-1$-dimensional mesh, there are $\frac{p}{p^{\frac{d-1}{d}}} = p^{1/d}$ total exchange operations to perform. Because of the induction hypothesis, each of the total exchange operations needs time $O\left(p^{\frac{d-1}{d}}\right)^{\frac{d}{d-1}} = O(p)$ and thus, the time $p^{1/d} \cdot O(p) = O(p^{\frac{d+1}{d}})$ for the first phase results. In the second phase, the messages between the different submeshes are exchanged. The d-dimensional mesh consists of $p^{\frac{d-1}{d}}$ meshes of dimension 1 with $\sqrt[d]{p}$ nodes each; these are linear arrays of size $\sqrt[d]{p}$. Each node of a one-dimensional mesh belongs to a different $(d-1)$-dimensional mesh and has already received $p^{\frac{d-1}{d}}$ messages in the first phase. Thus, each node of a one-dimensional mesh has $p^{\frac{d-1}{d}}$ messages different from the messages of the other nodes; these messages have to be exchanged between them. This takes time $O((\sqrt[d]{p})^2)$ for one message of each node and in total $p^{\frac{2}{d}} p^{\frac{d-1}{d}} = p^{\frac{d+1}{d}}$ time steps. Thus, the time complexity $\Theta(p^{\frac{d+1}{d}})$ results.

4.4.2 Communications Operations on a Hypercube

For a d-dimensional hypercube, we use the bit notation of the $p = 2^d$ nodes as d-bit words $\alpha = \alpha_1 \ldots \alpha_d \in \{0,1\}^d$ introduced in Section 2.7.2.

4.4.2.1 Single-broadcast operation

A single-broadcast operation can be implemented using a spanning tree rooted at a node α that is the root of the broadcast operation. We construct a spanning tree for $\alpha = 00\cdots0 = 0^d$ and then derive spanning trees for other root nodes. Starting with root node $\alpha = 00\cdots0 = 0^d$ the children of a node are chosen by inverting one of the zero bits that are right of the rightmost unity bit. For $d = 4$ the spanning tree in Figure 4.11 results.

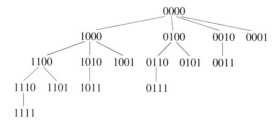

Fig. 4.11 Spanning tree for a single-broadcast operation on a hypercube for $d = 4$.

The spanning tree with root $\alpha = 00\cdots 0 = 0^d$ has the following properties: The bit names of two nodes connected by an edge differ in exactly one bit, i.e., the edges of the spanning tree correspond to hypercube links. The construction of the spanning tree creates all nodes of the hypercube. All leaf nodes end with a unity bit. The maximal degree of a node is d, since at most d bits can be inverted. Since a child node has one unity bit more than its parent node, an arbitrary path from the root to a leaf has a length not larger than d, i.e., the spanning tree has depth d, since there is one path from the root to node $11\ldots 1$ for which all d bits have to be inverted.

For a single-broadcast operation with an arbitrary root node z, a spanning tree T_z is constructed from the spanning tree T_0 rooted at node $00\cdots 0$ by keeping the structure of the tree but mapping the bit names of the nodes to new bit names in the following way. A node x of tree T_0 is mapped to node $x \oplus z$ of tree T_z, where $\oplus$ denotes the bitwise xor-operation (exclusive or operation), i.e.,

$$a_1 \cdots a_d \oplus b_1 \cdots b_d = c_1 \cdots c_d \text{ with } c_i = \begin{cases} 1 & \text{when } a_i \neq b_i \\ 0 & \text{otherwise} \end{cases} \text{ for } 1 \leq i \leq d.$$

Especially, node $\alpha = 00\cdots 0$ is mapped to node $\alpha \oplus z = z$. The tree structure of tree T_z remains the same as for tree T_0. Since the nodes v, w of T_0 connected by an edge (v, w) differ in exactly one bit position, the nodes $v \oplus z$ and $w \oplus z$ of tree T_z also differ in exactly one bit position and the edge $(v \oplus z, w \oplus z)$ is a hypercube link. Thus, a spanning tree of the d-dimensional hypercube with root z results.

The spanning tree can be used to implement a single-broadcast operation from the root node in d time steps. The messages are first sent from the root to all children and in the next time steps each node sends the message received to all its children. Since the diameter of a d-dimensional hypercube is d, the single-broadcast operation cannot be faster than d and the time $\Theta(d) = \Theta(\log(p))$ results.

4.4.2.2 Multi-broadcast operation on a hypercube

For a multi-broadcast operation, each node receives $p - 1$ messages from the other nodes. Since a node has $d = \log p$ incoming edges, which can receive messages simultaneously, an implementation of a multi-broadcast operation on a d-dimensional hypercube takes at least $\lceil (p - 1)/\log p \rceil$ time steps. There are algorithms that attain this lower bound and we construct one of them in the following according to [21].

The multi-broadcast operation is considered as a set of single-broadcast operations, one for each node in the hypercube. A spanning tree is constructed for each of these single-broadcast operations and the message is sent along the links of the tree in a sequence of time steps as described above for the single-broadcast in isolation. The idea of the algorithm for the multi-broadcast operation is to construct spanning trees for the single-broadcast operations such that the single-broadcast operations can be performed simultaneously. To achieve this, the links of the different spanning trees used for a transmission in the same time step have to be disjoint. This is the reason why the spanning trees for the single-broadcast in isolation cannot be used here, as will be seen later. We start by constructing the spanning tree T_0 for root node $00\cdots0$.

The spanning tree T_0 for root node $00\cdots0$ consists of disjoint sets of edges $A_1,\ldots,A_m$, where m is the number of time steps needed for a single-broadcast and A_i is the set of edges over which the messages are transmitted at time step i, $i = 1,\ldots,m$. The set of start nodes of the edges in A_i is denoted by S_i and the set of end nodes is denoted by E_i, $i = 1,\ldots,m$, with $S_1 = \{(00\cdots0)\}$ and $S_i \subset S_1 \cup \bigcup_{k=1}^{i-1} E_k$. The spanning tree T_t with root $t \in \{0,1\}^d$ is constructed from T_0 by mapping the edge sets of T_0 to edge sets $A_i(t)$ of T_t using the xor operation, i.e.,

$$A_i(t) = \{(x \oplus t, y \oplus t)|(x,y) \in A_i\} \quad \text{for } 1 \le i \le m. \tag{4.23}$$

If T_0 is a spanning tree, then T_t is also a spanning tree with root $T \in \{0,1\}^d$. The goal is to construct the sets $A_1,\ldots,A_m$ such that for each $i \in \{1,\ldots,m\}$ the sets $A_i(t)$ are pairwise disjoint for all $t \in \{0,1\}^d$ (with $A_i = A_i(0)$, $i = 1,\ldots,m$). This means that transmission of data can be performed simultaneously on those links. To get disjoint edges for the same transmission step i the sets A_i are constructed such that the following holds:

- For any two edges $(x,y) \in A_i$ and $(x',y') \in A_i$, the bit position in which the nodes x and y differ is **not** the same bit position in which the nodes x' and y' differ.

The reason for this requirement is that two edges whose start and end node differ in the same bit position can be mapped onto each other by an xor operation with an appropriate t. Thus, if such edges would be in set A_i for some $i \in \{1,\ldots,m\}$, then they would be in the set $A_i(t)$ and the sets A_i and $A_i(t)$ would not be disjoint. This is illustrated in Figure 4.12 for $d = 3$ using the spanning trees constructed earlier for the single-broadcast operations in isolation.

There are only d different bit positions so that each set A_i, $i = 1,\ldots,m$, can only contain at most d edges. Thus, the sets A_i are constructed such that $|A_i| = d$ for $1 \le i < m$ and $|A_m| \le d$. Since the sets $A_1,\ldots,A_m$ should be pairwise disjoint and the total number of edges in the spanning tree is $2^d - 1$ (there is an incoming edge for each node except the root node), we get

$$\left|\bigcup_{i=1}^{m} A_i\right| = 2^d - 1$$

and a first estimation for m:

$$m = \left\lceil \frac{2^d - 1}{d} \right\rceil.$$

Figure 4.13 shows the 8 spanning trees for $d = 3$ and edge sets A_1, A_2, A_3 with $|A_1| = |A_2| = 3$ and $|A_3| = 1$. In this example, there is no conflict in any of the three time steps $i = 1,2,3$. These spanning trees can be used simultaneously, and a multi-broadcast needs $m = \lceil (2^3 - 1)/3 \rceil = 3$ time steps.

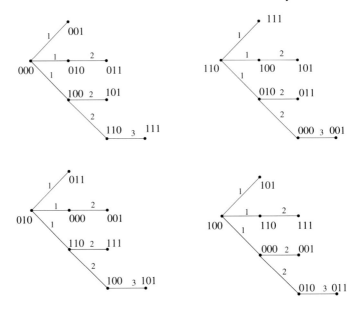

Fig. 4.12 Spanning tree for the single-broadcast operation in isolation. The start and end nodes of the edges $e_1 = ((010), (011))$ and $e_2 = ((100), (101))$ differ in the same bit position, which is the rightmost bit position (see tree top left). The xor operation with new root node $t = 110$ creates a tree that contains the same edges e_1 and e_2 for a data transmission in the second time step (tree top right). A delay of the transmission into the third time step would solve this conflict. However, a new conflict in time step 3 results with the spanning tree with root 010, which has edge e_2 in the third time step (tree bottom left), and with spanning tree with root 100, which has edge e_1 in the third time step (tree bottom right).

We now construct the edge sets A_i, $i = 1, \ldots, m$, for arbitrary d. The construction mainly consists of the following arrangement of the nodes of the d-dimensional hypercube. The set of nodes with k unity bits and $d - k$ zero bits is denoted as N_k, $k = 1, \ldots, d$, i.e.,

$$N_k = \{t \in \{0, 1\}^d \mid t \text{ has } k \text{ unity bits and } d - k \text{ zero bits }\}$$

for $0 \leq k \leq d$ with $N_0 = \{(00 \cdots 0)\}$ and $N_d = \{(11 \cdots 1)\}$. The number of elements in N_k is

$$|N_k| = \binom{d}{k} = \frac{d!}{k!(d-k)!} \, .$$

Each set N_k is further partitioned into disjoint sets $R_{k1}, \ldots, R_{kn_k}$, where one set R_{ki} contains all elements which result from a bit rotation to the left from each other. The sets R_{ki} are equivalence classes with respect to the relation *rotation to the left*. The first of these equivalence classes R_{k1} is chosen to be the set with the element $(0^{d-k}1^k)$, i.e., the rightmost bits are unity bits. Based on these sets, the following

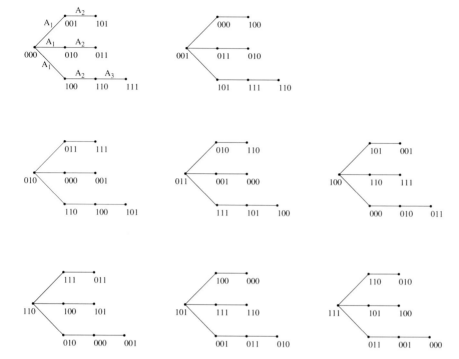

Fig. 4.13 Spanning trees for a multi-broadcast operation on a d-dimensional hypercube with $d = 3$. The sets A_1, A_2, A_3 for root 000 are $A_1 = \{(000, 001), (000, 010), (000, 100)\}$, $A_2 = \{(001, 101), (010, 011), (100, 110)\}$ and $A_3 = \{(110, 111)\}$ shown in the upper left corner. The other trees are constructed according to Formula (4.23).

assignment is made: Each node $t \in \{0, 1\}^d$ is assigned a number $n(t) \in \{0, \ldots, 2^d - 1\}$ corresponding to its position in the order

$$\{\alpha\} R_{11} R_{21} \ldots R_{2n_2} \ldots R_{k1} \ldots R_{kn_k} \ldots R_{(d-2)1} \ldots R_{(d-2)n_{d-2}} R_{(d-1)1} \{\beta\}, \tag{4.24}$$

with $\alpha = 00\ldots0$ and $\beta = 11\ldots1$ and position numbers $n(\alpha) = 0$ and $n(\beta) = 2^d - 1$. Each node $t \in \{0, 1\}^d$, except α, is also assigned a number $m(t)$ with

$$m(t) = 1 + [(n(t) - 1) \bmod d], \tag{4.25}$$

i.e., the nodes are numbered in a round-robin fashion by $1, \ldots, d$. So far, there is no specific order of the nodes within one of the equivalence classes R_{kj}, $k = 1, \ldots, d$, $j = 1, \ldots, n_k$. Using $m(t)$ we now specify the following order:

- the first element $t \in R_{kj}$ is chosen such that the following condition is satisfied

    The bit at position $m(t)$ from the right is 1. $\tag{4.26}$

- The subsequent elements of R_{kj} result from a single bit rotation to the left. Thus, property (4.26) is satisfied for all elements of R_{kj}.

For the first equivalence classes R_{k1}, $k = 1, \ldots, d$, we additionally require the following:

- the first element $t \in R_{k1}$ has a zero at the bit position right of position $m(t)$, i.e., when $m(t) > 1$, the bit at position $m(t) - 1$ is a zero, and when $m(t) = 1$, the bit at the leftmost position is a zero.
- The property holds for all elements in R_{k1}, since they result by a bit rotation to the left from the first element.

For the case $d = 4$, the following order of the nodes $t \in \{0, 1\}^4$ and $m(t)$ values result:

$$
\begin{array}{llllll}
 & \overset{0}{} & & & & \\
N_0 & (0000) & & & & \\
 & \overset{1}{} & \overset{2}{} & \overset{3}{} & \overset{4}{} & \\
N_1 & (0001) & (0010) & (0100) & (1000) & \\
 & & & \underbrace{\qquad R_{11} \qquad} & & \\
 & \overset{1}{} & \overset{2}{} & \overset{3}{} & \overset{4}{} & \overset{1}{}\quad\overset{2}{} \\
N_2 & (0011) & (0110) & (1100) & (1001) & (0101)\ (1010) \\
 & & \underbrace{\qquad R_{21} \qquad} & & & \underbrace{R_{22}} \\
 & \overset{3}{} & \overset{4}{} & \overset{1}{} & \overset{2}{} & \\
N_3 & (1101) & (1011) & (0111) & (1110) & \\
 & & \underbrace{\qquad R_{31} \qquad} & & & \\
 & \overset{3}{} & & & & \\
N_4 & (1111) & & & &
\end{array}
$$

Using the numbering $n(t)$ we now define the sets of end nodes $E_0, E_1, \ldots, E_m$ of the edge sets $A_1, \ldots, A_m$ as contiguous blocks of d nodes (or $< d$ nodes for the last set):

$$E_0 = \{(00 \cdots 0)\}$$
$$E_i = \{t \in \{0, 1\}^d \mid (i - 1)d + 1 \le n(t) \le i \cdot d\} \quad \text{for } 1 \le i < m$$
$$E_m = \{t \in \{0, 1\}^d \mid (m - 1)d + 1 \le n(t) \le 2^d - 1\} \quad \text{with } m = \left\lceil \frac{2^d - 1}{d} \right\rceil.$$

The sets of edges A_i, $1 \le i \le m$, are then constructed according to the following:

- The set of edges A_i, $1 \le i \le m$, consists of the edges that connect an end node $t \in E_i$ with the start node t' obtained from t by inverting the bit at position $m(t)$, which is alway a unity bit due to the construction.
- As an exception, the end node $t = (11 \cdots 1)$ for the case $m(11 \cdots 1) = d$ is connected to the start node $t' = (1011 \cdots 1)$ (and not $(011 \cdots 1)$).

Due to the construction the start nodes t' have one unity bit less than t and, thus, when $t \in N_k$, then $t' \in N_{k-1}$. Also the edges are links of the hypercube. Figure 4.14 shows the sets of end nodes and the sets of edges for $d = 4$.

Next, we show that these sets of edges define a spanning tree with root node $(00 \cdots 0)$ by showing that an end node $t \in E_i$ is connected to a start node $t' \in \bigcup_{k=1}^{i-1} E_k$,

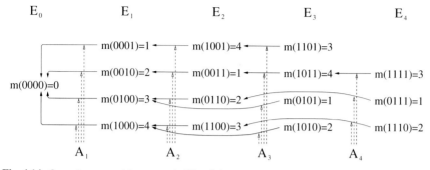

Fig. 4.14 Spanning tree with root node $00\ldots0$ for a multi-broadcast operation on a hypercube with $d = 4$. The sets of edges A_i, $i = 1,\ldots,4$ are indicated by dotted arrows.

i.e., that there exists $k < i$ with $t' \in E_k$. Since t' has one more zero than t by construction, $n(t') < n(t)$ and thus $k > i$ is not possible, i.e., $k \leq i$ holds. It remains to show that $k < i$.

- For $t = 11\cdots1$ and $m(t) = d$, the set E_m contains d nodes, which are node t and $d - 1$ other nodes from $R_{d-1,1}$. There is one node of $R_{d-1,1}$ left, which is in set E_{m-1}; this node has a 1 at position $m(t)$ from the right and a 0 left of it. Thus, this node is $(1011\cdots1)$ which has been chosen as the start node by exception.
- For $t = 11\cdots1$ and $m(t) = d - k < d$, with $1 \leq k < d$, the set E_m contains $d - k$ nodes s with numbers $n(s) < d - k$. The start node t' connected to t has a 0 at the position $d - k$ according to the construction and a 1 at the position $d - k - 1$ from the right. Thus, $m(t') = d - k + 1$. Since $m(t') > d - k$, the node t' cannot belong to the edge set E_m and thus $t' \in E_{m-1}$.

For the nodes $t \neq 11\cdots1$, we now show that $n(t) - n(t') \geq d$, i.e., t' belongs to a different set E_k than t, with $k < i$.

- For $t \in R_{kn}$ with $n > 1$, all elements of R_{k1} are between t and t', since $t' \in N_{k-1}$. This set R_{k1} is the equivalence class of nodes $(0^{d-k}1^k)$ and contains d elements. Thus, $n(t) - n(t') \geq d$.
- For $t \in R_{k1}$, the start node t' is an element of $R_{k-1,1}$, since it has one more zero bit (which is at position $m(t)$ and according to the internal order in the set $R_{k-1,1}$ all remaining unity bits are right of $m(t)$ in a contiguous block of bit positions. Therefore, all elements of $R_{k-1,2}, \ldots, R_{k-1,n_{k-1}}$ are between t and t'. These are $|N_{k-1}| - |R_{k-1,1}| = \binom{d}{k-1} - d$ elements. For $2 < k < d$ and $d \geq 5$, it can be shown by induction that $\binom{d}{k-1} - d \geq d$. For $k = 1, 2$, $R_{11} = E_1$ and $R_{21} = E_2$ for all d and $t' \in E_{k-1}$ holds. For $d = 3$ and $d = 4$, the estimation can be shown individually; Figure 4.13 shows the case $d = 3$ and Figure 4.14 shows the case $d = 4$.

Thus, the sets $A_i(t)$, $i = 1,\ldots,m$, can be used for one of the single-broadcast operations of the multi-broadcast operation. The other sets $A_i(t)$ are constructed using the xor operation as described above. The trees can be used simultaneously, since no conflicts result. This can be seen from the construction and the numbers

$m(t)$. The nodes in a set of end nodes E_i of edge set A_i have d different numbers $m(t) = 1, \ldots, d$ and, thus, for each of the nodes $t \in E_i$ a bit at a different bit position is inverted. Thus, the start and end nodes of the edges in A_i differ in different bit positions, which was required to get a conflict free transmission of messages in time step i. In summary, the single broadcast operations can be performed in parallel and the multi-broadcast operation can be performed in $m = \lceil (2^d - 1)/d \rceil$ time steps. This is the smallest number of time steps that is possible, since each node has to send $2^d - 1 = p - 1$ messages for a multi-broadcast operation and since only $d = \log p$ links are available for sending these messages. Thus, $\lceil (2^d - 1)/d \rceil$ is a lower bound for the number of time steps required and the multi-broadcast algorithm using the broadcast trees constructed is optimal.

4.4.2.3 Scatter operation

A scatter operation takes no more time than the multi-broadcast operation, i.e., it takes no more than $\lceil (2^d - 1)/d \rceil$ time steps. On the other hand, in a scatter operation $2^d - 1$ messages have to be sent out from the d outgoing edges of the root node, which needs at least $\lceil (2^d - 1)/d \rceil$ time steps. Thus, the time for a scatter operation on a d-dimensional hypercube is $\Theta(\lceil (p-1)/\log p \rceil)$. A gather operation requires the same time for the same reasons.

4.4.2.4 Total exchange

The total exchange on a d-dimensional hypercube has time $\Theta(p) = \Theta(2^d)$. The lower bound results from decomposing the hypercube into two hypercubes of dimension $d - 1$ with $p/2 = 2^{d-1}$ nodes each and 2^{d-1} edges between them. For a total exchange, each node of one of the $(d-1)$-dimensional hypercubes sends a message for each node of the other hypercube; these are $(2^{d-1})^2 = 2^{2d-2}$ messages, which have to be transmitted along the 2^{d-1} edges connecting both hypercubes. This takes at least $2^{2d-2}/2^{d-1} = 2^{d-1} = p/2$ time steps.

An algorithm implementing the total exchange in $p - 1$ steps can be built recursively. For $d = 1$, the hypercube consists of 2 nodes for which the total exchange can be done in one time step, which is $2^1 - 1$. Next, we assume that there is an implementation of the total exchange on a d-dimensional hypercube in time $2^d - 1$. A $(d+1)$-dimensional hypercube is decomposed into two hypercubes C_1 and C_2 of dimension d. The algorithm consists of the three phases:

1. Two separate total exchange operations within the hypercubes C_1 and C_2 are performed simultaneously. This takes time $2^d - 1$.
2. Each node in C_1 (or C_2) sends 2^d messages for the nodes in C_2 (or C_1) to its counterpart in the other hypercube. Since all nodes used different edges, this takes time 2^d.
3. A total exchange in each of the hypercubes is performed to distribute the messages received in phase 2. This again takes time $2^d - 1$.

The phases 1 and 2 can be done simultaneously and together take time 2^d. Phase 3 has to be performed after phase 2 and takes time $2^d - 1$. In summary, the time $2^d + 2^d - 1 = 2^{d+1} - 1$ results.

4.4.3 Communication Operations on a Complete Binary Tree

In a complete binary tree, all leaves have equal depth t and the number of nodes in the tree is $p = 2^{t+1} - 1$. This subsection investigates the implementation of the communication operations on a complete binary tree according to [216, 21].

A **single-broadcast operation** can be implemented as follows on a complete binary tree: In the first step, the root node of the single broadcast operation sends the broadcast message over all of its output links (at most three). In the following steps, each node sends the message received in the preceding step to all of its neighbors (parent node or child nodes) from which the message has not been received. In total, between t and $2t$ steps are required, depending on the position of the root node of the broadcast operation: If the root node of the tree network is the root of the broadcast operation, t steps are required. If a leaf node of the tree network is the root of the broadcast operation, $2t$ steps are required. Because of $p = 2^{t+1} - 1$, is is $t = log(p+1) - 1)$, i.e., the resulting asymptotic runtime is $\Theta(log p)$.

For the implementation of a **scatter operation**, the root node of the scatter operation sends consecutively its $p - 1$ messages over its outgoing edges. The messages are sorted according to their distance to the receiving node, i.e., for more distant receivers the messages are sent first. Messages are annotated with the receiver information to ensure a correct delivery. This allows intermediate nodes to forward a message received in the right direction. Figure 4.15 illustrates the implementation. After at most $p - 1$ steps the algorithm ends (upper limit), since the root node of the scatter operations sends a message in each step and since the messages are ordered appropriately. A smaller number of steps might be sufficient, depending on the position of the scatter root node in the tree network. For example, if the scatter root node is the root node of the tree network, only $\lceil \frac{p-1}{2} \rceil$ steps are required (lower limit). Thus, the resulting asymptotic runtime is $\Theta(p)$.

A **multi-broadcast operation** can also be implemented in $p - 1$ steps: In the first step, each node i sends its message to all its neighbors. In the following steps, each node i considers all outgoing edges (i, j). If node i has a message that has not yet been sent to node j and that has not been received from j in a preceding step, i sends this message to j using edge (i, j). If i has several such messages, one of these messages is chosen randomly.

To show that after $p - 1$ steps all messages have been delivered at their final destination, a set $R(i, x)$ is defined for each node i containing all messages that i has already received until time x. Similarly, a set $X(i, j)$ is defined for each node i containing all messages that i has already sent to its neighbor j using edge (i, j). At time $x = 0$, $R(i, 0)$ is initialized with the local message of node i. Furthermore, $X(i, j) = \emptyset$, since no messages have been sent so far. At each time $x \geq 1$, each node

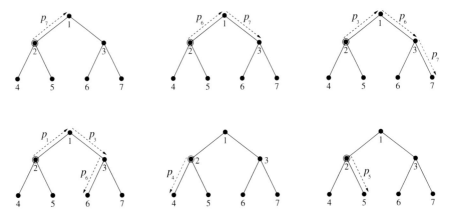

Fig. 4.15 Illustration of a scatter operation on a complete binary tree with seven nodes. The root node of the scatter operation is node 2. The message sent by node 2 for node i is denoted by p_i.

i investigates each adjacent edge (i, j). If $R(i,x) \setminus X(i, j) \neq \emptyset$, i.e., if node i has a message that has not yet been sent to node j, then i selects an arbitrary message from $R(i,x) \setminus X(i, j)$ sends it over edge (i, j) to node j, and inserts the message into $X(i, j)$. Messages arriving at node i at time x from neighboring nodes are inserted into $R(i, x+1)$. A message arriving from neighboring node j is inserted into set $X(i, j)$. The algorithm stops, if there are no messages left to be transmitted at any node i, i.e., if $R(i,x) = X(i, j)$.

It is now shown that the number of steps required by this algorithm is $p - 1$. To compute the number of steps, a value $b(i, j)$ is defined for each edge (i, j) counting the number of nodes k, whose path to i crosses edge (j, i). For these numbers $b(i, j)$ the following claim is made for arbitrary values $q \geq 0$:

- If $b(i, j) \geq q$, then a message is sent to node i in step q using the edge (j, i).

This means that there are no temporal gaps for the transmission over an edge of the tree network. The claim can be proven by induction over $b(i, j)$. For $b(i, j) = 1$, j is a leaf node and i is the parent node of a leaf node. In step 1, each node sends its message over all outgoing edges; especially the message from node j is sent over edge (j, i) to node i. Thus, the claim is true for $b(i, j) = 1$. For the induction step, it is assumed that the claim is true for all nodes i with neighbor j and $b(i, j) \leq f, f > 1$. Based on this assumption, it is shown that the claim is also true for all nodes i with neighbor j and $b(i, j) = f + 1$. To show this, edge (j, i) is considered. If there are $f + 1$ messages using edge (j, i) to arrive at node i, then there must be f messages arriving at node j over edges $(k, j) \neq (j, i)$, i.e.

$$\sum_{\substack{(k,j) \\ k \neq i}} b(j,k) = b(i, j) - 1 = f.$$

Therefore, the induction assumption can be applied to the edges (k, j) with $k \neq i$. According to this assumption, a message is sent over edge (k, j) to node j in each step q with $q \leq b(j,k)$, i.e., node j has received for each step $q \leq f$ at least q messages via edge (k, j) with $k \neq i$. Since node j also sends its own message to node i in the first step, at each point in time $1, 2, \ldots, b(i, j)$ a message is sent over edge (j, i) to node i.

Thus, it has been shown that there are no temporal gaps in the transmission of messages over an edge of the tree network. For the number of messages received by node i at time q it holds

$$|R(i,q)| = 1 + \sum_{(j,i)} \min(b(i,j), q),$$

i.e., additional to the own message of node i, $R(i, q)$ contains all messages arriving over edge (j, i) until time q. At most $b(i, j)$ messages can arrive over edge (j, i), since this is the number of nodes whose path to i crosses edge (j, i). Moreover, each edge can only transmit one message in each step, i.e., over edge (j, i) at most q messages can have been arrived until time q. Also the following holds:

$$Q(i) := \max\{b(i, j) \mid j \text{ is neighbor of } i\} \leq p - 1, \tag{4.27}$$

i.e., at most $p - 1$ nodes can be connected via node j with i. For times $q \geq b(i, j)$, it is $\min(b(i, j), q) = b(i, j)$. Since each node i has to receive $p - 1$ messages via its incoming edges, it is $\sum_{(j,i)} b(i, j) = p - 1$. For time $Q(i)$, it is therefore

$$|R(i, Q(i))| = 1 + \sum_{(j,i)} b(i, j) = p.$$

Because of $Q(i) \leq p - 1$ node i has received the messages of all other nodes after at most $p - 1$ steps. This is true for an arbitrary node i. Thus, the multi-broadcast operation is finished after $p - 1$ steps.

Figure 4.16 shows the communication steps of a multi-broadcast operation on a complete binary tree with seven nodes. The message sent by node i is denoted by p_i. Six steps are required to distribute all messages as shown in the figure. The corresponding sets $R(i, x)$ containing all messages that node i has received after step x are shown in Table 4.3. The set $C = \{p_1, p_2, p_3, p_4, p_5, p_6, p_7\}$ denotes all messages. The development of the sets of the seven nodes is shown in the seven rows of the table. The columns show the development of the sets $R(i, x)$ for the six steps $x = 1, \ldots 6$ that have to be executed. The content of $R(i, x)$ after step x is shown for each of the nodes $i = 1, \ldots 7$.

For a **total exchange** on a complete binary tree, a lower bound for the number of steps required is $\lceil p/2 \rceil \lfloor p/2 \rfloor \sim p^2/4$. This can be seen by considering the edges connecting the root node with the left or right subtree. For a total exchange, each of the $\lceil p/2 \rceil$ nodes in the left subtree has to send a message to each of the $\lceil p/2 \rceil$ nodes in the right subtree. These $\lceil p/2 \rceil \lfloor p/2 \rfloor$ messages have to be sent over a single edge one after another, requiring at least $\lceil p/2 \rceil \lfloor p/2 \rfloor$ steps.

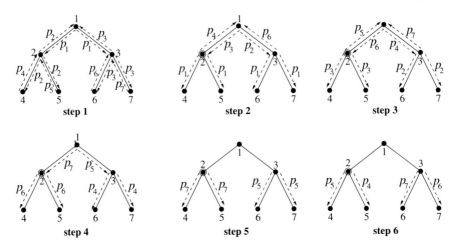

Fig. 4.16 Illustration of a multi-broadcast operation on a complete binary tree with seven nodes. The message sent by node i is denoted by p_i.

node	$x=1$	$x=2$	$x=3$	$x=4$	$x=5$	$x=6$
1	$\{p_1,p_2,p_3\}$	$\{p_1,p_2,p_3,p_4,p_6\}$	C	C	C	C
2	$\{p_1,p_2,p_4,p_5\}$	$\{p_1,p_2,p_3,p_4,p_5\}$	$\{p_1,p_2,p_3,p_4,p_5,p_6\}$	C	C	C
3	$\{p_1,p_3,p_6,p_7\}$	$\{p_1,p_2,p_3,p_6,p_7\}$	$\{p_1,p_2,p_3,p_4,p_6,p_7\}$	C	C	C
4	$\{p_2,p_4\}$	$\{p_1,p_2,p_4\}$	$\{p_1,p_2,p_3,p_4\}$	$\{p_1,p_2,p_3,p_4,p_6\}$	$\{p_1,p_2,p_3,p_4,p_6,p_7\}$	C
5	$\{p_2,p_5\}$	$\{p_1,p_2,p_5\}$	$\{p_1,p_2,p_3,p_5\}$	$\{p_1,p_2,p_3,p_5,p_6\}$	$\{p_1,p_2,p_3,p_5,p_6,p_7\}$	C
6	$\{p_3,p_6\}$	$\{p_1,p_3,p_6\}$	$\{p_1,p_2,p_3,p_6\}$	$\{p_1,p_2,p_3,p_4,p_6\}$	$\{p_1,p_2,p_3,p_4,p_5,p_6\}$	C
7	$\{p_3,p_7\}$	$\{p_1,p_3,p_7\}$	$\{p_1,p_2,p_3,p_7\}$	$\{p_1,p_2,p_3,p_4,p_7\}$	$\{p_1,p_2,p_3,p_4,p_5,p_7\}$	C

Table 4.3 Content of the sets $R(i,x)$ which contain all messages that node i has received after step x for the different steps $x = 1,\ldots 6$. These sets correspond to the six steps in Fig. 4.16. The six steps are shown in the six columns. C denotes the set of all messages.

An upper bound for the total exchange is obtained by mapping a ring network into the complete binary tree. The mapping of a ring network with p nodes into a complete binary tree is obtained by enumerating the tree nodes in a depth-first order from left to right starting from the root node of the tree. The nodes in this ring enumeration use all tree nodes exactly once in each direction. A total exchange can be performed in p^2 steps on a ring network, see Subsection 4.4.1.4. Thus, the total exchange can also be performed in p^2 steps on a complete binary tree.

4.5 Analysis of Parallel Execution Times

The time needed for the parallel execution of a parallel program depends on:

- the size of the input data n, and possibly further characteristics such as the number of iterations of an algorithm or the loop bounds,
- the number of processors p,

- the communication parameters, which describe the specifics of the communication of a parallel system or a communication library.

For a specific parallel program, the time needed for the parallel execution can be described as a function $T(p,n)$ depending on p and n. This function can be used to analyze the parallel execution time and its behavior depending on p and n. As an example, we consider the parallel implementations of a scalar product and of a matrix-vector product, presented in Sect. 3.7.

4.5.1 Parallel Scalar Product

The parallel scalar product of two vectors $\mathbf{a}, \mathbf{b} \in \mathbb{R}^n$ computes a scalar value which is the sum of the values $a_j \cdot b_j$, $j = 1, \ldots, n$. For a parallel computation on p processors, we assume that n is divisible by p with $n = r \cdot p$, $r \in \mathbb{N}$, and that the vectors are distributed in a blockwise way, see Section 3.5 for a description of data distributions. Processor P_k stores the elements a_j and b_j with $r \cdot (k-1) + 1 \leq j \leq r \cdot k$ and computes the partial scalar products

$$c_k = \sum_{j=r\cdot(k-1)+1}^{r\cdot k} a_j \cdot b_j \, ,$$

so that processor P_k stores value c_k. To get the final result $c = \sum_{k=1}^{p} c_k$, a single-accumulation operation is performed and one of the processors stores this value. The parallel execution time of the implementation depends on the computation time and the communication time. To build a function $T(p,n)$, we assume that the execution of an arithmetic operation needs α time units and that sending a floating point value to a neighboring processor in the interconnection network needs β time units. The parallel computation time for the partial scalar product is $2r\alpha$, since about r addition operations and r multiplication operations are performed.

The time for a single-accumulation operation depends on the specific interconnection network and we consider the linear array and the hypercube as examples. See also Sect. 2.7.2 for the definition of these direct networks.

4.5.1.1 Linear array

In the linear array, the optimal processor as root node for the single-accumulation operation is the node in the middle since it has a distance no more than $p/2$ from every other node. Each node gets a value from its left (or right) neighbor in time β, adds the value to the local value in time α, and sends the results to its right (or left) in the next step. This results in the communication time $\frac{p}{2}(\alpha + \beta)$. In total, the parallel execution time is

$$T(p,n) = 2\frac{n}{p}\alpha + \frac{p}{2}(\alpha+\beta). \tag{4.28}$$

The function $T(p,n)$ shows that the computation time decreases with increasing number of processors p but that the communication time increases with increasing number of processors. Thus, this function exhibits the typical situation in a parallel program that an increasing number of processors does not necessarily lead to faster programs since the communication overhead increases. Usually, the parallel execution time decreases for increasing p until the influence of the communication overhead is too large and then the parallel execution time increases again. The value for p at which the parallel execution time starts to increase again is the optimal value for p, since more processors do not lead to a faster parallel program

For the Function (4.28), we determine the optimal value of p which minimizes the parallel execution time for $T(p) \equiv T(p,n)$ using the derivatives of this function. The first derivative is

$$T'(p) = -\frac{2n\alpha}{p^2} + \frac{\alpha+\beta}{2} \,,$$

when considering $T(p)$ as a function of real values. For T'(p) = 0, we get $p^* = \pm\sqrt{\frac{4n\alpha}{\alpha+\beta}}$. The second derivative is $T''(p) = \frac{4n\alpha}{p^3}$ and $T''(p^*) > 0$, meaning that $T(p)$ has a minimum at p^*. From the formula for p^*, we see that the optimal number of processors increases with $\sqrt{n}$. We also see that $p^* = 2\sqrt{\frac{\alpha}{\alpha+\beta}}\sqrt{n} < 1$, if $\beta > (4n-1)\alpha$, so that the sequential program should be used in this case.

4.5.1.2 Hypercube

For the d-dimensional hypercube with $d = \log p$, the single-accumulation operation can be performed in $\log p$ time steps using a spanning tree, see Section 4.4.1. Again, each step for sending a data value to a neighboring node and the local addition takes time $\alpha+\beta$ so that the communication time $\log p(\alpha+\beta)$ results. In total, the parallel execution time is

$$T(n,p) = \frac{2n\alpha}{p} + \log p \cdot (\alpha+\beta) \,. \tag{4.29}$$

This function shows a slightly different behavior of the overhead than Function (4.28). The communication overhead increases with the factor $\log p$. The optimal number of processors is again determined by using the derivatives of $T(p) \equiv T(n,p)$. The first derivative (using $\log p = \ln p/\ln 2$ with the natural logarithm) is

$$T'(p) = -\frac{2n\alpha}{p^2} + (\alpha+\beta)\frac{1}{p}\frac{1}{\ln 2} \,.$$

For $T'(p) = 0$, we get the necessary condition $p^* = \frac{2n\alpha\ln 2}{\alpha+\beta}$. Since $T''(p) = \frac{4n\alpha}{p^3} -$ $\frac{1}{p^2}\frac{\alpha+\beta}{\ln 2} > 0$ for p^*, the function $T(p)$ has a minimum at p^*. This shows that the optimal number of processors increases with increasing n. This is faster than for the linear array and is caused by the faster implementation of the single-accumulation operation.

4.5.2 Parallel Matrix-vector Product

The parallel implementation of the matrix-vector product $\mathbf{A} \cdot \mathbf{b} = \mathbf{c}$ with $\mathbf{A} \in \mathbb{R}^{n \times n}$ and $\mathbf{b} \in \mathbb{R}^n$ can be performed with a row-oriented distribution of the matrix A or with a column-oriented distribution of matrix A, see Section 3.7. For deriving a function describing the parallel execution time, we assume that n is a multiple of the number of processors p with $r = \frac{n}{p}$ and that an arithmetic operation needs α time units.

- For an implementation using a row-oriented distribution of blocks of rows, processor P_k stores the rows i with $r \cdot (k-1) + 1 \leq i \leq r \cdot k$ of matrix $\mathbf{A}$ and computes the elements

$$c_i = \sum_{j=1}^{n} a_{ij} \cdot b_j$$

  of the result vector c. For each of these r values, the computation needs n multiplication and $n - 1$ addition operations so that approximately the computation time $2nr\alpha$ is needed. The vector $\mathbf{b}$ is replicated for this computation. If the result vector $\mathbf{c}$ has to be replicated as well, a multi-broadcast operation is performed, for which each processor P_k, $k = 1, \ldots, p$, provides $r = \frac{n}{p}$ elements.
- For an implementation with column-oriented distribution of blocks of columns, processor P_k stores the columns j with $r \cdot (k-1) + 1 \leq j \leq r \cdot k$ of matrix $\mathbf{A}$ as well as the corresponding elements of $\mathbf{b}$ and computes a partial linear combination, i.e. P_k computes n partial sums $d_{k1}, \ldots, d_{kn}$ with

$$d_{kj} = \sum_{l=r\cdot(k-1)+1}^{r\cdot k} a_{jl} b_l \; .$$

  The computation of each d_{kj} needs r multiplications and $r - 1$ additions so that for all n values the approximate computation time $n2r\alpha$ results. A final multi-accumulation operation with addition as reduction operation computes the final result $\mathbf{c}$. Each processor P_k adds the values $d_{1j}, \ldots, d_{nj}$ for $(k-1) \cdot r + 1 \leq j \leq k \cdot r$, i.e. P_k performs an accumulation with blocks of size r and vector $\mathbf{c}$ results in a blockwise distribution.

Thus, both implementation variants have the same execution time $2\frac{n^2}{p}\alpha$. Also, the communication time is asymptotically identical, since multi-broadcast and multi-accumulation are dual operations, see Section 3.6. For determining a function for the communication time, we assume that sending r floating-point values to a neighboring processor in the interconnection network needs $\beta + r \cdot \gamma$ time units and consider the two networks, a linear array and a hypercube.

4.5.2.1 Linear array

In the linear array with p processors, a multi-broadcast operation (or a multi-accumulation) operation can be performed in p steps in each of which messages of size r are sent. This leads to a communication time $p(\beta + r \cdot \gamma)$. Since the message size in this example is $r = \frac{n}{p}$, the following parallel execution time results

$$T(n,p) = \frac{2n^2}{p}\alpha + p \cdot (\beta + \frac{n}{p} \cdot \gamma) = \frac{2n^2}{p}\alpha + p \cdot \beta + n \cdot \gamma \,.$$

This function shows that the computation time decreases with increasing p but the communication time increases linearly with increasing p, which is similar as for the scalar product. But in contrast to the scalar product, the computation time increases quadratically with the system size n, whereas the communication time increases only linearly with the system size n. Thus, the relative communication overhead is smaller. Still, for a fixed number n, only a limited number of processors p leads to an increasing speedup.

To determine the optimal number p^* of processors, we again consider the derivatives of $T(p) \equiv T(n,p)$. The first derivative is

$$T'(p) = -\frac{2n^2\alpha}{p^2} + \beta \,,$$

for which $T'(p) = 0$ leads to $p^* = \sqrt{2\alpha n^2/\beta} = n \cdot \sqrt{2\alpha/\beta}$. Since $T''(p) = 4\alpha n^2/p^3$, we get $T''(n\sqrt{2\alpha/\beta}) > 0$ so that p^* is a minimum of $T(p)$. This shows that the optimal number of processors increases linearly with n.

4.5.2.2 Hypercube

In a $\log p$-dimensional hypercube, a multi-broadcast (or a multi-accumulation) operation needs $p/\log p$ steps, see Section 4.4, with $\beta + r \cdot \gamma$ time units in each step. This leads to a parallel execution time:

$$\begin{aligned}
T(n,p) &= \frac{2\alpha n^2}{p} + \frac{p}{\log p}(\beta + r \cdot \gamma) \\
&= \frac{2\alpha n^2}{p} + \frac{p}{\log p} \cdot \beta + \frac{\gamma n}{\log p} \,.
\end{aligned}$$

The first derivative of $T(p) \equiv T(n,p)$ is:

$$T'(p) = -\frac{2\alpha n^2}{p^2} + \frac{\beta}{\log p} - \frac{\beta}{\log^2 p \ln 2} - \frac{\gamma n}{p \cdot \log^2 p \ln 2} \ .$$

For $T'(p) = 0$ the equation

$$-2\alpha n^2 \log^2 p + \beta p^2 \log p - \beta p^2 \frac{1}{\ln 2} - \gamma n p \frac{1}{\ln 2} = 0$$

needs to be fulfilled. This equation cannot be solved analytically, so that the number of optimal processors p^* cannot be expressed in closed form. This is a typical situation for the analysis of functions for the parallel execution time, and approximations are used. In this specific case, the function for the linear array can be used since the hypercube can be embedded into a linear array. This means that the matrix-vector product on a hypercube is at least as fast as on the linear array.

4.6 Parallel Computational Models

A computational model of a computer system describes at an abstract level which basic operations can be performed when the corresponding actions take effect and how data elements can be accessed and stored [14]. This abstract description does not consider details of a hardware realization or a supporting runtime system. A computational model can be used to evaluate algorithms independently of an implementation in a specific programming language and of the use of a specific computer system. To be useful, a computational model must abstract from many details of a specific computer system while on the other hand it should capture those characteristics of a broad class of computer systems which have a larger influence on the execution time of algorithms.

To evaluate a specific algorithm in a computational model, its execution according to the computational model is considered and analyzed concerning a specific aspect of interest. This could, for example, be the number of operations that must be performed as a measure for the resulting execution time, or the number of data elements that must be stored as a measure for the memory consumption, both in relation to the size of the input data. In the following, we give a short overview of popular parallel computational models, including the PRAM model, the BSP model, and the LogP model. More information on computational models can be found in [197].

4.6.1 PRAM Model

The theoretical analysis of sequential algorithms is often based on the RAM (Random Access Machine) model, which captures the essential features of traditional sequential computers. The RAM model consists of a single processor and a memory with sufficient capacity. Each memory location can be accessed in a random (direct) way. In each time step, the processor performs one instruction as specified by a sequential algorithm. Instruction for (read or write) access to the memory as well as for arithmetic or logical operations are provided. Thus, the RAM model provides a simple model which abstracts from many details of real computers, like a fixed memory size, existence of a memory hierarchy with caches, complex addressing modes, or multiple functional units. Nevertheless, the RAM model can be used to perform a runtime analysis of sequential algorithms to describe their asymptotic behavior, which is also meaningful for real sequential computers.

The RAM model has been extended to the PRAM (Parallel Random Access Machine) model to analyze parallel algorithms [65, 121, 155]. A PRAM consists of a bounded set of identical processors $\{P_1, \ldots, P_n\}$, which are controlled by a global clock. Each processor is a RAM and can access the common memory to read and write data. All processors execute the same program synchronously. Besides the common memory of unbounded size, there is a local memory for each processor to store private data. Each processor can access any location in the common memory in unit time, which is the same time as needed for an arithmetic operation. The PRAM executes computation steps one after another. In each step, each processor (a) reads data from the common memory or its private memory (read phase), (b) performs a local computation, and (c) writes a result back into the common memory or into its private memory (write phase). It is important to note that there is no direct connection between the processors. Instead, communication can only be performed via the common memory.

Since each processor can access any location in the common memory, memory access conflicts can occur when multiple processors access the same memory location at the same time. Such conflicts can occur both in the read phase and the write phase of a computation step. Depending on how these read conflicts and write conflicts are handled, several variants of the PRAM model are distinguished. The **EREW** (*exclusive read, exclusive write*) PRAM model forbids simultaneous read accesses as well as simultaneous write accesses to the same memory location by more than one processor. Thus, in each step, each processor must read from and write into a different memory location as the other processors. The **CREW** (*concurrent read, exclusive write*) PRAM model allows simultaneous read accesses by multiple processors to the same memory location in the same step, but simultaneous write accesses are forbidden within the same step. The **ERCW** (*exclusive read, concurrent write*) PRAM model allows simultaneous write accesses, but forbids simultaneous read accesses within the same step. The **CRCW** (*concurrent read, concurrent write*) PRAM model allows both simultaneous read and write accesses within the same step. If simultaneous write accesses are allowed, write conflicts to the same memory location must be resolved to determine what happens if multiple processors try to write to the same memory location in the same step. Different resolution schemes have been proposed:

(1) The *common model* requires that all processors writing simultaneously to a common location write the same value.

(2) The *arbitrary model* allows an arbitrary value to be written by each processor; if multiple processors simultaneously write to the same location, an arbitrarily chosen value will succeed.
(3) The *combining model* assumes that the values written simultaneously to the same memory location in the same step are combined by summing them up and the combined value is written.
(4) The *priority model* assigns priorities to the processors and in the case of simultaneous writes the processor with the highest priority succeeds.

In the PRAM model, the cost of an algorithm is defined as the number of PRAM steps to be performed for the execution of an algorithm. As described above, each step consists of a read phase, a local computation, and a write phase. Usually, the costs are specified as asymptotic execution time with respect to the size of the input data. The theoretical PRAM model has been used as a concept to build the SB-PRAM as a real parallel machine which behaves like the PRAM model [2, 126]. This machine is an example for simultaneous multi-threading, since the unit memory access time has been reached by introducing logical processors which are simulated in a round-robin fashion and, thus, hide the memory latency.

A useful class of operations for PRAM models or PRAM-like machines is the multi-prefix operations which can be defined for different basic operations. We consider an MPADD operation as example. This operation works on a variable s in the common memory. The variable s is initialized with the value o. Each of the processors P_i, $i = 1, \ldots, n$, participating in the operation provides a value o_i. The operation is synchronously executed and has the effect that processor P_j obtains the value

$$o + \sum_{i=1}^{j-1} o_i$$

After the operation, the variable s has the value $o + \sum_{i=1}^{n} o_i$. Multi-prefix operations can be used for the implementation of synchronization operations and parallel data structures that can be accessed by multiple processors simultaneously without causing race conditions [93]. For an efficient implementation, hardware support or even a hardware implementation for multi-prefix operations is useful as it has been provided by the SB-PRAM prototype [2]. Multi-prefix operations are also useful for the implementation of a parallel task pool providing a dynamic load balancing for application programs with an irregular computational behavior, see [93, 127, 176, 185]. An example for such an application is the Cholesky factorization for sparse matrices for which the computational behavior depends on the sparsity structure of the matrix to be factorized. Section 8.5 gives a detailed description of this application. The implementation of task pools in Pthreads is considered in Section 6.2.1.

A theoretical runtime analysis based on the PRAM model provides useful information on the asymptotic behavior of parallel algorithms. But the PRAM model has its limitations concerning a realistic performance estimation of application programs on real parallel machines. One of the main reasons for these limitations is the assumption that each processor can access any location in the common memory in unit time. Real parallel machines do not provide memory access in unit time. Instead, large variations in memory access time often occur, and accesses to a global memory or to the local memory of other processors are usually much slower than accesses to the local memory of the accessing processor. Moreover, real parallel machines use a memory hierarchy with several levels of caches with different access times. This cannot be modeled with the PRAM model. Therefore, the PRAM model cannot be used to evaluate the locality behavior of the memory accesses of a parallel application program. Other unrealistic assumptions of the PRAM model are the synchronous execution of the processors and the absence of collisions when multiple processors access the common memory simultaneously. Because of these structures, several extensions of the original PRAM model have been proposed. The missing synchronicity of instruction execution in real parallel machines is addressed in the *phase PRAM* model [80], in which the computations are partitioned into phases such that the processors work asynchronously within the phases. At the end of each phase, a barrier synchronization is performed. The *delay PRAM* model [169] tries to model delays in memory access times by introducing a communication delay between the time at which a data element is produced by a processor and the time at which another processor can use this data element. A similar approach is used for the *local memory PRAM* and the *block PRAM* model [4, 5]. For the block PRAM, each access to the common memory takes time $l + b$, where l is a startup time and b is the size of the memory block addressed. A more detailed description of PRAM models can be found in [33].

4.6.2 BSP Model

None of the PRAM models proposed has really been able to capture the behavior of real parallel machines for a large class of application areas in a satisfactory way. One of the reasons is that there is a large variety of different architectures for parallel machines and the architectures are steadily evolving. To avoid that the computational model design constantly drags behind the development of parallel computer architecture, the BSP model (*bulk synchronously parallel*) has been proposed as a bridging model between hardware architecture and software development [218]. The idea is to provide a standard on which both hardware architects and software developers can agree. Thus, software development can be decoupled from the details of a specific architecture, and software does not have to be adapted when porting it to a new parallel machine.

The BSP model is an abstraction of a parallel machine with a physically distributed memory organization. Communication between the processors is not performed as separate point-to-point transfers, but is bundled in a step-oriented way. In the BSP model, a parallel computer consists of a number of *components* (processors), each of which can perform processing or memory functions. The components are connected by a *router* (interconnection network) which can send point-to-point messages between pairs of components. There is also a *synchronization unit*, which supports the synchronization of all or a subset of the components. A computation

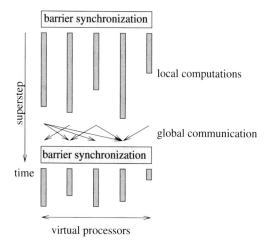

Fig. 4.17 In the BSP model, computations are performed in supersteps where each superstep consists of three phases: (1) simultaneous local computations of each processor, (2) communication operations for data exchange between processors, and (3) a barrier synchronization to terminate the communication operations and to make the data sent visible to the receiving processors. The communication pattern shown for the communication phase represents an h-relation with h= 3.

in the BSP model consists of a sequence of *supersteps*, see Figure 4.17 for an illustration. In each superstep, each component performs local computations and can participate in point-to-point message transmissions. A local computation can be performed in one time unit. The effect of message transmissions becomes visible in the next time step, i.e., a receiver of a message can use the received data not before the next superstep. At the end of each superstep, a **barrier synchronization** is performed. There is a *periodicity parameter L* which determines the length of the supersteps in time units. Thus, L determines the granularity of the computations. The BSP model allows that the value of L can be controlled by the program to be executed, even at runtime. There may be a lower bound for L given by the hardware. The parallel program to be executed should set an upper bound for L such that in each superstep, computations with approximately L steps can be assigned to each processor.

In each superstep, the router can implement arbitrary *h-relations* capturing communication patterns, where each processor sends or receives at most h messages. A computation in the BSP model can be characterized by four parameters [110]:

- p: the number of (virtual) processors used within the supersteps to perform computations;
- s: the execution speed of the processors expressed as the number of computation steps per second that each processor can perform, where each computation step performs an (arithmetic or logical) operation on a local data element;
- l: the number of steps that are required for the execution of a barrier synchronization;

- g: the number of steps that are required on the average for the transfer of a memory word in the context of an h-relation.

The parameter g is determined such that the execution of an h-relation with m words per message takes $l \cdot m \cdot g$ steps. For a real parallel computer, the value of g depends not only on the bisection bandwidth of the interconnection network, see page 39, but also on the communication protocol used and on the implementation of the communication library. The value of l is influenced by the diameter of the interconnection network, but also by the implementation of the communication library. Both l and g can be determined by suitable benchmark programs. Only p, l and g are independent parameters; the value of s is used for the normalization of the values of l and g.

The execution time of a BSP program is specified as the sum of the execution times of the supersteps which are performed for executing the program. The execution time $T_{superstep}$ of a single superstep consists of three terms: (1) the maximum of the execution time w_i for performing local computations of processor P_i, (2) the time for global communication for the implementation of an h-relation, and (3) the time for the barrier synchronization at the end of each superstep. This results in

$$T_{superstep} = \max_{processors} w_i + h \cdot g + l$$

The BSP model is a general model that can be used as a basis for different programming models. To support the development of efficient parallel programs with the BSP model, the BSPLib library has been developed [88, 110], which provides operations for the initialization of a superstep, for performing communication operations, and for participating in the barrier synchronization at the end of each superstep.

The BSP model has been extended to the Multi-BSP model, which extends the original BSP model to capture important characteristics of modern architectures, in particular multicore architectures [219]. In particular, the model is extended to a hierarchical model with an arbitrary number d of levels modeling multiple memory and cache levels. Moreover, at each level the memory size is incorporated as an additional parameter. The entire model is based on a tree of depth d with memory/caches at the internal nodes and processors at the leaves.

4.6.3 LogP Model

In [44], several concerns about the BSP model are formulated. First, the length of the supersteps must be sufficiently large to accommodate arbitrary h-relations. This has the effect that the granularity cannot be decreased below a certain value. Second, messages sent within a superstep can only be used in the next superstep, even if the interconnection network is fast enough to deliver messages within the same superstep. Third, the BSP model expects hardware support for synchronization at the end of each superstep. Such support may not be available for some parallel machines.

Because of these concerns, the BSP model has been extended to the *LogP model* to provide a more realistic modeling of real parallel machines.

Similar to the BSP model, the LogP model is based on the assumption that a parallel computer consists of a set of processors with local memory that can communicate by exchanging point-to-point messages over an interconnection network. Thus, the LogP model is also intended for the modeling of parallel computers with a distributed memory. The communication behavior of a parallel computer is described by four parameters:

- L (latency) is an upper bound on the latency of the network capturing the delay observed when transmitting a small message over the network;
- o (overhead) is the management overhead that a processor needs for sending or receiving a message; during this time, a processor cannot perform any other operation;
- g (gap) is the minimum time interval between consecutive send or receive operations of a processor;
- P (processors) is the number of processors of the parallel machine.

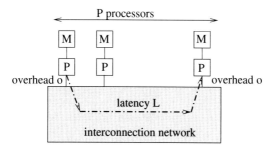

Fig. 4.18 Illustration of the parameters of the LogP model.

Figure 4.18 illustrates the meaning of these parameters [43]. All parameters except P are measured in time units or as multiples of the machine cycle time. Furthermore it is assumed that the network has a *finite capacity* which means that between any pair of processors at most $\lceil L/g \rceil$ messages are allowed to be in transmission at any time. If a processor tries to send a message that would exceed this limit, it is blocked until the message can be transmitted without exceeding the limit. The LogP model assumes that the processors exchange *small messages* that do not exceed a predefined size. Larger messages must be split into several smaller messages. The processors work *asynchronously* with each other. The latency of any single message cannot be predicted in advance, but is bounded by L if there is no blocking because of the finite capacity. This includes that messages do not necessarily arrive in the same order in which they have been sent. The values of the parameters L, o, and g depend not only on the hardware characteristics of the network, but also on the communication library and its implementation.

The execution time of an algorithm in the LogP model is determined by the maximum of the execution times of the participating processors. An access by a processor P_1 to a data element that is stored in the local memory of another processor P_2 takes time $2 \cdot L + 4 \cdot o$; half of this time is needed to bring the data element from P_2 to P_1, the other half is needed to bring the data element from P_1 back to P_2. A sequence of n messages can be transmitted in time $L + 2 \cdot o + (n-1) \cdot g$, see Figure 4.19.

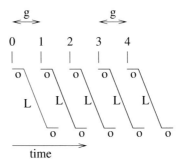

Fig. 4.19 Transmission of a larger message as a sequence of n smaller messages in the LogP model. The transmission of the last smaller message is started at time $(n-1) \cdot g$ and reaches its destination $2 \cdot o + L$ time units later.

A drawback of the original LogP model is that it is based on the assumption that the messages are small and that only point-to-point messages are allowed. More complex communication patterns must be assembled from point-to-point messages. To release the restriction to small messages, the LogP model has been extended to the LogGP model [10] which contains an additional parameter G (*Gap per byte*). This parameter specifies the transmission time per byte for long messages. $1/G$ is the bandwidth available per processor. The time for the transmission of a message with n bytes takes time $o + (n-1)G + L + o$ see Figure 4.20.

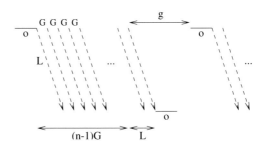

Fig. 4.20 Illustration of the transmission of a message with n bytes in the LogGP model. The transmission of the last byte of the message is started at time $o + (n-1) \cdot G$ and reaches its destination $o + L$ time units later. Between the transmission of the last byte of a message and the start of the transmission of the next message at least g time units must have been elapsed.

The LogGP model has been successfully used to analyze the performance of message-passing programs [9, 129]. The LogGP model has been further extended to the LogGPS model [117] by adding a parameter S to capture synchronization that must be performed when sending large messages. The parameter S is the threshold for the message length above which a synchronization between sender and receiver is performed before message transmission starts.

4.7 Loop Scheduling and Loop Tiling

Program loops are a rich source for parallelization, which can be exploited either by hand or by a compiler. The goal of a parallel execution of program loops is a low execution time, which can be achieved by assigning loop iterations to processors or cores such that a good load balancing is obtained, or by restructuring loops to improve data locality. An efficient execution of the internal loops can significantly reduce the execution times of the entire parallel program. A typical way to assess and to analyze the efficiency of the resulting program is to measure the resulting execution time. In this section, different loop scheduling and loop tiling or loop blocking techniques are introduced.

Loop blocking, also called strip mining, is a loop transformation that takes a single loop, cuts it into chunks of about equal size, and creates a doubly nested loop with an outer loop over the chunks. The goal is to improve data locality for caches or local memories, see Sect. 2.9, to facilitate vectorization, or to enable coarse-grain parallelism. **Loop tiling** is the corresponding transformation method for nested loops for which the partitioning of the iteration space results in so-called tiles. The potential for loop blocking or loop tiling is determined by the dependence structure of the loop iterations, see Sect. 3.3.1 for the definitions of data dependencies. If there are no dependencies between loop iterations, such as in parallel loop nests, any loop tiling is possible. Otherwise a dependence analysis is required so that tiling preserves the correct program behavior. Section 4.7.2 summarizes basic results for loop tiling according to [223] and [177]. An example for tiling is also given in Sect. 7.5, which presents GPU programming with CUDA.

A more fine-grained parallelism of loops is exploited by **loop scheduling**. Loop scheduling denotes the scheduling problem which assigns iterations of a loop, also called *iterates* in the following, to processors or cores with the goal to achieve a good load balance. This scheduling method is applied to parallel loops which are loops in which the iterates are independent of each other, see Sect. 3.3.3. Loop transformation may also play a role in loop scheduling when nested parallel loops are transformed into a single loop with the goal to treat all iterates in the same load balancing process. Due to the fine-grained parallelism, loop scheduling is typically applied for shared address space. Programming environments, such as OpenMP, see Sect. 6.5, provide different types of loop scheduling. Section 4.7.1 presents loop scheduling according to [177].

4.7.1 Loop Scheduling

Approaches to program **scheduling** deal with the assignment of computations or tasks of an application program to the compute resources of a parallel machine. The goal is to keep all available compute resources as busy as possible and to keep the overhead minimal. This should result in a good load balance and a small overall execution time. For the decision how to assign a computation or a task to a processor or core, a large variety of scheduling algorithms have been invented and studied.

A main distinction between scheduling algorithms is whether they are static or dynamic. **Static scheduling** is done before program execution and its decisions are solely based on static or global program information. Since scheduling has the goal to achieve a load balanced execution, some knowledge about the expected execution time is needed, which may be obtained by a prediction. If exact execution times are not available and execution times have to be predicted, the result of a static scheduling may not be optimal. However, static scheduling has the advantage of a low runtime overhead caused by the scheduling algorithm.

Dynamic scheduling approaches make scheduling decisions at runtime and can include dynamic or local information about the state of the program executed. Thus, more reliable information about the execution time of single computations or tasks and also about the current utilization of processors or cores is available. This can result in a better load balance and also a smaller program execution time. However, the runtime overhead caused by the scheduling algorithm and the assignment of tasks might be higher and, thus, might reduce the advantage over static scheduling.

A specific scheduling technique is loop scheduling exploiting loop level parallelism by assigning loop iterates to processing units. Loop scheduling is usually applied to **parallel loops**, i.e., those loops whose iterates are independent of each other. Also shared memory systems and thread-level parallelism are typically addressed, since the data for such a fine-grained parallelism have to be available and message-passing would be too expensive. An example is the programming library OpenMP, see Sect. 6.5, which supports the programming of a shared address space and provides the concept of parallel loops.

4.7.1.1 Static loop scheduling

As for all scheduling algorithms, there exists the distinction between static loop scheduling and dynamic loop scheduling. **Static loop scheduling** approaches determine how the iterates should be assigned to processors or cores before the execution of the program starts. Static approaches are reasonable in the case that all iterates of the parallel loop are expected to have about the same execution time. A parallel loop of the form is considered

```
for (i=1; i <= N;i ++) { //sequential specification
    DoWork(i); // same execution time T for all i
}
```

where the iterate DoWork(i) is independent from iterate DoWork(j) for $i \neq j$. The iterates DoWork(i), $i = 1,...n$, are assumed all have the same execution time. In this case, an exact prediction of the execution time is not required. For a parallel system with p processing units and a parallel loop of size N, each processor should execute about the same number of iterates, i.e. about $\lceil N/p \rceil$ iterates, to achieve a good load balance. The assignment can be done in a cyclic way in $\lceil N/p \rceil$ rounds or in a blockwise way so that each processor gets a consecutive block of $\lceil N/p \rceil$ iterates. A cyclic assignment results in the following code with $R = \lceil N/p \rceil$ rounds:

```
for (r=1; r <= R; r++) { // sequential
    forall (j=1; j <= p; j++) { // parallel
        processor P_j computes DoWork((r-1)p+j);
}}
```

The outer loop is a sequential loop which runs over the rounds. The inner forall loop specifies a parallel computation in which each processor P_j computes exactly one iterate DoWork((r-1)p+j) selected according to its processor number j. Thus, in each round the processors compute different iterates. In a programming environment with an SPMD programming model, the loop can be expressed as:

```
for (r=1; r <= R; r++) { // sequential
    DoWork((r-1)p+my_rank); // SPMD parallel
}
```

where my_rank denotes the rank of the processor, so that each processor executes a different iterate of round r in parallel with the other processors.

4.7.1.2 Dynamic loop scheduling

Dynamic loop scheduling approaches, also called self-scheduling, perform the decision about the assignment of loop iterates to processing units at runtime based on the utilization of processing units and the number of available iterates not yet executed. When a processing unit has finished its current work and becomes idle, a new iterate that has not yet been executed is assigned. In a shared address spaced, it has to be ensured that the access to iterates is synchronized so that a multiple assignment of the same loop iterate is avoided. Dynamic loop scheduling is advantageous, if the loop iterates are known to have different and varying execution times so that a static loop scheduling would very likely result in an unbalanced execution. However, dynamic loop scheduling has the disadvantage that N dispatch operations are required for accessing the loop iterate information, causing a so-called dispatch overhead, which includes the time for all executions and memory accesses related to the dispatch of a single loop iterate.

Several versions for dynamic loop scheduling have been introduced and the most common ones are self-scheduling (one iterate at a time), chunk scheduling, or guided self-scheduling, which are described in the following in more detail together with an investigation of the expected execution time. Loop scheduling algorithms typically consider a flat structure of independent iterates so that nested parallel loops are transformed into a single loop before the scheduling (which is possible due to the independence of the iterates). The transformation of a nested loop into a single loop (loop coalescing) is covered in Sect. 4.7.1.3.

Self-scheduling is the basic form of dynamic loop scheduling in which single iterates are assigned to processing units. Starting with a parallel loop of the form

```
for (i=1; i <= N; i ++) { //sequential specification
    DoWork(i); // different execution times T(i)
}
```

an idle processor or core picks a single iterate and executes it. This is expressed in the following SPMD code executed by each participating processor:

```
next = request_iteration() //synchronisation
while (next != EMPTY) {
    DoWork(next);
    next = request_iteration(); //synchronisation
}
```

where `request_iteration()` dynamically returns the index of the next iteration to be executed. Figure 4.21 shows two examples, each assuming different execution times for the iterates, and illustrates possible resulting mappings of iterates to processors.

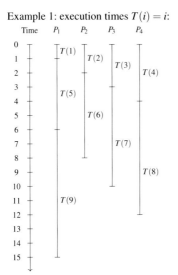

Example 1: execution times $T(i) = i$:

Example 2: execution times

$T(1) = 1; T(2) = 3; T(3) = 5;$
$T(4) = 1; T(5) = 6; T(6) = 3;$
$T(7) = 1; T(8) = 2; T(9) = 3;$

Fig. 4.21 Self-scheduling: Example using four processors $P_1, \ldots, P_4$ and loop iterates $1, \ldots, 9$ with execution time $T(1), \ldots, T(9)$. The left diagram shows the resulting mapping of iterates to processors for $T(i) = i$, the right diagram shows the resulting mapping for the execution times shown above the diagram.

The expected execution time for a loop with N iterates executed by p processors can be approximated by the formula $N/p \cdot (B + \sigma)$, where B denotes the average execution time of an iterate and σ denotes the dispatch overhead for one iterate. Self-scheduling is useful if B is much larger than σ and if the execution times of the iterates vary a lot. In cases with a large dispatch overhead the assignment of more than one iterate at a time may be more appropriate. This is the strategy of the next scheduling method, called chunk scheduling.

Chunk scheduling assigns a fixed number k of iterates to an idle processor or core, in contrast to only one iterate. The k iterates assigned to the same processor are called **chunk**. Again, a parallel loop of the form

```
for (i=1; i <= N; i ++) { //sequential specification
    DoWork(i); // each iterate has execution time B
}
```

is considered. The following parallel program implementing chunk scheduling contains three loops and is similar to basic self scheduling with the difference that there is an additional innermost loop for the execution of one chunk.

```
for (r=1; r <= N/(k*p); r++) { // sequential loop over rounds
    forall (j=1; j <= p; j++) { // parallel execution
        for (i=1; i <= k; i++) { // sequential loop over chunks
            processor P_j computes DoWork((r-1)*p*k + (j-1)*k + i);
}}}
```

The outer loop organizes the chunk execution in $\lceil N/(k \cdot p) \rceil$ rounds, where each processor executes one chunk in every round. The middle loop expresses the parallel execution of p processors specified by a forall loop over the processors P_j, $j = 1 \ldots p$. Each processor executes the inner loop, which is a sequential execution of the iterates of one specific chunk. In an SPMD programming style, the middle forall loop is replaced by an SPMD computation using the processor rank my_rank of the processor P_{my_rank}.

```
for (r=1; r <= N/(k*p); r++) { // sequential loop over rounds
    for (i=1; i <= k; i++) { // SPMD parallel
        DoWork((r-1)*p*k + (my_rank-1)*k + i);
}}
```

Chunk scheduling may lead to a less balanced execution caused by the coarser granularity but results in a smaller dispatch overhead compared to basic self-scheduling. Special cases are the chunk size $k = 1$ which corresponds to self-scheduling and the chunk size $k = \lceil N/p \rceil$ which corresponds to static scheduling. In general, the chunk size is between 1 and $\lceil N/p \rceil$, however the optimal chunk size (leading to the smallest execution time) is unknown, since the execution time B of the iterates is usually unknown.

The optimal chunk size leading to a minimum execution time can be calculated by a runtime modeling, in which it is assumed that N is a multiple of k. The overall parallel execution time achieved by chunk scheduling with chunk size k can be estimated by the following approximation for p processing units and a loop of N iterates, see [177]. The parallel execution time $T_p(k)$ is approximated by

$$T_p(k) = \frac{N}{k \cdot p}(k \cdot B + \sigma) = \frac{N \cdot B}{p} + \frac{N \cdot \sigma}{k \cdot p}$$

since N/k chunks are executed in parallel by p processors and each chunk has execution time $k \cdot B + \sigma$ where B denotes the execution of one iterate and σ denotes the dispatch overhead, now for one chunk. The first derivative of $T_p(k)$

$$T'_p(k) = -\frac{N\sigma}{k^2 \cdot p}$$

has no zero and, thus, even in this simple case, no optimal value for k can be calculated. A minor modification of the approximation to

$$\overline{T}_p(k) = \left(\frac{N}{k \cdot p} + 1\right)(k \cdot B + \sigma) = \frac{N \cdot B}{p} + k \cdot B + \frac{N \cdot \sigma}{k \cdot p} + \sigma$$

can be used. This has the advantage that the first derivative

$$\overline{T}'_p(k) = B - \frac{N \cdot \sigma}{k^2 \cdot p}$$

has a zero at $k = \sqrt{N \cdot \sigma / (p \cdot B)}$ which is the global minimum. However, since the value for B is unknown, even this value for k may not help to find the minimum of the actual loop execution, and chunk scheduling may lead to a slowdown (which means that the parallel execution time is greater than the sequential execution time). This leads to the related question of how large the chunk size k should be to avoid a slowdown, i.e., what is the minimum value for chunk size k for which $S_p(k) > 1$, where $S_p(k)$ denotes the speedup when chunk size k is used. More precisely, the value for k is to be determined, for which

$$S_p(k) = \frac{N \cdot B}{\frac{N}{k \cdot p}(k \cdot B + \sigma)} > 1.$$

This results in the following lower bound for k:

$$k > \frac{\sigma}{B \cdot (p - 1)}$$

for $p > 1$. The formula reflects that k can decrease for an increasing number of processors.

Guided self-scheduling is a scheduling scheme that dynamically assigns chunks of iterates to processors. However, this is done in a more flexible way starting with a maximal chunk size $k = \lceil N/p \rceil$ and then decreasing the chunk size continuously until all iterates have been assigned. Guided self-scheduling with bound k_b, abbreviated as GSS(k_b), denotes the scheduling scheme which allows the chunk size to decrease until the bound k_b is reached. For GSS(1), the smallest chunk size can be one.

The algorithm deciding on the number of iterates to be assigned to an idle processor assumes that each of the p processors starts the execution of some iterates of the loop at different times. This is reasonable, since the processors are executing other program parts preceding the loop and may not reach the loop synchronously at the same time. The first processor starting the loops gets a block of $k = \lceil N/p \rceil$ iterates. A next processor gets a chunk of size $k = \lceil R/p \rceil$, where R denotes the number of iterates not yet been executed. The value of R is updated in each assignment step, so that the iterates just assigned are not considered for the computation of the number of iterates to be assigned in the next step.

Guided self scheduling can be implemented in the following way. For p processors and R_i remaining iterates, the next processor that is idle gets $x_i = \left\lceil \frac{R_i}{p} \right\rceil$ iterates. The chunk size R_i is updated to $R_{i+1} = R_i - x_i$ to be used in the next step. The initialization is $R_1 = N$ where N is the total number of iterations. For a parallel loop with N iterations and p processors, the following algorithm computes the chunk sizes R_i:

$$
\begin{aligned}
&R_1 = N; \\
&\textbf{while } (R_i \neq 0) \textbf{ do } \{ \\
&\qquad \text{the next idle processor } P \text{ gets } x_i = \lceil R_i/p \rceil \text{ iterates}; \\
&\qquad R_{i+1} = R_i - x_i; \\
&\qquad \text{processor } P \text{ executes iterates } N - R_i + 1, \dots, N - R_i + x_i; \\
&\qquad i = i + 1; \\
&\}
\end{aligned}
$$

The algorithm is illustrated in Fig. 4.22 which shows the resulting chunk sizes for an example with $p = 5$ processors and $N = 100$ iterates.

4.7.1.3 Loop Coalescing

For loop scheduling, it is usually assumed that a singly nested parallel loop is scheduled. Thus before scheduling, a multiply nested loop is transformed into an equivalent singly nested loop executing exactly the same computations in the same order. The transformation restructuring multiply nested loops into a single parallel loop is called loop coalescing. For example, a nested loop of depth two with loop indices j and k and loop bound N in both loops is transformed into a single loop with loop index i and loop bound N^2 by choosing an appropriate index transformation. We first consider the following doubly nested loop accessing a two-dimensional array a in the loop body

$$
\begin{aligned}
&\textbf{forall } (j = 1; j \leq N; j++) \\
&\qquad \textbf{forall } (k = 1; k \leq N; k++) \\
&\qquad\qquad a[j][k] = (j-1) \cdot N + k;
\end{aligned} \tag{4.30}
$$

This loop is transformed into the coalesced loop

time	number of remaining iterations	next processor	number of iterations assigned
t_1	100	P_1	20
t_2	80	P_3	16
t_3	64	P_2	13
t_4	51	P_4	11
t_5	40	P_2	8
t_6	32	P_5	7
t_7	25	P_3	5
t_8	20	P_1	4
t_9	16	P_4	4
t_{10}	12	P_1	3
t_{11}	9	P_3	2
t_{12}	7	P_5	2
t_{13}	5	P_4	1
t_{14}	4	P_4	1
t_{15}	3	P_2	1
t_{16}	2	P_3	1
t_{17}	1	P_5	1

Fig. 4.22 Illustration of guided self-scheduling with selected chunk sizes.

$$\text{forall } (i = 1; i \leq N^2; i++) \qquad\qquad (4.31)$$
$$a\left[\lceil i/N \rceil\right]\left[i - N\lfloor (i-1)/N \rfloor\right] = i;$$

which differs in the index expressions used for the array access, since the array a is still stored in the original two-dimensional way in rowwise order, i.e., neighboring row elements are stored at consecutive memory locations. To access the correct array elements, the index expressions are transformed from $[j][k]$ to $[\lceil i/N \rceil][i - N\lfloor(i - 1)/N \rfloor]$. The mapping

$$i \mapsto (\lceil i/N \rceil, i - N\lfloor(i-1)/N \rfloor) \qquad\qquad (4.32)$$

always provides the correctly transformed index expressions. This mapping is specific for the rowwise ordering of the elements of the two-dimensional array a. Table 4.4 lists the indices j and k of the original loop and the corresponding index i of the transformed loop. In the example, we have assigned the value $(j - 1) \cdot N + k$ to $a[j][k]$, which corresponds to a numbering in a rowwise ordering of the array elements. The computation of this ordering also has to be transformed and is also shown in Table 4.4.

For a nested loop of depth three with loop indices j, k, and l and loop bound N in each sub-loop,

$$\text{forall } (j = 1; j \leq N; j++)$$
$$\quad \text{forall } (k = 1; k \leq N; k++) \qquad\qquad (4.33)$$
$$\quad\quad \text{forall } (l = 1; l \leq N; l++)$$
$$\quad\quad\quad a[j][k][l] = (j-1) \cdot N^2 + (k-1) \cdot N + l;$$

before transformation			after transformation		
j	k	$(j-1)\cdot N+k$	i	$\lceil i/N \rceil$	$i-N\cdot\lfloor \frac{(i-1)}{N} \rfloor$
1	1	1	1	1	1
1	2	2	2	1	2
$\vdots$	$\vdots$	$\vdots$	$\vdots$	$\vdots$	$\vdots$
1	N	N	N	1	N
2	1	$N+1$	$N+1$	2	1
2	2	$N+2$	$N+2$	2	2
$\vdots$	$\vdots$	$\vdots$	$\vdots$	$\vdots$	$\vdots$
2	N	$2\cdot N$	$2\cdot N$	2	N
$\vdots$	$\vdots$	$\vdots$	$\vdots$	$\vdots$	$\vdots$
N	1	$(N-1)\cdot N+1$	$(N-1)\cdot N+1$	N	1
N	2	$(N-1)\cdot N+2$	$(N-1)\cdot N+2$	N	2
$\vdots$	$\vdots$	$\vdots$	$\vdots$	$\vdots$	$\vdots$
N	N	$N\cdot N$	$N\cdot N$	N	N

Table 4.4 Index values j and k for the original loop and i for the coalesced loop for two dimensions and illustration of the transformation of the index expressions.

the corresponding transformation into a single loop results in the loop

$$
\begin{aligned}
&\text{forall } (i = 1; i \leq N^3; i{+}{+}) \\
&\quad a[\lceil i/N^2 \rceil][\lceil i/N \rceil - N \cdot \lfloor (i-1)/N^2 \rfloor][i - N \cdot \lfloor (i-1)/N \rfloor] = i;
\end{aligned}
\tag{4.34}
$$

The new loop index is i and the new loop bound is N^3. The array accesses use the index expressions according to the mapping $i \mapsto (f_3(i), f_2(i), f_1(i))$ with $f_3(i) = \lceil i/N^2 \rceil$, $f_2(i) = \lceil i/N \rceil - N \cdot \lfloor (i-1)/N^2 \rfloor$, and $f_1(i) = i - N \cdot \lfloor (i-1)/N \rfloor$. In the general case, a loop nest $L = (L_m, \dots, L_1)$ of parallel loops with index bounds $N_m, \dots, N_1$ of the form

$$
\begin{aligned}
L_m: &\quad \text{forall } (J_m = 1; J_m \leq N_m; J_m{+}{+}) \\
L_{m-1}: &\quad\quad \text{forall } (J_{m-1} = 1; J_{m-1} \leq N_m; J_{m-1}{+}{+}) \\
\vdots &\quad\quad \vdots \\
L_1: &\quad\quad\quad\quad \text{forall } (J_1 = 1; J_1 \leq N_1; J_1{+}{+}) \\
&\quad\quad\quad\quad\quad a[J_m, J_{m-1}, \dots, J_1] = \cdots
\end{aligned}
$$

is coalesced into loop L' with loop index I and loop bound $N = N_m \cdot N_{m-1} \cdots N_1$:

$$
\begin{aligned}
L': &\quad \text{forall } (J = 1; J \leq N; J{+}{+}) \\
&\quad\quad a[f_m(I), f_{m-1}(I), \dots, f_1(I)] = \cdots
\end{aligned}
$$

The array access $a[J_m, J_{m-1}, \dots, J_1]$ in the loop nest L is uniquely expressed by the array reference $a[f_m(I), f_{m-1}(I), \dots, f_1(I)]$ in loop L' with the new loop index I. The index expressions are:

$$f_k(I) = \left\lceil \frac{I}{\Pi_{i=1}^{k-1} N_i} \right\rceil - N_k \cdot \left\lfloor \frac{I-1}{\Pi_{i=1}^{k} N_i} \right\rfloor \quad \text{for } k = m, m-1, \ldots, 1. \tag{4.35}$$

For identical loop bounds $N_1 = \ldots N_m = N$, this is

$$f_k(I) = \left\lceil \frac{I}{N^{k-1}} \right\rceil - N \cdot \left\lfloor \frac{I-1}{N^k} \right\rfloor. \tag{4.36}$$

The proof for the correctness of this transformation is given in [177].

Example: The starting point is the following nested loop of depth three with loop indices j, k, and l and loop bounds $N_3 = 2$, $N_2 = 3$, and $N_1 = 6$. The loop body accesses elements of a three-dimensional array a.

> forall ($j = 1; j \leq 2; j++$)
>   forall ($k = 1; k \leq 3; k++$)
>   forall ($l = 1; l \leq 6; l++$)
> $a[j][k][l] = \ldots$

This loop is coalesced into a singly nested loop with loop index i.

> forall ($i = 1; i \leq 36; i++$)
> $a[\lceil \frac{i}{18} \rceil][\lceil \frac{i}{6} \rceil - 3 \cdot \lfloor \frac{i-1}{18} \rfloor][i - 6 \cdot \lfloor \frac{i-1}{6} \rfloor] = \ldots$

The index expressions for the array access in the loop body are built according to the index mapping (4.32), i.e.,

$$i \mapsto (f_3(i), f_2(i), f_1(i)) \tag{4.37}$$

with $f_3(i) = \lceil \frac{i}{18} \rceil$, $f_2(i) = \lceil \frac{i}{6} \rceil - 3 \cdot \lfloor \frac{i-1}{18} \rfloor$, and $f_1(i) = i - 6 \cdot \lfloor \frac{i-1}{6} \rfloor$. $\qquad \square$

4.7.1.4 Parallel execution of coalesced loops

The assignment of loop iterates to idle processors is based on the indices in the coalesced loop and the corresponding calculation of index expressions. The following example shows a static assignment in which each processor gets a consecutive block of loop iterates. Each processor q is assigned $r = \lceil N/p \rceil$ iterates of the coalesced loop L', which are the iterates with indices

$$I = (q-1) \cdot r + 1, \ldots, q \cdot r$$

so that the loop over I is cut into p sub-loops, one for each processor. In the loop body, each processor has to access the elements of the array a specified in the original nested loop. These array elements are determined by inserting the indices $I = (q-1) \cdot r + 1, \ldots, q \cdot r$ assigned to processor q into the index expressions $f_k(I)$, $k = m, m-1, \ldots, 1$, one after another. This results in r tuples of index expressions

$(f_m(I), \ldots, f_1(I))$ for $I = (q-1) \cdot r + 1, \ldots, q \cdot r$

and provides exactly q consecutive array elements in the rowwise storage scheme, i.e., $a[f_m(I)][\ldots][f_1(I)]$ for $I = (q-1) \cdot r + 1, \ldots, q \cdot r$. According to Eq. (4.35), the index expressions are

$$f_k((q-1) \cdot r + 1) = \left\lceil \frac{(q-1) \cdot r + 1}{\Pi_{j=1}^{k-1} N_j} \right\rceil - N_k \cdot \left\lfloor \frac{(q-1) \cdot r}{\Pi_{j=1}^{k} N_j} \right\rfloor$$

for the first and

$$f_k(q \cdot r) = \left\lceil \frac{q \cdot r}{\Pi_{j=1}^{k-1} N_j} \right\rceil - N_k \cdot \left\lfloor \frac{q \cdot r - 1}{\Pi_{j=1}^{k} N_j} \right\rfloor$$

for the last index assigned to processor q.

The following example shows the application of the mapping with $p = 5$ processors for the loop from the previous example.

Example: For an execution on $p = 5$ processors, the $N = N_3 \cdot N_2 \cdot N_1 = 36$ iterates are assigned such that each processor gets $r = \lceil N/p \rceil = \lceil 36/5 \rceil = 8$ iterates. For processor q, $1 \le q \le 5$, these are the indices $i = (q-1) \cdot r + 1, \ldots, q \cdot r$ of the singly nested loop. Written in an SPMD style, each of the processors q executes the loop

forall $(i = (q-1) \cdot r + 1; i \le \min(q \cdot r, 36); i++)$
$a[\lceil \frac{i}{18} \rceil][\lceil \frac{i}{6} \rceil - 3 \cdot \lfloor \frac{i-1}{18} \rfloor][i - 6 \cdot \lfloor \frac{i-1}{6} \rfloor] = \ldots$

The SPMD style means that the processors execute the same loop, but each processor uses its own private processor rank q so that different data is accessed. Processor $q = 3$, for example, accesses the array elements between

$$a\left[\lceil \frac{(q-1) \cdot r + 1}{18} \rceil\right]\left[\lceil \frac{(q-1) \cdot r + 1}{6} \rceil - 3 \cdot \lfloor \frac{(q-1)r}{18} \rfloor\right]\left[(q-1) \cdot r + 1 - 6 \cdot \lfloor \frac{(q-1) \cdot r}{6} \rfloor\right]$$

and

$$a\left[\lceil \frac{q \cdot r}{18} \rceil\right]\left[\lceil \frac{q \cdot r}{6} \rceil - 3 \cdot \lfloor \frac{q \cdot r - 1}{18} \rfloor\right]\left[q \cdot r - 6 \cdot \lfloor \frac{q \cdot r - 1}{6} \rfloor\right]$$

which are the elements between

$$a\left[\lceil \frac{17}{18} \rceil\right]\left[\lceil \frac{17}{6} \rceil - 3 \lfloor \frac{16}{18} \rfloor\right]\left[17 - 6 \lfloor \frac{16}{6} \rfloor\right] \text{ and } a\left[\lceil \frac{24}{18} \rceil\right]\left[\lceil \frac{24}{6} \rceil - 3 \lfloor \frac{23}{18} \rfloor\right]\left[24 - 6 \lfloor \frac{23}{6} \rfloor\right],$$

i.e., the elements

$$a[1,3,5], a[1,3,6], a[2,1,1], a[2,1,2], a[2,1,3], a[2,1,4], a[2,1,5], a[2,1,6]$$

The assignment of loop iterates for all processors is given in Table 4.5. □

Processor 1	Processor 2	Processor 3	Processor 4	Processor 5
$a[1,1,1]$	$a[1,2,3]$	$a[1,3,5]$	$a[2,2,1]$	$a[2,3,3]$
$a[1,1,2]$	$a[1,2,4]$	$a[1,3,6]$	$a[2,2,2]$	$a[2,3,4]$
$a[1,1,3]$	$a[1,2,5]$	$a[2,1,1]$	$a[2,2,3]$	$a[2,3,5]$
$a[1,1,4]$	$a[1,2,6]$	$a[2,1,2]$	$a[2,2,4]$	$a[2,3,6]$
$a[1,1,5]$	$a[1,3,1]$	$a[2,1,3]$	$a[2,2,5]$	
$a[1,1,6]$	$a[1,3,2]$	$a[2,1,4]$	$a[2,2,6]$	
$a[1,2,1]$	$a[1,3,5]$	$a[2,1,5]$	$a[2,3,1]$	
$a[1,2,2]$	$a[1,3,4]$	$a[2,1,6]$	$a[2,3,2]$	

Table 4.5 Example assignment of array elements, corresponding to loop iterates, to processors.

4.7.2 Loop Tiling

Loop transformations for vectorization or locality issues decompose loops into smaller parts so that data can be accessed from faster memory. For a single loop, the transformation called strip-mining decomposes the loop into two nested loops. The outer loop steps over the new strips of consecutive iterations and the inner loop steps between the consecutive iterations within the strips. This technique has originally been developed for vector computers, and the length of the strips is typically set to the length of the vector registers. For nested loops, a similar transformation is loop tiling. In the following, we give a short description, see [223] for a more detailed treatment.

Loop tiling decomposes the given iteration space into rectangular subspaces, called tiles, which are then computed one after another. For multiply nested loops, the tiling transformation is performed independently for each of the loops. In the following, we consider a specific loop dimension of a loop nest with loop index i, lower bound lo, and upper bound hi, i.e., there are $hi - lo + 1$ iterates:

$$\text{for } (i = lo; i \leq hi; i++) \atop \qquad statements; \tag{4.38}$$

The loop tiling transformation is described by a tile size ts and by a tile offset to with $(0 \leq to \leq ts)$. The tile offset to defines the starting points for the tiles in this dimension, which are $to, to + ts, to + 2 \cdot ts, \ldots$, i.e., the tiles start at an iteration i such that $i \bmod ts = to$. The use of the lower bound lo as tile offset is a special case and simplifies the resulting transformation. In the following, we cover the general case that $lo \neq to$ is possible. Each tile contains at most ts iterations of the original loop. The inner tiles cover exactly ts iterations, the first and the last tile may cover less than ts iterations. The tiles are numbered by a tile index t_n in increasing order of the iterations starting with $t_n = 0$. An inner tile with tile index t_n iterates from $t_n \cdot ts + to$ to $(t_n + 1) \cdot ts + to - 1$. There are at most $\lceil (hi - lo + 1)/ts \rceil + 1$ tiles.

The result of the transformation is a nested loop with an outer loop over the tiles and an inner loop covering a single tile. For the loop over the tiles, the minimum and maximum tile number is computed from the loop bounds of the original loop.

Given a loop with lower bound lo and upper bound hi, the first tiles starts at $\lfloor(lo - to)/ts\rfloor \cdot ts + to$ and the last tile starts at $\lfloor(hi - to)/ts\rfloor \cdot ts + to$. Applying the tiling transformation transforms the singly nested loop (4.38) into

$$\text{for } (it = \lfloor(lo-to)/ts\rfloor \cdot ts + to; it \leq \lfloor(hi-to)/ts\rfloor \cdot ts + to; it+ = ts)$$
$$\text{for } (i = \max(lo, it); i \leq \min(hi; it + ts - 1); i++)$$
$$statements;$$

$$(4.39)$$

The index expressions for the lower and upper bound of the transformed loop can be simplified for special cases such as $to = 0$ or $lo = 0$. The resulting outer loop (called *tile loop*) iterates over the tiles, starting with the first tile until the last tile is reached. The inner loop (called *element loop*) iterates over the elements of a single tile. For the correctness of the transformation, it is important that the loop resulting after the transformation covers exactly the iteration space of the original loop. Therefore, the max and min expressions in the element loop are required, when the first and the last tile should not be covered completely.

Loop tiling is typically applied to nested loops with the goal to interchange the element loop of the original outer loop with the tile loop of the original inner loop. This is demonstrated by the following example, see also [223].

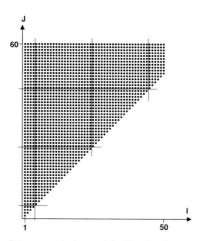

Fig. 4.23 Iteration space of the example loop with tiling borders.

Example: The following doubly nested loop is considered:

```
for(i = 1; i ≤ 50; i++)
    for(j = i; j ≤ 60; j++)
        a[i][j] = a[i][j] + 1;
```

The corresponding iteration space is shown in Fig. 4.23. For the tiling, the values $ts = 20$ and $to = 5$ are used, which is also shown in Fig. 4.23. Applying the tiling transformation to both the inner and the outer loop yields the following loop:

$$\begin{aligned}
&\text{for}(it = -15; it \leq 45; it+ = 20)\\
&\quad \text{for}(i = \max(1, it); i \leq \min(50, it + ts - 1); i++)\\
&\qquad \text{for}(jt = \lfloor(i-5)/20\rfloor \cdot 20 + 5; jt \leq 45; jt+ = 20)\\
&\qquad\quad \text{for}(j = \max(i, jt); j \leq \min(60, jt + ts - 1); j++)\\
&\qquad\qquad a[i][j] = a[i][j] + 1;
\end{aligned}$$

In this loop nest, the i loop and the jt loop can be interchanged without violating the dependencies. To perform the loop interchange, the loop boundaries of the two loops to be interchanged have to be re-computed, since the loop index of the i loop occurs in the loop bounds of the jt loop. This is done by taking the lower bound of the loop over i into consideration for each occurrence of i in the bounds of the jt loop. For $i = 1$, $jt = \lfloor(1-5)/20\rfloor \cdot 20 + 5 = -15$ results for the lower bound of the jt loop. For $i = it$, $jt = \lfloor\frac{it-5}{20}\rfloor \cdot 20 + 5 = it - 5 + 5 = it$ results, because it is a multiple of $20 + 5$. After the loop interchange and the adaptation of the loop bounds, the following loop results:

$$\begin{aligned}
&\text{for}(it = -15; it \leq 45; it+ = 20)\\
&\quad \text{for}(jt = \max(-15, it); jt \leq 45; jt+ = 20)\\
&\qquad \text{for}(i = (\max(1, it); i \leq \min(50, it + ts - 1); i++)\\
&\qquad\quad \text{for}(j = \max(i, jt); j \leq \min(60, jt + ts - 1); j++)\\
&\qquad\qquad a[i][j] = a[i][j] + 1;
\end{aligned}$$

The two inner loops now iterate over one two-dimensional tile.   □

In practice, the tile size is selected such that the data elements accessed when iterating over a single tile fit into a specific cache of the memory hierarchy of the execution platform. This can help to improve the spacial and temporal locality of the memory references of a program, which may reduce the resulting execution time considerably.

4.8 Exercises for Chapter 4

Exercise 4.1. We consider two processors P_1 and P_2 which have the same set of instructions. P_1 has a clock rate of 4 GHz, P_2 ha a clock rate of 2 GHz. The instructions of the processors can be partitioned into three classes A, B, and C. The following table specifies for each class the CPI values for both processors. We assume that there are three compilers C_1, C_2, and C_3 available for both processors. We consider a specific program X. All three compilers generate machine programs which lead to the execution of the same number of instructions. But the instruction classes are represented with different proportions according to the following table:

class	CPI for P_1	CPI for P_2	C_1	C_2	C_3
A	4	2	30%	30%	50%
B	6	4	50%	20%	30%
C	8	3	20%	50%	20%

(a) If C_1 is used for both processors, how much faster is P_1 than P_2?
(b) If C_2 is used for both processors, how much faster is P_2 than P_2?
(c) Which of the three compilers is best for P_1?
(d) Which of the three compilers is best for P_2?

Exercise 4.2. Consider the MIPS (Million Instructions Per Second) rate for estimating the performance of computer systems for a computer with instructions $I_1, \ldots, I_m$. Let p_k be the proportion with which instruction $I_k (1 \le k \le m)$ is represented in the machine program for a specific program X with $0 \le p_k \le 1$. Let CPI_k be the CPI value for I_k and let t_c be the cycle time of the computer system in nanoseconds (10^{-9}).

(a) Show that the MIPS rate for program X can be expressed as

$$MIPS(X) = \frac{1000}{(p_1 \cdot CPI_1 + \ldots + p_m CPI_m) \cdot t_c [ns]}$$

(b) Consider a computer with a clock rate of 3.3 GHz. The CPI values and proportion of occurrence of the different instructions for program X are given in the following table

instruction I_k	p_n	CPI_n
Load and store	20.4	2.5
Integer add and subtract	18.0	1
Integer multiply and divide	10.7	9
Floating-point add and subtract	3,5	7
Floating-point multiply and divide	4.6	17
Logical operations	6.0	1
Branch instruction	20.0	1.5
Compare and shift	16.8	2

Compute the resulting MIPS rate for program X.

Exercise 4.3. There is a SPEC benchmark suite MPI2007 for evaluating the MPI performance of parallel systems for floating-point, compute-intensive programs. Visit the SPEC webpage at www.spec.org and collect information on the benchmark programs included in the benchmark suite. Write a short summary for each of the benchmarks with computations performed, programming language used, MPI usage and input description. What criteria were used to select the benchmarks? Which information is obtained by running the benchmarks?

Exercise 4.4. There is a SPEC benchmark suite OMP2012 to evaluate the performance of parallel systems with a shared address space based on OpenMP applications. Visit the SPEC web page at www.spec.org and collect information about this benchmark suite. Which applications are included and what information is obtained by running the benchmark?

Exercise 4.5. There is a SPEC benchmark suite SPEChpc 2021 for High-Performance computing, combining MPI with OpenMP and OpenACC. Visit the SPEC web page at www.spec.org and collect information about this benchmark suite. Which applications are included and what information is obtained by running the benchmark? Compare SPEChpc 2021 with SPEC MPI2007 and OMP2012.

Exercise 4.6. The SPEC CPU2017 is the standard benchmark suite to evaluate the performance of computer systems. Visit the SPEC web page at www.spec.org and collect the following information:

(a) Which benchmark programs are used in SPEC CPU2017 to evaluate the integer performance? Give a short characteristic of each of the benchmarks.
(b) Which benchmark programs are used in SPEC CPU2017 to evaluate the floating-point performance? Give a short characteristic of each of the benchmarks.
(c) Which performance results have been submitted for your favorite desktop computer?

Exercise 4.7. Consider Amdahl's law and Gustafson's law. Assume the sequential fraction 0.1, i.e., $f = 0.1$ for Amdahl's law and $f_{seq} = 0.1$ for Gustafson's law. Compute the resulting $S_p(n)$ using Amdahl's law and Gustafson's law for $p = 100$ and $p = 1000$.

Exercise 4.8. Consider Equ. (4.14) and the following example with a strongly monotonic execution time $\tau_v(n, p) = n^2/p$. Use Equ. (4.13) and (4.14) to compute the minimum system size n for which an efficiency of at least 0.8 results when using different numbers $p = 100, 200, 500, 1000$ processors. Use the gnuplot tool to construct diagrams similar to Fig. 4.2.

Exercise 4.9. Consider a ring topology and assume that each processor can transmit at most one message at any time along an incoming or outgoing link (one-port communication). Show that the running time for a single-broadcast, a scatter operation, or a multi-broadcast take time $\Theta(p)$. Show that a total exchange needs time $\Theta(p^2)$.

Exercise 4.10. Give an algorithm for a scatter operation on a linear array which sends the message from the root node for more distant nodes first and determine the asymptotic running time.

Exercise 4.11. Given a 2-dimensional mesh with wraparound arrows forming a torus consisting of $n \times n$ nodes. Construct spanning trees for a multi-broadcast operation according to the construction in Section 4.4.2.2, page 208, and give an corresponding algorithm for the communication operation which takes time $(n^2 - 1)/4$ for n odd and $n^2/4$ for n even [21].

Exercise 4.12. Consider a d-dimensional mesh network with $\sqrt[d]{p}$ processors in each of the d dimensions. Show that a multi-broadcast operation requires at least $\lceil (p-1)/d \rceil$ steps to be implemented. Construct an algorithm for the implementation of a multi-broadcast that performs the operation with this number of steps.

Exercise 4.13. Consider the construction of a spanning tree in Section 4.4.2, page 207, and Fig. 4.11. Use this construction to determine the spanning tree for a 5-dimensional hypercube network.

Exercise 4.14. For the construction of the spanning trees for the realization of a multi-broadcast operation on a d-dimensional hypercube network we have used the relation

$$\binom{d}{k-1} - d \geq d$$

for $2 < k < d$ and $d \geq 5$, see Section 4.4.2, page 213. Show by induction that this relation is true.
Hint: it is $\binom{d}{k-1} = \binom{d-1}{k-1} + \binom{d-1}{k-2}$

Exercise 4.15. Consider a scalar product and a matrix-vector multiplication and derive the formula for the running time on a mesh topology.

Exercise 4.16. Develop a runtime function to capture the execution time of a parallel matrix-matrix computation $C = A \cdot B$ for a distributed address space. Assume a hypercube network as interconnection. Consider the following distributions for A and B:

(a) A is distributed in column-blockwise, B in row-blockwise order;
(b) both A and B are distributed in checkerboard order;

Compare the resulting runtime functions and try to identify situations in which one or the other distribution results in a faster parallel program.

Exercise 4.17. The multi-prefix operation leads to the effect that each participating processor P_j obtains the value $\sigma + \sum_{i=1}^{j-1} \sigma_i$ where processor P_i contributes values σ_i and σ is the initial value of the memory location used, see also page 225. Illustrate the effect of a multi-prefix operation with an exchange diagram similar to those used in Section 3.6.2. The effect of multi-prefix operations can be used for the implementation of parallel loops where each processor gets iterations to be executed. Explain this usage in more detail.

Exercise 4.18. Consider the guided self scheduling method for assigning independent loop iterates to processors. Compute the resulting chunk sizes for a parallel loop with $N = 200$ iterates and $p = 8$ processors assuming that all processors start execution at the same time.

Exercise 4.19. Consider the transformation needed for loop coalescing in Section 4.7.1.3 and the illustration of the index transformation in Table 4.4 for the two-dimensional case. Draw a similar table for the three-dimensional case and compute the transformed index expressions according to Eq. (4.35).

Exercise 4.20. The examples in Section 4.7.1.3 enumerate the array elements starting with 1 in increasing order, see Code sections (4.30) and (4.31) for the two-dimensional case and Code sections (4.33) and (4.34) for the three-dimensional case. Modify the original loop and the transformed loop such that the enumeration is in decreasing order starting with N^2 for the two-dimensional case and N^3 for the three-dimensional case.

Exercise 4.21. Apply the loop coalescing transformation to the following loop nest:

```
forall (j = 0; j < 5; j++)
    forall (k = 0; k < 3; k++)
        forall (l = 0; l < 5; l++)
            a[j][k][l] = a[j-1][k+1][l] + 1;
```

(a) Transform this loop nest into a singly nested loop with loop index i and transform the index expressions in the loop body by applying Eq. (4.35).
(b) Consider a parallel execution of the resulting singly nested loop on $p = 8$ processors. Compute for each processor the array elements that are modified by this processor by a calculation according to Section 4.7.1.4.

Exercise 4.22. Apply the tiling loop transformation from Section 4.7.2 to the following loop nests:

(a)
```
for (i = 1; i ≤ 10; i++)
    for (j = 1; j ≤ 10; j++)
        a[i][j] = a[i][j-1] + b[i][j];
```

with tile offset $to = 2$ and tile size $ts = 5$ in each dimension.

(b)
```
for (i = 0; i ≤ 50; i++)
    for (j = 0; j ≤ i; j++)
        a[i][j] = a[i-1][j] + a[i][j-1] + 1;
```

with tile offset $to = 0$ and tile size $ts = 20$ in each dimension.

For each of these loop nests, draw the iteration space with the resulting tiles and apply the tiling transformation (4.39). If possible, exchange the tile loop of the originally inner loop with the element loop of the originally outer loop and recompute loop bounds if necessary.

Chapter 5
Message-Passing Programming

About this Chapter

The message-passing programming model is based on the abstraction of a parallel computer with a distributed address space where each processor has a local memory to which it has exclusive access. This chapter introduces the message-passing interface (MPI), which is the most widely used standard for message-passing programming. Topics covered include point-to-point operations, collective communication operations, process groups and communicators as well as dynamic process creation and one-sided communication.

In a distributed address space, there is no global memory. Thus, data exchange must be performed by explicit message-passing: to transfer data from the local memory of one processor A to the local memory of another processor B, A must send a message containing the data to B, and B must receive the data in a buffer in its local memory. To guarantee portability of programs, no assumptions on the topology of the interconnection network is made. Instead, it is assumed that each processor can send a message to any other processor.

A message-passing program is executed by a set of processes where each process has its own local data. Usually, one process is executed on one processor or core of the execution platform. The number of processes is often fixed when starting the program. Each process can access its local data and can exchange information and data with other processes by sending and receiving messages. In principle, each of the processes could execute a different program (MPMD, *multiple program multiple data*). But to make program design easier, it is usually assumed that each of the processes executes the same program (SPMD, *single program, multiple data*), see also Sect. 2.4. In practice, this is not really a restriction, since each process can still execute different parts of the program, selected, for example, by its process rank.

The processes executing a message-passing program can exchange local data by using communication operations. These could be provided by a communication library. To activate a specific communication operation, the participating processes call the corresponding communication function provided by the library. In the simplest case, this could be a point-to-point transfer of data from a process A to a process B. In this case, A calls a send operation, and B calls a corresponding receive operation. Communication libraries often provide a large set of communication functions

© The Author(s), under exclusive license to Springer Nature Switzerland AG 2023
T. Rauber, G. Rünger, *Parallel Programming*, https://doi.org/10.1007/978-3-031-28924-8_5

to support different point-to-point transfers and also global communication operations like broadcast in which more than two processes are involved, see Sect. 3.6.2 for a typical set of global communication operations.

A communication library could be vendor or hardware specific, but in most cases portable libraries are used which define syntax and semantics of communication functions and which are supported for a large class of parallel computers. By far the most popular portable communication library is MPI (*Message-Passing Interface*), but PVM (*Parallel Virtual Machine*) is also used, see [77]. In this chapter, we give an introduction to MPI and show how parallel programs with MPI can be developed. The description includes point-to-point and global communication operations, but also more advanced features like process groups and communicators are covered.

5.1 Introduction to MPI

The Message-Passing Interface (MPI) is a standardization of a message-passing library interface specification. MPI defines the syntax and semantics of library routines for standard communication patterns as they have been considered in Section 3.6.2. Language bindings for C, C++, Fortran-77 and Fortran-95 are supported. In the following, we concentrate on the interface for C and describe the most important features. For a detailed description, we refer to the official MPI documents, see www.mpi-forum.org. The MPI standard has first been introduced in 1994, now referred to as MPI-1, and there have been several minor and major extensions since then. The current version of the MPI standard is MPI-4, introduced in 2021 and discussions for MPI-5 are underway. MPI-1 defines standard MPI communication operations and is based on a static process model. MPI-2 from 2008 extends MPI-1 and provides additional support for dynamic process management, one-sided communication, and parallel I/O. MPI-3 from 2012 adds non-blocking collective communication operations as well as new one-sided communication operations. MPI-4 adds persistent collectives, partitioned communications, as well as large-count versions of many MPI functions.

MPI is an interface specification for the syntax and semantics of communication operations, but leaves the details of the implementation open. Thus, different MPI libraries can use different implementations, possibly using specific optimizations for specific hardware platforms. For the programmer, MPI provides a standard interface, thus ensuring the portability of MPI programs. Freely available MPI libraries are OpenMPI (see www.open-mpi.org and [75]), MPICH (see www.mpich.org), and MVAPICH (see mvapich.cse.ohio-state.edu and [168]).

In this section, we give an overview of MPI. An MPI program consists of a collection of processes that can exchange messages. For MPI-1, a static process model is used, which means that the number of processes is set when starting the MPI program and cannot be changed during program execution. Thus, MPI-1 does not support dynamic process creation during program execution. Such a feature is added by MPI-2. Normally, each processor of a parallel system executes one MPI process, and the number of MPI processes started should be adapted to the number of processors that are available. Typically, all MPI processes execute the same program in an SPMD style. In principle, each process can read and write data from/into files. For a coordinated I/O behavior, it is essential that only one specific process performs the input or output operations. To support portability, MPI programs should be written for an arbitrary number of processes. The actual number of processes used for a specific program execution is set when starting the program.

On many parallel systems, an MPI program can be started from the command line. The following two commands are common or widely used:

```
mpiexec -n 4 programname programarguments

mpirun -np 4 programname programarguments
```

This call starts the MPI program `programname` with $p = 4$ processes. The specific command to start an MPI program on a parallel system can differ.

A significant part of the operations provided by MPI are operations for the exchange of data between processes. In the following, we describe the most important MPI operations. For a more detailed description of all MPI operations, we refer to [167, 203, 204, 90]. In particular the official description of the MPI standard provides many more details that cannot be covered in our short description, see [69]. Most examples given in this chapter are taken from these sources. Before describing the individual MPI operations, we first introduce some semantic terms that are used for the description of MPI operations:

- **blocking operation**: An MPI communication operation is *blocking*, if return of control to the calling process indicates that all resources, such as buffers, specified in the call can be reused, e.g., for other operations. In particular, all state transitions initiated by a blocking operation are completed before control returns to the calling process.
- **non-blocking operation**: An MPI communication operation is *non-blocking*, if the corresponding call may return before all effects of the operation are completed, and before the resources used by the call can be reused. Thus, a call of a non-blocking operation only **starts** the operation. The operation itself is completed not before all state transitions caused are completed and the resources specified can be reused.

The terms *blocking* and *non-blocking* describe the behavior of operations from the *local* view of the executing process, without taking the effects on other processes into account. But it is also useful to consider the effect of communication operations from a *global* viewpoint. In this context, it is reasonable to distinguish between *synchronous* and *asynchronous* communication:

- **synchronous communication**: For a synchronous MPI operation, the communication between a sending process and a receiving process is performed such that the communication operation does not complete before both processes have started their communication operation. This means in particular that the completion of a synchronous send indicates not only that the send buffer can be reused, but also that the receiving process has started the execution of the corresponding receive operation.

- **asynchronous communication**: Using asynchronous communication, the sender can execute its communication operation without any coordination with the receiving process.

In the next section, we consider single-transfer operations provided by MPI, which are also called point-to-point communication operations.

5.1.1 MPI point-to-point communication

In MPI, all communication operations are executed using a **communicator**. A communicator represents a communication domain which is essentially a set of processes that exchange messages between each other. In this section, we assume that the MPI default communicator MPI_COMM_WORLD is used for the communication. This communicator captures all processes executing a parallel program. In Section 5.3, the grouping of processes and the corresponding communicators are considered in more detail.

The most basic form of data exchange between processes is provided by point-to-point communication. Two processes participate in this communication operation: a sending process executes a send operation and a receiving process executes a corresponding receive operation. The send operation is *blocking* and has the syntax:

```
int MPI_Send(void *smessage,
             int count,
             MPI_Datatype datatype,
             int dest,
             int tag,
             MPI_Comm comm).
```

The parameters have the following meaning:

- smessage specifies a send buffer which contains the data elements to be sent in successive order;
- count is the number of elements to be sent from the send buffer;
- datatype is the data type of each entry of the send buffer; all entries have the same data type;
- dest specifies the rank of the target process which should receive the data; each process of a communicator has a unique rank; the ranks are numbered from 0 to the number of processes minus one;
- tag is a message tag which can be used by the receiver to distinguish different messages from the same sender;
- comm specifies the communicator used for the communication.

The size of the message in bytes can be computed by multiplying the number count of entries with the number of bytes used for type datatype. The tag parameter should be an integer value between 0 and 32767. Larger values can be permitted by specific MPI libraries.

To receive a message, a process executes the following operation:

```
int MPI_Recv(void *rmessage,
             int count,
             MPI_Datatype datatype,
             int source,
             int tag,
             MPI_Comm comm,
             MPI_Status *status).
```

This operation is also blocking. The parameters have the following meaning:

- rmessage specifies the receive buffer in which the message should be stored;
- count is the maximum number of elements that should be received;
- datatype is the data type of the elements to be received;
- source specifies the rank of the sending process which sends the message;
- tag is the message tag that the message to be received must have;
- comm is the communicator used for the communication;
- status specifies a data structure which contains information about a message after the completion of the receive operation.

The most important predefined MPI data types and the corresponding C data types are shown in Table 5.1. There is no corresponding C data type to MPI_PACKED and MPI_BYTE. The type MPI_BYTE represents a single byte value. The type MPI_PACKED is used by special MPI pack operations.

MPI Data type	C Data type
MPI_CHAR	signed char
MPI_SHORT	signed short int
MPI_INT	signed int
MPI_LONG	signed long int
MPI_LONG_LONG_INT	signed long long int
MPI_UNSIGNED_CHAR	unsigned char
MPI_UNSIGNED_SHORT	unsigned short int
MPI_UNSIGNED	unsigned int
MPI_UNSIGNED_LONG	unsigned long int
MPI_UNSIGNED_LONG_LONG	unsigned long long int
MPI_FLOAT	float
MPI_DOUBLE	double
MPI_LONG_DOUBLE	long double
MPI_WCHAR	wide char
MPI_PACKED	special data type for packing
MPI_BYTE	single byte value

Table 5.1 Predefined data types for MPI.

By using source = MPI_ANY_SOURCE, a process can receive a message from an arbitrary other process. Similarly, by using tag = MPI_ANY_TAG, a process can receive a message with an arbitrary tag. In both cases, the status data structure contains the information, from which process the message received has been sent and which tag has been used by the sender. After completion of MPI_Recv(), status contains the following information:

- status.MPI_SOURCE specifies the rank of the sending process;
- status.MPI_TAG specifies the tag of the message received;
- status.MPI_ERROR contains an error code.

The status data structure also contains information about the length of the message received. This can be obtained by calling the MPI function

```
int MPI_Get_count (MPI_Status *status,
                   MPI_Datatype datatype,
                   int *count_ptr)
```

where status is a pointer to the data structure status returned by MPI_Recv(). The function returns the number of elements received in the variable pointed to by count_ptr. Some programs are constructed in such a way that it is not required to examine the status data structure. For these programs, the predefined constant MPI_STATUS_IGNORE can be used instead of a status data structure, informing the runtime system that the status fields do not need to be filled.

Internally a message transfer in MPI is usually performed in three steps:

1. The data elements to be sent are copied from the send buffer smessage specified as parameter into a system buffer of the MPI runtime system. The message is assembled by adding a header with information on the sending process, the receiving process, the tag, and the communicator used.
2. The message is sent via the network from the sending process to the receiving process.
3. At the receiving side, the data entries of the message are copied from the system buffer into the receive buffer rmessage specified by MPI_Recv().

Both MPI_Send() and MPI_Recv() are *blocking, asynchronous* operations. This means that an MPI_Recv() operation can be started also when the corresponding MPI_Send() operation has not yet been started. The process executing the MPI_Recv() operation is blocked until the specified receive buffer contains the data elements sent. Similarly, an MPI_Send() operation can be started also when the corresponding MPI_Recv() operation has not yet been started. The process executing the MPI_Send() operation is blocked until the specified send buffer can be reused. The exact behavior depends on the specific MPI library used. The following two behaviors can often be observed:

- If the message is sent directly from the send buffer specified without using an internal system buffer, then the MPI_Send() operation is blocked until the entire message has been copied into a receive buffer at the receiving side. In particular, this requires that the receiving process has started the corresponding MPI_Recv() operation.

- If the message is first copied into an internal system buffer of the runtime system, the sender can continue its operations as soon as the copy operation into the system buffer is completed. Thus, the corresponding MPI_Recv() operation does not need to be started. This has the advantage that the sender is not blocked for a long period of time. The drawback of this version is that the system buffer needs additional memory space and that the copying into the system buffer requires additional execution time.

```
#include <stdio.h>
#include <string.h>
#include "mpi.h"

int main (int argc, char *argv[]) {
    int my_rank, p, tag=0;
    char msg [20];
    MPI_Status status;

    MPI_Init (&argc, &argv);
    MPI_Comm_rank (MPI_COMM_WORLD, &my_rank);
    MPI_Comm_size (MPI_COMM_WORLD, &p);
    if (my_rank == 0) {
        strcpy (msg, "Hello ");
        MPI_Send (msg, strlen(msg)+1, MPI_CHAR, 1, tag, MPI_COMM_WORLD);
    }
    if (my_rank == 1)
        MPI_Recv (msg, 20, MPI_CHAR, 0, tag, MPI_COMM_WORLD, &status);
    MPI_Finalize ();
}
```

Fig. 5.1 A first MPI program: message passing from process 0 to process 1.

Example: Fig. 5.1 shows a first MPI program in which the process with rank 0 uses MPI_Send() to send a message to the process with rank 1. This process uses MPI_Recv() to receive a message. The MPI program shown is executed by all participating processes, i.e., each process executes the same program. But different processes may execute different program parts, e.g., depending on the values of local variables. The program defines a variable status of type MPI_Status which is used for the MPI_Recv() operation. Any MPI program must include <mpi.h>. The MPI function MPI_Init() must be called before any other MPI function to initialize the MPI runtime system. The call MPI_Comm_rank(MPI_COMM_WORLD, &my_rank) returns the rank of the calling process in the communicator specified, which is MPI_COMM_WORLD here. The rank is returned in the variable my_rank. The function MPI_Comm_size(MPI_COMM_WORLD, &p) returns the total number of processes in the specified communicator in variable p. In the example program, different processes execute different parts of the program depending on their rank stored in my_rank: process 0 executes a string copy and an MPI_Send() operation; process 1 executes a corresponding MPI_Recv() operation. The MPI_Send() operation specifies in its fourth parameter that the receiving process has rank 1. The MPI_Recv() operation specifies in its fourth parameter that the sending process should have rank 0. The last operation in the example program is MPI_Finalize() which should be the last MPI operation in any MPI program.   □

An important property to be fulfilled by any MPI library is that messages are delivered in the order in which they have been sent. If a sender sends two messages one after another to the same receiver and both messages fit to the first MPI_Recv() called by the receiver, the MPI runtime system ensures that the first message sent will always be received first. But this order can be disturbed if more than two processes are involved. This can be illustrated with the following program fragment:

```
/* example to demonstrate the order of receive operations */
MPI_Comm_rank (comm, &my_rank);
if (my_rank == 0) {
  MPI_Send (sendbuf1, count, MPI_INT, 2, tag, comm);
  MPI_Send (sendbuf2, count, MPI_INT, 1, tag, comm);
}
else if (my_rank == 1) {
  MPI_Recv (recvbuf1, count, MPI_INT, 0, tag, comm, &status);
  MPI_Send (recvbuf1, count, MPI_INT, 2, tag, comm);
}
else if (my_rank == 2) {
  MPI_Recv (recvbuf1, count, MPI_INT, MPI_ANY_SOURCE, tag, comm,
      &status);
  MPI_Recv (recvbuf2, count, MPI_INT, MPI_ANY_SOURCE, tag, comm,
      &status);
}
```

Process 0 first sends a message to process 2 and then to process 1. Process 1 receives a message from process 0 and forwards it to process 2. Process 2 receives two messages in the order in which they arrive using MPI_ANY_SOURCE. In this scenario, it can be expected that process 2 first receives the message that has been sent by process 0 directly to process 2, since process 0 sends this message first and since the second message sent by process 0 has to be forwarded by process 1 before arriving at process 2. But this must not necessarily be the case, since the first message sent by process 0 might be delayed because of a collision in the network whereas the second message sent by process 0 might be delivered without delay. Therefore it can happen that process 2 first receives the message of process 0 that has been forwarded by process 1. Thus, if more than two processes are involved, there is no guaranteed delivery order. In the example, the expected order of arrival can be ensured if process 2 specifies the expected sender in the MPI_Recv() operation instead of MPI_ANY_SOURCE.

5.1.2 Deadlocks with Point-to-point Communications

Send and receive operations must be used with care, since **deadlocks** can occur in ill-constructed programs. This can be illustrated by the following example:

```
/* program fragment which always causes a deadlock */
MPI_Comm_rank (comm, &my_rank);
if (my_rank == 0) {
  MPI_Recv (recvbuf, count, MPI_INT, 1, tag, comm, &status);
  MPI_Send (sendbuf, count, MPI_INT, 1, tag, comm);
}
else if (my_rank == 1) {
  MPI_Recv (recvbuf, count, MPI_INT, 0, tag, comm, &status);
  MPI_Send (sendbuf, count, MPI_INT, 0, tag, comm);
}
```

Both processes 0 and 1 execute an MPI_Recv() operation before an MPI_Send() operation. This leads to a deadlock because of mutual waiting: for process 0, the MPI_Send() operation can be started not before the preceding MPI_Recv() operation has been completed. This is only possible when process 1 executes its MPI_Send() operation. But this cannot happen because process 1 also has to complete its preceding MPI_Recv() operation first which can happen only if process 0 executes its MPI_Send() operation. Thus, cyclic waiting occurs, and this program always leads to a deadlock.

The occurrence of a deadlock might also depend on the question whether the runtime system uses internal system buffers or not. This can be illustrated by the following example:

```
/* program fragment for which the occurrence of a deadlock
   depends on the implementation */
MPI_Comm_rank (comm, &my_rank);
if (my_rank == 0) {
  MPI_Send (sendbuf, count, MPI_INT, 1, tag, comm);
  MPI_Recv (recvbuf, count, MPI_INT, 1, tag, comm, &status);
}
else if (my_rank == 1) {
  MPI_Send (sendbuf, count, MPI_INT, 0, tag, comm);
  MPI_Recv (recvbuf, count, MPI_INT, 0, tag, comm, &status);
}
```

Message transmission is performed correctly here without deadlock, if the MPI runtime system uses system buffers. In this case, the messages sent by process 0 and 1 are first copied from the specified send buffer sendbuf into a system buffer before the actual transmission. After this copy operation, the MPI_Send() operation is completed because the send buffers can be reused. Thus, both process 0 and 1 can execute their MPI_Recv() operation and no deadlock occurs. But a deadlock occurs, if the runtime system does not use system buffers or if the system buffers used are too small. In this case, none of the two processes can complete its MPI_Send() operation, since the corresponding MPI_Recv() cannot be executed by the other process.

A **secure implementation** which does not cause deadlocks even if no system buffers are used is the following:

```
/* program fragment that does not cause a deadlock */
MPI_Comm_rank (comm, &myrank);
if (my_rank == 0) {
  MPI_Send (sendbuf, count, MPI_INT, 1, tag, comm);
  MPI_Recv (recvbuf, count, MPI_INT, 1, tag, comm, &status);
}
else if (my_rank == 1) {
  MPI_Recv (recvbuf, count, MPI_INT, 0, tag, comm, &status);
  MPI_Send (sendbuf, count, MPI_INT, 0, tag, comm);
}
```

An MPI program is called **secure** if the correctness of the program does not depend on assumptions about specific properties of the MPI runtime system, like the existence of system buffers or the size of system buffers. Thus, secure MPI programs work correctly even if no system buffers are used. If more than two processes exchange messages such that each process sends and receives a message, the program must exactly specify in which order the send and receive operations are executed to avoid deadlocks. As example, we consider a program with p processes where process i sends a message to process $(i + 1) \bmod p$ and receives a message from process $(i - 1) \bmod p$ for $0 \leq i \leq p - 1$. Thus, the messages are sent in a logical ring. A secure implementation can be obtained if processes with an even rank first execute their send and then their receive operation, whereas processes with an odd rank first execute their receive and then their send operation. This leads to a communication with two phases. The corresponding exchange scheme when using four processes is the following:

phase	process 0	process 1	process 2	process 3
1	MPI_Send() to 1	MPI_Recv() from 0	MPI_Send() to 3	MPI_Recv() from 2
2	MPI_Recv() from 3	MPI_Send() to 2	MPI_Recv() from 1	MPI_Send() to 0

The described execution order leads to a secure implementation also for an odd number of processes. For three processes, the following exchange scheme results:

phase	process 0	process 1	process 2
1	MPI_Send() to 1	MPI_Recv() from 0	MPI_Send() to 0
2	MPI_Recv() from 2	MPI_Send() to 2	-wait-
3		-wait-	MPI_Recv() from 1

In this scheme, some communication operations such as the MPI_Send() operation of process 2 can be delayed because the receiver calls the corresponding MPI_Recv() operation at a later time. But a deadlock cannot occur.

In many situations, processes both send and receive data. MPI provides the following operations to support this behavior:

```
int MPI_Sendrecv (void *sendbuf,
                  int sendcount,
                  MPI_Datatype sendtype,
                  int dest,
                  int sendtag,
                  void *recvbuf,
                  int recvcount,
                  MPI_Datatype recvtype,
                  int source,
                  int recvtag,
                  MPI_Comm comm,
                  MPI_Status *status).
```

This operation is blocking and combines a send and a receive operation in one call. The parameters have the following meaning:

- sendbuf specifies a send buffer in which the data elements to be sent are stored;
- sendcount is the number of elements to be sent;
- sendtype is the data type of the elements in the send buffer;
- dest is the rank of the target process to which the data elements are sent;
- sendtag is the tag for the message to be sent;
- recvbuf is the receive buffer for the message to be received;
- recvcount is the maximum number of elements to be received;
- recvtype is the data type of elements to be received;
- source is the rank of the process from which the message is expected;
- recvtag is the expected tag of the message to be received;
- comm is the communicator used for the communication;
- status is the data structure storing information about the message received.

Using MPI_Sendrecv(), the programmer does not need to worry about the order of the send and receive operations. The MPI runtime system guarantees deadlock freedom, also for the case that no internal system buffers are used. The parameters sendbuf and recvbuf, specifying the send and receive buffer of the executing process, must be disjoint, non-overlapping memory locations. But the buffers may have different lengths, and the entries stored may even contain elements of different data types. A variant of MPI_Sendrecv() has identical send and receive buffers. This operation is also blocking and has the following syntax:

```
int MPI_Sendrecv_replace (void *buffer,
                          int count,
                          MPI_Datatype type,
                          int dest,
                          int sendtag,
                          int source,
                          int recvtag,
                          MPI_Comm comm,
                          MPI_Status *status).
```

Here, buffer specifies the buffer that is used as both, send and receive buffer. For this function, count is the number of elements to be sent and to be received; these elements now have to have identical type type.

5.1.3 Nonblocking Point-to-Point Operations

The use of blocking communication operations can lead to waiting times in which the blocked process does not perform useful work. For example, a process executing a blocking send operation must wait until the user send buffer has been copied into a system buffer or even until the message has completely arrived at the receiving process if no system buffers are used. Often, it is desirable to fill the waiting times with useful operations of the waiting process, e.g., by overlapping communications and computations. This can be achieved by using *non-blocking communication operations*.

A **non-blocking send operation** initiates the sending of a message and returns control to the sending process as soon as possible. Upon return, the send operation has been started, but the send buffer specified cannot be reused safely, i.e., the transfer into an internal system buffer may still be in progress. A separate completion operation is provided to test whether the send operation has been completed locally. A non-blocking send has the advantage that control is returned as fast as possible to the calling process which can then execute other useful operations. A non-blocking send is performed by calling the following MPI function:

```
int MPI_Isend (void *buffer,
               int count,
               MPI_Datatype type,
               int dest,
               int tag,
               MPI_Comm comm,
               MPI_Request *request).
```

The parameters have the same meaning as for MPI_Send(). There is an additional parameter of type MPI_Request which denotes an opaque object that can be used for the identification of a specific communication operation. This request object is also used by the MPI runtime system to report information on the status of the communication operation.

A **non-blocking receive operation** initiates the receiving of a message and returns control to the receiving process as soon as possible. Upon return, the receive operation has been started and the runtime system has been informed that the receive buffer specified is ready to receive data. But the return of the call does not indicate that the receive buffer already contains the data, i.e., the message to be received cannot be used yet. A non-blocking receive is provided by MPI using the function

```
int MPI_Irecv (void *buffer,
               int count,
               MPI_Datatype type,
               int source,
               int tag,
               MPI_Comm comm,
               MPI_Request *request)
```

where the parameters have the same meaning as for MPI_Recv(). Again, a request object is used for the identification of the operation. Before reusing a send or receive buffer specified in a non-blocking send or receive operation, the calling process must test the completion of the operation. The request objects returned are used for the identification of the communication operations to be tested for completion. The following MPI function can be used to test for completion of a non-blocking communication operation:

```
int MPI_Test (MPI_Request *request,
              int *flag,
              MPI_Status *status).
```

The call returns flag = 1 (true), if the communication operation identified by request has been completed. Otherwise, flag = 0 (false) is returned. If request denotes a receive operation and flag = 1 is returned, the parameter status contains information on the message received as described for MPI_Recv(). The parameter status is undefined if the specified receive operation has not yet been completed. If request denotes a send operation, all entries of status except status.MPI_ERROR are undefined. The MPI function

```
int MPI_Wait (MPI_Request *request, MPI_Status *status)
```

can be used to wait for the completion of a non-blocking communication operation. When calling this function, the calling process is blocked until the operation identified by request has been completed. For a non-blocking send operation, the send buffer can be reused after MPI_Wait() returns. Similarly for a non-blocking receive, the receive buffer contains the message after MPI_Wait() returns.

MPI ensures also for non-blocking communication operations that messages are non-overtaking. Blocking and non-blocking operations can be mixed, i.e., data sent by MPI_Isend() can be received by MPI_Recv() and data sent by MPI_Send() can be received by MPI_Irecv().

Example: As example for the use of non-blocking communication operations, we consider the collection of information from different processes such that each process gets all available information [167]. We consider p processes and assume that each process has computed the same number of floating-point values. These values should be communicated such that each process gets the values of all other processes. To reach this goal, $p - 1$ steps are performed and the processes are logically arranged in a ring. In the first step, each process sends its local data to its successor process in the ring. In the following steps, each process forwards the data that it has received in the previous step from its predecessor to its successor. After $p - 1$ steps, each process has received all the data.

The steps to be performed are illustrated in Fig. 5.2 for four processes. For the implementation, we assume that each process provides its local data in an array x and that the entire data is collected in an array y of size p times the size of x.

Fig. 5.3 shows an implementation with blocking send and receive operations. The size of the local data blocks of each process is given by parameter blocksize. First, each process copies its local block x into the corresponding position in its local buffer y and determines its predecessor process pred as well as its successors process succ in the ring. Then, a loop with $p - 1$ steps is executed. In each step, the

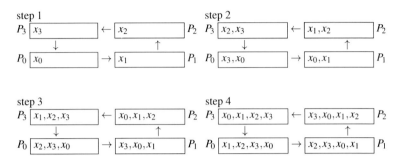

Fig. 5.2 Illustration for the collection of data in a logical ring structure for $p = 4$ processes.

data block received in the previous step is sent to the successor process, and a new block is received from the predecessor process and stored in the next block position to the left in y. To do so, each process computes the send position in y of the block to be send next to its successor process in its local variable send_offset. Similarly, the position in y for the next block to be received from its predecessor process is computed in recv_offset. Each iteration of the loop is executed by all processes in parallel in SPMD style with each process working on its local buffer y. It should be noted that this implementation requires the use of system buffers that are large enough to store the data blocks to be sent. Figure 5.4 illustrates the communication for $p = 4$ processes with $p - 1 = 3$ iteration steps and shows explicitly the specific send and receive offsets for each step, resulting in the same order of the data blocks in the local y arrays.

An implementation with non-blocking communication operations is shown in Fig. 5.5. This implementation allows an overlapping of communication with local computations. In this example, the local computations overlapped are the computations of the positions of send_offset and recv_offset of the next blocks to be sent or to be received in array y. The send and receive operations are started with MPI_Isend() and MPI_Irecv(), respectively. After control returns from these operations, send_offset and recv_offset are re-computed and MPI_Wait() is used to wait for the completion of the send and receive operations. According to [167], the non-blocking version leads to a smaller execution time than the blocking version.   □

5.1.4 Communication modes

MPI provides different **communication modes** for both blocking and non-blocking communication operations. These communication modes determine the coordination between a send and its corresponding receive operation. The following three modes are available.

```
void Gather_ring (float x[], int blocksize, float y[])
{
    int i, p, my_rank, succ, pred;
    int send_offset, recv_offset;
    MPI_Status status;

    MPI_Comm_size (MPI_COMM_WORLD, &p);
    MPI_Comm_rank (MPI_COMM_WORLD, &my_rank);
    for (i=0; i<blocksize; i++)
        y[i+my_rank * blocksize] = x[i];
    succ = (my_rank+1) % p;
    pred = (my_rank-1+p) % p;
    for (i=0; i<p-1; i++) { // loop of communication steps
        send_offset = ((my_rank-i+p) % p) * blocksize;
        recv_offset = ((my_rank-i-1+p) % p) * blocksize;
        MPI_Send (y+send_offset, blocksize, MPI_FLOAT, succ, 0,
                MPI_COMM_WORLD);
        MPI_Recv (y+recv_offset, blocksize, MPI_FLOAT, pred, 0,
                MPI_COMM_WORLD, &status);
    }
}
```

Fig. 5.3 MPI program for the collection of distributed data blocks. The participating processes are logically arranged as a ring. The communication is performed with *blocking* point-to-point operations. Deadlock freedom is ensured only if the MPI runtime system uses system buffers that are large enough.

5.1.4.1 Standard mode

The communication operations described until now use the standard mode of communication. In this mode, the MPI runtime system decides whether outgoing messages are buffered in a local system buffer or not. The runtime system could, for example, decide to buffer small messages up to a predefined size, but not large messages. For the programmer, this means that he cannot rely on a buffering of messages. Hence, programs should be written in such a way that they also work if no buffering is used.

5.1.4.2 Synchronous mode

In the standard mode, a send operation can also be completed if the corresponding receive operation has not yet been started (if system buffers are used). In contrast, in synchronous mode, a send operation will be completed not before the corresponding receive operation has been started and the receiving process has started to receive the data sent. Thus, the execution of a send and receive operation in synchronous mode leads to a form of synchronization between the sending and the receiving process: the return of a send operation in synchronous mode indicates that the receiver has started to store the message in its local receive buffer. A blocking send operation in

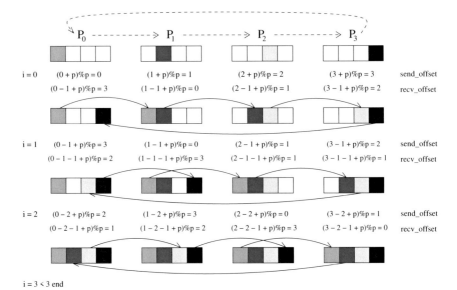

Fig. 5.4 Illustration of the ring communication from the MPI program in Figure 5.3 for $p = 4$ processes. The data blocks of the different processes are shown in different colors. For each process, the fill state of the local arrays y is shown for the different iteration steps. The computation of the position of the send and receive buffers for the next communication is also illustrated. After the last iteration step, each process stores the data blocks of all processes at the same position.

synchronous mode is provided in MPI by the function MPI_Ssend() which has the same parameters as MPI_Send() with the same meaning. A non-blocking send operation in synchronous mode is provided by the MPI function MPI_Issend() which has the same parameters as MPI_Isend() with the same meaning. Similar to a non-blocking send operation in standard mode, control is returned to the calling process as soon as possible, i.e., in synchronous mode there is **no synchronization** between MPI_Issend() and MPI_Irecv(). Instead, synchronization between sender and receiver is performed when the sender calls MPI_Wait(). When calling MPI_Wait() for a non-blocking send operation in synchronous mode, control is returned to the calling process not before the receiver has called the corresponding MPI_Recv() or MPI_Irecv() operation.

5.1.4.3 Buffered mode

In buffered mode, the local execution and termination of a send operation is not influenced by non-local events as this is the case for the synchronous mode and can be the case for standard mode if no or too small system buffers are used. Thus, when starting a send operation in buffered mode, control will be returned to the

```
void Gather_ring_nb (float x[], int blocksize, float y[])
{
    int i, p, my_rank, succ, pred;
    int send_offset, recv_offset;
    MPI_Status status;
    MPI_Request send_request, recv_request;

    MPI_Comm_size (MPI_COMM_WORLD, &p);
    MPI_Comm_rank (MPI_COMM_WORLD, &my_rank);
    for (i=0; i<blocksize; i++)
        y[i+my_rank * blocksize] = x[i];
    succ = (my_rank+1) % p;
    pred = (my_rank-1+p) % p;
    send_offset = my_rank * blocksize;
    recv_offset = ((my_rank-1+p) % p) * blocksize;
    for (i=0; i<p-1; i++) {
        MPI_Isend (y+send_offset, blocksize, MPI_FLOAT, succ, 0,
            MPI_COMM_WORLD, &send_request);
        MPI_Irecv (y+recv_offset, blocksize, MPI_FLOAT, pred, 0,
            MPI_COMM_WORLD, &recv_request);
        send_offset = ((my_rank-i-1+p) % p) * blocksize;
        recv_offset = ((my_rank-i-2+p) % p) * blocksize;
        MPI_Wait (&send_request, &status);
        MPI_Wait (&recv_request, &status);
    }
}
```

Fig. 5.5 MPI program for the collection of distributed data blocks, see Fig. 5.3. Nonblocking communication operations are used instead of blocking operations.

calling process even if the corresponding receive operation has not yet been started. Moreover, the send buffer can be reused immediately after control returns, even if a non-blocking send is used. If the corresponding receive operation has not yet been started, the runtime system must buffer the outgoing message. A blocking send operation in buffered mode is performed by calling the MPI function MPI_Bsend() which has the same parameters as MPI_Send() with the same meaning. A non-blocking send operation in buffered mode is performed by calling MPI_Ibsend() which has the same parameters as MPI_Isend(). In buffered mode, the buffer space to be used by the runtime system must be provided by the programmer. Thus, it is the programmer who is responsible that a sufficiently large buffer is available. In particular, a send operation in buffered mode may fail if the buffer provided by the programmer is too small to store the message. The buffer for the buffering of messages by the sender is provided by calling the MPI function

```
int MPI_Buffer_attach (void *buffer, int buffersize)
```

where buffersize is the size of the buffer buffer in bytes. Only one buffer can be attached by each process at a time. A buffer previously been provided can be detached again by calling the function

```
int MPI_Buffer_detach (void *buffer, int *buffersize)
```

where `buffer` is the *address* of the buffer pointer used in `MPI_Buffer_attach()`; the size of the buffer detached is returned in the parameter `buffersize`. A process calling `MPI_Buffer_detach()` is blocked until all messages that are currently stored in the buffer have been transmitted.

For receive operations, MPI provides the standard mode only.

5.2 Collective Communication Operations

A communication operation is called *collective* or *global* if all or a subset of the processes of a parallel program are involved. In Sect. 3.6.2, we have shown global communication operations which are often used. In this section, we show how these communication operations can be used in MPI. The following table gives an overview of operations supported:

global communication operation	MPI function
Broadcast operation	`MPI_Bcast()`
Accumulation operation	`MPI_Reduce()`
Gather operation	`MPI_Gather()`
Scatter operation	`MPI_Scatter()`
Multi-broadcast operation	`MPI_Allgather()`
Multi-accumulation operation	`MPI_Allreduce()`
Total exchange	`MPI_Alltoall()`

5.2.1 Collective Communication in MPI

5.2.1.1 Broadcast operation

For a broadcast operation, one specific process of a group of processes sends the same data block to all other processes of the group, see Sect. 3.6.2. In MPI, a broadcast is performed by calling the following MPI function

```
int MPI_Bcast (void *message,
               int count,
               MPI_Datatype type,
               int root,
               MPI_Comm comm)
```

where `root` denotes the process which sends the data block. This process provides the data block to be sent in parameter `message`. The other processes specify in `message` their receive buffer. The parameter `count` denotes the number of elements in the data block, `type` is the data type of the elements of the data block. `MPI_Bcast()` is a *collective* communication operation, i.e., each process of the communicator `comm` must call the `MPI_Bcast()` operation. Each process must specify the same `root` process and must use the same communicator. Similarly, the type

type and number `count` specified by any process including the root process must be the same for all processes. Data blocks sent by `MPI_Bcast()` *cannot* be received by an `MPI_Recv()` operation.

As can be seen in the parameter list of `MPI_Bcast()`, no tag information is used as this is the case for point-to-point communication operations. Thus, the receiving processes cannot distinguish between different broadcast messages based on tags.

The MPI runtime system guarantees that broadcast messages are received in the same order in which they have been sent by the root process, even if the corresponding broadcast operations are not executed at the same time. Figure 5.6 shows as example a program part in which process 0 sends two data blocks x and y by two successive broadcast operations to process 1 and process 2 [167].

```
if (my_rank == 0) {
    MPI_Bcast (&x, 1, MPI_INT, 0, comm);
    MPI_Bcast (&y, 1, MPI_INT, 0, comm);
    local_work ();
}
else if (my_rank == 1) {
    local_work ();
    MPI_Bcast (&y, 1, MPI_INT, 0, comm);
    MPI_Bcast (&x, 1, MPI_INT, 0, comm);
}
else if (my_rank == 2) {
    local_work ();
    MPI_Bcast (&x, 1, MPI_INT, 0, comm);
    MPI_Bcast (&y, 1, MPI_INT, 0, comm);
}
```

Fig. 5.6 Example for the receive order with several broadcast operations.

Process 1 first performs local computations by `local_work()` and then stores the first broadcast message in its local variable y, the second one in x. Process 2 stores the broadcast messages in the same local variables from which they have been sent by process 0. Thus, process 1 will store the messages in other local variables as process 2. Although there is no explicit synchronization between the processes executing `MPI_Bcast()`, synchronous execution semantics is used, i.e., the order of the `MPI_Bcast()` operations is such as if there were a synchronization between the executing processes.

Collective MPI communication operations are always *blocking*; no non-blocking versions are provided as it is the case for point-to-point operations. The main reason for this is to avoid a large number of additional MPI functions. For the same reason, only the standard modus is supported for collective communication operations. A process participating in a collective communication operation can complete the operation and return control as soon as its local participation has been completed, no matter what the status of the other participating processes is. For the root process, this means that control can be returned as soon as the message has been copied into a system buffer and the send buffer specified as parameter can be reused. The other processes need not have to have received the message before the root process can continue its computations. For a receiving process, this means that control can be returned as soon as the message has been transferred into the local receive buffer, even if other receiving processes do not even have started their corresponding MPI_Bcast() operation. Thus, the execution of a collective communication operation does not involve a synchronization of the participating processes.

5.2.1.2 Reduction operation

An *accumulation* operation is also called *global reduction* operation. For such an operation, each participating process provides a block of data that is combined with the other blocks using a binary reduction operation. The accumulated result is collected at a root process, see also Sect. 3.6.2. In MPI, a global reduction operation is performed by letting each participating process call the function

```
int MPI_Reduce (void *sendbuf,
                void *recvbuf,
                int count,
                MPI_Datatype type,
                MPI_Op op,
                int root,
                MPI_Comm comm)
```

where sendbuf is a send buffer in which each process provides its local data for the reduction. The parameter recvbuf specifies the receive buffer which is provided by the root process root. The parameter count specifies the number of elements that are provided by each process, type is the data type of each of these elements. The parameter op specifies the reduction operation to be performed for the accumulation. This must be an *associative* operation. MPI provides a number of predefined reduction operations which are also *commutative*:

representation	operation
MPI_MAX	maximum
MPI_MIN	minimum
MPI_SUM	sum
MPI_PROD	product
MPI_LAND	logical and
MPI_BAND	bit-wise and
MPI_LOR	logical or
MPI_BOR	bit-wise or
MPI_LXOR	logical exclusive or
MPI_BXOR	bit-wise exclusive or
MPI_MAXLOC	maximum value and corresponding index
MPI_MINLOC	minimum value and corresponding index

The predefined reduction operations MPI_MAXLOC and MPI_MINLOC can be used to determine a global maximum or minimum value and also an additional index attached to this value. This will be used in Chapter 8 in Gaussian elimination to determine a global pivot element of a row as well as the process which owns this pivot element and which is then used as root of a broadcast operation. In this case, the additional index value is a process rank. Another use could be to determine the maximum value of a distributed array as well as the corresponding index position. In this case, the additional index value is an array index. The operation defined by MPI_MAXLOC is

$$(u,i) \circ_{max} (v,j) = (w,k)$$

$$\text{where } w = max(u,v) \text{ and } k = \begin{cases} i, & \text{if } u > v \\ min(i,j), & \text{if } u = v \ . \\ j, & \text{if } u < v \end{cases}$$

Analogously, the operation defined by MPI_MINLOC is

$$(u,i) \circ_{min} (v,j) = (w,k)$$

$$\text{where } w = min(u,v) \text{ and } k = \begin{cases} i, & \text{if } u < v \\ min(i,j), & \text{if } u = v \ . \\ j, & \text{if } u > v \end{cases}$$

Thus, both operations work on pairs of values, consisting of a value and an index. Therefore the data type provided as parameter of MPI_Reduce() must represent such a pair of values. MPI provides the following pairs of data types:

MPI_FLOAT_INT	(float,int)
MPI_DOUBLE_INT	(double,int)
MPI_LONG_INT	(long,int)
MPI_SHORT_INT	(short,int)
MPI_LONG_DOUBLE_INT	(long double,int)
MPI_2INT	(int,int)

For an MPI_Reduce() operation, all participating processes must specify the same values for the parameters count, type, op, and root. The send buffers sendbuf and the receive buffer recvbuf must have the same size. At the root process, they must denote disjoint memory areas. An in-place version can be activated by passing MPI_IN_PLACE for sendbuf at the root process. In this case, the input data block is taken from the recvbuf parameter at the root process, and the resulting accumulated value then replaces this input data block after the completion of MPI_Reduce().

Example: As example, we consider the use of a global reduction operation using MPI_MAXLOC, see Fig. 5.7. Each process has an array of 30 values of type double, stored in array ain of length 30. The program part computes the maximum value for each of the 30 array positions as well as the rank of the process that stores this maximum value. The information is collected at process 0: the maximum values are stored in array aout and the corresponding process ranks are stored in array ind. For the collection of the information based on value pairs, a data structure is defined for the elements of array in and out, consisting of a double and an int value. □

```
double ain[30], aout[30];
int ind[30];
struct {double val; int rank;} in[30], out[30];
int i, my_rank, root=0;

MPI_Comm_rank (MPI_COMM_WORLD, &my_rank);
for (i=0; i<30; i++) {
   in[i].val = ain[i];
   in[i].rank = my_rank;
}
MPI_Reduce(in,out,30,MPI_DOUBLE_INT,MPI_MAXLOC,root,MPI_COMM_WORLD);
if (my_rank == root)
   for (i=0; i<30; i++) {
      aout[i] = out[i].val;
      ind[i] = out[i].rank;
   }
```

Fig. 5.7 Example for the use of MPI_Reduce() using MPI_MAXLOC for the reduction operator.

MPI supports the definition of user-defined reduction operations using the following MPI function:

```
int MPI_Op_create (MPI_User_function *function,
                   int commute,
                   MPI_Op *op).
```

The parameter function specifies a user-defined function which must declare the following four parameters:

void *in, void *out, int *len, MPI_Datatype *type.

The user-defined function must be associative. The parameter commute specifies whether the function is also commutative (commute=1) or not (commute=0). The call of MPI_Op_create() returns a reduction operation op which can then be used as parameter of MPI_Reduce().

Example: We consider the parallel computation of the scalar product of two vectors x and y of length m using p processes. Both vectors are partitioned into blocks of size local_m = m/p. Each block is stored by a separate process such that each process stores its local blocks of x and y in local vectors local_x and local_y. Thus, the process with rank my_rank stores the following parts of x and y

```
local_x[j] = x[j + my_rank * local_m];
local_y[j] = y[j + my_rank * local_m];
```

for $0 \leq j <$ local_m.

Fig. 5.8 shows a program part for the computation of a scalar product. Each process executes this program part and computes a scalar product for its local blocks in local_x and local_y. The result is stored in local_dot. An MPI_Reduce() operation with reduction operation MPI_SUM is then used to add up the local results. The final result is collected at process 0 in variable dot.   □

```
int j, m, p, local_m;
float local_dot, dot;
float local_x[100], local_y[100];
MPI_Status status;

MPI_Comm_rank( MPI_COMM_WORLD, &my_rank);
MPI_Comm_size( MPI_COMM_WORLD, &p);
if (my_rank == 0) scanf("%d",&m);
local_m = m/p;
local_dot = 0.0;
for (j=0; j < local_m; j++)
    local_dot = local_dot + local_x[j] * local_y[j];
MPI_Reduce(&local_dot, &dot,1, MPI_FLOAT, MPI_SUM,0, MPI_COMM_WORLD);
```

Fig. 5.8 MPI program for the parallel computation of a scalar product.

5.2.1.3 Scan operation as further reduction operation

MPI supports a further reduction operation, which is referred to as prefix sum, inclusive scan, or simply scan. In general, a scan operation is based on a binary associative reduction operation $\oplus$ and is defined as follows, see [36]. The input for a scan operation is a list $[x_1, x_2, \ldots, x_n]$. The result of the scan operation is a new list $[y_1, y_2, \ldots, y_n]$ of the same length with the entries defined as follows: $y_1 = x_1$ and

$$y_k = y_{k-1} \oplus x_k = x_1 \oplus x_2 \oplus \ldots \oplus x_k \qquad (5.1)$$

for $2 \leq k \leq n$. MPI supports a scan operation with two operations: MPI_Scan() and MPI_Exscan(). For both operations, the length of the lists corresponds to the number of participating processes. MPI_Scan() has the following syntax:

```
int MPI_Scan (void *sendbuf,
              void *recvbuf,
              int count,
              MPI_Datatype type,
              MPI_Op op,
              MPI_Comm comm)
```

The parameters have the same meaning as for MPI_Reduce(): sendbuf is a send buffer in which each process provides its local data block for the scan operation. The parameter recvbuf specifies the receive buffer provided by each of the participating processes. This buffer is used to store the result for the corresponding process, which is obtained by a element-wise reduction. In particular, there is no root process, since each process receives a partial result. The parameter count specifies the number of elements that are provided by each process in sendbuf, type is the data type of each of these elements. The parameter op specifies the reduction operation to be used for the element-wise reduction. This must be an *associative* operation as described above. The same predefined reduction operations can be used as for

MPI_Reduce(). Moreover, the use of user-defined reduction operations is also possible. All buffers must have the same size, since the reduction is performed element-wise. MPI_Scan() is an inclusive scan operation, i.e., the result buffer for process i also includes the values in the local sendbuf buffer of i. This corresponds to the definition of scan in Equation (5.1).

The MPI_Exscan() operation has the same parameters as the MPI_Scan() operation. In contrast to MPI_Scan(), MPI_Exscan() is an exclusive scan, i.e., the result for process i does not include the values in the local sendbuf buffer of i. Thus, for processes with rank $i > 0$, the elements in recvbuf result by the element-wise reduction of the elements in the sendbuf buffers of all processes j with $j < i$. The recvbuf of the process with rank 0 contains undefined values.

5.2.1.4 Gather operation

For a gather operation, each process provides a block of data which are collected at a root process, see Sect. 3.6.2. In contrast to MPI_Reduce(), no reduction operation is applied. Thus, for p processes, the data block collected at the root process is p times larger than the individual blocks provided by each process. A gather operation is performed by calling the following MPI function

```
int MPI_Gather(void *sendbuf,
               int sendcount,
               MPI_Datatype sendtype,
               void *recvbuf,
               int recvcount,
               MPI_Datatype recvtype,
               int root,
               MPI_Comm comm).
```

The parameter sendbuf specifies the send buffer which is provided by each participating process. Each process provides sendcount elements of type sendtype. The parameter recvbuf is the receive buffer that is provided by the root process. No other process has to allocate a receive buffer. The root process receives recvcount elements of type recvtype from each process of communicator comm and stores them in the order of the ranks of the processes according to comm. For p processes the effect of the MPI_Gather() call can also be achieved if each process, including the root process, calls a send operation

```
MPI_Send (sendbuf, sendcount, sendtype, root, my_rank, comm)
```

and the root process executes p receive operations

```
MPI_Recv (recvbuf+i*recvcount*extent,
          recvcount, recvtype, i, i, comm, &status)
```

where i enumerates all processes of comm. The number of bytes used for each element of the data blocks is stored in extend and can be determined by calling the function MPI_Type_extent(recvtype, &extent). For a correct execution of MPI_Gather(), each process must specify the same root process root. Moreover, each process must specify the same element data type and the same number of elements to be sent. Fig. 5.9 shows a program part in which process 0 collects 100 integer values from each process of a communicator.

```
MPI_Comm comm;
int sendbuf[100], my_rank, root = 0, gsize, *rbuf;
MPI_Comm_rank (comm, &my_rank);
if (my_rank == root) {
    MPI_Comm_size (comm, &gsize);
    rbuf = (int *) malloc (gsize*100*sizeof(int));
}
MPI_Gather(sendbuf, 100, MPI_INT, rbuf, 100, MPI_INT, root, comm);
```

Fig. 5.9 Example for the application of MPI_Gather().

MPI provides a variant of MPI_Gather() for which each process can provide a *different* number of elements to be collected. The variant is MPI_Gatherv() which uses the same parameters as MPI_Gather() with the following two changes:

• the integer parameter recvcount is replaced by an integer array recvcounts of length p where recvcounts[i] denotes the number of elements provided by process i;

• there is an additional parameter displs after recvcounts. This is also an integer array of length p and displs[i] specifies at which position of the receive buffer of the root process the data block of process i is stored. Only the root process must specify the array parameters recvcounts and displs.

The effect of an MPI_Gatherv() operation can also be achieved if each process executes the send operation described above and the root process executes the following p receive operations

```
MPI_Recv(recvbuf+displs[i]*extent, recvcounts[i], recvtype, i, i,
         comm, &status).
```

For a correct execution of MPI_Gatherv(), the parameter sendcount specified by process i must be equal to the value of recvcounts[i] specified by the root process. Moreover, the send and receive types must be identical for all processes. The array parameters recvcounts and displs specified by the root process must be chosen such that no location in the receive buffer is written more than once, i.e., an overlapping of received data blocks is not allowed.

Fig. 5.10 shows an example for the use of MPI_Gatherv() which is a generalization of the example in Fig. 5.9: each process provides 100 integer values, but the blocks received are stored in the receive buffer in such a way that there is a free gap between neighboring blocks; the size of the gaps can be controlled by parameter displs. In Fig. 5.10, stride is used to define the size of the gap, and the gap size is set to 10. An error occurs for stride $<$ 100, since this would lead to an overlapping in the receive buffer.

```
MPI_Comm comm;
int sbuf[100];
int my_rank, root = 0, gsize, *rbuf, *displs, *rcounts, stride=110;
MPI_Comm_rank (comm, &my_rank);
if  (my_rank == root) {
    MPI_Comm_size (comm, &gsize);
    rbuf = (int *) malloc(gsize*stride*sizeof(int));
    displs = (int *) malloc(gsize*sizeof(int));
    rcounts = (int *) malloc(gsize*sizeof(int));
    for (i = 0; i < gsize; i++) {
        displs[i] = i*stride;
        rcounts[i] = 100;
    }
}
MPI_Gatherv(sbuf,100,MPI_INT,rbuf,rcounts,displs,MPI_INT,root,comm);
```

Fig. 5.10 Example for the use of MPI_Gatherv().

5.2.1.5 Scatter operation

For a scatter operation, a root process provides a different data block for each participating process. By executing the scatter operation, the data blocks are distributed to
these processes, see Sect. 3.6.2. In MPI, a scatter operation is performed by calling

```
int MPI_Scatter (void *sendbuf,
                 int sendcount,
                 MPI_Datatype sendtype,
                 void *recvbuf,
                 int recvcount,
                 MPI_Datatype recvtype,
                 int root,
                 MPI_Comm comm)
```

where sendbuf is the send buffer provided by the root process root which contains
a data block for each process of the communicator comm. Each data block contains
sendcount elements of type sendtype. In the send buffer, the blocks are ordered
in rank order of the receiving process. The data blocks are received in the receive
buffer recvbuf provided by the corresponding process. Each participating process
including the root process must provide such a receive buffer. For p processes, the
effects of MPI_Scatter() can also be achieved by letting the root process execute
p send operations

```
MPI_Send (sendbuf+i*sendcount*extent, sendcount, sendtype, i, i, comm)
```

for $i = 0, \ldots, p - 1$. Each participating process executes the corresponding receive
operation

```
MPI_Recv (recvbuf, recvcount, recvtype, root, my_rank, comm, &status)
```

For a correct execution of MPI_Scatter(), each process must specify the same
root, the same data types, and the same number of elements.

Similar to MPI_Gather(), there is a generalized version MPI_Scatterv() of MPI_Scatter() for which the root process can provide data blocks of different size. MPI_Scatterv() uses the same parameters as MPI_Scatter() with the following two changes:

- the integer parameter sendcount is replaced by the integer array sendcounts where sendcounts[i] denotes the number of elements sent to process i for i $= 0, \ldots, p - 1$.
- There is an additional parameter displs after sendcounts which is also an integer array with p entries, displs[i] specifies from which position in the send buffer of the root process the data block for process i should be taken.

The effect of an MPI_Scatterv() operation can also be achieved by point-to-point operations: the root process executes p send operations

```
MPI_Send (sendbuf+displs[i]*extent,sendcounts[i],sendtype,i,i,comm)
```

and each process executes the receive operation described above.

For a correct execution of MPI_Scatterv(), the entry sendcounts[i] specified by the root process for process i must be equal to the value of recvcount specified by process i. In accordance to MPI_Gatherv(), it is required that the arrays sendcounts and displs are chosen such that no entry of the send buffer is sent to more than one process. This restriction is imposed for symmetry reasons with MPI_Gatherv() although this is not essential for a correct behavior. The program in Fig. 5.11 illustrates the use of a scatter operation. Process 0 distributes 100 integer values to each other process such that there is a gap of 10 elements between neighboring send blocks.

```
MPI_Comm comm;
int rbuf[100];
int my_rank, root = 0, gsize, *sbuf, *displs, *scounts, stride=110;
MPI_Comm_rank (comm, &my_rank);
if (my_rank == root) {
    MPI_Comm_size (comm, &gsize);
    sbuf = (int *) malloc(gsize*stride*sizeof(int));
    displs = (int *) malloc(gsize*sizeof(int));
    scounts = (int *) malloc(gsize*sizeof(int));
    for (i=0; i<gsize; i++) {
        displs[i] = i*stride; scounts[i]=100;
    }
}
MPI_Scatterv(sbuf,scounts,displs,MPI_INT,rbuf,100,MPI_INT,root,comm);
```

Fig. 5.11 Example for the use of an MPI_Scatterv() operation.

5.2.1.6 Multi-broadcast operation

For a multi-broadcast operation, each participating process contributes a block of data which could, for example, be a partial result from a local computation. By executing the multi-broadcast operation, all blocks will be provided to all processes. There is no distinguished root process, since each process obtains all blocks provided. In MPI, a multi-broadcast operation is performed by calling the function

```
int MPI_Allgather (void *sendbuf,
                    int sendcount,
                    MPI_Datatype sendtype,
                    void *recvbuf,
                    int recvcount,
                    MPI_Datatype recvtype,
                    MPI_Comm comm)
```

where sendbuf is the send buffer provided by each process containing the block of data. The send buffer contains sendcount elements of type sendtype. Each process also provides a receive buffer recvbuf in which all received data blocks are collected in the order of the ranks of the sending processes. The values of the parameters sendcount and sendtype must be the same as the values of recvcount and recvtype. In the following example, each process contributes a send buffer with 100 integer values which are collected by a multi-broadcast operation at each process:

```
int sbuf[100], gsize, *rbuf;
MPI_Comm_size (comm, &gsize);
rbuf = (int*) malloc (gsize*100*sizeof(int));
MPI_Allgather (sbuf, 100, MPI_INT, rbuf, 100, MPI_INT, comm);
```

For an MPI_Allgather() operation, each process must contribute a data block of the same size. There is a vector version of MPI_Allgather() which allows each process to contribute a data block of a different size. This vector version is obtained as a similar generalization as MPI_Gatherv() and is performed by calling the following function:

```
int MPI_Allgatherv (void *sendbuf,
                     int sendcount,
                     MPI_Datatype sendtype,
                     void *recvbuf,
                     int *recvcounts,
                     int *displs,
                     MPI_Datatype recvtype,
                     MPI_Comm comm).
```

The parameters have the same meaning as for MPI_Gatherv().

Figure 5.12 shows an example for the use of an MPI_Allgatherv() operation for the implementation of a matrix-vector multiplication $A \cdot b = c$ based on a row-wise block distribution of the matrix, see also Figure 3.12 in Sect. 3.7. The program illustrates the case that the number n of rows of the matrix is not necessarily a multiple of the number p of processes employed. Each process obtains about the same

```
MPI_Comm comm;
int me, p, n, i, j, vec_start, vec_end, elem_per_p, elem_rem, local_size;
int vec_lens[MAX_PROCS], vec_offsets[MAX_PROCS];
double local_c[SIZE], b[SIZE], c[SIZE], local_A[SIZE][SIZE];

MPI_Comm_size (comm, &p);
MPI_Comm_rank (comm, &me);
n = SIZE;
elem_per_p = n / p; elem_rem = n % p;
if (me < n % p) {
  local_size = n / p + 1;
  vec_start = me * (elem_per_p + 1);
  vec_end = vec_start + elem_per_p + 1;
}
else {
  local_size = n / p;
  vec_start = me * (elem_per_p);
  vec_end = vec_start + elem_per_p;
}
for (i = 0; i < local_size; i++) local_c[i] = 0.0;
for (i = 0; i < local_size; i++)
    for (j = 0; j < SIZE; j++)
        local_c[i] = local_c[i] + local_A[i][j] * b[j];
for (i = 0; i < p; ++i)
    vec_lens[i] = i < elem_rem ? elem_per_p + 1 : elem_per_p;
vec_offsets[0] = 0;
for (i = 1; i < p; ++i)
    vec_offsets[i] = vec_offsets[i - 1] + vec_lens[i - 1];
MPI_Allgatherv(local_c, vec_lens[me], MPI_DOUBLE, c, vec_lens,
        vec_offsets, MPI_DOUBLE, MPI_COMM_WORLD);
```

Fig. 5.12 MPI program implementing a parallel matrix-vector multiplication using an MPI_Allgatherv() operation to obtain a replicated distribution of the result vector.

number of rows. If n is not a multiple of p, the first processes obtain an additional matrix row until no more rows are available. The variable local_size is used to compute the exact number of rows assigned to each of the processes. The vector b is assumed to be available in a replicated distribution. The same notation as in Figure 3.12 is used: each process stores local_size rows of the matrix in its local array local_A. Each process uses its local array local_c to store the locally computed part of the result vector. A replicated distribution of the result vector c is established by using a multi-broadcast operation. The use of the vector version MPI_Allgatherv() guarantees that the program also works if the number of rows of the matrix is not a multiple of the number of MPI processes employed. For this operation, the array vec_lens[] of length p collects the number of elements contributed by each of the processes. The array vec_offsets[] stores the positions of the blocks provided by the different processes in the replicated result.

5.2.1.7 Multi-accumulation operation

For a multi-accumulation operation, each participating process performs a separate single-accumulation operation for which each process provides a different block of data, see Sect. 3.6.2. MPI provides a version of a multi-accumulation with a restricted functionality: each process provides the same data block for each single-accumulation operation. This can be illustrated by the following diagram:

$$
\begin{array}{ll}
P_0 : x_0 & P_0 : x_0 + x_1 + \ldots + x_{p-1} \\
P_1 : x_1 & P_1 : x_0 + x_1 + \ldots + x_{p-1} \\
\vdots \quad \overset{MPI-accumulation(+)}{\Longrightarrow} & \vdots \\
P_{p-1} : x_n & P_{p-1} : x_0 + x_1 + \ldots + x_{p-1}
\end{array}
$$

In contrast to the general version described in Sect. 3.6.2, each of the processes $P_0, \ldots, P_{p-1}$ only provides one data block for $k = 0, \ldots, p-1$, expressed as $P_k : x_k$. After the operation, each process has accumulated the *same* result block, represented by $P_k : x_0 + x_1 + \ldots + x_{p-1}$. Thus, a multi-accumulation operation in MPI has the same effect as a single-accumulation followed by a single-broadcast operation which distributes the accumulated data block to all processes. The MPI operation provided has the following syntax:

```
int MPI_Allreduce (void *sendbuf,
                   void *recvbuf,
                   int count,
                   MPI_Datatype type,
                   MPI_Op op,
                   MPI_Comm comm)
```

where `sendbuf` is the send buffer in which each process provided its local data block. The parameter `recvbuf` specifies the receive buffer in which each process of the communicator `comm` collects the accumulated result. Both buffer contain `count` elements of type `type`. The reduction operation `op` is used. Each process must specify the same size and type for the data block.

Example: A multi-accumulation operation is used for the parallel computation of a matrix-vector multiplication $c = A \cdot b$ of an $n \times m$ matrix A with an m-dimensional vector b. The result is stored in the n-dimensional vector c. The matrix A is assumed to be distributed in a column-oriented blockwise way such that each of the p processes stores `local_m` = `m/p` contiguous columns of A in its local memory, see also Sect. 3.5 on data distributions. Correspondingly, vector b is distributed in a blockwise way among the processes. The matrix-vector multiplication is performed in parallel as described in Sect. 3.7, see also Fig. 3.15. Figure 5.13 shows an outline of an MPI implementation. The blocks of columns stored by each process are stored in the two-dimensional array a which contains n rows and `local_m` columns. Each process stores its local columns consecutively in this array. The one-dimensional array `local_b` contains for each process its block of b of length `local_m`. Each process computes n partial scalar products for its local block of columns using par-

```
int m, local_m, n, p;
float a[MAX_N][MAX_LOC_M], local_b[MAX_LOC_M];
float c[MAX_N], sum[MAX_N];
local_m = m/p;
for (i=0; i<n; i++) {
   sum[i] = 0;
   for (j=0; j<local_m; j++)
      sum[i] = sum[i] + a[i][j]*local_b[j];
}
MPI_Allreduce (sum, c, n, MPI_FLOAT, MPI_SUM, comm);
```

Fig. 5.13 MPI program fragment to compute a matrix-vector multiplication with a column-blockwise distribution of the matrix using an MPI_Allreduce() operation.

tial vectors of length local_m. The global accumulation to the final result is performed with an MPI_Allreduce() operation, providing the result to all processes in a replicated way. □

5.2.1.8 Total exchange

For a total exchange operation, each process provides a different block of data for each other process, see Sect. 3.6.2. The operation has the same effect as if each process performs a separate scatter operation (sender view), or as if each process performs a separate gather operation (receiver view). In MPI, a total exchange is performed by calling the function

```
int MPI_Alltoall (void *sendbuf,
                  int sendcount,
                  MPI_Datatype sendtype,
                  void *recvbuf,
                  int recvcount,
                  MPI_Datatype recvtype,
                  MPI_Comm comm)
```

where sendbuf is the send buffer in which each process provides for each process (including itself) a block of data with sendcount elements of type sendtype. The blocks are arranged in rank order of the target process. Each process also provides a receive buffer recvbuf in which the data blocks received from the other processes are stored. Again, the blocks received are stored in rank order of the sending processes. For p processes, the effect of a total exchange can also be achieved if each of the p processes executes p send operations

```
MPI_Send (sendbuf+i*sendcount*extent, sendcount, sendtype,i, my_rank, comm)
```

as well as p receive operations

```
MPI_Recv (recvbuf+i*recvcount*extent, recvcount, recvtype,i, i, comm, &status)
```

where i is the rank of one of the p processes and therefore lies between 0 and $p-1$.

For a correct execution, each participating process must provide for each other process data blocks of the same size and must also receive from each other process data blocks of the same size. Thus, all processes must specify the same values for sendcount and recvcount. Similarly, sendtype and recvtype must be the same for all processes. If data blocks of different size should be exchanged, the vector version must be used. This has the following syntax

```
int MPI_Alltoallv (void *sendbuf,
                    int *scounts,
                    int *sdispls,
                    MPI_Datatype sendtype,
                    void *recvbuf,
                    int *rcounts,
                    int *rdispls,
                    MPI_Datatype recvtype,
                    MPI_Comm comm).
```

For each process i, the entry scounts[j] specifies how many elements of type sendtype process i sends to process j. The entry sdispls[j] specifies the start position of the data block for process j in the send buffer of process i. The entry rcounts[j] at process i specifies how many elements of type recvtype process i receives from process j. The entry rdispls[j] at process i specifies at which position in the receive buffer of process i the data block from process j is stored.

For a correct execution of MPI_Alltoallv(), scounts[j] at process i must have the same value as rcounts[i] at process j. For p processes, the effect of Alltoallv() can also be achieved, if each of the processes executes p send operations

```
MPI_Send (sendbuf+sdispls[i]*sextent, scounts[i],
          sendtype, i, my_rank, comm)
```

and p receive operations

```
MPI_Recv (recvbuf+rdispls[i]*rextent, rcounts[i],
          recvtype, i, i, comm, &status)
```

where i is the rank of one of the p processes and therefore lies between 0 and $p-1$.

5.2.2 Deadlocks with Collective Communication

Similar to single transfer operations, different behavior can be observed for collective communication operations, depending on the use of internal system buffers by the MPI implementation. A careless use of collective communication operations may lead to **deadlocks** see also Section 3.8.4 (Page 162) for the occurrence of deadlocks with single transfer operations. This can be illustrated for MPI_Bcast() operations: we consider two MPI processes which execute two MPI_Bcast() operations in opposite order

```
switch (my_rank) {
case 0: MPI_Bcast (buf1, count, type, 0, comm);
        MPI_Bcast (buf2, count, type, 1, comm);
        break;
case 1: MPI_Bcast (buf2, count, type, 1, comm);
        MPI_Bcast (buf1, count, type, 0, comm);
}
```

Executing this piece of program may lead to two different error situations:

1. The MPI runtime system may match the first MPI_Bcast() call of each process. Doing this results in an error, since the two processes specify different roots.
2. The runtime system may match the MPI_Bcast() calls with the same root, as it has probably been intended by the programmer. Then a **deadlock** may occur if no system buffers are used or if the system buffers are too small. Collective communication operations are always **blocking**; thus, the operations are *synchronizing* if no or too small system buffers are used. Therefore, the first call of MPI_Bcast() blocks the process with rank 0 until the process with rank 1 has called the corresponding MPI_Bcast() with the same root. But his cannot happen, since process 1 is blocked due to its first MPI_Bcast() operation, waiting for process 0 to call its second MPI_Bcast(). Thus, a classical deadlock situation with cyclic waiting results.

The error or deadlock situation can be avoided in this example by letting the participating processes call the matching collective communication operations in the same order.

Deadlocks can also occur when mixing collective communication and single-transfer operations. This can be illustrated by the following example:

```
switch (my_rank) {
case 0: MPI_Bcast (buf1, count, type, 0, comm);
        MPI_Send (buf2, count, type, 1, tag, comm);
        break;
case 1: MPI_Recv (buf2, count, type, 0, tag, comm, &status);
        MPI_Bcast (buf1, count, type, 0, comm);
}
```

If no system buffers are used by the MPI implementation, a deadlock because of cyclic waiting occurs: process 0 blocks when executing MPI_Bcast(), until process 1 executes the corresponding MPI_Bcast() operation. Process 1 blocks when executing MPI_Recv(), until process 0 executes the corresponding MPI_Send() operation, resulting in cyclic waiting. This can be avoided if both processes execute their corresponding communication operations in the same order.

The **synchronization behavior** of collective communication operations depends on the use of system buffers by the MPI runtime system. If no internal system buffers are used or if the system buffers are too small, collective communication operations may lead to the synchronization of the participating processes. If system buffers are used, there is not necessarily a synchronization. This can be illustrated by the following example:

```
switch (my_rank) {
case 0: MPI_Bcast (buf1, count, type, 0, comm);
        MPI_Send (buf2, count, type, 1, tag, comm);
        break;
case 1: MPI_Recv (buf2, count, type, MPI_ANY_SOURCE, tag,
            comm, &status);
        MPI_Bcast (buf1, count, type, 0, comm);
        MPI_Recv (buf2, count, type, MPI_ANY_SOURCE, tag,
            comm, &status);
        break;
case 2: MPI_Send (buf2, count, type, 1, tag, comm);
        MPI_Bcast (buf1, count, type, 0, comm);
}
```

After having executed MPI_Bcast(), process 0 sends a message to process 1 using MPI_Send(). Process 2 sends a message to process 1 before executing an MPI_Bcast() operation. Process 1 receives two messages from MPI_ANY_SOURCE, one before and one after the MPI_Bcast() operation. The question is which message will be received from process 1 by which MPI_Recv(). Two execution orders are possible:

1. Process 1 first receives the message from process 2:

process$_0$	process$_1$	process$_2$
	MPI_Recv() $\Longleftarrow$	MPI_Send()
MPI_Bcast()	MPI_Bcast()	MPI_Bcast()
MPI_Send() $\Longrightarrow$	MPI_Recv()	

This execution order may occur independently from the fact that system buffers are used or not. In particular, this execution order is possible, also if the the calls of MPI_Bcast() are synchronizing.

2. Process 1 first receives the message from process 0:

process$_0$	process$_1$	process$_2$
MPI_Bcast()		
MPI_Send() $\Longrightarrow$	MPI_Recv()	
	MPI_Bcast()	
	MPI_Recv() $\Longleftarrow$	MPI_Send()
		MPI_Bcast()

This execution order can only occur, if large-enough system buffers are used, because otherwise process 0 cannot finish its MPI_Bcast() call before process 1 has started its corresponding MPI_Bcast().

Thus, a non-deterministic program behavior results depending on the use of system buffers. Such a program is correct only if both execution orders lead to the intended result. The previous examples have shown that collective communication operations are synchronizing only if the MPI runtime system does not use system buffers to store messages locally before their actual transmission. Thus, when writing a parallel program, the programmer cannot rely on the expectation that collective communication operations lead to a synchronization of the participating processes.

To synchronize a group of processes, MPI provides the operation

```
MPI_Barrier (MPI_Comm comm).
```

The effect of this operation is that all processes belonging to the group of communicator comm are blocked until all other processes of this group also have called this operation.

5.2.3 Nonblocking Collective Communication Operations

Similar to nonblocking point-to-point operations, MPI also supports nonblocking collective communication operations to make an overlapping of communication and computation possible. The usage model for nonblocking collective communication operations is the same as for nonblocking point-to-point operations, see Section 5.1.3: A call of a nonblocking collective operation initiates the corresponding operation locally and a separate completion operation is required to test whether the operation has been completed locally. In particular, the call of a nonblocking collective communication operation immediately returns control to the calling process and the specified communication operation is executed in the background. During this background communication, the calling process can execute other local computation operations, thus an overlapping communication and computation is achieved. However, only after the return of the completion operation, it is safe to overwrite the buffer resources that have been provided by the calling process for the communication operation. A process can issue multiple nonblocking collective operations one after another and multiple nonblocking collective operations can be outstanding. For the nonblocking collective communication operations, the same completion operations, such as MPI_Test() or MPI_Wait(), are used as for nonblocking point-to-point operations, see Section 5.1.3. After completion, a nonblocking collective operation has the same effect on data movements between processes as its blocking counterpart. For a full coverage of nonblocking collective operations, we refer to [69]. Here we only give two examples, a nonblocking broadcast and a nonblocking multi-broadcast operation in vector version. A nonblocking broadcast operation is started by the following MPI call:

```
int MPI_Ibcast (void *message,
                int count,
                MPI_Datatype type,
                int root,
                MPI_Comm comm,
                MPI_Request *request).
```

As for the blocking counterpart, root denotes the process which sends the data block. This process provides the data block to be sent in the parameter message. The other processes specify message as their receive buffer. The parameter count denotes the number of elements in the data block, type is the data type of the elements of the data block, comm is the communicator handle. Compared to the blocking counterpart, there is an additional parameter of type MPI_Request which can be used for the identification of the specific communication operation by a following completion operations. The usage model is the same as for nonblocking point-to-point operations. Figure 5.14 shows an example for the usage of MPI_Ibcast(): After initiating the nonblocking broadcast operation by MPI_Ibcast(), the participating processes execute local computations in the function do_local_computations() and then use MPI_Wait() as completion operation to wait for the broadcast operation to be finished locally. The local computations in do_local_computations() should not involve the buffer commbuf used by MPI_Ibcast(): for the root process, this is the send buffer for the broadcast that should not be modified before the return of the completion operation; for the other processes, this is the receive buffer that is not filled with the broadcast message before the return of the completion operation. MPI_Wait() is blocking and returns control to the calling process only after the MPI_Ibcast() has been terminated locally. The program in Figure 5.14 has the property that the local computations in do_local_computations() are overlapped with the broadcast operation performed in the background. Ideally, the broadcast operation would be finished when control returns from do_local_computations(). In this case, the following MPI_Wait() would return immediately and would not cause any waiting times.

```
MPI_Comm comm;
int commbuf[100], my_rank, root = 0;
MPI_Request request;
MPI_Comm_rank (comm, &my_rank);
if (my_rank == root)   /* only root process fills buffer */
    fill_buf (commbuf);
/* now follows nonblocking communication and overlapping computation */
MPI_Ibcast(commbuf, 100, MPI_INT, root, comm, &request);
do_local_computations(); /* local computations not involving commbuf */
MPI_Wait(&request,MPI_STATUS_IGNORE);
```

Fig. 5.14 Program fragment illustrating the usage of MPI_Ibcast() and the overlapping of communication and computation.

There are also nonblocking versions for the vector versions of the blocking collective communication operations described in Section 5.2.1. As example, we consider a nonblocking multi-broadcast operation in vector version. This operation is started by the following MPI call:

```
int MPI_Iallgatherv (void *sendbuf,
                     int sendcount,
                     MPI_Datatype sendtype,
                     void *recvbuf,
                     int *recvcounts,
                     int *displs,
                     MPI_Datatype recvtype,
                     MPI_Comm comm,
                     MPI_Request *request).
```

The first eight parameters have the same meaning as for `MPI_Allgatherv()` and the additional parameter of type `MPI_Request` can again be used for the identification of the specific communication operation by a following completion operations.

5.3 Process Groups and Communicators

MPI allows the construction of subsets of processes by defining *groups* and *communicators*. A **process group** (or **group** for short) is an ordered set of processes of an application program. Each process of a group gets an uniquely defined process number which is also called **rank**. The ranks of a group always start with 0 and continue consecutively up to the number of processes minus one. A process may be a member of multiple groups and may have different ranks in each of these groups. The MPI system handles the representation and management of process groups. For the programmer, a group is an object of type `MPI_Group` which can only be accessed via a **handle** which may be internally implemented by the MPI system as an index or a reference. Process groups are useful for the implementation of **task parallel programs** and are the basis for the communication mechanism of MPI.

In many situations, it is useful to partition the processes of a parallel program into disjoint subsets (groups) which perform independent tasks of the program. This is called **task parallelism**, see also Section 3.3.4. The execution of task parallel program parts can be obtained by letting the processes of a program call different functions or communication operations, depending on their process numbers. But task parallelism can be implemented much easier using the group concept.

5.3.1 Process Groups in MPI

MPI provides a lot of support for process groups. In particular, collective communication operations can be restricted to process groups by using the corresponding communicators. This is important for program libraries where the communication operations of the calling application program and the communication operations of functions of the program library must be distinguished. If the same communicator is used, an error may occur, e.g., if the application program calls `MPI_Irecv()` with communicator `MPI_COMM_WORLD` using source `MPI_ANY_SOURCE` and tag `MPI_ANY_TAG` immediately before calling a library function. This is dangerous, if the library functions also use `MPI_COMM_WORLD` and if the library function called sends data to the process which executes `MPI_Irecv()` as mentioned above, since this process may then receive library-internal data. This can be avoided by using separate communicators.

In MPI, each point-to-point communication and each collective communication is executed in a **communication domain**. There is a separate communication domain for each process group using the ranks of the group. For each process of a group, the corresponding communication domain is *locally* represented by a **communicator**. In MPI, there is a communicator for each process group and each communicator defines a process group. A communicator knows all other communicators of the same communication domain. This may be required for the internal implementation of communication operations. Internally, a group may be implemented as an array of process numbers where each array entry specifies the global process number of one process of the group.

For the programmer, an MPI communicator is an opaque data object of type `MPI_Comm`. MPI supports **intra-communicators** and **inter-communicators**. Intra-communicators support the execution of arbitrary collective communication operations on a single group of processes. Inter-communicators support the execution of point-to-point communication operations between two process groups. In the following, we only consider intra-communicators which we call communicators for short.

In the preceding sections, we have always used the predefined communicator `MPI_COMM_WORLD` for communication. This communicator comprises all processes participating in the execution of a parallel program. MPI provides several operations to build additional process groups and communicators. These operations are all based on existing groups and communicators. The predefined communicator `MPI_COMM_WORLD` and the corresponding group are normally used as starting point. The process group to a given communicator can be obtained by calling

```
int MPI_Comm_group (MPI_Comm comm, MPI_Group *group),
```

where `comm` is the given communicator and `group` is a pointer to a previously declared object of type `MPI_Group` which will be filled by the MPI call. A predefined group is `MPI_GROUP_EMPTY` which denotes an empty process group.

5.3.1.1 Operations on Process Groups

MPI provides operations to construct new process groups based on existing groups. The predefined empty group `MPI_GROUP_EMPTY` can also be used. The **union** of two existing groups `group1` and `group2` can be obtained by calling

```
int MPI_Group_union (MPI_Group group1,
                     MPI_Group group2,
                     MPI_Group *new_group).
```

The ranks in the new group `new_group` are set such that the processes in `group1` keep their ranks. The processes from `group2` which are not in `group1` get subsequent ranks in consecutive order. The **intersection** of two groups is obtained by calling

```
int MPI_Group_intersection (MPI_Group group1,
                            MPI_Group group2,
                            MPI_Group *new_group)
```

where the process order from group1 is kept for new_group. The processes in new_group get successive ranks starting from 0. The **set difference** of two groups is obtained by calling

```
int MPI_Group_difference (MPI_Group group1,
                          MPI_Group group2,
                          MPI_Group *new_group).
```

Again, the process order from group1 is kept. A sub_group of an existing group can be obtained by calling

```
int MPI_Group_incl (MPI_Group group,
                    int p,
                    int *ranks,
                    MPI_Group *new_group)
```

where ranks is an integer array with p entries. The call of this function creates a new group new_group with p processes which have ranks from 0 to p-1. Process i is the process which has rank ranks[i] in the given group group. For a correct execution of this operation, group must contain at least p processes, and for $0 \leq i < p$, the values ranks[i] must be valid process numbers in group which are different from each other. Processes can be deleted from a given group by calling

```
int MPI_Group_excl (MPI_Group group,
                    int p,
                    int *ranks,
                    MPI_Group *new_group).
```

This function call generates a new group new_group which is obtained from group by deleting the processes with ranks ranks[0],..., ranks[p-1]. Again, the entries ranks[i] must be valid process ranks in group which are different from each other.

Example: The following program fragment splits the group large_group with p processes into two groups: the group even_group that contains all processes with even process numbers in large_group and the group odd_group that contains all processes with odd process numbers in large_group:

```
for (i=0; i<(p+1)/2; i++) members[i] = 2 * i;
N_even = (p+1)/2;
N_odd = p - N_even;
MPI_Comm_group (MPI_COMM_WORLD, &large_group);
MPI_Group_incl (large_group, N_even, members, &even_group);
MPI_Group_excl (large_group, N_even, members, &odd_group);
```

The function MPI_Comm_group is used to create the group large_group to contain all processes the program has been started with. The array members is used to collect the processes that should be included in the new group even_group. N_even is the number of processes in even_group. □

Data structures of type MPI_Group cannot be directly accessed by the programmer. But MPI provides operations to obtain information about process groups. The **size** of a process group can be obtained by calling

```
int MPI_Group_size (MPI_Group group, int *size)
```

where the size of the group is returned in parameter `size`. The **rank** of the calling process in a group can be obtained by calling

```
int MPI_Group_rank (MPI_Group group, int *rank)
```

where the rank is returned in parameter `rank`. The function

```
int MPI_Group_compare (MPI_Group group1, MPI_Group group2, int *res)
```

can be used to check whether two group representations `group1` and `group2` describe the same group. The parameter value `res = MPI_IDENT` is returned if both groups contain the same processes in the same order. The parameter value `res = MPI_SIMILAR` is returned if both groups contain the same processes, but `group1` uses a different order than `group2`. The parameter value `res = MPI_UNEQUAL` means that the two groups contain different processes. The function

```
int MPI_Group_free (MPI_Group *group)
```

can be used to free a group representation if it is no longer needed. The group handle is set to `MPI_GROUP_NULL`.

5.3.1.2 Operations on Communicators

A new intra-communicator to a given group of processes can be generated by calling

```
int MPI_Comm_create (MPI_Comm comm,
                     MPI_Group group,
                     MPI_Comm *new_comm)
```

where `comm` specifies an existing communicator. The parameter `group` must specify a process group which is a subset of the process group associated with `comm`. For a correct execution, it is required that all processes of `comm` perform the call of `MPI_Comm_create()` and that each of these processes specifies the same group argument. As result of this call, each calling process which is a member of `group` obtains a pointer to the new communicator in `new_comm`. Processes not belonging to `group` get `MPI_COMM_NULL` as return value in `new_comm`.

MPI also provides functions to get information about communicators. These functions are implemented as local operations which do not involve communication to be executed. The size of the process group associated with a communicator `comm` can be requested by calling the function

```
int MPI_Comm_size (MPI_Comm comm, int *size).
```

The size of the group is returned in parameter `size`. For `comm = MPI_COMM_WORLD` the total number of processes executing the program is returned. The rank of a process in a particular group associated with a communicator `comm` can be obtained by calling

```
int MPI_Comm_rank (MPI_Comm comm, int *rank).
```

The group rank of the calling process is returned in `rank`. In previous examples, we have used this function to obtain the global rank of processes with `MPI_COMM_WORLD`. Two communicators `comm1` and `comm2` can be compared by calling

```
int MPI_Comm_compare (MPI_Comm comm1, MPI_Comm comm2, int *res)
```

The result of the comparison is returned in parameter `res`; `res = MPI_IDENT` is returned, if `comm1` and `comm2` denote the same communicator data structure. The value `res = MPI_CONGRUENT` is returned, if the associated groups of `comm1` and `comm2` contain the same processes with the same rank order. If the two associated groups contain the same processes in different rank order, `res = MPI_SIMILAR` is returned. If the two groups contain different processes, `res = MPI_UNEQUAL` is returned.

For the direct construction of communicators, MPI provides operations for the duplication, deletion and splitting of communicators. A communicator can be **duplicated** by calling the function

```
int MPI_Comm_dup (MPI_Comm comm, MPI_Comm *new_comm)
```

which creates a new intra-communicator `new_comm` with the same characteristics (assigned group and topology) as `comm`. The new communicator `new_comm` represents a new distinct communication domain. Duplicating a communicator allows the programmer to separate communication operations executed by a library from communication operations executed by the application program itself, thus avoiding any conflict. A communicator can be **deallocated** by calling the MPI operation

```
int MPI_Comm_free (MPI_Comm *comm).
```

This operation has the effect that the communicator data structure `comm` is freed, as soon as all pending communication operations performed with this communicator are completed. This operation could, e.g., be used to free a communicator which has previously been generated by duplication to separate library communication from communication of the application program. Communicators should not be assigned by simple assignments of the form `comm1 = comm2`, since a deallocation of one of the two communicators involved with `MPI_Comm_free()` would have a side-effect on the other communicator, even if this is not intended. A **splitting** of a communicator can be obtained by calling the function

```
int MPI_Comm_split (MPI_Comm comm,
                    int color,
                    int key,
                    MPI_Comm *new_comm).
```

The effect is that the process group associated with `comm` is partitioned into disjoint subgroups. The number of subgroups is determined by the number of different values of `color`. Each subgroup contains all processes which specify the same value for `color`. Within each subgroup, the processes are ranked in the order defined by argument value `key`. If two processes in a subgroup specify the same value for key, the order in the original group is used. If a process of comm specifies color = MPI_UNDEFINED, it is not a member of any of the subgroups generated. The subgroups are not directly provided in the form of an MPI_GROUP representation. Instead, each process of comm gets a pointer new_comm to the communicator of that subgroup which the process belongs to. For color = MPI_UNDEFINED, MPI_COMM_NULL is returned as new_comm.

Example: We consider a group of 10 processes each of which calls the operation MPI_Comm_split() with the following argument values [204]:

process	a	b	c	d	e	f	g	h	i	j
rank	0	1	2	3	4	5	6	7	8	9
color	0	⊥	3	0	3	0	0	5	3	⊥
key	3	1	2	5	1	1	1	2	1	0

This call generates three subgroups {f, g, a, d}, {e, i, c} und {h} which contain the processes in this order. In the table, the entry ⊥ represents color = MPI_UNDEFINED.   □

The operation MPI_Comm_split() can be used to prepare a task parallel execution. The different communicators generated can be used to perform communication within the task parallel parts, thus separating the communication domains.

5.3.2 *Process Topologies*

Each process of a process group has a unique rank within this group which can be used for the communication with this process. Although a process is uniquely defined by its group rank, it is often useful to have an alternative representation and access. This is the case if an algorithm performs computations and communication on a two-dimensional or a three-dimensional grid where grid points are assigned to different processes and the processes exchange data with their neighboring processes in each dimension by communication. In such situations, it is useful if the processes can be arranged according to the communication pattern in a grid structure such that they can be addressed via two-dimensional or three-dimensional coordinates. Then each process can easily address its neighboring processes in each dimension. MPI supports such a logical arrangement of processes by defining **virtual topologies** for intra-communicators, which can be used for communication within the associated process group. A virtual Cartesian grid structure of arbitrary dimension can be generated by calling

```
int MPI_Cart_create (MPI_Comm comm,
                     int ndims,
                     int *dims,
                     int *periods,
                     int reorder,
                     MPI_Comm *new_comm)
```

where comm is the original communicator without topology, ndims specifies the number of dimensions of the grid to be generated, dims is an integer array of size ndims such that dims[i] is the number of processes in dimension i. The entries of dims must be set such that the product of all entries is the number of processes contained in the new communicator new_comm. In particular, this product must not exceed the number of processes of the original communicator comm. The boolean array periods of size ndims specifies for each dimension whether the grid is periodic (entry 1 or true) or not (entry 0 or false) in this dimension. For reorder = false, the processes in new_comm have the same rank as in comm. For reorder = true, the runtime system is allowed to reorder processes, e.g., to obtain a better mapping of the process topology to the physical network of the parallel machine.

Example: We consider a communicator with 12 processes [204]. For ndims=2, using the initializations dims[0]=3, dims[1]=4, periods[0]=periods[1]=0, reorder=0, the call

```
MPI_Cart_create (comm, ndims, dims, periods, reorder, &new_comm)
```

generates a virtual 3×4 grid with the following group ranks and coordinates:

0 (0,0)	1 (0,1)	2 (0,2)	3 (0,3)
4 (1,0)	5 (1,1)	6 (1,2)	7 (1,3)
8 (2,0)	9 (2,1)	10 (2,2)	11 (2,3)

The Cartesian coordinates are represented in the form (row, column). In the communicator, the processes are ordered according to their rank row-wise in increasing order. □

To help the programmer to select a balanced distribution of the processes for the different dimensions, MPI provides the function

```
int MPI_Dims_create (int nnodes, int ndims, int *dims)
```

where ndims is the number of dimensions in the grid and nnodes is the total number of processes available. The parameter dims is an integer array of size ndims. After the call, the entries of dims are set such that the nnodes processes are balanced as close as possible among the different dimensions, i.e., each dimension has about equal size. But the size of a dimension i is set only if dims[i] = 0 when calling MPI_Dims_create(). The number of processes in a dimension j can be fixed by setting dims[j] to a positive value before the call. This entry is then not modified by this call and the other entries of dims are set by the call accordingly.

When defining a virtual topology, each process has a group rank, but also a position in the virtual grid topology which can be expressed by its Cartesian coordinates. For the translation between group ranks and Cartesian coordinates, MPI provides two operations. The operation

```
int MPI_Cart_rank (MPI_Comm comm, int *coords, int *rank)
```

translates the Cartesian coordinates provided in the integer array `coords` into a group rank and returns it in parameter `rank`. The parameter `comm` specifies the communicator with Cartesian topology. For the opposite direction, the operation

```
int MPI_Cart_coords (MPI_Comm comm,
                     int rank,
                     int ndims,
                     int *coords)
```

translates the group rank provided in `rank` to Cartesian coordinates, returned in integer array `coords`, for a virtual grid; `ndims` is the number of dimensions of the virtual grid defined for communicator `comm`.

Virtual topologies are typically defined to facilitate the determination of communication partners of processes. A typical communication pattern in many grid-based algorithms is that processes communicate with their neighboring processes in a specific dimension. To determine these neighboring processes, MPI provides the operation

```
int MPI_Cart_shift (MPI_Comm comm,
                    int dir,
                    int displ,
                    int *rank_source,
                    int *rank_dest).
```

where `dir` specifies the dimension for which the neighboring process should be determined. The parameter `displ` specifies the displacement, i.e., the distance to the neighbor. Positive values of `displ` request the neighbor in upward direction, negative values request for downward direction. Thus, `displ = -1` requests the neighbor immediately preceding, `displ = 1` requests the neighboring process which follows directly. The result of the call is that `rank_dest` contains the group rank of the neighboring process in the specified dimension and distance. The rank of the process for which the calling process is the neighboring process in the specified dimension and distance is returned in `rank_source`. Thus, the group ranks returned in `rank_dest` and `rank_source` can be used as parameters for `MPI_Sendrecv()`, as well as for separate `MPI_Send()` and `MPI_Recv()`, respectively.

Example: As example, we consider 12 processes that are arranged in a 3×4 grid structure with periodic connections [204]. Each process stores a floating-point value which is exchanged with the neighboring process in dimension 0, i.e., within the columns of the grid. In this example, the specification `displs = coord[1]` is used as displacement for `MPI_Cart_shift()`, i.e., the position in dimension 1 is used as displacement. Thus, the displacement increases with the column position, and in each column of the grid, a different exchange distance is used. `MPI_Cart_shift()` is used to determine the communication partners `dest` and `source` for each process. These are then used as parameters for `MPI_Sendrecv()`. The resulting program code is as follows:

```
int coords[2], dims[2], periods[2], source, dest, my_rank, reorder;
MPI_Comm comm_2d;
MPI_status status;
float a, b;
MPI_Comm_rank (MPI_COMM_WORLD, &my_rank);
dims[0] = 3; dims[1] = 4;
periods[0] = periods[1] = 1;
reorder = 0;
MPI_Cart_create (MPI_COMM_WORLD, 2, dims, periods, reorder, &comm_2d);
MPI_Cart_coords (comm_2d, my_rank, 2, coords);
MPI_Cart_shift (comm_2d, 0, coords[1], &source, &dest);
a = my_rank;
MPI_Sendrecv (&a, 1, MPI_FLOAT, dest, 0, &b, 1, MPI_FLOAT,
    source, 0, comm_2d, &status);
```

The following diagram illustrates the exchange. For each process, its rank, its Cartesian coordinates, and its communication partners in the form source/dest are given in this order. For example, for the process with rank=5, it is coords[1]=1, and therefore source=9 (lower neighbor in dimension 0) and dest=1 (upper neighbor in dimension 0).

0 (0,0) 0|0	1 (0,1) 9|5	2 (0,2) 6|10	3 (0,3) 3|3
4 (1,0) 4|4	5 (1,1) 1|9	6 (1,2) 10|2	7 (1,3) 7|7
8 (2,0) 8|8	9 (2,1) 5|1	10 (2,2) 2|6	11 (2,3) 11|11

□

If a virtual topology has been defined for a communicator, the corresponding grid can be partitioned into subgrids by using the MPI function

```
int MPI_Cart_sub (MPI_Comm comm,
                  int *remain_dims,
                  MPI_Comm *new_comm).
```

The parameter comm denotes the communicator for which the virtual topology has been defined. The subgrid selection is controlled by the integer array remain_dims which contains an entry for each dimension of the original grid.

Setting remain_dims[i] = 1 means that the ith dimension is kept in the subgrid; remain_dims[i] = 0 means that the ith dimension is dropped in the subgrid. In this case, the size of this dimension determines the number of subgrids generated in this dimension. A call of MPI_Cart_sub() generates a new communicator new_comm for each calling process, representing the corresponding subgroup of the subgrid to which the calling process belongs. The dimensions of the different subgrids result from the dimensions for which remain_dims[i] has been set to 1.

The total number of subgrids generated is defined by the product of the number of processes in all dimensions i for which remain_dims[i] has been set to 0.

Example: We consider a communicator comm for which a $2 \times 3 \times 4$ virtual grid topology has been defined. Calling

```
int MPI_Cart_sub (comm_3d, remain_dims, &new_comm)
```

with remain_dims=(1,0,1) generates three 2×4 grids and each process gets a communicator for its corresponding subgrid, see Fig. 5.15 for an illustration. □

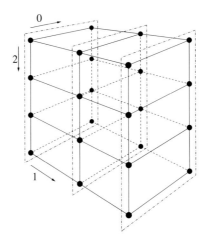

Fig. 5.15 Partitioning of a $(2 \times 3 \times 4)$ grid into three (2×4) grids.

MPI also provides functions to inquire information about a virtual topology that has been defined for a communicator. The MPI function

```
int MPI_Cartdim_get (MPI_Comm comm,int *ndims)
```

returns in parameter ndims the number of dimensions of the virtual grid associated with communicator comm. The MPI function

```
int MPI_Cart_get (MPI_Comm comm,
                  int maxdims,
                  int *dims,
                  int *periods,
                  int *coords)
```

returns information on the virtual topology defined for communicator comm. This virtual topology should have maxdims dimensions, and the arrays dims, periods, and coords should have this size. The following information is returned by this call: integer array dims contains the number of processes in each dimension of the virtual grid, the boolean array periods contains the corresponding periodicity information. The integer array coords contains the Cartesian coordinates of the calling process.

5.3.3 Timings and aborting processes

To measure the parallel execution times of program parts, MPI provides the function

```
double MPI_Wtime (void)
```

which returns as a floating-point value the number of seconds elapsed since a fixed point in time in the past. A typical usage for timing would be:

```
start = MPI_Wtime();
part_to_measure();
end = MPI_Wtime();
```

MPI_Wtime() does not return a system time, but the absolute time elapsed between the start and the end of a program part, including times at which the process executing part_to_measure() has been interrupted. The resolution of MPI_Wtime() can be requested by calling

```
double MPI_Wtick (void)
```

which returns the time between successive clock ticks in seconds as floating-point value. If the resolution is a microsecond, MPI_Wtick() will return 10^{-6}. The execution of all processes of a communicator can be aborted by calling the MPI function

```
int MPI_Abort (MPI_Comm comm, int error_code)
```

where error_code specifies the error code to be used, i.e., the behavior is as if the main program has been terminated with return error_code.

5.4 Advanced topics

For a continuous development of MPI, the MPI Forum has defined extensions to MPI as it has been described in the previous sections. These extensions are often referred to as MPI-2. The original MPI standard is referred to as MPI-1. The current version of MPI-1 is described in the MPI document, version 1.3 [67]. Since MPI-2 comprises all MPI-1 operations, each correct MPI-1 program is also a correct MPI-2 program. The most important extensions contained in MPI-2 are dynamic process management, one-sided communications, parallel I/0, and extended collective communications. In the following, we give a short overview of the most important extensions. For a more detailed description, we refer to the current version of the MPI-2 document, version 2.1, see [68].

5.4.1 Dynamic Process Generation and Management

MPI-1 is based on a **static process model**: the processes used for the execution of
a parallel program are implicitly created before starting the program. No processes
can be added during program execution. Inspired by PVM [77], MPI-2 extends this
process model to a **dynamic process model** which allows the creation and deletion
of processes at any time during program execution. MPI-2 defines the interface for
dynamic process management as a collection of suitable functions and gives some
advice for an implementation. But not all implementation details are fixed to support
an implementation for different operating systems.

5.4.1.1 MPI_Info objects

Many MPI-2 functions use an additional argument of type MPI_Info which allows
the provision of additional information for the function, depending on the specific
operating system used. But using this feature may lead to non-portable MPI programs. MPI_Info provides opaque objects where each object can store arbitrary
(key, value) pairs. In C, both entries are strings of type char, terminated with
\0. Since MPI_Info objects are opaque, their implementation is hidden from the
user. Instead, some functions are provided for access and manipulation. The most
important ones are described in the following. The function

```
int MPI_Info_create (MPI_Info *info)
```

can be used to generate a new object of type MPI_Info. Calling the function

```
int MPI_Info_set (MPI_Info info, char *key, char *value)
```

adds a new (key, value) pair to the MPI_Info structure info. If a value for the
same key was previously stored, the old value is overwritten. The function

```
int MPI_Info_get (MPI_Info info,
                  char *key,
                  int valuelen,
                  char *value,
                  int *flag)
```

can be used to retrieve a stored pair (key, value) from info. The programmer
specifies the value of key and the maximum length valuelen of the value entry.
If the specified key exists in info, the associated value is returned in parameter
value. If the associated value string is longer than valuelen, the returned string
is truncated after valuelen characters. If the specified key exists in info, true is
returned in parameter flag; otherwise, false is returned. The function

```
int MPI_Info_delete(MPI_Info info, char *key)
```

can be used to delete an entry (key, value) from info. Only the key has to be
specified.

5.4.1.2 Process creation and management

A number of separate MPI processes can dynamically be created within an MPI program by calling the function

```
int MPI_Comm_spawn (char *command,
                    char *argv[],
                    int maxprocs,
                    MPI_Info info,
                    int root,
                    MPI_Comm comm,
                    MPI_Comm *intercomm,
                    int errcodes[])
```

The parameter command specifies the name of the program to be executed by each of the processes, argv[] contains the arguments for this program. In contrast to the standard C convention, argv[0] is not the program name but the first argument for the program. An empty argument list is specified by MPI_ARGV_NULL. The parameter maxprocs specifies the number of processes to be started. If the MPI runtime system is not able to start maxprocs processes, an error message is generated. The parameter info specifies an MPI_Info data structure with (key, value) pairs providing additional instructions for the MPI runtime system on how to start the processes. This parameter could be used to specify the path of the program file as well as its arguments, but this may lead to non-portable programs. Portable programs should use MPI_INFO_NULL.

The parameter root specifies the number of the root process from which the new processes are spawned. Only this root process provides values for the preceding parameters. But the function MPI_Comm_spawn() is a collective operation, i.e., all processes belonging to the group of the communicator comm must call the function. The parameter intercomm contains an intercommunicator after the successful termination of the function call. This intercommunicator can be used for communication between the original group of comm and the group of processes just spawned.

The parameter errcodes is an array with maxprocs entries in which the status of each process to be spawned is reported. When a process could be spawned successfully, its corresponding entry in errcodes will be set to MPI_SUCCESS. Otherwise, an implementation-specific error code will be reported.

A successful call of MPI_Comm_spawn() starts maxprocs identical copies of the specified program and creates an intercommunicator which is provided to all calling processes. The new processes belong to a separate group and have a separate MPI_COMM_WORLD communicator comprising all processes spawned. The spawned processes can access the intercommunicator created by MPI_Comm_spawn() by calling the function

```
int MPI_Comm_get_parent(MPI_Comm *parent)
```

The requested intercommunicator is returned in parameter parent. Multiple MPI programs or MPI programs with different argument values can be spawned by calling the function

```
int MPI_Comm_spawn_multiple (int count,
                             char *commands[],
                             char **argv[],
                             int maxprocs[],
                             MPI_Info infos[],
                             int root,
                             MPI_Comm comm,
                             MPI_Comm *intercomm,
                             int errcodes[])
```

where count specifies the number of different programs to be started. Each of the following four arguments specifies an array with count entries where each entry has the same type and meaning as the corresponding parameters for MPI_Comm_spawn():
The argument commands[] specifies the names of the programs to be started, argv[] contains the corresponding arguments, maxprocs[] defines the number of copies to be started for each program, and infos[] provides additional instructions for each program. The other arguments have the same meaning as for MPI_comm_spawn().

After the call of MPI_Comm_spawn_multiple() has been terminated, the array errcodes[] contains an error status entry for each process created. The entries are arranged in the order given by the commands[] array. In total, errcodes[] contains

$$\sum_{i=0}^{count-1} \texttt{maxprocs[i]}$$

entries. There is a difference between calling MPI_Comm_spawn() multiple times and calling MPI_Comm_spawn_multiple() with the same arguments. A call of the function MPI_Comm_spawn_multiple() creates one communicator MPI_COMM_-WORLD for all newly created processes. Multiple calls of MPI_Comm_spawn() generate separate communicators MPI_COMM_WORLD, one for each process group created.

The attribute MPI_UNIVERSE_SIZE specifies the maximum number of processes that can be started in total for a given application program. The attribute is initialized by MPI_Init().

5.4.2 One-sided communication

MPI provides single-transfer and collective communication operations as described in the previous sections. For collective communication operations, each process of a communicator calls the communication operation to be performed. For single-transfer operations, a sender and a receiver process must cooperate and actively execute communication operations: in the simplest case, the sender executes an MPI_Send() operation and the receiver executes an MPI_Recv() operation. Therefore, this form of communication is also called *two-sided communication*. The position of the MPI_Send() operation in the sender process determines at which time the data are sent. Similarly, the position of the MPI_Recv() operation in the receiver process determines at which time the receiver stores the received data in its local address space.

In addition to two-sided communication, MPI-2 supports *one-sided communication*. Using this form of communication, a source process can access the address space at a target process without an active participation of the target process. This form of communication is also called Remote Memory Access (RMA). RMA facilitates communication for applications with dynamically changing data access patterns by supporting a flexible dynamic distribution of program data among the address spaces of the participating processes. But the programmer is responsible for the coordinated memory access. In particular, a concurrent manipulation of the same address area by different processes at the same time must be avoided to inhibit race conditions. Such race conditions cannot occur for two-sided communications.

5.4.2.1 Window objects

If a process A should be allowed to access a specific memory region of a process B using one-sided communication, process B must expose this memory region for external access. Such a memory region is called *window*. A window can be exposed by calling the function

```
int MPI_Win_create (void *base,
                    MPI_Aint size,
                    int displ_unit,
                    MPI_Info info,
                    MPI_Comm comm,
                    MPI_Win *win)
```

This is a collective call which must be executed by each process of the communicator comm. Each process specifies a window in its local address space that it exposes for RMA by other processes of the same communicator.

The starting address of the window is specified in parameter base. The size of the window is given in parameter size as number of bytes. For the size specification the predefined MPI type MPI_Aint is used instead of int to allow window sizes of more than 2^{32} bytes. The parameter displ_unit specifies the displacement (in bytes) between neighboring window entries used for one-sided memory accesses. Typically, displ_unit is set to 1 if bytes are used as unit, or to sizeof(type) if the window consists of entries of type type. The parameter info can be used to provide additional information for the runtime system. Usually, info=MPI_INFO_NULL is used. The parameter comm specifies the communicator of the processes which participate in the MPI_Win_create() operation. The call of MPI_Win_create() returns a window object of type MPI_Win in parameter win to the calling process. This window object can then be used for RMA to memory regions of other processes of comm.

A window exposed for external accesses can be closed by letting all processes of the corresponding communicator call the function

```
int MPI_Win_free (MPI_Win *win)
```

thus freeing the corresponding window object win. Before calling MPI_Win_free(), the calling process must have finished all operations on the specified window.

5.4.2.2 RMA operations

For the actual one-sided data transfer, MPI provides three *non-blocking* RMA operations: MPI_Put() transfers data from the memory of the calling process into the window of another process; MPI_Get() transfers data from the window of a target process into the memory of the calling process; MPI_Accumulate() supports the accumulation of data in the window of the target process. These operations are *non-blocking*: when control is returned to the calling process, this does not necessarily mean that the operation is completed. To test for the completion of the operation, additional synchronization operations like MPI_Win_fence() are provided as described below. Thus, a similar usage model as for non-blocking two-sided communication can be used. The local buffer of a RMA communication operation should not be updated or accessed until the subsequent synchronization call returns.

The transfer of a data block into the window of another process can be performed by calling the function

```
int MPI_Put (void *origin_addr,
             int origin_count,
             MPI_Datatype origin_type,
             int target_rank,
             MPI_Aint target_displ,
             int target_count,
             MPI_Datatype target_type,
             MPI_Win win)
```

where origin_addr specifies the start address of the data buffer provided by the calling process and origin_count is the number of buffer entries to be transferred. The parameter origin_type defines the type of the entries. The parameter target_rank specifies the rank of the target process which should receive the data block. This process must have created the window object win by a preceding MPI_Win_create() operation, together with all processes of the communicator group which the process calling MPI_Put() also belongs to. The remaining parameters define the position and size of the target buffer provided by the target process in its window: target_displ defines the displacement from the start of the window to the start of the target buffer, target_count specifies the number of entries in the target buffer, target_type defines the type of each entry in the target buffer. The data block transferred is stored in the memory of the target process at position target_addr := window_base + target_displ * displ_unit where window_base is the start address of the window in the memory of the target process and displ_unit is the distance between neighboring window entries as defined by the target process when creating the window with MPI_Win_create(). The execution of an MPI_Put() operation by a process source has the same effect as a two-sided communication for which process source executes the send operation

```
int MPI_Isend (origin_addr, origin_count, origin_type,
               target_rank, tag, comm)
```

and the target process executes the receive operation

```
int MPI_Recv (target_addr, target_count, target_type,
              source, tag, comm, &status)
```

where comm is the communicator for which the window object has been defined. For a correct execution of the operation, some constraints must be satisfied: The target buffer defined must fit in the window of the target process and the data block provided by the calling process must fit into the target buffer. In contrast to MPI_Isend() operations, the send buffers of multiple successive MPI_Put() operations may overlap, even if there is no synchronization in between. Source and target process of an MPI_Put() operation may be identical.

To transfer a data block from the window of another process into a local data buffer, the MPI function

```
int MPI_Get (void *origin_addr,
             int origin_count,
             MPI_Datatype origin_type,
             int target_rank,
             MPI_Aint target_displ,
             int target_count,
             MPI_Datatype target_type,
             MPI_Win win)
```

is provided. The parameter origin_addr specifies the start address of the receive buffer in the local memory of the calling process; origin_count defines the number of elements to be received; origin_type is the type of each of the elements. Similar to MPI_Put(), target_rank specifies the rank of the target process which provides the data and win is the window object previously created. The remaining parameters define the position and size of the data block to be transferred out of the window of the target process. The start address of the data block in the memory of the target process is given by target_addr := window_base + target_displ * displ_unit.

For the accumulation of data values in the memory of another process, MPI provides the operation

```
int MPI_Accumulate (void *origin_addr,
                    int origin_count,
                    MPI_Datatype origin_type,
                    int target_rank,
                    MPI_Aint target_displ,
                    int target_count,
                    MPI_Datatype target_type,
                    MPI_Op op,
                    MPI_Win win)
```

The parameters have the same meaning as for MPI_Put(). The additional parameter op specifies the reduction operation to be applied for the accumulation. The same predefined reduction operations as for MPI_Reduce() can be used, see Section 5.2, page 268. Examples are MPI_MAX and MPI_SUM. User-defined reduction operations cannot be used. The execution of an MPI_Accumulate() has the effect that the specified reduction operation is applied to corresponding entries of the source buffer and the target buffer and that the result is written back into the target buffer. Thus, data values can be accumulated in the target buffer provided by another process. There is an additional reduction operation MPI_REPLACE which allows the replacement of buffer entries in the target buffer, without taking the previous values of the entries into account. Thus, MPI_Put() can be considered as a special case of MPI_Accumulate() with reduction operation MPI_REPLACE.

There are some constraints for the execution of one-sided communication operations by different processes to avoid race conditions and to support an efficient implementation of the operations. Concurrent conflicting accesses to the same memory location in a window are not allowed. At each point in time during program execution, each memory location of a window can be used as target of at most one one-sided communication operation. Exceptions are accumulation operations: multiple concurrent MPI_Accumulate() operations can be executed at the same time for the same memory location. The result is obtained by using an arbitrary order of the executed accumulation operations. The final accumulated value is the same for all orders, since the predefined reduction operations are commutative. A window of a process P cannot be used concurrently by an MPI_Put() or MPI_Accumulate() operation of another process and by a local store operation of P, even if different locations in the window are addressed.

MPI provides three synchronization mechanisms for the coordination of one-sided communication operations executed on the windows of a group of processes. These three mechanisms are described in the following.

5.4.2.3 Global synchronization

A global synchronization of all processes of the group of a window object can be obtained by calling the MPI function

```
int MPI_Win_fence (int assert, MPI_Win win)
```

where win specifies the window object. MPI_Win_fence() is a collective operation to be performed by all processes of the group of win. The effect of the call is that all RMA operations originating from the calling process and started before the MPI_Win_fence() call are locally completed at the calling process before control is returned to the calling process. RMA operations started after the MPI_Win_fence() call access the specified target window only after the corresponding target process has called its corresponding MPI_Win_fence() operation. The intended use of MPI_Win_fence() is the definition of program areas in which one-sided communication operations are executed. Such program areas are surrounded by calls of MPI_Win_fence() thus establishing communication phases which can be mixed with computation phases during which no communication is required. Such communication phases are also referred to as **access epochs** in MPI. The parameter `assert` can be used to specify assertions on the context of the call of `MPI_Win_fence()` which can be used for optimizations by the MPI runtime system. Usually, `assert=0` is used, not providing additional assertions.

Global synchronization with `MPI_Win_fence()` is useful in particular for applications with regular communication pattern in which computation phases alternate with communication phases.

Example: As example, we consider an iterative computation of a distributed data structure A. In each iteration step, each participating process updates its local part of the data structure using the function `update()`. Then, parts of the local data structure are transferred into the windows of neighboring processes using `MPI_Put()`. Before the transfer, the elements to be transferred are copied into a contiguous buffer. This copy operation is performed by `update_buffer()`. The communication operations are surrounded by `MPI_Win_fence()` operations to separate the communication phases of successive iterations from each other. This results in the following program structure:

```
while (!converged(A)) {
    update(A);
    update_buffer(A, from_buf);
    MPI_Win_fence(0, win);
    for (i=0; i<num_neighbors; i++)
        MPI_Put(&from_buf[i], size[i], MPI_INT, neighbor[i], to_disp[i],
                size[i], MPI_INT, win);
    MPI_Win_fence(0, win);
}
```

The iteration is controlled by the function `converged()`.   □

5.4.2.4 Loose synchronization

MPI also supports a loose synchronization which is restricted to pairs of communicating processes. To perform this form of synchronization, an accessing process defines the start and the end of an **access epoch** by a call to `MPI_Win_start()` and `MPI_Win_complete()`, respectively. The target process of the communication defines a corresponding **exposure epoch** by calling `MPI_Win_post()` to start the exposure epoch and `MPI_Win_wait()` to end the exposure epoch. A synchronization is established between `MPI_Win_start()` and `MPI_Win_post()` in the sense that all RMA which the accessing process issues after its `MPI_Win_start()` call are executed not before the target process has completed its `MPI_Win_post()` call. Similarly, a synchronization between `MPI_Win_complete()` and `MPI_Win_wait()` is established in the sense that the `MPI_Win_wait()` call is completed at the target process not before all RMA of the accessing process in the corresponding access epoch are terminated.

To use this form of synchronization, before performing an RMA, a process defines the start of an access epoch by calling the function

```
int MPI_Win_start (MPI_Group group,
                   int assert,
                   MPI_Win win)
```

where `group` is a group of target processes. Each of the processes in `group` must issue a matching call of `MPI_Win_post()`. The parameter `win` specifies the window object to which the RMA is made. MPI supports a blocking and a non-blocking behavior of `MPI_Win_start()`:

- blocking behavior: the call of `MPI_Win_start()` is blocked until all processes of group have completed their corresponding calls of `MPI_Win_post()`;
- non-blocking behavior: the call of `MPI_Win_start()` is completed at the accessing process without blocking, even if there are processes in `group` which have not yet issued or finished their corresponding call of `MPI_Win_post()`. Control is returned to the accessing process and this process can issue RMA operations like `MPI_Put()` or `MPI_Get()`. These calls are then delayed until the target process has finished its `MPI_Win_post()` call.

The exact behavior depends on the MPI implementation. The end of an access epoch is indicated by the accessing process by calling

```
int MPI_Win_complete (MPI_Win win)
```

where `win` is the window object which has been accessed during this access epoch. Between the call of `MPI_Win_start()` and `MPI_Win_complete()`, only RMA operations to the window `win` of processes belonging to `group` are allowed. When calling `MPI_Win_complete()`, the calling process is blocked until all RMA operations to `win` issued in the corresponding access epoch have been completed at the accessing process. An `MPI_Put()` call issued in the access epoch can be completed at the calling process as soon as the local data buffer provided can be reused. But this does not necessarily mean that the data buffer has already been stored in the window of the target process. It might as well have been stored in a local system buffer of the MPI runtime system. Thus, the termination of `MPI_Win_complete()` does not imply that all RMA operations have taken effect at the target processes.

A process indicates the start of an RMA exposure epoch for a local window `win` by calling the function

```
int MPI_Win_post (MPI_Group group,
                  int assert,
                  MPI_Win win).
```

Only processes in `group` are allowed to access the window during this exposure epoch. Each of the processes in `group` must issue a matching call of the function `MPI_Win_start()`. The call of `MPI_Win_post()` is non-blocking. A process indicates the end of an RMA exposure epoch for a local window `win` by calling the function

```
int MPI_Win_wait (MPI_Win win).
```

This call blocks until all processes of the group defined in the corresponding MPI_Win_post() call have issued their corresponding MPI_Win_complete() calls. This ensures that all these processes have terminated the RMA operations of their corresponding access epoch to the specified window. Thus, after the termination of MPI_Win_wait(), the calling process can reuse the entries of its local window, e.g., by performing local accesses. During an exposure epoch, indicated by surrounding MPI_Win_post() and MPI_Win_wait() calls, a process should not perform local operations on the specified window to avoid access conflicts with other processes.

By calling the function

```
int MPI_Win_test (MPI_Win win, int *flag)
```

a process can test whether the RMA operation of other processes to a local window have been completed or not. This call can be considered as the non-blocking version of MPI_Win_wait(). The parameter flag=1 is returned by the call if all RMA operations to win have been terminated. In this case, MPI_Win_test() has the same effect as MPI_Win_wait() and should not be called again for the same exposure epoch. The parameter flag=0 is returned if not all RMA operations to win have been finished yet. In this case, the call has no further effect and can be repeated later.

The synchronization mechanism described can be used for arbitrary communication patterns on a group of processes. The communication pattern can be described by a directed graph $G = (V, E)$ where V is the set of participating processes. There exists an edge $(i, j) \in E$ from process i to process j, if i accesses the window of j by an RMA operation. Assuming that the RMA operations are performed on window win, the required synchronization can be reached by letting each participating process execute MPI_Win_start(target_group,0,win) followed by MPI_Win_post(source_group,0,win) where source_group= $\{i; (i, j) \in E\}$ denotes the set of accessing processes and target_group= $\{j; (i, j) \in E\}$ denotes the set of target processes.

Example: This form of synchronization is illustrated by the following example which is a variation of the previous example describing the iterative computation of a distributed data structure:

```
while (!converged (A)) {
    update(A);
    update_buffer(A, from_buf);
    MPI_Win_start(target_group, 0, win);
    MPI_Win_post(source_group, 0, win);
    for (i=0; i<num_neighbors; i++)
        MPI_Put(&from_buf[i], size[i], MPI_INT, neighbor[i], to_disp[i],
                size[i], MPI_INT, win);
    MPI_Win_complete(win);
    MPI_Win_wait(win);
}
```

In the example, it is assumed that source_group and target_group have been
defined according to the communication pattern used by all processes as described
above. An alternative would be that each process defines a set source_group of
processes which are allowed to access its local window and a set target_group
of processes whose window the process is going to access. Thus, each process potentially defines different source and target groups, leading to a weaker form of
synchronization as for the case that all processes define the same source and target
groups.   □

5.4.2.5 Lock synchronization

To support the model of a shared address space, MPI provides a synchronization
mechanism for which only the accessing process actively executes communication
operations. Using this form of synchronization, it is possible that two processes exchange data via RMA operations executed on the window of a third process without
an active participation of the third process. To avoid access conflicts, a lock mechanism is provided as typically used in programming environments for shared address
spaces, see Chapter 6. This means that the accessing process locks the accessed
window before the actual access and releases the lock again afterwards. To lock a
window before an RMA operation, MPI provides the operation

```
int MPI_Win_lock (int lock_type,
                  int rank,
                  int assert,
                  MPI_Win win).
```

A call of this function starts an RMA access epoch for the window win at the
process with rank rank. Two lock types are supported as specified by parameter
lock_type. An *exclusive lock* is indicated by lock_type=MPI_LOCK_EXCLUSIVE.
This lock type guarantees that the following RMA operation executed by the calling
process are protected from RMA operations of other processes, i.e., exclusive access
to the window is ensured. Exclusive locks should be used if the executing process
will change the value of window entries using MPI_Put() and if these entries could
also be accessed by other processes.

A shared lock is indicated by lock_type=MPI_LOCK_SHARED. This lock type
guarantees that the following RMA operations of the calling process are protected
from *exclusive* RMA operations of other processes, i.e., other processes are not allowed to change entries of the window via RMA operations that are protected by an
exclusive lock. But other processes are allowed to perform RMA operations on the
same window that are also protected by a shared lock.

Shared locks should be used if the executing process accesses window entries
only by MPI_Get() or MPI_Accumulate(). When a process wants to read or manipulate entries of its local window using local operations, it must protect these
local operations with a lock mechanism, if these entries can also be accessed by
other processes.

An access epoch started by MPI_Win_lock() for a window win can be terminated by calling the MPI function

```
int MPI_Win_unlock (int rank,
                    MPI_Win win)
```

where rank is the rank of the target process. The call of this function blocks until all RMA operations issued by the calling process on the specified window have been completed both at the calling process and the target process. This guarantees that all manipulations of window entries issued by the calling process have taken effect at the target process.

Example: The use of lock synchronization for the iterative computation of a distributed data structure is illustrated in the following example, which is a variation of the previous examples. Here, an exclusive lock is used to protect the RMA operations:

```
while (!converged (A)) {
    update(A);
    update_buffer(A, from_buf);
    for (i=0; i<num_neighbors; i++) {
        MPI_Win_lock(MPI_LOCK_EXCLUSIVE, neighbor[i], 0, win);
        MPI_Put(&from_buf[i], size[i], MPI_INT, neighbor[i], to_disp[i],
                size[i], MPI_INT, win);
        MPI_Win_unlock(neighbor[i], win);
    }
}
```

$\square$

5.5 Exercises for Chapter 5

Exercise 5.1. Consider the following incomplete piece of an MPI program:

```
int rank, p, size=8;
int left, right;
char send_buffer1[8], recv_buffer1[8];
char send_buffer2[8], recv_buffer2[8];
  :
MPI_Comm_rank(MPI_COMM_WORLD, 8 rank);
MPI_Comm_size(MPI_COMM_WORLD, & p);
left = (rank-1 + p) % p;
right = (rank+1) % p;
  :
MPI_Send(send_buffer1, size, MPI_CHAR, left, ...);
MPI_Recv(recv_buffer1, size, MPI_CHAR, right, ...);

MPI_Send(send_buffer2, size, MPI_CHAR, right, ...);
MPI_Recv(recv_buffer2, size, MPI_CHAR, left, ...);
  :
```

(a) In the program, the processors are arranged in a logical ring and each processor should exchange its name with its neighbor to the left and its neighbor to the right. Assign a unique name to each MPI process and fill out the missing pieces of the program such that each process prints its own name as well as its neighbors' names.

(b) In the given program piece, the MPI_Send() and MPI_Recv() operations are arranged such that depending on the implementation a deadlock can occur. Describe how a deadlock may occur.

(c) Change the implementation such that no deadlock is possible by arranging the order of the MPI_Send() and MPI_Recv() operations appropriately.

(d) Change the program such that MPI_Sendrecv() is used to avoid deadlocks.

(e) Change the program such that MPI_Isend() and MPI_Irecv() are used.

Exercise 5.2. Consider the MPI program in Fig. 5.3 for the collection of distributed data block with point-to-point messages. The program assumes that all data blocks have the same size blocksize. Generalize the program such that each process can contribute a data block of a size different from the data blocks of the other processes. To do so, assume that each process has a local variable which specifies the size of its data block.

Hint: First make the size of each data block available to each process in a precollection phase with a similar communication pattern as in Fig. 5.3 and then perform the actual collection of the data blocks.

Exercise 5.3. Modify the program from the previous exercise for the collection of data blocks of different size such that no pre-collection phase is used. Instead, use `MPI_Get_count()` to determine the size of the data block received in each step. Compare the resulting execution time with the execution time of the program from the previous exercise for different data block sizes and different numbers of processors. Which of the programs is faster?

Exercise 5.4. Consider the program `Gather_ring()` from Fig. 5.3. As described in the text, this program does not avoid deadlocks if the runtime system does not use internal system buffers. Change the program such that deadlocks are avoided in any case by arranging the order of the `MPI_Send()` and `MPI_Recv()` operations appropriately.

Exercise 5.5. The program in Fig. 5.3 arranges the processors logically in a ring to perform the collection. Modify the program such that the processors are logically arranged in a logical two-dimensional torus network. For simplicity, assume that all data blocks have the same size. Develop a mechanism with which each processor can determine its predecessor and successor in x and y direction. Perform the collection of the data blocks in two phases, the first phase with communication in x direction, the second phase with communication in y direction.
In both directions, communication in different rows or columns of the processor torus can be performed concurrently. For the communication in y direction, each process distributes all blocks that it has collected in the x direction phase. Use the normal blocking send and receive operations for the communication. Compare the resulting execution time with the execution time of the ring implementation from Fig. 5.3 for different data block sizes and different numbers of processors. Which of the programs is faster?

Exercise 5.6. Modify the program from the previous exercise such that non-blocking communication operations are used.

Exercise 5.7. Consider the parallel computation of a matrix-vector multiplication $A \cdot b$ using a distribution of the scalar products based on a row-wise distribution of A, see Figure 3.12, page 148 for a sketch of a parallel pseudo program. Transform this program into a running MPI program. Select the MPI communication operations for the multi-broadcast operations used appropriately.

Exercise 5.8. Similar to the preceding exercise, consider a matrix-vector multiplication using a distribution of the linear combinations based on a column-wise distribution of the matrix. Transform the pseudo program from Figure 3.14, page 151 to a running MPI program. Use appropriate MPI operations for the single-accumulation and single-broadcast operations. Compare the execution time with the execution time of the MPI program from the preceding exercise for different sizes of the matrix.

Exercise 5.9. For a broadcast operation a root process sends the same data block to all other processes. Implement a broadcast operation by using point-to-point send and receive operations (`MPI_Send()` and `MPI_Recv()`) such that the same effect as MPI_Bcast() is obtained. For the processes, use a logical ring arrangement similar to Fig. 5.3.

Exercise 5.10. Modify the program from the previous exercise such that two other logical arrangements are used for the processes: a two-dimensional mesh and a three-dimensional hypercube. Measure the execution time of the three different versions (ring, mesh, hypercube) for eight processors for different sizes of the data block and make a comparison by drawing a diagram. Use MPI_Wtime() for the timing.

Exercise 5.11. Consider the construction of conflict-free spanning trees in a d-dimensional hypercube network for the implementation of a multi-broadcast operation, see Section 4.4.2, page 211, and Fig. 4.13. For $d = 3$, $d = 4$, and $d = 5$ write a MPI program with 8, 16, and 32 processes, respectively that uses these spanning trees for a multi-broadcast operation.

(a) Implement the multi-broadcast by concurrent single-to-single transfers along the spanning trees and measure the resulting execution time for different message sizes.

(b) Implement the multi-broadcast by using multiple broadcast operations where each broadcast operation is implemented by single-to-single transfers along the usual spanning trees for hypercube networks as defined on page 207, see Fig. 4.11. These spanning trees do not avoid conflicts in the network. Measure the resulting execution time for different message sizes and compare them with the execution times from (a).

(c) Compare the execution times from (a) and (b) with the execution time of an MPI_Allgather() operation to perform the same communication.

Exercise 5.12. For a global exchange operation, each process provides a potentially different block of data for each other process, see pages 142 and 279 for a detailed explanation. Implement a global exchange operation by using point-to-point send and receive operations (MPI_Send() and MPI_Recv()) such that the same effect as MPI_Alltoall() is obtained. For the processes, use a logical ring arrangement similar to Fig. 5.3.

Exercise 5.13. Modify the program Gather_ring() from Fig. 5.3 such that synchronous send operations (MPI_Send() and MPI_Recv()) are used. Compare the resulting execution time with the execution time obtained for the standard send and receive operations from Fig. 5.3.

Exercise 5.14. Repeat the previous exercise with buffered send operations.

Exercise 5.15. Modify the program Gather_ring() from Fig. 5.3 such that the MPI operation MPI_Test() is used instead of MPI_Wait(). When a nonblocking receive operation is found by MPI_Test() to be completed, the process sends the received data block to the next process.

Exercise 5.16. Write an MPI program which implements a broadcast operation with MPI_Send() and MPI_Recv() operations. The program should use $n = 2^k$

processes which should logically be arranged as a hypercube network. Based on this arrangement the program should define a spanning tree in the network with root 0, see Figure 3.10 and page 144, and should use this spanning tree to transfer a message step-wise from the root along the tree edges up to the leaves. Each node in the tree receives the message from its parent node and forwards it to its child nodes. Measure the resulting runtime for different message sizes up to 1 MB for different numbers of processors using MPI_Wtime() and compare the execution times with the execution times of MPI_Bcast() performing the same operation.

Exercise 5.17. The execution time of point-to-point communication operations between two processors can normally be described by a linear function of the form

$$t_{s2s}(m) = \tau + t_c \cdot m$$

where m is the size of the message, τ is a startup time, which is independent of the message size, and t_c is the inverse of the network bandwidth. Verify this functions by measuring the time for a ping-pong message transmission where process A sends a message to process B, and B sends the same message back to A. Use different message sizes and draw a diagram which shows the dependence of the communication time on the message size. Determine the size of τ and t_c on your parallel computer.

Exercise 5.18. The program fragment in Figure 5.12 implements a matrix-vector multiplication $A \cdot b = c$ based on a row-wise block distribution of the matrix A using an MPI_Allgatherv() operation to obtain a replicated provision of the result vector c. The same matrix-vector multiplication can also be implemented using a column-wise block distribution of A, see the pseudo-code in Figure 3.14, using a single-accumulation operation followed by a single-broadcast operation. Modify the program in Figure 5.12 to perform this modified implementation using a column-wise block distribution of A. Use the appropriate MPI operations for the communication. Assume that the number of columns is not necessarily a multiple of the number of processes used. Is for this modified implementations the use of vector versions of the communication operations required?

Exercise 5.19. Write an MPI program which arranges 24 processes in a (periodic) Cartesian grid structure of dimension $2 \times 3 \times 4$ using MPI_Cart_create(). Each process should determine and print the process rank of its two neighbors in x, y, and z directions.

For each of the three sub-grids in y-direction, a communicator should be defined. This communicator should then be used to determine the maximum rank of the processes in the sub-grid by using an appropriate MPi_Reduce() operation. This maximum rank should be printed out.

Exercise 5.20. Write an MPI program which arranges the MPI processes in a two-dimensional torus of size $\sqrt{p} \times \sqrt{p}$ where p is the number of processes. Each process exchanges its rank with its two neighbors in x and y dimension. For the exchange, one-sided communication operations should be used. Implement three different schemes for the exchange with the following one-sided communication operations:

- global synchronization with `MPI_Win_fence()`;
- loose synchronization by using `MPI_Win_start()`, `MPi_Win_post()`, `MPI_Win_complete()`, and `MPI_Win_wait()`;
- lock synchronization with `MPI_Win_lock()` and `MPI_Win_unlock()`.

Test your program for $p = 16$ processors, i.e., for a 4×4 torus network.

Chapter 6
Thread Programming

About this Chapter

Many parallel computing platforms, in particular multicore platforms, offer a shared address space. A natural programming model for these architectures is a thread-based model in which all threads have access to shared variables. Several programming environments support the programming with threads. This chapter introduces the programming with Pthreads, Java threads, as well as OpenMP and explains the synchronization and coordination mechanisms provided by these environments. Implementations of parallel programming patterns, such as reader-writer locks, pipelining, client-server, and task pools are also provided.

For shared address spaces, shared variables are used for information and data exchange. To coordinate the access to shared variables, synchronization mechanisms have to be used to avoid race conditions in case of concurrent accesses. Basic synchronization mechanisms are lock synchronization and condition synchronization, see Section 3.8 for an overview.

In this chapter, thread programming is studied in more detail. In particular, synchronization problems such as deadlocks or priority inversion are considered and programming techniques to avoid such problems are presented. Moreover, it is demonstrated how basic synchronization mechanisms such as lock synchronization or condition synchronization can be used to build more complex synchronization mechanisms such as read/write locks. A set of parallel patterns such as task-based or pipelined processing that can be used to structure a parallel application are addressed. These issues are considered in the context of popular programming environments for thread-based programming to directly show the usage of the mechanisms in practice. The programming environments Pthreads, Java threads, and OpenMP are introduced in detail. For Java, also an overview of the package `java.util.concurrent` is contained. The package provides many advanced synchronization mechanisms as well as a task-based execution environment. The goal of this chapter is to enable the reader to develop correct and efficient thread programs that can be used, for example, on multicore architectures.

T. Rauber, G. Rünger, *Parallel Programming*, https://doi.org/10.1007/978-3-031-28924-8_6

6.1 Programming with Pthreads

The POSIX threads model (also called Pthreads) defines a standard for the programming with threads, based on the programming language C. The threads of a process
share a common address space. Thus, the global variables and dynamically generated data objects can be accessed by all threads of a process. In addition, each thread
has a separate runtime stack which is used to control the functions activated and to
store their local variables. These variables declared locally within the functions are
local data of the executing thread and cannot be accessed directly by other threads.
Since the runtime stack of a thread is deleted after a thread is terminated, it is dangerous to pass a reference to a local variable in the runtime stack of a thread A to
another thread B.

The data types, interface definitions and macros of Pthreads are usually available
via the header file <pthread.h>. This header file must therefore be included into
a Pthreads program. The functions and data types of Pthreads are defined according
to a naming convention. According to this convention, Pthreads functions are named
in the form

$$\texttt{pthread[\_<object>]\_<operation> ()}$$

where <operation> describes the operation to be performed and the optional
<object> describes the object to which this operation is applied. For example,
pthread_mutex_init() is a function for the initialization of a mutex variable;
thus, the <object> is mutex and the <operation> is init; we give a more
detailed description later.

For functions which are involved in the manipulation of threads, the specification
of <object> is omitted. For example, the function for the generation of a thread
is pthread_create(). All Pthreads functions yield a return value 0, if they are
executed without failure. In case of a failure, an error code from <error.h> will
be returned. Thus, this header file should also be included in the program. Pthreads
data types describe, similarly to MPI, opaque objects whose exact implementation
is hidden from the programmer. Data types are named according to the syntax form

$$\texttt{pthread\_<object>\_t}$$

where <object> specifies the specific data object. For example, a mutex variable
is described by the data type pthread_mutex_t. If <object> is omitted, the data
type pthread_t for threads results. The following table contains important Pthread
data types which will be described in more detail later.

Pthreads data types	meaning
pthread_t	Thread ID
pthread_mutex_t	mutex variable
pthread_cond_t	condition variable
pthread_key_t	access key
pthread_attr_t	thread attributes object
pthread_mutexattr_t	mutex attributes object
pthread_condattr_t	condition variable attributes object
pthread_once_t	*one time initialization* control context

For the execution of threads, we assume a two-step scheduling method according to Fig. 3.18 in Section 3, as this is the most general case. In this model, the programmer has to partition the program into a suitable number of user threads which can be executed concurrently with each other. The user threads are mapped by the library scheduler to system threads which are then brought to execution on the processors of the computing system by the scheduler of the operating system. The programmer cannot control the scheduler of the operating system and has only little influence on the library scheduler. Thus, the programmer cannot directly perform the mapping of the user-level threads to the processors of the computing system, e.g., by a scheduling at program level. This facilitates program development, but also prevents an efficient mapping directly by the programmer according to his specific needs. It should be noted that there are operating system specific extensions that allow thread execution to be bound to specific processors. But in most cases, the scheduling provided by the library and the operating system leads to good results and relieves the programmer from additional programming effort, thus providing more benefits than drawbacks.

In this section, we give an overview of the programming with Pthreads. Section 6.1.1 describes thread generation and management in Pthreads. Section 6.1.2 describes the lock mechanism for the synchronization of threads accessing shared variables. The Sections 6.1.3 and 6.1.4 introduce Pthreads condition variables and an extended lock mechanism using condition variables. Sections 6.2.1 - 6.2.3 describe the use of the basic synchronization techniques in the context of more advanced synchronization patterns, such as task pools, pipelining, and client-server coordination. Section 6.3.1 discusses additional mechanisms for the control of threads, including scheduling strategies. We describe in Section 6.3.2 how the programmer can influence the scheduling controlled by the library. The phenomenon of *priority inversion* is then explained in Section 6.3.3 and finally thread-specific data is considered in Section 6.3.4. Only the most important mechanisms of the Pthreads standard are described; for a more detailed description, we refer to [27, 131, 146, 160, 178].

6.1.1 Creating and Merging Threads

When a Pthreads program is started, a single *main thread* is active, executing the main() function of the program. The main thread can generate more threads by calling the function

```
int pthread_create (pthread_t *thread,
                    const pthread_attr_t *attr,
                    void *(*start_routine)(void *),
                    void *arg)
```

The first argument is a pointer to an object of type pthread_t which is also referred to as *thread identifier* (TID); this TID is generated by pthread_create() and can later be used by other Pthreads functions to identify the generated thread.

The second argument is a pointer to a previously allocated and initialized attributes object of type pthread_attr_t, defining the desired attributes of the generated thread. The argument value NULL causes the generation of a thread with default attributes. If different attribute values are desired, an attribute data structure has to be created and initialized before calling pthread_create(); this mechanism is described in more detail in Section 6.3.1. The third argument specifies the function start_routine() which will be executed by the generated thread. The specified function should expect a single argument of type void * and should have a return value of the same type. The fourth argument is a pointer to the argument value with which the thread function start_routine() will be executed.

To execute a thread function with more than one argument, all arguments must be put into a single data structure; the address of this data structure can then be specified as argument of the thread function. If several threads are started by a parent-thread using the same thread function but different argument values, *separate* data structures should be used for each of the threads to specify the arguments. This avoids situations where argument values are overwritten too early by the parent thread before they are read by the child threads or where different child threads manipulate the argument values in a common data structure concurrently.

A thread can determine its own thread identifier by calling the function

```
pthread_t pthread_self()
```

This function returns the thread ID of the calling thread. To compare the thread ID of two threads, the function

```
int pthread_equal (pthread_t t1, pthread_t t2)
```

can be used. This function returns the value 0 if t_1 and t_2 do not refer to the same thread. Otherwise, a non-zero value is returned. Since pthread_t is an opaque data structure, only pthread_equal should be used to compare thread IDs. The number of threads that can be generated by a process is typically limited by the system. The Pthreads standard determines that at least 64 threads can be generated by any process. But depending on the specific system used, this limit may be higher. For most systems, the maximum number of threads that can be started can be determined by calling

```
maxThreads = sysconf (_SC_THREAD_THREADS_MAX)
```

in the program. Knowing this limit, the program can avoid to start more than maxThreads threads. If the limit is reached, a call of the pthread_create() function returns the error value EAGAIN. A thread is terminated if its thread function terminates, e.g., by calling return. A thread can terminate itself explicitly by calling the function

```
void pthread_exit (void *valuep)
```

The argument `valuep` specifies the value that will be returned to another thread which waits for the termination of this thread using `pthread_join()`. When a thread terminates its thread function, the function `pthread_exit()` is called *implicitly*, and the return value of the thread function is used as argument of this implicit call of `pthread_exit()`. After the call to `pthread_exit()`, the calling thread is terminated, and its runtime stack is freed and can be used by other threads. Therefore, the return value of the thread should not be a pointer to a local variable of the thread function or another function called by the thread function. These local variables are stored on the runtime stack and may not exist any longer after the termination of the thread. Moreover, the memory space of local variables can be reused by other threads, and it can usually not be determined when the memory space is overwritten, thereby destroying the original value of the local variable. Instead of a local variable, a global variable or a variable that has been dynamically allocated should be used.

A thread can wait for the termination of another thread by calling the function

```
int pthread_join (pthread_t thread, void **valuep)
```

The argument `thread` specifies the thread ID of the thread for which the calling thread waits to be terminated. The argument `valuep` specifies a memory address where the return value of this thread should be stored. The thread calling `pthread_join()` is blocked until the specified thread has terminated. Thus, `pthread_join()` provides a possibility for the *synchronization* of threads. After the thread with TID `thread` has terminated, its return value is stored at the specified memory address. If several threads wait for the termination of the same thread, using `pthread_join()`, all waiting threads are blocked until the specified thread has terminated. But only one of the waiting threads successfully stores the return value. For all other waiting threads, the return value of `pthread_join()` is the error value ESRCH. The runtime system of the Pthreads library allocates for each thread an internal data structure to store information and data needed to control the execution of the thread. This internal data structure is preserved by the runtime system also after the termination of the thread to ensure that another thread can later successfully access the return value of the terminated thread using `pthread_join()`.

After the call to `pthread_join()`, the internal data structure of the terminated thread is released and can no longer be accessed. If there is no `pthread_join()` for a specific thread, its internal data structure is not released after its termination and occupies memory space until the complete process is terminated. This can be a problem for large programs with many thread creations and terminations without corresponding calls to `pthread_join()`. The preservation of the internal data structure of a thread after its termination can be avoided by calling the function

```
int pthread_detach (pthread_t thread)
```

This function notifies the runtime system that the internal data structure of the thread with TID `thread` can be detached as soon as the thread has terminated. A thread may detach itself, and any thread may detach any other thread. After a thread has been set into a detached state, calling `pthread_join()` for this thread returns the error value `EINVAL`.

Example: We give a first example for a Pthreads program; Figure 6.1 shows a program fragment for the multiplication of two matrices, see also [160]. The matrices MA and MB to be multiplied have a fixed size of eight rows and eight columns. For each of the elements of the result matrix MC, a separate thread is created. The IDs of these threads are stored in the array `thread`. Each thread obtains a separate data structure of type `matrix_type_t` which contains pointers to the input matrices MA and MB, the output matrix MC, and the row and column position of the entry of MC to be computed by the corresponding thread. Each thread executes the same thread function `thread_mult()` which computes the scalar product of one row of MA and one column of MB. After creating a new thread for each of the 64 elements of MC to be computed, the main thread waits for the termination of each of these threads using `pthread_join()`. The program in Fig. 6.1 creates 64 threads which is exactly the limit defined by the Pthreads standard for the number of threads that must be supported by each implementation of the standard. Thus, the given program works correctly. But it is not scalable in the sense that it can be extended to the multiplication of matrices of any size. Since a separate thread is created for each element of the output matrix, it can be expected that the upper limit for the number of threads that can be generated will be reached even for matrices of moderate size. Therefore, the program should be re-written when using larger matrices such that a fixed number of threads is used and each thread computes a block of entries of the output matrix; the size of the blocks increases with the size of the matrices.   □

6.1.2 Thread Coordination with Pthreads

The threads of a process share a common address space. Therefore, they can concurrently access shared variables. To avoid race conditions, these concurrent accesses must be coordinated. To perform such coordinations, Pthreads provides *mutex variables* and *condition variables*.

6.1.2.1 Mutex variables

In Pthreads, a **mutex variable** denotes a data structure of the predefined opaque type `pthread_mutex_t`. Such a mutex variable can be used to ensure *mutual exclusion* when accessing common data, i.e., it can be ensured that only one thread at a time has exclusive access to a common data structure, all other threads have to wait. A mutex variable can be in one of two states: *locked* and *unlocked*. To ensure mutual exclusion when accessing a common data structure, a separate mutex variable is assigned to the data structure. All accessing threads must behave as follows: *Before* an access to the common data structure, the accessing thread locks the corresponding

```
#include <pthread.h>

typedef struct {
  int size, row, column;
  double (*MA)[8], (*MB)[8], (*MC)[8];
} matrix_type_t;

void *thread_mult (void *w) {
  matrix_type_t *work = (matrix_type_t *) w;
  int i, row = work->row, column = work->column;
  work -> MC[row][column] = 0;
  for (i=0; i < work->size; i++)
    work->MC[row][column] += work->MA[row][i] * work->MB[i][column];
  return NULL;
}

int main() {
  int row, column, size = 8, i;
  double MA[8][8], MB[8][8], MC[8][8];
  matrix_type_t *work;
  pthread_t thread[8*8];
  for (row=0; row<size; row++)
    for (column=0; column<size; column++) {
      work = (matrix_type_t *) malloc (sizeof (matrix_type_t));
      work->size = size;
      work->row = row;
      work->column = column;
      work->MA = MA; work->MB = MB; work->MC = MC;
      pthread_create (&(thread[column + row*8]), NULL,
                      thread_mult, (void *) work);
    }

  for (i=0; i<size*size; i++)
    pthread_join (thread[i], NULL);
}
```

Fig. 6.1 Pthreads program for the multiplication of two matrices MA and MB. A separate thread is created for each element of the output matrix MC. A separate data structure work is provided for each of the threads created.

mutex variable using a specific Pthreads function. When this is successful, the thread is the *owner* of the mutex variable. *After* each access to the common data structure, the accessing thread unlocks the corresponding mutex variable. After the unlocking, it is no longer owner of the mutex variable and another thread can become owner and is allowed to access the data structure.

When a thread A tries to lock a mutex variable that is already owned by another thread B, thread A is blocked until thread B unlocks the mutex variable. The Pthreads runtime system ensures that only one thread at a time is the owner of a specific mutex variable. Thus, a conflicting manipulation of a common data structure is avoided if each thread uses the described behavior. But if a thread accesses the data structure without locking the mutex variable before, mutual exclusion is no longer guaranteed.

The assignment of mutex variables to data structures is done implicitly by the programmer by protecting accesses to the data structure with locking and unlocking operations of a specific mutex variable. There is no explicit assignment of mutex variables to data structures. The programmer can improve the readability of Pthreads programs by grouping a common data structure and the protecting mutex variable into a new structure.

In Pthreads, mutex variables have the predefined type `pthread_mutex_t`. Like normal variables, they can be statically declared or dynamically generated. Before a mutex variable can be used, it must be initialized. For a mutex variable `mutex` that is allocated statically, this can be done by

`mutex = PTHREAD_MUTEX_INITIALIZER`

where `PTHREAD_MUTEX_INITIALIZER` is a predefined macro. For arbitrary mutex variables (statically allocated or dynamically generated), an initialization can be performed dynamically by calling the function

```
int pthread_mutex_init (pthread_mutex_t *mutex,
                        const pthread_mutexattr_t *attr)
```

For `attr = NULL`, a mutex variable with default properties results. The properties of mutex variables can be influenced by using different attribute values, see Section 6.3.1. If a mutex variable that has been initialized dynamically is no longer needed, it can be destroyed by calling the function

```
int pthread_mutex_destroy (pthread_mutex_t *mutex)
```

A mutex variable should only be destroyed if none of the threads is waiting for the mutex variable to become owner and if there is currently no owner of the mutex variable. A mutex variable that has been destroyed can later be re-used after a new initialization. A thread can lock a mutex variable `mutex` by calling the function

```
    int pthread_mutex_lock (pthread_mutex_t *mutex)
```

If another thread B is owner of the mutex variable `mutex` when a thread A issues the call of `pthread_mutex_lock()`, then thread A is blocked until thread B unlocks `mutex`. When several threads $T_1, \ldots, T_n$ try to lock a mutex variable which is owned by another thread, all threads $T_1, \ldots, T_n$ are blocked and are stored in a waiting queue for this mutex variable. When the owner releases the mutex variable, one of the blocked threads in the waiting queue is unblocked and becomes the new owner of the mutex variable. Which one of the waiting threads is unblocked may depend on their priorities and the scheduling strategies used, see Section 6.3.1 for more information. The order in which waiting threads become owner of a mutex variable is not defined in the Pthreads standard and may depend on the specific Pthreads library used.

A thread should not try to lock a mutex variable when it is already the owner. Depending on the specific runtime system, this may lead to an error return value

EDEADLK or may even cause a self-deadlock. A thread which is owner of a mutex variable mutex can unlock mutex by calling the function

```
int pthread_mutex_unlock (pthread_mutex_t *mutex)
```

After this call, mutex is in the state *unlocked*. If there is no other thread waiting for mutex, there is no owner of mutex after this call. If there are threads waiting for mutex, one of these threads is woken up and becomes the new owner of mutex. In some situations, it is useful that a thread can check without blocking whether a mutex variable is owned by another thread. This can be achieved by calling the function

```
int pthread_mutex_trylock (pthread_mutex_t *mutex)
```

If the specified mutex variable is currently not held by another thread, the calling thread becomes the owner of the mutex variable. This is the same behavior as for pthread_mutex_lock(). But different from pthread_mutex_lock(), the calling thread is *not blocked* if another thread already holds the mutex variable. Instead, the call returns with error return value EBUSY without blocking. The calling thread can then perform other computations and can later retry to lock the mutex variable. The calling thread can also repeatedly try to lock the mutex variable until it is successful (*spinlock*).

Example: Fig. 6.2 shows a simple program fragment to illustrate the use of mutex variables to ensure mutual exclusion when concurrently accessing a common data structure, see also [160]. In the example, the common data structure is a linked list. The nodes of the list have type node_t. The complete list is protected by a single mutex variable. To indicate this, the pointer to the first element of the list (first) is combined with the mutex variable (mutex) into a data structure of type list_t. The linked list will be kept sorted according to increasing values of the node entry index. The function list_insert() inserts a new element into the list while keeping the sorting. Before the first call to list_insert(), the list must be initialized by calling list_init(), e.g., in the main thread. This call also initializes the mutex variable. In list_insert(), the executing thread first locks the mutex variable of the list before performing the actual insertion. After the insertion, the mutex variable is released again using pthread_mutex_unlock(). This procedure ensures that it is not possible for different threads to insert new elements at the same time. Hence, the list operations are *sequentialized*. The function list_insert() is a *thread-safe* function, since a program can use this function without performing additional synchronization.

In general, a (library) function is thread-safe if it can be called by different threads concurrently, without performing additional operations to avoid race conditions.   □

In Fig. 6.2, a single mutex variable is used to control the complete list. This results in a *coarse-grain* lock granularity. Only a single insert operation can happen at a time, independently from the length of the list. Alternatively, the list could be partitioned into fixed-size areas and protect each area with a mutex variable, or even

```
typedef struct node {
  int index;
  void *data;
  struct node *next;
} node_t;

typedef struct list {
  node_t *first;
  pthread_mutex_t mutex;
} list_t;

void list_init (list_t *listp)
{
  listp->first = NULL;
  pthread_mutex_init (&(listp->mutex), NULL);
}

void list_insert (int newindex, void *newdata, list_t *listp)
{
  node_t *current, *previous, *new;
  int found = FALSE;

  pthread_mutex_lock (&(listp->mutex));
  for (current = previous = listp->first; current != NULL;
       previous = current, current = current->next)
  {
    if (current->index == newindex) {
      found = TRUE; break;
    }
    else
      if (current->index > newindex) break;
  }
  if (!found) {
    new = (node_t *) malloc (sizeof (node_t));
    new->index = newindex;
    new->data = newdata;
    new->next = current;
    if (current == listp->first) lstp->first = new;
    else previous->next = new;
  }
  pthread_mutex_unlock (&(lstp->mutex));
}
```

Fig. 6.2 Pthread implementation of a linked list. The function list_insert() can be called by different threads concurrently to insert new elements into the list. In the form presented, list_insert() cannot be used as the start function of a thread, since the function has more than one argument. To be used as start function, the arguments of list_insert() have to be put into a new data structure which is then passed as argument. The original arguments could then be extracted from this data structure at the beginning of list_insert().

to protect each single element of the list with a separate mutex variable. In this case, the granularity would be **fine-grained**, and several threads could access different parts of the list concurrently. But this also requires a substantial re-organization of the synchronization, possibly leading to a larger overhead.

6.1.2.2 Mutex variables and deadlocks

When multiple threads work with different data structures each of which is protected by a separate mutex variable, caution has to be taken to avoid deadlocks. A deadlock may occur, if the threads use a different order for locking the mutex variables. This can be seen for two threads T_1 and T_2 and two mutex variables ma and mb as follows:

- thread T_1 first locks ma and then mb;
- thread T_2 first locks mb and then ma.

If T_1 is interrupted by the scheduler of the runtime system after locking ma such that T_2 is able to successfully lock mb, a deadlock occurs:
T_2 will be blocked when it is trying to lock ma, since ma is already locked by T_1; similarly, T_1 will be blocked when it is trying to lock mb after it has been woken up again, since mb has already been locked by T_2. In effect, both threads are blocked forever and are mutually waiting for each other. The occurrence of deadlocks can be avoided by using a *fixed locking order* for all threads or by employing a *backoff strategy*.

When using a **fixed locking order**, each thread locks the critical mutex variables always in the same predefined order. Using this approach for the example above, thread T_2 must lock the two mutex variables ma and mb in the same order as T_1, e.g., both threads must first lock ma and then mb. The deadlock described above cannot occur now, since T_2 cannot lock mb if ma has previously been locked by T_1. To lock mb, T_2 must first lock ma. If ma has already been locked by T_1, T_2 will be blocked when trying to lock ma and, hence, cannot lock mb. The specific locking order used can in principle be arbitrarily selected, but to avoid deadlocks it is important that the order selected is used throughout the entire program. If this does not conform to the program structure, a backoff strategy should be used.

When using a **backoff strategy**, each participating thread can lock the mutex variables in its individual order, and it is not necessary to use the same predefined order for each thread. But a thread must back off when its attempt to lock a mutex variable fails. In this case, the thread must release all mutex variables that it has previously locked successfully. After the backoff, the thread starts the entire lock procedure from the beginning by trying to lock the first mutex variable again. To implement a backoff strategy, each thread uses `pthread_mutex_lock()` to lock its first mutex variable, and `pthread_mutex_trylock()` to lock the remaining mutex variables needed. If `pthread_mutex_trylock()` returns EBUSY, this means that this mutex variable is already locked by another thread. In this case, the calling thread releases all mutex variables that it has previously locked successfully using `pthread_mutex_unlock()`.

Example: Backoff strategy (see Fig. 6.3 and 6.4):
The use of a backoff strategy is demonstrated in Fig. 6.3 for two threads f and b which lock three mutex variables m[0], m[1], and m[2] in different orders, see [27]. The thread f (forward) locks the mutex variables in the order m[0], m[1], and m[2] by calling the function lock_forward(). The thread b (backward) locks the mutex variables in the opposite order m[2], m[1], and m[0] by calling the function lock_backward(), see Fig. 6.4. Both threads repeat the locking 10 times. The main program in Fig. 6.3 uses two control variables backoff and yield_flag which are read in as arguments. The control variable backoff determines whether a backoff strategy is used (value 1) or not (value 0). For backoff = 1, no deadlock occurs when running the program because of the backoff strategy. For backoff = 0, a deadlock occurs in most cases, in particular if f succeeds in locking m[0] and b succeeds in locking m[2].

But depending on the specific scheduling situation concerning f and b, no deadlock may occur even if no backoff strategy is used. This happens when both threads succeed in locking all three mutex variables, before the other thread is executed. To illustrate this dependence of deadlock occurrence from the specific scheduling situation, the example in Fig. 6.3 and 6.4 contains a mechanism to influence the scheduling of f and b. This mechanism is activated by using the control variable yield_flag. For yield_flag = 0, each thread tries to lock the mutex variables without interruption. This is the behavior described so far. For yield_flag = 1, each thread calls sched_yield() after having locked a mutex variable, thus transferring control to another thread with the same priority. Therefore, the other thread has a chance to lock a mutex variable. For yield_flag = -1, each thread calls sleep(1) after having locked a mutex variable, thus waiting for 1 s. In this time, the other thread can run and has a chance to lock another mutex variable. In both cases, a deadlock will likely occur if no backoff strategy is used.

Calling pthread_exit() in the main thread causes the termination of the main thread, but not of the entire process. Instead, using a normal return would terminate the entire process, including the threads f and b.   □

Compared to a fixed locking order, the use of a backoff strategy typically leads to larger execution times, since threads have to back off when they do not succeed in locking a mutex variable. In this case, the locking of the mutex variables has to be started from the beginning.

But using a backoff strategy leads to an increased flexibility, since no fixed locking order has to be ensured. Both techniques can also be used in combination by using a fixed locking order in code regions where this is not a problem, and using a backoff strategy where the additional flexibility is beneficial.

```
#include <pthread.h>
#include <sched.h>
#include <stdlib.h>
#include <stdio.h>

pthread_mutex_t m[3] = {
    PTHREAD_MUTEX_INITIALIZER,
    PTHREAD_MUTEX_INITIALIZER,
    PTHREAD_MUTEX_INITIALIZER
};
int backoff = 1; // == 1: with backoff strategy
int yield_flag = 0; // > 0: use sched_yield, < 0: sleep

int main(int argc, char *argv[]) {
    pthread_t f, b;
    if (argc > 1) backoff = atoi(argv[1]);
    if (argc > 2) yield_flag = atoi(argv[2]);
    pthread_create(&f, NULL, lock_forward, NULL);
    pthread_create(&b, NULL, lock_backward, NULL);
    pthread_exit(NULL); // both threads continue execution
}
```

Fig. 6.3 Control program to illustrate the use of a backoff strategy.

6.1.3 Condition Variables

Mutex variables are typically used to ensure mutual exclusion when accessing global data structures concurrently. But mutex variables can also be used to wait for the occurrence of a specific condition which depends on the state of a global data structure and which has to be fulfilled before a certain operation can be applied. An example might be a shared buffer from which a consumer thread can remove entries only if the buffer is not empty. To apply this mechanism, the shared data structure is protected by one or several mutex variables, depending on the specific situation. To check whether the condition is fulfilled, the executing thread locks the mutex variable(s) and then evaluates the condition. If the condition is fulfilled, the intended operation can be performed. Otherwise, the mutex variable(s) are released again and the thread repeats this procedure again at a later time. This method has the drawback that the thread which is waiting for the condition to be fulfilled may have to repeat the evaluation of the condition quite often before the condition becomes true. This consumes execution time (*active waiting*), in particular because the mutex variable(s) have to be locked before the condition can be evaluated. To enable a more efficient method for waiting for a condition, Pthreads provides condition variables.

A **condition variable** is an opaque data structure which enables a thread to wait for the occurrence of an arbitrary condition without active waiting. Instead, a signaling mechanism is provided which blocks the executing thread during the waiting

```
void *lock_forward(void *arg) {
  int iterate, i, status;
  for (iterate = 0; iterate < 10; iterate++) {
    for (i = 0; i < 3; i++) { // lock order forward
      if (i == 0 || !backoff)
        status = pthread_mutex_lock(&m[i]);
      else status = pthread_mutex_trylock(&m[i]);
      if (status == EBUSY)
        for (--i; i >= 0; i--) pthread_mutex_unlock(&m[i]);
      else printf("forward locker got mutex %d\n", i);
      if (yield_flag) {
        if (yield_flag > 0) sched_yield(); // switch threads
        else sleep(1); // block executing thread
      }
    }
    for (i = 2; i >= 0; i--)
      pthread_mutex_unlock(&m[i]);
    sched_yield(); // new trial with potentially different order
  }
}
void *lock_backward(void *arg) {
  int iterate, i, status;
  for (iterate = 0; iterate < 10; iterate++) {
    for (i = 2; i >= 0; i--) { // lock order backward
      if (i == 2 || !backoff)
        status = pthread_mutex_lock(&m[i]);
      else status = pthread_mutex_trylock(&m[i]);
      if (status == EBUSY)
        for (++i; i < 3; i++) pthread_mutex_unlock(&m[i]);
      else printf("backward locker got mutex %d\n", i);
      if (yield_flag)
        if (yield_flag > 0) sched_yield();
        else sleep(1);
    }
    for (i = 0; i < 3; i++)
      pthread_mutex_unlock(&m[i]);
    sched_yield();
  }
}
```

Fig. 6.4 Functions lock_forward and lock_backward to lock mutex variables in opposite directions.

time, so that it does not consume CPU time. The waiting thread is woken up again as soon as the condition is fulfilled. To use this mechanism, the executing thread must define a condition variable and a mutex variable. The mutex variable is used to protect the evaluation of the specific condition which is waited for to be fulfilled. The use of the mutex variable is necessary, since the evaluation of a condition usually requires to access shared data which may be modified by other threads concurrently.

A condition variable has type pthread_cond_t. After the declaration or the dynamic generation of a condition variable, it must be initialized before it can be used. This can be done dynamically by calling the function

```
int pthread_cond_init (pthread_cond_t *cond,
                       const pthread_condattr_t *attr)
```

where cond is the address of the condition variable to be initialized and attr is the address of an attribute data structure for condition variables. Using attr=NULL leads to an initialization with the default attributes. For a condition variable cond that has been declared statically, the initialization can also be obtained by using the PTHREAD_COND_INITIALIZER initialization macro. This can also be done directly with the declaration

```
pthread_cond_t cond = PTHREAD_COND_INITIALIZER.
```

The initialization macro cannot be used for condition variables that have been generated dynamically using, e.g., malloc(). A condition variable cond that has been initialized with pthread_cond_init() can be destroyed by calling the function

```
int pthread_cond_destroy (pthread_cond_t *cond)
```

if it is no longer needed. In this case, the runtime system can free the information stored for this condition variable. Condition variables that have been initialized statically with the initialization macro, do not need to be destroyed.

Each condition variable must be uniquely associated with a specific mutex variable. All threads which wait for a condition variable at the same time must use the same associated mutex variable. It is not allowed that different threads associate different mutex variables with a condition variable at the same time. But a mutex variable can be associated to different condition variables. A condition variable should only be used for a single condition to avoid deadlocks or race conditions [27]. A thread must first lock the associated mutex variable mutex with pthread_mutex_lock() before it can wait for a specific condition to be fulfilled using the function

```
int pthread_cond_wait (pthread_cond_t *cond,
                       pthread_mutex_t *mutex)
```

where cond is the condition variable used and mutex is the associated mutex variable. The condition is typically embedded into a surrounding control statement. A standard usage pattern is:

```
pthread_mutex_lock (&mutex);
while (!condition())
  pthread_cond_wait (&cond, &mutex);
compute_something();
pthread_mutex_unlock (&mutex);
```

The evaluation of the condition and the call of pthread_cond_wait() are protected by a mutex variable mutex to ensure that the condition does not change between the evaluation and the call of pthread_cond_wait(), e.g., because another thread changes the value of a variable that is used within the condition. Therefore, each thread must use this mutex variable mutex to protect the manipulation of each variable that is used within the condition. Two cases can occur for this usage pattern for condition variables:

- If the specified condition is fulfilled when executing the code segment from above, the function pthread_cond_wait() is **not** called. The executing thread releases the mutex variable and proceeds with the execution of the succeeding program part.
- If the specified condition is not fulfilled, pthread_cond_wait() is called. This call has the effect that the specified mutex variable mutex is implicitly released and that the executing thread is blocked, waiting for the condition variable until another thread sends a signal using pthread_cond_signal() to notify the blocked thread that the condition may now be fulfilled. When the blocked thread is woken up again in this way, it implicitly tries to lock the mutex variable mutex again. If this is owned by another thread, the woken-up thread is blocked again, now waiting for the mutex variable to be released. As soon as the thread becomes owner of the mutex variable mutex, it continues the execution of the program. In the context of the usage pattern from above, this results in a new evaluation of the condition because of the while loop.

In a Pthreads program, it should be ensured that a thread which is waiting for a condition variable is woken up only if the specified condition is fulfilled. Nevertheless, it is useful to evaluate the condition again after the wake up because there are other threads working concurrently. One of these threads might become owner of the mutex variable before the woken-up thread. Thus the woken-up thread is blocked again. During the blocking time, the owner of the mutex variable may modify common data such that the condition is no longer fulfilled. Thus, from the perspective of the executing thread, the state of the condition may change in the time interval between being woken up and becoming owner of the associated mutex variable. Therefore, the thread must again evaluate the condition to be sure that it is still fulfilled. If the condition is fulfilled, it cannot change before the executing thread calls pthread_mutex_unlock() or pthread_cond_wait() for the same condition variable, since each thread must be owner of the associated mutex variable to modify a variable used in the evaluation of the condition.

Pthreads provides two functions to wake up (*signal*) a thread waiting on a condition variable:

```
int pthread_cond_signal (pthread_cond_t *cond)
int pthread_cond_broadcast (pthread_cond_t *cond).
```

A call of pthread_cond_signal() wakes up a *single* thread waiting on the condition variable cond. A call of this function has no effect, if there are no threads waiting for cond. If there are several threads waiting for cond, one of them is selected to be woken up. For the selection, the priorities of the waiting threads and the scheduling method used are taken into account. A call of pthread_cond_broadcast() wakes up *all* threads waiting on the condition variable cond. If several threads are woken up, only one of them can become owner of the associated mutex variable. All other threads that have been woken up are blocked on the mutex variable.

The functions pthread_cond_signal() and pthread_cond_broadcast() should only be called if the condition associated with cond is fulfilled. Thus, before calling one of these functions, a thread should evaluate the condition. To do so safely, it must first lock the mutex variable associated with the condition variable to ensure a consistent evaluation of the condition. The actual call of pthread_cond_signal() or pthread_cond_broadcast() does not need to be protected by the mutex variable. Issuing a call without protection by the mutex variable has the drawback that another thread may become owner of the mutex variable when it has been released after the evaluation of the condition, but before the signaling call. In this situation, the new owner thread can modify shared variables such that the condition is no longer fulfilled. This does not lead to an error, since the woken-up thread will again evaluate the condition. The advantage of not protecting the call of pthread_cond_signal() or pthread_cond_broadcast() by the mutex variable is the chance that the mutex variable may not have an owner when the waiting thread is woken up. Thus, there is a chance that this thread becomes owner of the mutex variable without waiting. If mutex protection is used, the signaling thread is owner of the mutex variable when the signal arrives, so the woken-up thread must block on the mutex variable immediately after being woken up.

To wait for a condition, Pthreads also provides the function

```
int pthread_cond_timedwait (pthread_cond_t *cond,
                            pthread_mutex_t *mutex,
                            const struct timespec *time)
```

The difference to pthread_cond_wait() is that the blocking on the condition variable cond is ended with return value ETIMEDOUT after the specified time interval time has elapsed. This maximum waiting time is specified using type

```
struct timespec {
  time_t tv_sec;
  long tv_nsec;
}
```

where tv_sec specifies the number of seconds and tv_nsec specifies the number of additional nanoseconds. The time parameter of pthread_cond_timedwait() specifies an absolute clock time rather than a time interval. A typical use may look as follows:

```
pthread_mutex_t m = PTHREAD_MUTEX_INITIALIZER;
pthread_cond_t c = PTHREAD_COND_INITIALIZER;
struct timespec time;
pthread_mutex_lock (&m);
time.tv_sec = time (NULL) + 10;
time.tv_nsec = 0;
while (!Bedingung)
  if (pthread_cond_timedwait (&c, &m, &time) == ETIMEDOUT)
    timed_out_work();
pthread_mutex_unlock (&m);
```

In this example, the executing thread waits at most 10 s for the condition to be fulfilled. The function `time()` from `<time.h>` is used to define `time.tv_sec`. The call `time(NULL)` yields the absolute time in seconds elapsed since Jan 1, 1970. If no signal arrives after 10 s, the function `timed_out_work()` is called before the condition is evaluated again.

6.1.4 Extended Lock Mechanism

Condition variables can be used to implement more complex synchronization mechanisms that are not directly supported by Pthreads. In the following, we consider a *read/write lock* mechanism as an example for an extension of the standard lock mechanism provided by normal mutex variables. If we use a normal mutex variable to protect a shared data structure, only one thread at a time can access (read or write) the shared data structure. The following user-defined read/write locks extend this mechanism by allowing an arbitrary number of reading threads at a time. But only one thread at a time is allowed to write to the data structure. In the following, we describe a simple implementation of this extension, see also [160]. For more complex and more efficient implementations, we refer to [27, 131].

For the implementation of read/write locks, we define read/write lock variables (r/w lock variables) by combining a mutex variable and a condition variable as follows:

```
typedef struct rw_lock {
  int num_r, num_w;
  pthread_mutex_t mutex;
  pthread_cond_t cond;
} rw_lock_t;
```

Here, `num_r` specifies the current number of read permits, and `num_w` specifies the current number of write permits; `num_w` should have a maximum value of 1. The mutex variable `mutex` is used to protect the access to `num_r` and `num_w`. The condition variable `cond` coordinates the access to the r/w lock variable.

Fig. 6.5 shows the functions that can be used to implement the read/write lock mechanism. The function `rw_lock_init()` initializes a read/write lock variable.

```
int rw_lock_init (rw_lock_t *rwl) {
  rwl->num_r = rwl->num_w = 0;
  pthread_mutex_init (&(rwl->mutex),NULL);
  pthread_cond_init (&(rwl->cond),NULL);
  return 0;
}

int rw_lock_rlock (rw_lock_t *rwl) {
  pthread_mutex_lock (&(rwl->mutex));
  while (rwl->num_w > 0)
    pthread_cond_wait (&(rwl->cond), &(rwl->mutex));
  rwl->num_r ++;
  pthread_mutex_unlock (&(rwl->mutex));
  return 0;
}

int rw_lock_wlock (rw_lock_t *rwl) {
  pthread_mutex_lock (&(rwl->mutex));
  while ((rwl->num_w > 0) || (rwl->num_r > 0))
    pthread_cond_wait (&(rwl->cond), &(rwl->mutex));
  rwl->num_w ++;
  pthread_mutex_unlock (&(rwl->mutex));
  return 0;
}

int rw_lock_runlock (rw_lock_t *rwl) {
  pthread_mutex_lock (&(rwl->mutex));
  rwl->num_r --;
  if (rwl->num_r == 0) pthread_cond_signal (&(rwl->cond));
  pthread_mutex_unlock (&(rwl->mutex));
  return 0;
}

int rw_lock_wunlock (rw_lock_t *rwl) {
  pthread_mutex_lock (&(rwl->mutex));
  rwl->num_w --;
  pthread_cond_broadcast (&(rwl->cond));
  pthread_mutex_unlock (&(rwl->mutex));
  return 0;
}
```

Fig. 6.5 Funktion for the control of read/write lock variables.

The function rw_lock_rlock() requests a read permit to the common data structure. The read permit is granted only if there is no other thread that currently has
a write permit. Otherwise the calling thread is blocked until the write permit is returned. The function rw_lock_wlock() requests a write permit to the common data
structure. The write permit is granted only if there is no other thread that currently
has a read or write permit.

The function `rw_lock_runlock()` is used to return a read permit. This may cause that the number of threads with a read permit decreases to zero. In this case, a thread which is waiting for a write permit is woken up by `pthread_cond_signal()`. The function `rw_lock_wunlock()` is used to return a write permit. Since only one thread with a write permit is allowed, there cannot be a thread with a write permit after this operation. Therefore, all threads waiting for a read or write permit can be woken up using `pthread_cond_broadcast()`.

The implementation sketched in Fig. 6.5 favors read requests over write requests: If a thread A has a read permit and a thread B waits for a write permit, then other threads will obtain a read permit without waiting, even if they put their read request long after B has put its write request. Thread B will get a write permit only if there are no other threads requesting a read permit. Depending on the intended usage, it might also be useful to give write requests priority over read requests to keep a data structure up to date. An implementation for this is given in [27].

The r/w lock mechanism can be used for the implementation of a shared linked list, see Fig. 6.2, by replacing the mutex variable `mutex` by a r/w lock variable. In the `list_insert()` function, the list access will then be protected by `rw_lock_wlock()` and `rw_lock_wunlock()`. A function to search for a specific entry in the list could use `rw_lock_rlock()` and `tw_lock_runlock()`, since no entry of the list will be modified when searching.

6.1.5 One-time initialization

In some situations, it is useful to perform an operation only once, no matter how many threads are involved. This is useful for initialization operations or opening a file. If several threads are involved, it sometimes cannot be determined in advance which of the threads is first ready to perform an operation. A one-time initialization can be achieved using a boolean variable initialized to 0 and protected by a mutex variable. The first thread arriving at the critical operation sets the boolean variable to 1, protected by the mutex variable, and then performs the one-time operation. If a thread arriving at the critical operation finds that the boolean variable has value 1, it does not perform the operation. Pthreads provides another solution for one-time operations by using a control variable of the predefined type `pthread_once_t`. This control variable must be statically initialized using the initialization macro `PTHREAD_ONCE_INIT`:

```
pthread_once_t once_control = PTHREAD_ONCE_INIT
```

The code to perform the one-time operation must be put into a separate function without parameter. We call this function `once_routine()` in the following. The one-time operation is then performed by calling the function

```
pthread_once (pthread_once_t *once_control,
              void (*once_routine)(void)).
```

This function can be called by several threads. If the execution of once_routine()
has already been completed, then control is directly returned to the calling thread.
If the execution of once_routine() has not yet been started, once_routine() is
executed by the calling thread. If the execution of the function once_routine()
has been started by another thread, but is not finished yet, then the thread executing pthread_once() waits until the other thread has finished its execution of
once_routine().

6.2 Parallel programming patterns with Pthreads

Parallel programming patterns are an essential feature to organize the parallelism
in parallel programs. They are especially useful in thread-based parallel programs
due to their irregular structure. Several parallel programming patterns have been
introduced in Section 3.3.6. This section provides Pthreads implementations for the
patterns task pool, pipelining and client-server. All Pthreads programming features
needed for this implementation have been introduced in the preceding section.

6.2.1 Implementation of a Task Pool

A thread program usually has to perform several operations or tasks. A simple approach would be to put each task into a separate function which is then called by a
separate thread which executes exactly this function and then terminates. Depending on the granularity of the tasks, this may lead to the generation and termination
of a large number of threads, causing a significant overhead. For many applications,
a more efficient implementation can be obtained by using a **task pool** (also called
work crew). The idea of a task pool is to use a specific data structure (called task
pool) to store the tasks that are ready for execution. For task execution, a fixed number of worker threads is used which are generated by the main thread at program
start and exist until the program terminates. The worker threads access the task pool
to retrieve tasks for execution. During the execution of a task, new tasks may be
generated which are then inserted into the task pool. The execution of the parallel
program is terminated, if the task pool is empty and each worker thread has finished the execution of its task. The advantage of the task pool execution scheme
is that a fixed number of threads is used, no matter how many tasks are generated.
This keeps the overhead for thread management small, independent of the number
of tasks. Moreover, tasks can be generated dynamically, thus enabling the realization of adaptive and irregular applications. In the following, we describe a simple
implementation of a task pool, see also [160]. More advanced implementations are
described in [27, 131].

Figure 6.6 presents the data structure of type tpool_t that can be used for the
task pool, and a function tpool_init for the initialization of the task pool. The data

type work_t represents a single task. It contains a reference routine to the task function containing the code of the task and the argument arg of this function. The tasks are organized as a linked list, and next is a pointer to the next task element. The data type tpool_t represents the actual task pool. It contains pointers head and tail to the first and last element of the linked list of tasks, respectively. The entry num_threads specifies the fixed number of worker threads used for executing the tasks. The array threads with entries of type pthread_t contains the references to the thread IDs of these threads. The entry max_size specifies the maximum number of tasks that can be stored in the task pool during execution and current_size specifies the varying number of tasks in the task pool during execution.

The data structure tpool_t contains one mutex variable and two condition variables to organize the insertion and extraction of tasks by concurrently running worker threads. The mutex variable lock is used to ensure mutual exclusion for threads accessing the task pool. The condition variable not_empty ensures the correct behavior when the task pool is empty. Similarly, the condition variable not_full organizes the concurrent behavior when the task pool is full, i.e., contains the maximum number max_size of tasks. If a worker thread attempts to retrieve a task from an empty task pool, it is blocked on the condition variable not_empty. If another thread inserts a task into an empty task pool, it wakes up a worker thread that is blocked on not_empty. If a thread attempts to insert a task into a full task pool, it is blocked on the condition variable not_full. If another thread retrieves a task from a full task pool so that there is space for a new task, it wakes up a thread that is blocked on not_full. The integer variable finished is used to control the termination of the task pool execution.

The function tpool_init() in Fig. 6.6 initializes the task pool by allocating the data structure and initializing it with the argument values provided. Moreover, the worker threads used for the execution of the tasks are generated and their IDs are stored in tpl->threads[i] for i=0,...,num_threads-1. Each of these worker threads uses the function tpool_thread() as start function.

Figure 6.7 shows the thread function tpool_thread(). This function has one argument specifying the task pool data structure to be used (called tpl inside tpool_thread()). Task execution is performed in an infinite loop. In each iteration of the loop, a task is retrieved from the head of the task list in tpl. If the task list is empty, the executing thread is blocked on the condition variable not_empty as described above. In the Fig. 6.7, the corresponding program position is denoted as position NE. Otherwise, a task wl is retrieved from the task list. If the task pool has been full before the retrieval of the task wl, all threads blocked on not_full and waiting to insert a task, are woken up using pthread_cond_broadcast(). In Fig. 6.7, the corresponding program code has the comment wake up threads at position NF. The access to the entire task pool structure is protected by the mutex variable tpl->lock. Finally, the retrieved task wl is executed by calling the task function wl->routine() using the argument wl->arg. The execution of the retrieved task wl may lead to the generation of new tasks which are then inserted into the task pool using tpool_insert() by the executing thread.

```
typedef struct work { // structure for a single task
  void (*routine)();
  void *arg;
  struct work *next;
} work_t;

typedef struct tpool { // entire task pool with tasks and threads
  int num_threads, max_size, current_size;
  pthread_t *threads;
  work_t *head, *tail;
  pthread_mutex_t lock;
  pthread_cond_t not_empty, not_full;
  int finished;
} tpool_t;

tpool_t *tpool_init (int num_threads, int max_size) {
  // initialization of task pool with fixed number num_threads of threads
  int i;
  tpool_t *tpl;

  // initialization of data structure
  tpl = (tpool_t *) malloc (sizeof (tpool_t));
  tpl->num_threads = num_threads;
  tpl->max_size = max_size;
  tpl->current_size = 0;
  tpl->head = tpl->tail = NULL;

  pthread_mutex_init (&(tpl->lock), NULL);
  pthread_cond_init (&(tpl->not_empty), NULL);
  pthread_cond_init (&(tpl->not_full), NULL);
  tpl->threads = (pthread_t *)malloc(sizeof(pthread_t)*num_threads);

  for (i=0; i<num_threads; i++) // creation of threads
    pthread_create (&(tpl->threads[i]), NULL,
                    tpool_thread, (void *) tpl);
  return tpl;
}
```

Fig. 6.6 Implementation of a task pool (Part 1): The data structure work_t represents a task to be executed. The task pool data structure tpool_t contains a list of tasks with head pointing to the first element and tail pointing to the last element, as well as a set of threads threads to execute the tasks. The function tpool_init() is used to initialize a task pool data structure tpl.

```
void *tpool_thread (void *tpl_arg) {
  work_t *wl;
  tpool_t *tpl = (tpool_t *) tpl_arg;

  for( ; ; ) { // each loop iteration extracts one task
    pthread_mutex_lock (&(tpl->lock));
    while (tpl->current_size == 0) { // task pool empty
      pthread_cond_wait (&(tpl->not_empty), &(tpl->lock)); // position NE
      if (tpl->finished) { pthread_mutex_unlock(&(tpl -> lock)); return 0; }
    }
    wl = tpl->head;
    tpl->current_size--;
    if (tpl->current_size == 0)
      tpl->head = tpl->tail = NULL;
    else tpl->head = wl->next;
    if (tpl->current_size == tpl->max_size - 1)
      pthread_cond_broadcast (&(tpl->not_full));
      // wake up threads waiting at position NF in function tpool_insert()
    pthread_mutex_unlock (&(tpl->lock));
    (*(wl->routine))(wl->arg); // execution of task
    free(wl);
  }
}

void tpool_insert (tpool_t *tpl, void (*routine)(), void *arg) {
  // insertion of one new task into task pool tpl
  work_t *wl;

  pthread_mutex_lock (&(tpl->lock));
  while (tpl->current_size == tpl->max_size) // task pool full
    pthread_cond_wait (&(tpl->not_full), &(tpl->lock)); // position NF
  wl = (work_t *) malloc (sizeof (work_t));
  wl->routine = routine;
  wl->arg = arg;
  wl->next = NULL;
  if (tpl->current_size == 0) {
    tpl->tail = tpl->head = wl;
    pthread_cond_broadcast (&(tpl->not_empty));
    // wake up threads waiting at position NE in function tpool_thread()
  }
  else {
    tpl->tail->next = wl;
    tpl->tail = wl;
  }
  tpl->current_size++;
  pthread_mutex_unlock (&(tpl->lock));
}
```

Fig. 6.7 Implementation of a task pool (Part 2): The function tpool_thread() extracts a task wl and executes the task by calling wl->routine with argument wl->arg. The function tpool_insert() inserts tasks into the task pool.

Situation	State of Task Pool	Thread A		Thread B
Situation 1	empty	executes `tpool_thread()`; blocks at **NE**;		. . .
		. . .		executes `tpool_insert()`; inserts task `w1`;
	not empty	. . .		state of task pool changes; broadcasts to thread that is
		wakes up; executes a task;	←—— *broadcast*	blocked at **NE**;
Situation 2	full	executes `tpool_thread()`; . . .		executes `tpool_insert()`; tries to insert a task `w12`; blocks at **NF**;
	not full	extracts task `w1`; state of task pool changes; broadcast to thread that is		. . .
		blocked at **NF**; executes task `w1`;	*broadcast* ——→	wakes up; inserts task `w12`;

Fig. 6.8 Illustration of two threads executing the function `tpool_thread()` and two selected situations. Situation 1: Thread A blocks at **NE** because of an empty task pool and Thread B inserts task `w1` and signals to wake up Thread A. Situation 2: Thread B blocks at **NF** because it tries to insert a task into a full task pool and Thread A extracts a task from a full pool and broadcasts to wake up Thread B. The program position **NE** is given in program code `tpool_thread()` and the program position **NF** is given in program code `tpool_insert()`, see Fig.6.7.

The function `tpool_insert()`, see Fig. 6.7, is used to insert tasks into the task pool. If the task pool is full when calling this function, the executing thread is blocked on the condition variable `not_full`. In Fig. 6.7, the corresponding program code is denoted as `position NF`. If the task pool is not full, a new task structure is generated and filled, and is inserted at the end of the task list. If the task pool has been empty before the insertion, all threads blocked on the condition variable `not_empty` are woken up using `pthread_cond_broadcast()`. In the Fig. 6.7, the corresponding program code has the comment `wake up threads at position NE`. The access to the task pool `tpl` is protected by the mutex variable `tpl->lock` to ensure a correct retrieval and insertion of tasks by concurrently working threads.

Figure 6.8 illustrates the interaction of concurrently working threads executing the functions `tpool_thread()` and `tpool_insert()`. Two situations are described in the figure in which the activity of the threads is given in the text column below the thread. Situation 1 illustrates the blocking of a Thread A at program `position NE` with respect to condition variable `not_empty` and its waking up by a signal from another Thread B which has inserted a new task such that the task pool is no longer empty. Situation 2 illustrates the blocking of a Thread B at program `position NF` with respect to condition variable `not_full` and its waking up by a signal from another Thread A which has extracted a task such that the task pool is no longer filled up to the maximum size.

```
int main (int argc, char *argv[]) {
  int i;
  int a[100];
  tpool_t *tpl;

  // create task pool tpl with 100 entries for 4 threads
  tpl = tpool_init(4,100);
  for (i=0; i<100; i++) a[i] = i;
  for (i=0; i<100; i++)
    tpool_insert(tpl, do_work, (void *) &a[i]);
  sleep(1);
  tpl -> finished = 1;
  pthread_cond_broadcast(&(tpl -> not_empty));
  for (i=0; i<4; i++) pthread_join(tpl->threads[i], NULL);
  printf("tpool work finished \n");
  pthread_exit(NULL);
}
void do_work (void * arg) {
  pthread_t my_id = pthread_self();
  int j = * (int *) arg;
  printf("Thread %d executes task %d computing %d \n",my_id,j,j+1);
}
```

Fig. 6.9 Implementation of a task pool (Part 3): example main program with 100 tasks and four worker threads. The tasks do_work() add integer value 1 to the integer argument value and print the thread ID executing the task.

Using pthread_cond_broadcast() in function tpool_thread() is essential, since using pthread_cond_signal() could lead to the following scenario with four threads: Thread A and B execute tpool_thread() and the task pool is full. There are two other threads C and D executing tpool_insert() which are blocked on not_full, since the task pool is full. Thread A obtains the mutex variable tpl->lock first, retrieves a task from the task pool, such that its size is tpl->max_size - 1, and wakes up thread C using pthread_cond_signal(). Thread C is then trying to obtain the mutex variable tpl->lock and is blocked on tpl->lock, since it is owned by thread A. However, after thread A releases tpl->lock, thread B becomes the owner of tpl->lock and retrieves another task, such that the task pool size is tpl->max_size - 2. Thus, thread B does not execute pthread_cond_signal() and no other thread is woken up, i.e., thread D is still blocked on not_full. If the task pool never becomes full again, thread D will never be woken up. Using pthread_cond_broadcast() would prevent this situation, since it ensures that thread D is also woken up by thread A. Similarly, pthread_cond_broadcast() is used in function tpool_insert() and not pthread_cond_signal() to keep all threads busy in all situations.

An example usage of the task pool functions is shown in Fig. 6.9. The main thread creates a task pool for 100 tasks with four worker threads using tpool_init(). The tasks are represented by the function do_work(), which obtains an integer argument, increments it, and prints the result along with the task number and the thread number. After having inserted the 100 tasks into the task pool, the main thread waits for one second, sets the finished signal and wakes up all worker threads waiting on the empty task pool. The main thread then uses `pthread_join()` to wait for the termination of the worker threads and then terminates the execution with `pthread_exit()`.

The described implementation is especially suited for a master-worker model. A master thread uses `tpool_init()` to generate a number of worker threads each of which executes the function `tpool_thread()`. The tasks to be executed are defined according to the specific requirements of the application problem and are inserted in the task pool by the master thread using `tpool_insert()`. Tasks can also be inserted by the worker threads when the execution of a task leads to the generation of new tasks. After the execution of all tasks is completed, the master thread terminates the worker threads. To do so, the master thread wakes up all threads blocked on the condition variables `not_full` and `not_empty` and terminates them. Threads that are currently executing a task are terminated as soon as they have finished the execution of this task.

6.2.2 Parallelism by Pipelining

In the pipelining model, a stream of data items is processed one after another by a sequence of threads $T_1, \ldots, T_n$ where each thread T_i performs a specific operation on each element of the data stream and passes the element onto the next thread T_{i+1}:

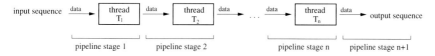

This results in an input/output relation between the threads: thread T_i receives the output of thread T_{i+1} as input and produces data elements for thread $T_{i+1}, 1 < i < n$. Thread T_1 reads the sequence of input elements, thread T_n produces the sequence of output elements. After a start-up phase with n steps, all threads can work in parallel and can be executed by different processors in parallel. The pipeline model requires some coordination between the cooperating threads: thread T_i can start the computation of its corresponding stage only if the predecessor thread T_{i-1} has provided the input data element. Moreover, thread T_i can forward its output element to the successor thread T_{i+1}, if T_{i+1} has finished its computation of the previous data item and is ready to receive a new data element.

The coordination of the threads of the pipeline stages can be organized with the help of condition variables. This will be demonstrated in the following for a simple example in which a sequence of integer values is incremented step by step in each pipeline stage, see also [27]. Thus, in each pipeline stage, the same computation is performed. But the coordination mechanism can also be applied if each pipeline stage performs a different computation.

```
typedef struct stage { // pipeline stage
  pthread_mutex_t m;
  pthread_cond_t avail; // input data available for this stage?
  pthread_cond_t ready; // stage ready to receive new data?
  int data_ready; // != 0, if other data is currently computed
  long data; // data element
  pthread_t thread; // Thread ID
  struct stage *next;
} stage_t;
typedef struct pipe { // pipeline
  pthread_mutex_t m;
  stage_t *head, *tail; // first/last stage of the pipeline
  int stages; // number of stages of the pipeline
  int active; // number of active data elements in the pipeline
} pipe_t;
const N_STAGES = 10;
```

Fig. 6.10 Implementation of a pipeline (part 1): Data structures for the implementation of a pipeline model in Pthreads.

For each stage of the pipeline, a data structure of type stage_t is used, see Fig. 6.10. This data structure contains a mutex variable m for synchronizing the access to the stage and two condition variables avail and ready for synchronizing the threads of neighboring stages. The condition variable avail is used to notify a thread that a data element is available to be processed by its pipeline stage. Thus, the thread can start the computation. A thread is blocked on the condition variable avail if no data element from the predecessor stage is available. The condition variable ready is used to notify the thread of the preceding pipeline stage that it can forward its output to the next pipeline stage. The thread of the preceding pipeline stage is blocked on this condition variable if it cannot directly forward its output data element to the next stage. The entry data_ready in the data structure for a stage is used to record whether a data element is currently available (value 1) for this pipeline stage or not (value 0). The entry data contains the actual data element to be processed. For the simple example discussed here, this is a single integer value, but this could be any data element for more complex applications. The entry thread is the TID of the thread used for this stage, and next is a reference to the next pipeline stage.

The entire pipeline is represented by the data structure pipe_t containing a mutex variable m and two pointers head and tail to the first and the last stages of the pipeline, respectively. The last stage of the pipeline is used to store the final result of the computation performed for each data element. There is no computation performed in this last stage, and there is no corresponding thread associated to this stage.

The function pipe_send(), shown in Fig. 6.11 is used to send a data element to a stage of the pipeline. This function is used to send a data element to the first stage of the pipeline, and it is also used to pass a data element to the next stage of the

```
int pipe_send(stage_t *nstage, long data) {
  //parameter: target stage and data element to be processed
  pthread_mutex_lock(&nstage->m);
  {
      while (nstage->data_ready)
        pthread_cond_wait(&nstage->ready, &nstage->m);
      nstage->data = data;
      nstage->data_ready = 1;
      pthread_cond_signal(&nstage->avail);
  }
  pthread_mutex_unlock(&nstage->m);
}
void *pipe_stage(void *arg) {
  stage_t *stage = (stage_t*)arg;
  long result_data;
  pthread_mutex_lock(&stage->m);
  {
      for ( ; ; ) {
        while (!stage->data_ready) { // wait for data
          pthread_cond_wait(&stage->avail, &stage->m);
        }
        // process data element and forward to next stage :
        result_data = stage->data + 1; // compute result data element
        pipe_send(stage->next, result_data);
        stage->data_ready = 0; // processing finished
        pthread_cond_signal(&stage->ready);
      }
  }
  pthread_mutex_unlock(&stage->m); //this unlock will never be reached
}
```

Fig. 6.11 Implementation of a pipeline (part 2): Functions to forward data elements to a pipeline stage and thread functions for the pipeline stages.

pipeline after the computation of a stage has been completed. The stage receiving the data element is identified by the parameter nstage. Before inserting the data element, the mutex variable m of the receiving stage is locked to ensure that only one thread at a time is accessing the stage. A data element can be written into the receiving stage only if the computation of the previous data element in this stage has been finished. This is indicated by the condition data_ready=0. If this is not the case, the sending thread is blocked on the condition variable ready of the receiving stage. If the receiving stage is ready to receive the data element, the sending thread writes the element into the stage and wakes up the thread of the receiving stage if it is blocked on the condition variable avail.

Each of the threads participating in the pipeline computation executes the function pipe_stage(), see Fig. 6.11. The same function can be used for each stage for our example, since each stage performs the same computations. The function receives a pointer to its corresponding pipeline stage as an argument. A thread exe-

```
int pipe_create(pipe_t *pipe, int stages) { // creation of the pipeline
  int pi;
  stage_t **link = &pipe->head;
  stage_t *new_stage, *stage;
  pthread_mutex_init(&pipe->m, NULL);
  pipe->stages = stages;
  pipe->active = 0;
  // create stages+1 stages, last stage is for the result
  for (pi = 0; pi <= stages; pi++) {
    new_stage = (stage_t *)malloc(sizeof(stage_t));
    pthread_mutex_init(&new_stage->m, NULL);
    pthread_cond_init(&new_stage->avail, NULL);
    pthread_cond_init(&new_stage->ready, NULL);
    new_stage->data_ready = 0;
    *link = new_stage; // make list link
    link = &new_stage->next;
  }
  *link = (stage_t *) NULL;
  pipe->tail = new_stage;
  // create a thread for each stage except the last one
  for (stage = pipe_head; stage->next != NULL; stage = stage->next) {
    pthread_create(&stage->thread, NULL, pipe_stage, (void*)stage);
  }
}
int pipe_start(pipe_t *pipe, long v) { // start pipeline computation
  pthread_mutex_lock(&pipe->m);
  {
    pipe->active++;
  }
  pthread_mutex_unlock(&pipe->m);
  pipe_send(pipe->head, v);
}
```

Fig. 6.12 Implementation of a pipeline (part 3): Pthreads functions to generate and start a pipeline computation.

cuting the function performs an infinite loop waiting for the arrival of data elements to be processed. The thread blocks on the condition variable avail if there is currently no data element available. If a data element is available, the thread performs its computation (increment by 1) and sends the result to the next pipeline stage stage->next using pipe_send(). Then it sends a notification to the thread associated with the next stage, which may be blocked on the condition variable ready. The notified thread can then continue its computation.

Thus, the synchronization of two neighboring threads is performed by using the condition variables avail and ready of the corresponding pipeline stages. The entry data_ready is used for the condition and determines which of the two threads is blocked or woken up. The entry of a stage is set to 0 if the stage is ready to receive a new data element to be processed by the associated thread. The entry data_ready of the next stage is set to 1 by the associated thread of the preceding stage if a new data element has been put into the next stage and is ready to be processed. In the simple example given here, the same computations are performed in each stage, i.e., all corresponding threads execute the same function `pipe_stage()`. For more complex scenarios, it is also possible that the different threads execute different functions, thus performing different computations in each pipeline stage.

The generation of a pipeline with a given number of stages can be achieved by calling the function `pipe_create()`, see Fig. 6.12. This function generates and initializes the data structures for the representation of the different stages. An additional stage is generated to hold the final result of the pipeline computation, i.e., the total number of stages is `stages+1`. For each stage except for the last additional stage, a thread is created. Each of these threads executes the function `pipe_stage()`.

The function `pipe_start()` is used to transfer a data element to the first stage of the pipeline, see Fig. 6.12. The actual transfer of the data element is done by calling the function `pipe_send()`. The thread executing `pipe_start()` does not wait for the result of the pipeline computation. Instead, `pipe_start()` returns control immediately. Thus, the pipeline works asynchronously to the thread which transfers data elements to the pipeline for computation. The synchronization between this thread and the thread of the first pipeline stage is performed within the function `pipe_send()`.

The function `pipe_result()` is used to take a result value out of the last stage of the pipeline, see Fig. 6.13. The entry `active` in the pipeline data structure `pipe_t` is used to count the number of data elements that are currently stored in the different pipeline stages. For `pipe->active = 0`, no data element is stored in the pipeline. In this case, `pipe_result()` immediately returns without providing a data element. For `pipe->active > 0`, `pipe_result()` is blocked on the condition variable `avail` of the last pipeline stage until a data element arrives at this stage. This happens if the thread associated with the next to the last stage uses `pipe_send()` to transfer a processed data element to the last pipeline stage, see Fig. 6.11. By doing so, this thread wakes up a thread that is blocked on the condition variable `avail` of the last stage, if there is a thread waiting. If so, the woken-up thread is the one which tries to take a result value out of the last stage using `pipe_result()`.

The main program of the pipeline example is given in Fig. 6.13. It first uses `pipe_create()` to generate a pipeline with a given number of stages. Then it reads from `stdin` lines with numbers, which are the data elements to be processed. Each such data element is forwarded to the first stage of the pipeline using `pipe_start()`. Doing so, the executing main thread may be blocked on the condition variable `ready` of the first stage until the stage is ready to receive the data element. An input line with a single character `'='` causes the main thread to call `pipe_result()` to take a result element out of the last stage, if present.

Figure 6.14 illustrates the synchronization between neighboring pipeline threads as well as between the main thread and the threads of the first or the next to last stage for a pipeline with three stages and two pipeline threads T_1 and T_2. The figure shows the relevant entries of the data structure `stage_t` for each stage. The order of the access and synchronization operations performed by the pipeline threads is

```
int pipe_result(pipe_t *pipe, long *result) {
  stage_t *tail = pipe->tail;
  int empty = 0;
  pthread_mutex_lock(&pipe->m);
  {
     if (pipe->active <= 0) empty = 1; // empty pipeline
     else pipe->active--; // remove a data element
  }
  pthread_mutex_unlock(&pipe->m);
  if (empty) return 0;
  pthread_mutex_lock(&tail->m);
  {
     while (!tail->data_ready) // wait for data element
       pthread_cond_wait(&tail->avail, &tail->m);
     *result = tail->data;
     tail->data_ready = 0;
     pthread_cond_signal(&tail->ready);
  }
  pthread_mutex_unlock(&tail->m);
  return 1;
}
int main(int argc, char *argv[]) {
  pipe_t pipe;
  long value, result;
  char line[128];
  pipe_create(&pipe, N_STAGES);
  // terminate program for input error or EOF
  while (fgets(line, sizeof(line), stdin)) {
     if (*line == '\0')
       continue; // ignore empty input lines
     if (!strcmp(line, "=")) {
       if (pipe_result(&pipe, &result))
         printf("%ld\n", result);
       else printf("ERROR: Pipe empty\n");
     }
     else {
       if (sscanf(line, "%ld", &value) < 1)
         printf("ERROR: Not an int\n");
       else pipe_start(&pipe, value);
     }
  }
  return 0;
}
```

Fig. 6.13 Implementation of a pipeline (part 4): Main program and Pthreads function pipe_result which extracts a result element from the pipeline.

determined by the statements in `pipe_stage()` and is illustrated by circled numbers. The access and synchronization operations of the main thread result from the statements in `pipe_start()` and `pipe_result()`.

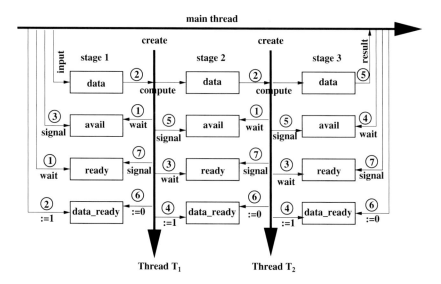

Fig. 6.14 Illustration of the synchronization between the pipeline threads for a pipeline with two pipeline threads and three stages, from the view of the data structures used. The circled numbers describe the order in which the synchronization steps are executed by the different threads according to the corresponding thread functions.

6.2.3 Implementation of a Client-Server Model

In a client-server system, we can distinguish between client threads and server threads. In a typical scenario, there are several server threads and several client threads. Server threads process requests that have been issued by the client threads. Client threads often represent the interface to the users of a system. During the processing of a request by a server thread, the issuing client thread can either wait for the request to be finished or can perform other operations, working concurrently with the server, and can collect the result at a later time when it is required. In the following, we illustrate the implementation of a client-server model for a simple example, see also [27].

Several threads repeatedly read input lines from `stdin` and output result lines to `stdout`. Before reading, a thread outputs a prompt to indicate which input is expected from the user. Server threads can be used to ensure the synchronization between the output of a prompt and the reading of the corresponding input line so that

```
#define REQ_READ 1
#define REQ_WRITE 2
#define REQ_QUIT 3
#define PROMPT_SIZE 32
#define TEXT_SIZE 128
typedef struct request {
  struct _request_t *next; // linked list
  int op;
  int synchronous; // 1 iff client waits for server
  int done_flag;
  pthread_cond_t done;
  char prompt[PROMPT_SIZE], text[TEXT_SIZE];
} request_t;
typedef struct tty_server { // data structure for server context
  request_t *first, *last;
  int running; // != 0, if server is running
  pthread_mutex_t m;
  pthrad_cond_t request;
} tty_server_t;
#define TTY_SERVER_INITIALIZER { NULL, NULL, 0,
  PTHREAD_MUTEX_INITIALIZER,
  PTHREAD_COND_INITIALIZER }
tty_server_t tty_server = TTY_SERVER_INITIALIZER;
int client_threads;
pthread_mutex_t client_mutex = PTHREAD_MUTEX_INITIALIZER;
pthread_cond_t client_done = PTHREAD_COND_INITIALIZER;
pthread_t server_thread;
```

Fig. 6.15 Implementation of a client-server system (part 1): Data structure for the implementation of a client-server model with Pthreads.

no output of another thread can occur in between. Client threads forward requests to the server threads to output a prompt or to read an input line. The server threads are terminated by a specific QUIT command. Fig. 6.15 shows the data structures used for an implementation with Pthreads. The data structure request_t represents requests from the clients for the servers. The entry op specifies the requested operation to be performed (REQ_READ, REQ_WRITE, or REQ_QUIT). The entry synchronous indicates whether the client waits for the termination of the request (value 1) or not (value 0). The condition variable done is used for the synchronization between client and server, i.e., the client thread is blocked on done to wait until the server has finished the execution of the request. The entries prompt and text are used to store a prompt to be output or a text read in by the server, respectively. The data structure tty_server_t is used to store the requests sent to a server. The requests are stored in a FIFO (*first-in, first-out*) queue which can be accessed by first and last. The server thread is blocked on the condition variable request if the request queue is empty. The entry running indicates whether the corresponding server is running (value 1) or not (value 0). The program described in the following works with a single server thread, but can in principle be extended to an arbitrary number of servers.

```
void *tty_server_routine(void *arg) {
  static pthread_mutex_t prompt_mutex = PTHREAD_MUTEX_INITIALIZER;
  request_t *request;
  int op, len;
  for (;;) {
    pthread_mutex_lock(&tty_server.m);
    {
        while (tty_server.first == NULL)
          pthread_cond_wait(&tty_server.request, &tty_server.m);
        request = tty_server.first;
        tty_server.first = request->next;
        if (tty_server.first == NULL)
          tty_server.last = NULL;
    }
    pthread_mutex_unlock(&tty_server.m);
    switch (request->op) {
    case REQ_READ:
      puts(request->prompt);
      if (fgets(request->text, TEXT_SIZE, stdin) == NULL)
        request->text[0] = '\0';
      len = strlen(request->text);
      if (len > 0 && request->text[len - 1] == '\n')
        request->text[len - 1] = '\0';
      break;
    case REQ_WRITE:
      puts(request->text); break;
    default: // auch REQ_QUIT
      break;
    }
    op = request->op;
    if (request->synchronous) {
      pthread_mutex_lock(&tty_server.m);
      request->done_flag = 1;
      pthread_cond_signal(&request->done);
      pthread_mutex_unlock(&tty_server.m);
    }
    else free (request);
    if (op == REQ_QUIT) break;
  }
  return NULL;
}
```

Fig. 6.16 Implementation of a client-server system (part 2): Server thread to process client requests.

The server thread executes the function tty_server_routine(), see Fig. 6.16. The server is blocked on the condition variable request as long as there are no

```
void tty_server_request(int op, int sync, char *prompt, char *string) {
  request_t *request;
  pthread_mutex_lock(&tty_server.m);
  {
    if (!tty_server.running) {
      pthread_create(&server_thread,NULL,tty_server_routine,NULL);
      tty_server.running = 1;
    }
    request = (request_t *) malloc(sizeof(request_t));
    request->op = op;
    request->synchronous = sync;
    request->next = NULL;
    if (sync) {
      request->done_flag = 0;
      pthread_cond_init(&request->done, NULL);
    }
    if (prompt != NULL) {
      strncpy(request->prompt, prompt, PROMPT_SIZE);
      request->prompt[PROMPT_SIZE - 1] = '\0';
    }
    else request->prompt[0] = '\0';
    if (op == REQ_WRITE && string != NULL) {
      strncpy(request->text, string, TEXT_SIZE);
      request->text[TEXT_SIZE - 1] = '\0';
    }
    else request->text[0] = '\0';
    if (tty_server.first == NULL)
      tty_server.first = tty_server.last = request;
    else {
      tty_server.last->next = request;
      tty_server.last = request;
    }
    pthread_cond_signal(&tty_server.request);
    if (sync) {
      while (!request->done_flag)
        pthread_cond_wait(&request->done, &tty_server.m);
      if (op == REQ_READ)
        strcpy(string, request->text);
      pthread_cond_destroy(&request->done);
      free(request);
    }
  }
  pthread_mutex_unlock(&tty_server.m);
}
```

Fig. 6.17 Implementation of a client-server system (part 3): Forwarding of a request to the server thread.

requests to be processed. If there are requests, the server removes the first request from the queue and executes the operation (REQ_READ, REQ_WRITE, or REQ_QUIT) specified in the request. For the REQ_READ operation, the prompt specified with the request is output and a line is read in and stored into the text entry of the request structure. For a REQ_WRITE operation, the line stored in the text entry is written to stdout. The operation REQ_QUIT causes the server to finish its execution. If an issuing client waits for the termination of a request (entry synchronous), it is blocked on the condition variable done in the corresponding request structure. In this case, the server thread wakes up the blocked client thread using pthread_cond_signal() after the request has been processed. For asynchronous requests, the server thread is responsible for freeing the request data structure.

The client threads use the function tty_server_request() to forward a request to the server, see Fig. 6.17. If the server thread is not running yet, it will be started in tty_server_request(). The function allocates a request structure of type request_t and initializes it according to the requested operation. The request structure is then inserted into the request queue of the server. If the server is blocked waiting for requests to arrive, it is woken up using pthread_cond_signal(). If the client wants to wait for the termination of the request by the server, it is blocked on the condition variable done in the request structure, waiting for the server to wake it up again. The client threads execute the function client_routine(), see Fig. 6.18. Each client sends read and write requests to the server using the function tty_server_request() until the user terminates the client thread by specifying an empty line as input. When the last client thread has been terminated, the main thread which is blocked on the condition variable client_done is woken up again. The main thread generates the client threads and then waits until all client threads have been terminated. The server thread is not started by the main thread, but by the client thread which sends the first request to the server using tty_server_routine(). After all client threads are terminated, the server thread is terminated by the main thread by sending a REQ_QUIT request.

6.3 Advanced Pthread features

6.3.1 Thread Attributes and Cancellation

Threads are created using pthread_create(). In the previous sections, we have specified NULL as the second argument, thus leading to the generation of threads with default characteristics. These characteristics can be changed with the help of attribute objects. To do so, an attribute object has to be allocated and initialized before using the attribute object as parameter of pthread_create(). An attribute object for threads has type pthread_attr_t. Before an attribute object can be used, it first must be initialized by calling the function

```
void *client_routine(void *arg) {
  int my_nr = *(int*)arg, loops;
  char prompt[PROMPT_SIZE], string[TEXT_SIZE];
  char format[TEXT_SIZE + 64];
  sprintf(prompt, "Client %d>", my_nr);
  for (;;) {
    tty_server_request(REQ_READ, 1, prompt, string);
    // synchronized input
    if (string[0] == '\0') break; // program exit
    for (loops = 0; loops < 4; loops++) {
        sprintf(format, "(%d # %d)%s", my_nr, loops, string);
        tty_server_request(REQ_WRITE, 0, NULL, format);
        sleep(1);
    }
  }
  pthread_mutex_lock(&client_mutex);
  client_threads--;
  if (client_threads == 0) pthread_cond_signal(&client_done);
  pthread_mutex_unlock(&client_mutex);
  return NULL;
}
#define N_THREADS 4
int main(int argc, char *argv[]) {
  pthread_t thread;
  int i;
  int args[N_THREADS];
  client_threads = N_THREADS;
  pthread_mutex_lock(&client_mutex);
  {
    for (i = 0; i < N_THREADS; i++) {
        args[i] = i;
        pthread_create(&thread, NULL, client_routine, &args[i]);
    }
    while (client_threads > 0)
        pthread_cond_wait(&client_done, &client_mutex);
  }
  pthread_mutex_unlock(&client_mutex);
  printf("All clients done\n");
  tty_server_request(REQ_QUIT, 1, NULL, NULL);
  return 0;
}
```

Fig. 6.18 Implementation of a client-server system (part 4): Client thread and main thread.

```
int pthread_attr_init (pthread_attr_t *attr).
```

This leads to an initialization with the default attributes, corresponding to the default characteristics. By changing an attribute value, the characteristics can be changed. Pthreads provides attributes to influence the return value of threads, setting the size and address of the runtime stack, or the cancellation behavior of the thread. For each attribute, Pthreads defines functions to get and set the current attribute value. But Pthreads implementations are not required to support the modification of all attributes. In the following, the most important aspects are described.

6.3.1.1 Return value

An important property of a thread is its behavior concerning thread termination. This is captured by the attribute detachstate. This attribute can be influenced by all Pthreads libraries. By default, the runtime system assumes that the return value of a thread T_1 may be used by another thread after the termination of T_1. Therefore, the internal data structure maintained for a thread will be kept by the runtime system after the termination of a thread until another thread retrieves the return value using pthread_join(), see Section 6.1.1. Thus, a thread may bind resources even after its termination. This can be avoided if the programmer knows in advance that the return value of a thread will not be needed. If so, the thread can be generated such that its resources are immediately returned to the runtime system after its termination. This can be achieved by changing the detachstate attribute. The following two functions are provided to get or set this attribute value:

```
int pthread_attr_getdetachstate (const pthread_attr_t *attr,
                                 int *detachstate)
int pthread_attr_setdetachstate (pthread_attr_t *attr,
                                 int detachstate).
```

The attribute value detachstate=PTHREAD_CREATE_JOINABLE means that the return value of the thread is kept until it is joined by another thread. The attribute value detachstate=PTHREAD_CREATE_DETACHED means that the thread resources are freed immediately after thread termination.

6.3.1.2 Stack characteristics

The different threads of a process have a shared program and data memory and a shared heap, but each thread has its own runtime stack. For most Pthreads libraries, the size and address of the local stack of a thread can be changed, but it is not required that a Pthreads library supports this option. The local stack of a thread is used to store local variables of functions whose execution has not yet been terminated. The size required for the local stack is influenced by the size of the local variables and the nesting depth of function calls to be executed. This size may be large for re-

cursive functions. If the default stack size is too small, it can be increased by changing the corresponding attribute value. The Pthreads library that is used supports this
if the macro

```
_POSIX_THREAD_ATTR_STACKSIZE
```

is defined in <unistd.h>. This can be checked by

```
#ifdef _POSIX_THREAD_ATTR_STACKSIZE        or
if (sysconf (_SC_THREAD_ATTR_STACKSIZE) == -1)
```

in the program. If it is supported, the current stack size stored in an attribute object
can be retrieved or set by calling the functions

```
int pthread_attr_getstacksize (const pthread_attr_t *attr,
                               size_t *stacksize)
int pthread_attr_setstacksize (pthread_attr_t *attr,
                               size_t stacksize).
```

Here, size_t is a data type defined in <unistd.h> which is usually implemented
as unsigned int. The parameter stacksize is the size of the stack in bytes. The
value of stacksize should be at least PTHREAD_STACK_MIN which is predefined by
Pthreads as the minimum stack size required by a thread. Moreover, if the macro

```
_POSIX_THREAD_ATTR_STACKADDR
```

is defined in <unistd.h>, the address of the local stack of a thread can also be
influenced. The following two functions

```
int pthread_attr_getstackaddr (const pthread_attr_t *attr,
                               size_t **stackaddr)
int pthread_attr_setstackaddr (pthread_attr_t *attr,
                               size_t *stackaddr)
```

are provided to get or set the current stack address stored in an attribute object. The
modification of stack-related attributes should be used with caution, since such modification can result in non-portable programs. Moreover, the option is not supported
by all Pthreads libraries.

After the modification of specific attribute values in an attribute object a thread
with the chosen characteristics can be generated by specifying the attribute object
as second parameter of pthread_create(). The characteristics of the new thread
are defined by the attribute values stored in the attribute object at the time at which
pthread_create() is called. These characteristics cannot be changed at a later
time by changing attribute values in the attribute object.

6.3.1.3 Thread Cancellation

In some situations, it is useful to stop the execution of a thread from outside, e.g., if
the result of the operation performed is no longer needed. An example could be an

application where several threads are used to search in a data structure for a specific entry. As soon as the entry is found by one of the threads, all other threads can stop execution to save execution time. This can be reached by sending a cancellation request to these threads.

In Pthreads, a thread can send a cancellation request to another thread by calling the function

```
int pthread_cancel (pthread_t thread)
```

where `thread` is the thread ID of the thread to be terminated. A call of this function does not necessarily lead to an immediate termination of the specified target thread. The exact behavior depends on the cancellation type of this thread. In any case, control immediately returns to the calling thread, i.e., the thread issuing the cancellation request does not wait for the cancelled thread to be terminated. By default, the cancellation type of the thread is **deferred**. This means that the thread can only be cancelled at specific **cancellation points** in the program. After the arrival of a cancellation request, thread execution continues until the next cancellation point is reached. The Pthreads standard defines obligatory and optional cancellation points. Obligatory cancellation points typically include all functions at which the executing thread may be blocked for a substantial amount of time. Examples are `pthread_cond_wait()`, `pthread_cond_timedwait()`, `open()`, `read()`, `wait()` or `pthread_join()`, see [27] for a complete list. Optional cancellation points include many file and I/O operations. The programmer can insert additional cancellation points into the program by calling the function

```
void pthread_testcancel().
```

When calling this function, the executing thread checks whether a cancellation request has been sent to it. If so, the thread is terminated. If not, the function has no effect. Similarly, at predefined cancellation points the executing thread also checks for cancellation requests. A thread can set its cancellation type by calling the function

```
int pthread_setcancelstate (int state, int *oldstate).
```

A call with `state = PTHREAD_CANCEL_DISABLE` disables the cancelability of the calling thread. The previous cancellation type is stored in `*oldstate`. If the cancelability of a thread is disabled, it does not check for cancellation requests when reaching a cancellation point or when calling `pthread_testcancel()`, i.e., the thread cannot be cancelled from outside. The cancelability of a thread can be enabled again by calling `pthread_setcancelstate()` with the parameter value `state = PTHREAD_CANCEL_ENABLE`.

By default, the cancellation type of a thread is deferred. This can be changed to **asynchronous cancellation** by calling the function

```
int pthread_setcanceltype (int type, int *oldtype)
```

with type=PTHREAD_CANCEL_ASYNCHRONOUS. This means that this thread can be cancelled not only at cancellation points. Instead, the thread is terminated immediately after the cancellation request arrives, even if the thread is just performing computations within a critical section. This may lead to inconsistent states causing errors for other threads. Therefore, asynchronous cancellation may be harmful and should be avoided. Calling pthread_setcanceltype() with type = PTHREAD_CANCEL_DEFERRED sets a thread to the usual deferred cancellation type.

6.3.1.4 Cleanup Stack

In some situations, a thread may need to restore a certain state when it is cancelled. For example, a thread may have to release a mutex variable when it is the owner before being cancelled. To support such state restorations, a cleanup stack is associated with each thread, containing function calls to be executed just before thread cancellation. These function calls can be used to establish a consistent state at thread cancellation, e.g., by unlocking mutex variables that have previously been locked. This is necessary if there is a cancellation point between acquiring and releasing a mutex variable. If a cancellation happens at such a cancellation point without releasing the mutex variable, another thread might wait forever to become the owner. To avoid such situations, the cleanup stack can be used: when acquiring the mutex variable, a function call (cleanup handler) to release it is put onto the cleanup stack. This function call is executed when the thread is cancelled. A cleanup handler is put onto the cleanup stack by calling the function

```
void pthread_cleanup_push (void (*routine) (void *), void *arg)
```

where routine is a pointer to the function used as cleanup handler and arg specifies the corresponding argument values. The cleanup handlers on the cleanup stack are organized in LIFO (*last-in, first-out*) order, i.e., the handlers are executed in the opposite order of their placement, beginning with the most recently added handler. The handlers on the cleanup stack are automatically executed when the corresponding thread is cancelled or when it exits by calling pthread_exit(). A cleanup handler can be removed from the cleanup stack by calling the function

```
void pthread_cleanup_pop (int execute).
```

This call removes the most recently added handler from the cleanup stack. For execute≠0, this handler will be executed when it is removed. For execute=0, this handler will be removed without execution. To produce portable programs, corresponding calls of pthread_cleanup_push() and pthread_cleanup_pop() should be organized in pairs within the same function.

Example: To illustrate the use of cleanup handlers, we consider the implementation of a semaphore mechanism in the following. A *(counting) semaphore* is a data type with a counter which can have non-negative integer values and which can be modified by two operations: a *signal* operation increments the counter and wakes up a thread which is blocked on the semaphore, if there is such a thread; a *wait* operation blocks the executing thread until the counter has a value > 0, and then decrements the counter. Counting semaphores can be used for the management of limited resources. In this case, the counter is initialized with the number of available resources. *Binary semaphores*, on the other hand, can only have value 0 or 1. They can be used to ensure mutual exclusion when executing critical sections.

Fig. 6.19 illustrates the use of cleanup handlers to implement a semaphore mechanism based on condition variables, see also [178]. A semaphore is represented by the data type `sema_t`. The function `AcquireSemaphore()` waits until the counter has values > 0, before decrementing the counter. The function `ReleaseSemaphore()` increments the counter and then wakes up a waiting thread using `pthread_cond_signal()`. The access to the semaphore data structure is protected by a mutex variable in both cases, to avoid inconsistent states by concurrent accesses. At the beginning, both functions call `pthread_mutex_lock()` to lock the mutex variable. At the end, the call `pthread_cleanup_pop(1)` leads to the execution of `pthread_mutex_unlock()`, thus releasing the mutex variable again. If a thread is blocked in `AcquireSemaphore()` when executing the function `pthread_cond_wait(&(ps->cond),&(ps->mutex))` it implicitly releases the mutex variable `ps->mutex`. When the thread is woken up again, it first tries to become owner of this mutex variable again. Since `pthread_cond_wait()` is a cancellation point, a thread might be cancelled while waiting for the condition variable `ps->cond`. In this case, the thread first becomes the owner of the mutex variable before termination. Therefore, a cleanup handler is used to release the mutex variable again. This is obtained by the function `Cleanup_Handler()` in Fig. 6.19.   ☐

6.3.1.5 Producer-Consumer threads

The semaphore mechanism from Fig. 6.19 can be used for the synchronization between producer and consumer threads, see Fig. 6.20. A producer thread inserts entries into a buffer of fixed length. A consumer thread removes entries from the buffer for further processing. A producer can insert entries only if the buffer is not full. A consumer can remove entries only if the buffer is not empty. To control this, two semaphores `full` and `empty` are used. The semaphore `full` counts the number of occupied entries in the buffer. It is initialized with 0 at program start. The semaphore `empty` counts the number of free entries in the buffer. It is initialized with the buffer capacity. In the example, the buffer is implemented as an array of length 100, storing entries of type ENTRY. The corresponding data structure `buffer` also contains the two semaphores `full` and `empty`.

As long as the buffer is not full, a producer thread produces entries and inserts them into the shared buffer using `produce_item()`. For each insert operation, `empty` is decremented by using `AcquireSemaphore()` and `full` is incremented by using `ReleaseSemaphore()`. If the buffer is full, a producer thread will be blocked when calling `AcquireSemaphore()` for `empty`. As long as the buffer is

```
typedef struct Sema {
  pthread_mutex_t mutex;
  pthread_cond_t cond;
  int count;
} sema_t;

void CleanupHandler (void *arg)
{ pthread_mutex_unlock ((pthread_mutex_t *) arg);}

void AquireSemaphore (sema_t *ps)
{
  pthread_mutex_lock (&(ps->mutex));
  pthread_cleanup_push (CleanupHandler, &(ps->mutex));
  while (ps->count == 0)
    pthread_cond_wait (&(ps->cond), &(ps->mutex));
  --ps->count;
  pthread_cleanup_pop (1);
}

void ReleaseSemaphore (sema_t *ps)
{
  pthread_mutex_lock (&(ps->mutex));
  pthread_cleanup_push (CleanupHandler, &(ps->mutex));
  ++ps->count;
  pthread_cond_signal (&(ps->cond));
  pthread_cleanup_pop (1);
}
```

Fig. 6.19 Use of a cleanup handler for the implementation of a semaphore mechanism. The function AquireSemaphore() implements the access to the semaphore. The call of pthread_cond_wait() ensures that the access is performed not before the value count of the semaphore is larger than zero. The function ReleaseSemaphore() implements the release of the semaphore.

not empty, a consumer thread removes entries from the buffer and processes them using comsume_item(). For each remove operation, full is decremented using AcquireSemaphore() and empty is incremented using ReleaseSemaphore(). If the buffer is empty, a consumer thread will be blocked when calling the function AcquireSemaphore() for full. The internal buffer management is hidden in the functions produce_item() and consume_item().

After a producer thread has inserted an entry into the buffer, it wakes up a consumer thread which is waiting for the semaphore full by calling the function ReleaseSemaphore(&buffer.full), if there is such a waiting consumer. After a consumer has removed an entry from the buffer, it wakes up a producer which is waiting for empty by calling ReleaseSemaphore(&buffer.empty), if there is such a waiting producer. The program in Fig. 6.20 uses one producer and one consumer thread, but it can easily be generalized to an arbitrary number of producer and consumer threads.

```
struct linebuf {
  ENTRY line[100];
  sema_t full, empty;
} buffer;

void *Producer(void *arg)
{
  while (1) {
    AquireSemaphore (&buffer.empty);
    produce_item();
    ReleaseSemaphore (&buffer.full);
  }
}

void *Consumer(void *arg)
{
  while (1) {
    AquireSemaphore (&buffer.full);
    consume_item();
    ReleaseSemaphore (&buffer.empty);
  }
}

void CreateSemaphore (sema_t *ps, int count)
{
  ps->count = count;
  pthread_mutex_init (&ps->mutex, NULL);
  pthread_cond_init (&ps->cond, NULL);
}

int main()
{
  pthread_t threadID[2];
  int i;
  void *status;

  CreateSemaphore (&buffer.empty, 100);
  CreateSemaphore (&buffer.full, 0);

  pthread_create (&threadID[0], NULL, Consumer, NULL);
  pthread_create (&threadID[1], NULL, Producer, NULL);

  for (i=0; i<2; i++)
    pthread_join (threadID[i], &status);
}
```

Fig. 6.20 Implementation of producer-consumer threads using the semaphore operations from Fig. 6.19.

6.3.2 Thread Scheduling with Pthreads

The user threads defined by the programmer for each process are mapped to kernel threads by the library scheduler. The kernel threads are then brought to execution on the available processors by the scheduler of the operating system. For many Pthreads libraries, the programmer can influence the mapping of user threads to kernel threads using **scheduling attributes**. The Pthreads standard specifies a scheduling interface for this, but this is not necessarily supported by all Pthreads libraries. A specific Pthreads library supports the scheduling programming interface, if the macro POSIX_THREAD_PRIORITY_SCHEDULING is defined in <unistd.h>. This can also be checked dynamically in the program using sysconf() with parameter _SC_THREAD_PRIORITY_SCHEDULING. If the scheduling programming interface is supported and shall be used, the header file <sched.h> must be included into the program.

Scheduling attributes are stored in data structures of type struct sched_param which must be provided by the Pthreads library if the scheduling interface is supported. This type must at least have the entry

```
int sched_priority;
```

The scheduling attributes can be used to assign scheduling priorities to threads and to define scheduling policies and scheduling scopes. This can be set when a thread is created, but it can also be changed dynamically during thread execution.

6.3.2.1 Explicit setting of scheduling attributes

In the following, we first describe how scheduling attributes can be set explicitly at thread creation.

The **scheduling priority** of a thread determines how privileged the library scheduler treats the execution of a thread compared to other threads. The priority of a thread is defined by an integer value which is stored in the sched_priority entry of the sched_param data structure and which must lie between a minimum and a maximum value. These minimum and maximum values allowed for a specific scheduling policy can be determined by calling the functions

```
int sched_get_priority_min (int policy)
int sched_get_priority_max (int policy) ,
```

where policy specifies the scheduling policy. The minimum or maximum priority values are given as return value of these functions. The library scheduler maintains for each priority value a separate queue of threads with this priority that are ready for execution. When looking for a new thread to be executed, the library scheduler accesses the thread queue with the highest priority that is not empty. If this queue contains several threads, one of them is selected for execution according to the scheduling policy. If there are always enough executable threads available at each point in program execution, it can happen that threads of low priority are not executed for quite a long time. The two functions

```
int pthread_attr_getschedparam (const pthread_attr_t *attr,
                                struct sched_param *param)
int pthread_attr_setschedparam (pthread_attr_t *attr,
                                const struct sched_param *param)
```

can be used to extract or set the priority value of an attribute data structure `attr`. To set the priority value, the entry `param->sched_priority` must be set to the chosen priority value before calling `pthread_attr_setschedparam()`.

The scheduling policy of a thread determines how threads of the same priority are executed and share the available resources. In particular, the scheduling policy determines how long a thread is executed if it is selected by the library scheduler for execution. Pthreads supports three different scheduling policies:

- **SCHED_FIFO** (*first-in, first-out*): The executable threads of the same priority are stored in a FIFO queue. A new thread to be executed is selected from the beginning of the thread queue with the highest priority. The selected thread is executed until it either exits or blocks, or until a thread with a higher priority becomes ready for execution. In the latter case, the currently executed thread with lower priority is interrupted and is stored at the *beginning* of the corresponding thread queue. Then, the thread of higher priority starts execution. If a thread that has been blocked, e.g., by waiting on a condition variable, becomes ready for execution again, it is stored at the *end* of the thread queue of its priority. If the priority of a thread is dynamically changed, it is stored at the *end* of the thread queue with the new priority.
- **SCHED_RR** (*round-robin*): The thread management is similar as for the policy SCHED_FIFO. The difference is that each thread is allowed to run for only a fixed amount of time, given by a predefined timeslice interval. After the interval has elapsed, and another thread of the same priority is ready for execution, the running thread will be interrupted and put at the end of the corresponding thread queue. The timeslice intervals are defined by the library scheduler. All threads of the same process use the same timeslice interval. The length of a timeslice interval of a process can be queried with the function

```
int sched_rr_get_interval (pid_t pid, struct timespec *quantum)
```

where `pid` is the process ID of the process. For `pid=0`, the information for that process is returned to which the calling thread belongs. The data structure of type `timespec` is defined as

```
struct timespec { time_t tv_sec; long tv_nsec; } .
```

- **SCHED_OTHER**: Pthreads allows an additional scheduling policy the behavior of which is not specified by the standard, but completely depends on the specific Pthreads library used. This allows the adaptation of the scheduling to a specific operating system. Often, a scheduling strategy is used which adapts the priorities of the threads to their I/O behavior, such that interactive threads get a higher priority as compute-intensive threads. This scheduling policy is often used as default for newly created threads.

The scheduling policy used for a thread is set when the thread is created. If the programmer wants to use a scheduling policy other than the default he can achieve this by creating an attribute data structure with the appropriate values and providing this data structure as argument for `pthread_create()`. The two functions

```
int pthread_attr_getschedpolicy (const pthread_attr_t *attr,
                                  int *schedpolicy)
int pthread_attr_setschedpolicy (pthread_attr_t *attr,
                                  int schedpolicy)
```

can be used to extract or set the scheduling policy of an attribute data structure `attr`. On some Unix systems, setting the scheduling policy may require superuser rights.

The **contention scope** of a thread determines which other threads are taken into consideration for the scheduling of a thread. Two options are provided: The thread may compete for processor resources with the threads of the corresponding process (*process contention scope*) or with the threads of all processes on the system (*system contention scope*). Two functions can be used to extract or set the contention scope of an attribute data structure `attr`:

```
int pthread_attr_getscope (const pthread_attr_t *attr,
                           int *contentionscope)
int pthread_attr_setscope (pthread_attr_t *attr,
                           int contentionscope)
```

The parameter value `contentionscope=PTHREAD_SCOPE_PROCESS` corresponds to a process contention scope, whereas a system contention scope can be obtained by the parameter value `contentionscope=PTHREAD_SCOPE_SYSTEM`. Typically, using a process contention scope leads to better performance than a system contention scope, since the library scheduler can switch between the threads of a process without calling the operating system, whereas switching between threads of different processes usually requires a call of the operating system, and this is usually relatively expensive [27]. A Pthreads library only needs to support one of the two contention scopes. If a call of `pthread_attr_setscope()` tries to set a contention scope that is not supported by the specific Pthreads library, the error value ENOT-SUP is returned.

6.3.2.2 Implicit setting of scheduling attributes

Some application codes create a lot of threads for specific tasks. To avoid setting the scheduling attributes before each thread creation, Pthreads supports the inheritance of scheduling information from the creating thread. The two functions

```
int pthread_attr_getinheritsched (const pthread_attr_t *attr,
                                  int *inheritsched)
int pthread_attr_setinheritsched (pthread_attr_t *attr,
                                  int inheritsched)
```

can be used to extract or set the inheritance status of an attribute data structure attr. Here, inheritsched=PTHREAD_INHERIT_SCHED means that a thread creation with this attribute structure generates a thread with the scheduling attributes of the creating thread, ignoring the scheduling attributes in the attribute structure. The parameter value inheritsched=PTHREAD_EXPLICIT_SCHED disables the inheritance, i.e., the scheduling attributes of the created thread must be set explicitly if they should be different from the default setting. The Pthreads standard does not specify a default value for the inheritance status. Therefore, if a specific behavior is required, the inheritance status must be set explictly.

6.3.2.3 Dynamic setting of scheduling attributes

The priority of a thread and the scheduling policy used can also be changed dynamically during the execution of a thread. The two functions

```
int pthread_getschedparam (pthread_t thread, int *policy,
                           struct sched_param *param)
int pthread_setschedparam (pthread_t thread, int policy,
                           const struct sched_param *param)
```

can be used to dynamically extract or set the scheduling attributes of a thread with TID thread. The parameter policy defines the scheduling policy, param contains the priority value.

Fig. 6.21 illustrates how the scheduling attributes can be set explicitly before the creation of a thread. In the example, SCHED_RR is used as scheduling policy. Moreover, a medium priority value is used for the thread with ID thread_id. The inheritance status is set to PTHREAD_EXPLICIT_SCHED to transfer the scheduling attributes from attr to the newly created thread thread_id.

6.3.3 Priority Inversion

When scheduling several threads with different priorities, an unsuitable order of synchronization operations can lead to situations in which a thread of lower priority prevents a thread of higher priority from being executed. This phenomenon is called **priority inversion**, indicating that a thread of lower priority is running although a thread of higher priority is ready for execution. This phenomenon is illustrated in the following example, see also [160].

```
#include <unistd.h>
#include <pthread.h>
#include <sched.h>

void *thread_routine (void *arg) {return NULL;}

int main()
{
  pthread_t thread_id;
  pthread_attr_t attr;
  struct sched_param param;
  int policy, min_prio, max_prio;

  pthread_attr_init (&attr);
  if (sysconf (_SC_THREAD_PRIORITY_SCHEDULING) != -1) {
    pthread_attr_getschedpolicy (&attr, &policy);
    pthread_attr_getschedparam (&attr, &param);
    printf ("Default: Policy %d, Priority %d \n", policy,
            param.sched_priority);
    pthread_attr_setschedpolicy (&attr, SCHED_RR);
    min_prio = sched_get_priority_min (SCHED_RR);
    max_prio = sched_get_priority_max (SCHED_RR);
    param.sched_priority = (min_prio + max_prio)/2;
    pthread_attr_setschedparam (&attr, &param);
    pthread_attr_setinheritsched (&attr, PTHREAD_EXPLICIT_SCHED);
  }
  pthread_create (&thread_id, &attr, thread_routine, NULL);
  pthread_join (thread_id, NULL);
  return 0;
}
```

Fig. 6.21 Use of scheduling attributes to define the scheduling behavior of a generated thread.

Example: We consider the execution of three threads A, B, C with high, medium and low priority, respectively, on a single processor competing for a mutex variable m. The threads perform at program points $t_1, \ldots, t_6$ the following actions, see Fig. 6.22 for an illustration. After the start of the program at time t_1, thread C of low priority is started at time t_2. At time t_3, thread C calls pthread_mutex_lock(m) to lock m. Since m has not been locked before, C becomes owner of m and continues execution. At time t_4, thread A of high priority is started. Since A has a higher priority than C, C is blocked and A is executed. The mutex variable m is still locked by C. At time t_5, thread A tries to lock m using pthread_mutex_lock(m). Since m has already been locked by C, A blocks on m. The execution of C resumes. At time t_6, thread B of medium priority is started. Since B has a higher priority than C, C is blocked and B is executed. C is still owner of m. If B does not try to lock m, it may be executed for quite some time, even if there is a thread A of higher priority. But A cannot be executed, since it waits for the release of m by C. But C cannot release m, since C is not executed. Thus, the processor is continuously executing B and not A, although A has a higher priority than B. □

point in time	event	thread A high priority	thread B medium priority	thread C low priority	mutex variable m
t_1	start	/	/	/	free
t_2	start C	/	/	running	free
t_3	C locks m	/	/	running	locked by C
t_4	start A	running	/	ready for execution	locked by C
t_5	A locks m	blocked	/	running	locked by C
t_6	start B	blocked	running	ready for execution	locked by C

Fig. 6.22 Illustration of a priority inversion.

Pthreads provides two mechanisms to avoid priority inversion: priority ceiling and priority inheritance. Both mechanisms are optional, i.e., they are not necessarily supported by each Pthreads library. We describe both mechanisms in the following.

6.3.3.1 Priority ceiling

The mechanism of priority ceiling is available for a specific Pthreads library if the macro

$$\text{\_POSIX\_THREAD\_PRIO\_PROTECT}$$

is defined in `<unistd.h>`. If priority ceiling is used, each mutex variable gets a priority value. The priority of a thread is automatically raised to this *priority ceiling value* of a mutex variable, whenever the thread locks the mutex variable. The thread keeps this priority as long as it is the owner of the mutex variable. Thus, a thread X cannot be interrupted by another thread Y with a lower priority than the priority of the mutex variable as long as X is the owner of the mutex variable. The owning thread can therefore work without interruption and can release the mutex variable as soon as possible.

In the example given above, priority inversion is avoided by using priority ceiling if a priority ceiling value is used which is equal to or larger than the priority value of thread A. In the general case, priority inversion is avoided if the highest priority at which a thread will ever be running is used as priority ceiling value.

To use priority ceiling for a mutex variable, it must be initialized appropriately using a mutex attribute data structure of type `pthread_mutex_attr_t`. This data structure must first be declared and initialized using the function

```
int pthread_mutex_attr_init(pthread_mutex_attr_t attr)
```

where `attr` is the mutex attribute data structure. The default priority protocol used for `attr` can be extracted by calling the function

```
int pthread_mutexattr_getprotocol(const pthread_mutex_attr_t
                           *attr, int *prio)
```

which returns the protocol in the parameter `prio`. The following three values are possible for `prio`:

- `PTHREAD_PRIO_PROTECT`: the priority ceiling protocol is used;
- `PTHREAD_PRIO_INHERIT`: the priority inheritance protocol is used;
- `PTHREAD_PRIO_NONE`: none of the two protocols is used, i.e., the priority of a thread does not change if it locks a mutex variable.

The function

```
int pthread_mutexattr_setprotocol(pthread_mutex_attr_t *attr,
                           int prio)
```

can be used to set the priority protocol of a mutex attribute data structure `attr` where `prio` has one of the three values just described. When using the priority ceiling protocol, the two functions

```
int pthread_mutexattr_getprioceiling(const pthread_mutex_attr_t
                           *attr, int *prio)
int pthread_mutexattr_setprioceiling(pthread_mutex_attr_t *attr,
                           int prio)
```

can be used to extract or set the priority ceiling value stored in the attribute structure `attr`. The ceiling value specified in `prio` must be a valid priority value. After a mutex attributed data structure `attr` has been initialized and possibly modified, it can be used for the initialization of a mutex variable with the specified properties, using the function

```
pthread_mutex_init (pthread_mutex_t *m, pthread_mutexattr_t
                           *attr)
```

see also Sect. 6.1.2.

6.3.3.2 Priority inheritance

When using the priority inheritance protocol, the priority of a thread which is the owner of a mutex variable is automatically raised, if a thread with a higher priority tries to lock the mutex variable and is therefore blocked on the mutex variable. In this situation, the priority of the owner thread is raised to the priority of the blocked thread. Thus, the owner of a mutex variable always has the maximum priority of all threads waiting for the mutex variable. Therefore, the owner thread cannot be interrupted by one of the waiting threads, and priority inversion cannot occur. When the owner thread releases the mutex variable again, its priority is decreased again to the original priority value.

The priority inheritance protocol can be used if the macro

_POSIX_THREAD_PRIO_INHERIT

is defined in <unistd.h>. If supported, priority inheritance can be activated by calling the function pthread_mutexattr_setprotocol() with parameter value prio = PTHREAD_PRIO_INHERIT as described above. Compared to priority ceiling, priority inheritance has the advantage that no fixed priority ceiling value has to be specified in the program. Priority inversion is avoided also for threads with unknown priority values. But the implementation of priority inheritance in the Pthreads library is more complicated and expensive and therefore usually leads to a larger overhead than priority ceiling.

6.3.4 Thread-specific Data

The threads of a process share a common address space. Thus, global and dynamically allocated variables can be accessed by each thread of a process. For each thread, a private stack is maintained for the organization of function calls performed by the thread. The local variables of a function are stored in the private stack of the calling thread. Thus, they can only be accessed by this thread, if this thread does not expose the address of a local variable to another thread. But the lifetime of local variables is only as long as the lifetime of the corresponding function activation. Thus, local variables do not provide a persistent thread-local storage. To use the value of a local variable throughout the lifetime of a thread, it has to be declared in the start function of the thread and passed as parameter to all functions called by this thread. But depending on the application, this would be quite tedious and would artificially increase the number of parameters. Pthreads supports the use of thread-specific data with an additional mechanism.

To generate thread-specific data, Pthreads provides the concept of *keys* that are maintained in a process-global way. After the creation of a key it can be accessed by each thread of the corresponding process. Each thread can associate thread-specific data to a key. If two threads associate different data to the same key, each of the two threads gets only its own data when accessing the key. The Pthreads library handles the management and storage of the keys and their associated data.

In Pthreads, keys are represented by the predefined data type pthread_key_t. A key is generated by calling the function

```
int pthread_key_create (pthread_key_t *key,
                        void (*destructor)(void *))
```

The generated key is returned in the parameter key. If the key is used by several threads, the address of a global variable or a dynamically allocated variable must be passed as key. The function pthread_key_create() should only be called once for each pthread_key_t variable. This can be ensured with the pthread_once() mechanism, see Section 6.1.4. The optional parameter destructor can be used to assign a deallocation function to the key to clean up the data stored when the thread terminates. If no deallocation is required, NULL should be specified. A key can be deleted by calling the function

```
int pthread_key_delete (pthread_key_t key)
```

After the creation of a key, its associated data is initialized with NULL. Each thread can associate new data value to the key by calling the function

```
int pthread_setspecific (pthread_key_t key, void *value)
```

Typically, the address of a *dynamically* generated data object will be passed as value. Passing the address of a local variable should be avoided, since this address is no longer valid after the corresponding function has been terminated. The data associated with a key can be retrieved by calling the function

```
void *pthread_getspecific (pthread_key_t key)
```

The calling thread always obtains the data value that it has previously associated with the key using pthread_setspecific(). When no data has been associated yet, NULL is returned. NULL is also returned, if another thread has associated data with the key, but not the calling thread. When a thread uses the function pthread_setspecific() to associate new data to a key, data that has previously been associated to this key by this thread will be overwritten and is lost.

An alternative to thread-specific data is the use of thread-local storage (TLS) which is provided since the C99 standard. This mechanism allows the declaration of variables with the storage class keyword __thread with the effect that each thread gets a separate instance of the variable. The instance is deleted as soon as the thread terminates. The __thread storage class keyword can be applied to global variables and static variables. It cannot be applied to block-scoped automatic or non-static variables.

6.4 Java Threads

Java supports the development of multi-threaded programs at the language level. Java provides language constructs for the synchronized execution of program parts and supports the creation and management of threads by predefined classes. In this chapter, we demonstrate the use of Java threads for the development of parallel programs for a shared address space. We assume that the reader knows the principles of object-oriented programming as well as the standard language elements of Java. We concentrate on the mechanisms for the development of multi-threaded programs and describe the most important elements. We refer to [162, 142] for a more detailed description. For a detailed description of Java, we refer to [63].

6.4.1 Thread Generation in Java

Each Java program in execution consists of at least one thread of execution, the *main thread*. This is the thread which executes the main() method of the class which has been given to the Java Virtual Machine (JVM) as start argument.

More user threads can be created explicitly by the main thread or other user threads that have been started earlier. The creation of threads is supported by the predefined class Thread from the standard package java.lang. This class is used for the representation of threads and provides methods for the creation and management of threads.

The interface Runnable from java.lang is used to represent the program code executed by a thread; this code is provided by a run() method and is executed asynchronously by a separate thread. There are two possibilities to arrange this: inheriting from the Thread class or using the interface Runnable.

6.4.1.1 Inheriting from the Thread class

One possibility to obtain a new thread is to define a new class NewClass which inherits from the predefined class Thread and which defines a method run() containing the statements to be executed by the new thread. The run() method defined in NewClass overwrites the predefined run() method from Thread.

The Thread class also contains a method start() which creates a new thread executing the given run() method.

The newly created thread is executed asynchronously to the generating thread. After the execution of start() and the creation of the new thread, the control will be immediately returned to the generating thread. Thus, the generating thread resumes execution usually before the new thread has terminated, i.e., the generating thread and the new thread are executed concurrently to each other.

```
import java.lang.Thread;
public class NewClass extends Thread { // inheritance
  public void run() {
  // overwriting method run() of class Thread
    System.out.println("hello from new thread");
  }
  public static void main (String args[]) {
    NewClass nc = new NewClass();
    nc.start();
  }
}
```

Fig. 6.23 Thread creation by overwriting the run() method of the Thread class.

The new thread is terminated when the execution of the run() method has been finished. This mechanism for thread creation is illustrated in Fig. 6.23 with a class NewClass whose main() method generates an object of NewClass and whose run() method is activated by calling the start() method of the newly created object. Thus, thread creation can be performed in two steps:

1. definition of a class NewClass which inherits from Thread and which defines a run() method for the new thread;
2. instantiation of an object nc of class NewClass and activation of nc.start().

The creation method just described requires that the class NewClass inherits from Thread. Since Java does not support multiple inheritance, this method has the drawback that NewClass cannot be embedded into another inheritance hierarchy. Java provides interfaces to obtain a similar mechanism as multiple inheritance. For thread creation, the interface Runnable is used.

6.4.1.2 Using the interface Runnable

The interface Runnable defines an abstract run() method as follows:

```
public interface Runnable {
  public abstract void run();
}
```

The predefined class Thread implements the interface Runnable. Therefore, each class which inherits from Thread, also implements the interface Runnable. Hence, instead of inheriting from Thread the newly defined class NewClass can directly implement the interface Runnable.

This way, objects of class NewClass are not thread objects. The creation of a new thread requires the generation of a new Thread object to which the object NewClass is passed as parameter. This is obtained by using the constructor

$$\text{public Thread (Runnable target).}$$

Using this constructor, the start() method of Thread activates the run() method of the Runnable object which has been passed as argument to the constructor.

This is obtained by the run() method of Thread which is specified as follows:

```
public void run() {
    if (target != null) target.run();
}
```

After activating start(), the run() method is executed by a separate thread which runs asynchronously to the calling thread. Thus, thread creation can be performed by the following steps:

1. definition of a class NewClass which implements Runnable and which defines a run() method with the code to be executed by the new thread;

2. instantiation of a Thread object using the constructor Thread (Runnable target) and of an object of NewClass which is passed to the Thread constructor;
3. activation of the start() method of the Thread object.

This is illustrated in Fig. 6.24 for a class NewClass. An object of this class is passed to the Thread constructor as parameter.

```java
import java.lang.Thread;
public class NewClass implements Runnable {
  public void run() {
    System.out.println("hello from new thread");
  }
  public static void main (String args[]) {
    NewClass nc = new NewClass();
    Thread th = new Thread(nc);
    th.start(); // start() activates nc.run() in a new thread
  }
}
```

Fig. 6.24 Thread creation by using the interface Runnable based on the definition of a new class NewClass.

6.4.1.3 Further methods of the Thread class

A Java thread can wait for the termination of another Java thread t by calling t.join(). This call blocks the calling thread until the execution of t is terminated. There are three variants of this method:

- void join(): the calling thread is blocked until the target thread is terminated;
- void join (long timeout): the calling thread is blocked until the target thread is terminated or the given time interval timeout has passed; the time interval is given in milliseconds;
- void join (long timeout, int nanos): the behavior is similar to void join (long timeout); the additional parameter allows a more exact specification of the time interval using an additional specification in nanoseconds.

The calling thread will not be blocked, if the target thread has not yet been started. The method

boolean isAlive()

of the Thread class gives information about the execution status of a thread: the method returns true, if the target thread has been started but has not yet been terminated; otherwise, false is returned. The join() and isAlive() methods have no effect on the calling thread. A name can be assigned to a specific thread by using the methods:

```
void setName (String name);
String getName();
```

An assigned name can later be used to identify the thread. A name can also be assigned at thread creation by using the constructor `Thread (String name)`. The `Thread` class defines static methods which affect the calling thread or provide information about the program execution:

```
static Thread currentThread();
static void sleep (long milliseconds);
static void yield();
static int enumerate (Thread[] th_array);
static int activeCount();
```

Since these methods are static, they can be called without using a target `Thread` object. The call of `currentThread()` returns a reference to the `Thread` object of the calling thread. This reference can later be used to call non-static methods of the `Thread` object. The method `sleep()` blocks the execution of the calling thread until the specified time interval has passed; at this time, the thread again becomes ready for execution and can be assigned to an execution core or processor. The method `yield()` is a directive to the Java Virtual Machine (JVM) to assign another thread with the same priority to the processor. If such a thread exists, then the scheduler of the JVM can bring this thread to execution. The use of `yield()` is useful for JVM implementations without a time-sliced scheduling, if threads perform long-running computations which do not block. The method `enumerate()` yields a list of all active threads of the program. The return value specifies the number of `Thread` objects collected in the parameter array `th_array`. The method `activeCount()` returns the number of active threads in the program. The method can be used to determine the required size of the parameter array before calling `enumerate()`.

Example: Fig. 6.25 gives an example of a class for performing a matrix multiplication with multiple threads. The input matrices are read in into `in1` and `in2` by the main thread using the static method `ReadMatrix()`. The thread creation is performed by the constructor of the `MatMult` class such that each thread computes one row of the result matrix. The corresponding computations are specified in the `run()` method. All threads access the same matrices `in1`, `in2` and `out` that have been allocated by the main thread. No synchronization is required, since each thread writes to a separate area of the result matrix `out`.   □

6.4.2 Synchronization of Java Threads

The threads of a Java program access a shared address space. Suitable synchronisation mechanisms have to be applied to avoid race conditions when a variable is accessed by several threads concurrently. Java provides `synchronized` blocks and methods to guarantee mutual exclusion for threads accessing shared data. A

```
import java.lang.*;
import java.io.*;
class MatMult extends Thread {
  static int in1[][]; static int in2[][]; static int out[][];
  static int n=3; int row;
  MatMult (int i) {
    row=i;
    this.start();
  }
  public void run() {
  //compute a row of the result matrix
    int i,j;
    for(i=0;i<n;i++) {
      out[row][i]=0;
      for (j=0;j<n;j++)
        out[row][i]=out[row][i]+in1[row][j]*in2[j][i];
    }
  }
  public static void ReadMatrix (int in[][]) {
  //read the input matrix
    int i,j;
    BufferedReader br=new BufferedReader(new
      InputStreamReader(System.in));
    System.out.println("Enter the Matrix : ");
    for(i=0;i<n;i++)
      for(j=0;j<n;j++)
        try {
          in[i][j]=Integer.parseInt(br.readLine());
        } catch(Exception e){ }
  }
  public static void PrintMatrix (int out[][]) {
  //print the result matrix
    int i,j;
    System.out.println("OUTPUT :");
    for(i=0;i<n;i++)
      for(j=0;j<n;j++)
        System.out.println(out[i][j]);
  }
  public static void main(String args[]) {
    int i,j;
    in1=new int[n][n]; in2=new int[n][n];
    out=new int[n][n];
    ReadMatrix(in1); ReadMatrix(in2);
    MatMult mat[] = new MatMult[n];
    for(i=0;i<n;i++)
      mat[i]=new MatMult(i);
    try {
      for(i=0;i<n;i++)
        mat[i].join();
    } catch(Exception e){ }
    PrintMatrix(out);
  }
}
```

Fig. 6.25 Parallel matrix multiplication in Java.

synchronized block or method avoids a concurrent execution of the block or method by two or more threads. A data structure can be protected by putting all accesses to it into synchronized blocks or methods, thus ensuring mutual exclusion. A synchronized increment operation of a counter can be realized by the following method incr():

```
public class Counter {
    private int value = 0;
    public synchronized int incr() {
        value = value + 1;
        return value;
    }
}
```

Java implements the synchronization by assigning to each Java object an implicit mutex variable. This is achieved by providing the general class Object with an implicit mutex variable. Since each class is directly or indirectly derived from the class Object, each class inherits this implicit mutex variable, and every object instantiated from any class implicitly possesses its own mutex variable. The activation of a synchronized method of an object Ob by a thread t has the following effects:

- When starting the synchronized method, t implicitly tries to lock the mutex variable of Ob. If the mutex variable is already locked by another thread s, thread t is blocked. The blocked thread becomes ready for execution again when the mutex variable is released by the locking thread s. The called synchronized method will only be executed after successfully locking the mutex variable of Ob.
- When t leaves the synchronized method called, it implicitly releases the mutex variable of Ob so that it can be locked by another thread.

A synchronized access to an object can be realized by declaring all methods accessing the object as synchronized. The object should only be accessed with these methods to guarantee mutual exclusion.

In addition to synchronized methods, Java provides synchronized blocks: such a block is started with the keyword synchronized and the specification of an arbitrary object that is used for the synchronization in parenthesis. Instead of an arbitrary object, the synchronization is usually performed with the object whose method contains the synchronized block. The above method for the incrementation of a counter variable can be realized using a synchronized block as follows:

```
public int incr() {
    synchronized (this) {
        value = value + 1; return value;
    }
}
```

The synchronization mechanism of Java can be used for the realization of **fully synchronized objects** (also called **atomar objects**); these can be accessed by an arbitrary number of threads without any additional synchronization. To avoid race conditions, the synchronization has to be performed within the methods of the corresponding class of the objects. This class must have the following properties:

- all methods must be declared synchronized;
- no public entries are allowed that can be accessed without using a local method;
- all entries are consistently initialized by the constructors of the class;
- the objects remain in a consistent state also in case of exceptions.

```java
import java.lang.*;
import java.util.*;
public class ExpandableArray {
  private Object[] data;
  private int size = 0;
  public ExpandableArray(int cap) {
    data = new Object[cap];
  }
  public synchronized int size() {
    return size;
  }
  public synchronized Object get(int i)
    throws NoSuchElementException {
      if (i < 0 || i >= size)
        throw new NoSuchElementException();
      return data[i];
  }
  public synchronized void add(Object x) {
    if (size == data.length) { // array too small
      Object[] od = data;
      data = new Object[3 * (size + 1) / 2];
      System.arraycopy(od, 0, data, 0, od.length);
    }
    data[size++] = x;
  }
  public synchronized void removeLast()
    throws NoSuchElementException {
      if (size == 0)
        throw new NoSuchElementException();
      data[--size] = null;
  }
}
```

Fig. 6.26 Example for a fully synchronized class.

Fig. 6.26 demonstrates the concept of fully synchronized objects for the example of a class ExpandableArray; this is a simplified version of the predefined synchronized class java.util.Vector, see also [142]. The class implements an adaptable array of arbitrary objects, i.e., the size of the array can be increased or decreased according to the number of objects to be stored. The adaptation is realized by the method add(): if the array data is fully occupied when trying to add a new object, the size of the array will be increased by allocating a larger array and using the method arraycopy() from the java.lang.System class to copy the content of the old array into the new array. Without the synchronization included, the class could not be used concurrently by more than one thread safely. A conflict could occur, if, e.g., two threads tried to perform an add() operation at the same time.

6.4.2.1 Deadlocks

The use of fully synchronized classes avoids the occurrence of race conditions, but may lead to deadlocks when threads are synchronized with different objects. This is illustrated in Fig. 6.27 for a class Account which provides a method swapBalance() to swap account balances, see [142]. A deadlock can occur when swapBalance() is executed by two threads A and B concurrently: For two account objects a and b, if A calls a.swapBalance(b) and B calls b.swapBalance(a) and A and B are executed on different processors or cores, a deadlock occurs with the following execution order:

- **time** T_1: thread A calls a.swapBalance(b) and locks the mutex variable of object a;
- **time** T_2: thread A calls getBalance() for object a and executes this function;
- **time** T_2: thread B calls b.swapBalance(a) and locks the mutex variable of object b;
- **time** T_3: thread A calls b.getBalance() and blocks because the mutex variable of b has previously been locked by thread B;
- **time** T_3: thread B calls getBalance() for object b and executes this function;
- **time** T_4: thread B calls a.getBalance() and blocks because the mutex variable of a has previously been locked by thread A.

```
pulic class Account {
  private long balance;
  synchronized long getBalance() {return balance;}
  synchronized void setBalance(long v) {
    balance = v;
  }
  synchronized void swapBalance(Account other) {
    long t = getBalance();
    long v = other.getBalance();
    setBalance(v);
    other.setBalance(t);
  }
}
```

Fig. 6.27 Example for a deadlock situation in Java.

The execution order is illustrated in Fig. 6.28. After time T_4, both threads are blocked: thread A is blocked, since it could not acquire the mutex variable of object b. This mutex variable is owned by thread B and only B can free it. Thread B is blocked, since it could not acquire the mutex variable of object a. This mutex variable is owned by thread A and only A can free it. Thus, both threads are blocked and none of them can proceed; a deadlock has occurred.

time	operation Thread A	operation Thread B	owner mutex a	owner mutex b
T_1	a.swapBalance(b)		A	–
T_2	t = getBalance()	b.swapBalance(a)	A	B
T_3	blocked with respect to b	t = getBalance()	A	B
T_4		blocked with respect to a	A	B

Fig. 6.28 Execution order to cause a deadlock situation for the class in Fig. 6.27.

Deadlocks typically occur, if different threads try to lock the mutex variables of the same objects in different orders. For the example in Fig. 6.27, thread A tries to lock first a and then b, whereas thread B tries to lock first b and then a. In this situation, a deadlock can be avoided by a backoff strategy or by using the same locking order for each thread, see also Section 6.1.2. A unique ordering of objects can be obtained by using the Java method System.identityHashCode() which refers to the default implementation Object.hashCode(), see [142]. But any other unique object ordering can also be used. Thus, we can give an alternative formulation of swapBalance() which avoids deadlocks, see Fig. 6.29. The new formulation also contains an alias check to ensure that the operation is only executed if different objects are used. The method swapBalance() is not declared synchronized any more.

```
public void swapBalance(Account other) {
    if (other == this) return;
    else if (System.identityHashCode(this) <
        System.identityHashCode(other))
      this.doSwap(other);
    else other.doSwap(this);
}
protected synchronized void doSwap(Account other) {
    long t = getBalance();
    long v = other.getBalance();
    setBalance(v);
    other.setBalance(t);
}
```

Fig. 6.29 Deadlock-free implementation of swapBalance() from Fig. 6.27.

For the synchronization of Java methods, several issues should be considered to make the resulting programs efficient and safe:

- Synchronization is expensive. Therefore, `synchronized` methods should only be used for methods that can be called concurrently by several threads and that may manipulate common object data.
  If an application ensures that a method is always executed by a single thread at each point in time, then a synchronization can be avoided to increase efficiency.
- Synchronization should be restricted to critical regions to reduce the time interval of locking. For larger methods, the use of `synchronized` blocks instead of `synchronized` methods should be considered.
- To avoid unnecessary sequentializations, the mutex variable of the same object should not be used for the synchronization of different, non-contiguous critical sections.
- Several Java classes are internally synchronized; Examples are `Hashtable`, `Vector`, and `StringBuffer`. No additional synchronization is required for objects of these classes.
- If an object requires synchronization, the object data should be put into `private` or `protected` instance fields to inhibit non-synchronized accesses from outside. All object methods accessing the instance fields should be declared as `synchronized`.
- For cases in which different threads access several objects in different orders, deadlocks can be prevented by using the same lock order for each thread.

6.4.2.2 Synchronization with variable lock granularity

To illustrate the use of the synchronization mechanism of Java, we consider a synchronization class with a variable lock granularity, which has been adapted from [162].

The new class `MyMutex` allows the synchronization of arbitrary object accesses by explicitly acquiring and releasing objects of the class `MyMutex`, thus realizing a lock mechanism similar to mutex variables in Pthreads, see Section 6.1.2, page 318. The new class also enables the synchronization of threads accessing different objects. The class `MyMutex` uses an instance field `OwnerThread` which indicates which thread has currently acquired the synchronization object. Fig. 6.30 shows a first draft of the implementation of `MyMutex`.

The method `getMyMutex` can be used to acquire the explicit lock of the synchronization object for the calling thread. The lock is given to the calling thread by assigning `Thread.currentThread()` to the instance field `OwnerThread`. The synchronized method `freeMyMutex()` can be used to release a previously acquired explicit lock; this is implemented by assigning `null` to the instance field `OwnerThread`. If a synchronization object has already been locked by another thread, `getMyMutex()` repeatedly tries to acquire the explicit lock after a fixed time interval of 100 ms. The method `getMyMutex()` is not declared `synchronized`. The synchronized method `tryGetMyMutex()` is used to access the instance field

```
public class MyMutex {
  protected Thread OwnerThread = null;
  public void getMyMutex() {
    while (!tryGetMyMutex()) {
      try { Thread.sleep(100); }
      catch (InterruptedException e) { }
    }
  }
  public synchronized boolean tryGetMyMutex() {
    if (OwnerThread == null) {
      OwnerThread = Thread.currentThread();
      return true;
    }
    else return false;
  }
  public synchronized void freeMyMutex() {
    if (OwnerThread == Thread.currentThread())
      OwnerThread = null;
  }
}
```

Fig. 6.30 Synchronization class with variable lock granularity.

OwnerThread. This protects the critical section for acquiring the explicit lock by using the implicit mutex variable of the synchronization object. This mutex variable is used both for tryGetMyMutex() and freeMyMutex().

If getMyMutex() would have been declared synchronized, the activation of getMyMutex() by a thread T_1 would lock the implicit mutex variable of the synchronization object of the class MyMutex before entering the method. If another thread T_2 holds the explicit lock of the synchronization object, T_2 cannot release this lock with freeMyMutex() since this would require to lock the implicit mutex variable which is held by T_1. Thus, a deadlock would result. The use of an additional method tryGetMyMutex() can be avoided by using a synchronized block within getMyMutex(), see Fig. 6.31.

Objects of the new synchronization class MyMutex can be used for the explicit protection of critical sections. This can be illustrated for a counter class Counter to protect the counter manipulation, see Fig. 6.32.

6.4.2.3 Synchronization of static methods

The implementation of synchronized blocks and methods based on the implicit object mutex variables works for all methods that are activated with respect to an object. Static methods of a class are not activated with respect to an object and thus, there is no implicit object mutex variable. Nevertheless, static methods can also be declared synchronized. In this case, the synchronization is implemented by using the implicit mutex variable of the corresponding class object of the class

```
public void getMyMutex() {
  for ( ; ; ) {
    synchronized(this) {
      if (OwnerThread == null) {
        OwnerThread = Thread.currentThread();
        break;
      }
    }
    try { Thread.sleep(100); }
    catch (InterruptedException e) { }
  }
}
```

Fig. 6.31 Implementation variant of getMyMutex().

```
public class Counter {
  private int value;
  private MyMutex flag = new MyMutex();
  public int incr() {
    int res;
    flag.getMyMutex();
    value = value + 1;
    res = value;
    flag.freeMyMutex();
    return res;
  }
}
```

Fig. 6.32 Implementation of a counter class with synchronization by an object of class MyMutex.

java.lang.Class (Class mutex variable). An object of this class is automatically generated for each class defined in a Java program.

Thus, static and non-static methods of a class are synchronized by using different implicit mutex variables. A static synchronized method can acquire both the mutex variable of the Class object and of an object of this class by using an object of this class for a synchronized block or by activating a synchronized non-static method for an object of this class. This is illustrated in Fig. 6.33. Similarly, a synchronized non-static method can also acquire both the mutex variables of the object and of the Class object by calling a synchronized static method. For an arbitrary class Cl, the Class mutex variable can be directly used for a synchronized block by using

$$synchronized\ (Cl.class)\ \{/*Code*/\}$$

```
public class MyStatic {
  public static synchronized void staticMethod(MyStatic obj) {
    // here, the class mutex is used
    obj.nonStaticMethod();
    synchronized(obj) {
      // here, additionally, the object mutex is used
    }
  }
  public synchronized void nonStaticMethod() {
    // using the object mutex
  }
}
```

Fig. 6.33 Synchronization of static methods.

6.4.3 Wait and Notify

In some situations, it is useful for a thread to wait for an event or condition. As soon as the event occurs, the thread executes a predefined action. The thread waits as long as the event does not occur or the condition is not fulfilled. The event can be signaled by another thread; similarly, another thread can make the condition to be fulfilled. Pthreads provides condition variables for these situations. Java provides a similar mechanism via the methods wait() and notify() of the predefined Object class. These methods are available for each object of any class which is explicitly or implicitly derived from the Object class. Both methods can only be used within synchronized blocks or methods. A typical usage pattern for wait() is:

```
synchronized (lockObject) {
    while (!condition) { lockObject.wait(); }
    Action();
}
```

The call of wait() blocks the calling thread until another thread calls notify() for the same object. When a thread blocks by calling wait(), it releases the implicit mutex variable of the object used for the synchronization of the surrounding synchronized method or block. Thus, this mutex variable can be acquired by another thread.

Several threads may block waiting for the same object. Each object maintains a list of waiting threads. When another thread calls the notify() method of the same object, one of the waiting threads of this object is woken up and can continue running. Before resuming its execution, this thread first acquires the implicit mutex variable of the object. If this is successful, the thread performs the action specified in the program. If this is not successful, the thread blocks and waits until the implicit mutex variable is released by the owning thread by leaving a synchronized method or block.

The methods `wait()` and `notify()` work similarly as the operations `pthread_cond_wait()` and `pthread_cond_signal()` for condition variables in Pthreads, see Section 6.1.3, page 325. The methods `wait()` and `notify()` are implemented using an implicit waiting queue for each object; this waiting queue contains all blocked threads waiting to be woken up by a `notify()` operation. The waiting queue does not contain those threads that are blocked waiting for the implicit mutex variable of the object.

The Java language specification does not specify which of the threads in the waiting queue is woken up if `notify()` is called by another thread. The method `notifyAll()` can be used to wake up all threads in the waiting queue; this has a similar effect as `pthread_cond_broadcast()` in Pthreads. The method `notifyAll()` also has to be called in a `synchronized` block or method.

6.4.3.1 Producer-Consumer pattern

The Java waiting and notification mechanism described above can be used for the implementation of a producer-consumer pattern using an item buffer of fixed size. Producer threads can put new items into the buffer and consumer threads can remove items from the buffer. Figure 6.34 shows a thread-safe implementation of such a buffer mechanism adapted from [142] using the `wait()` and `notify()` methods of Java. When creating an object of the class `BoundedBufferSignal`, an array `array` of a given size `capacity` is generated; this array is used as buffer. The indices `putptr` and `takeptr` indicate the next position in the buffer to put or take the next item, respectively; these indices are used in a circular way.

The class provides a `put()` method to enable a producer to enter an item into the buffer and a `take()` method to enable a consumer to remove an item from the buffer. A buffer object can have one of three states: full, partially full and empty. Figure 6.35 illustrates the possible transitions between the states when calling `take()` or `put()`. The states are characterized by the following conditions:

state	condition	put possible	take possible
full	`size == capacity`	no	yes
partially full	`0 < size < capacity`	yes	yes
empty	`size == 0`	yes	no

If the buffer is full, the execution of the `put()` method by a producer thread will block the executing thread; this is implemented by a `wait()` operation. If the `put()` method is executed for a previously empty buffer, all waiting (consumer) threads will be woken up using `notifyAll()` after the item has been entered into the buffer. If the buffer is empty, the execution of the `take()` method by a consumer thread will block the executing thread using `wait()`. If the `take()` method is executed for a previously full buffer, all waiting (producer) threads will be woken up using `notifyAll()` after the item has been removed from the buffer. The implementation

```
public class BoundedBufferSignal {
  private final Object[] array;
  private int putptr = 0;
  private int takeptr = 0;
  private int numel = 0; // number of items in buffer
  public BoundedBufferSignal (int capacity)
    throws IllegalArgumentException {
      if (capacity <= 0)
          throw new IllegalArgumentException();
        array = new Object[capacity];
  }

  public synchronized int size() {return numel; }
  public int capacity() {return array.length;}
  public synchronized void put(Object obj)
    throws InterruptedException {
    while (numel == array.length)
      wait(); // buffer full
    array [putptr] = obj;
    putptr = (putptr +1) % array.length;
    if (numel++ == 0)
      notifyAll(); // wake up all threads
  }

  public synchronized Object take()
    throws InterruptedException {
    while (numel == 0)
      wait(); // buffer empty
    Object x = array [takeptr];
    takeptr = (takeptr +1) % array.length;
    if (numel-- == array.length)
      notifyAll(); // wake up all threads
    return x;
  }
}
```

Fig. 6.34 Implementation of a thread-safe buffer mechanism using the wait() and notify() methods of Java.

of put() and take() ensures that each object of the class BoundedBufferSignal can be accessed concurrently by an arbitrary number of threads without race conditions.

Fig. 6.35 Illustration of the states of a thread-safe buffer mechanism.

6.4.3.2 Modification of the MyMutex class

The methods wait() and notify() can be used to improve the synchronization class MyMutex from Fig. 6.30 by avoiding the active waiting in the method
getMyMutex(), see Fig. 6.36 (according to [162]).

```
public synchronized void getMyMutex() {
   while (!tryGetMyMutex()) {
      try { wait(); }
      catch (InterruptedException e) { }
   }
}
public synchronized boolean tryGetMyMutex() {
   if (OwnerThread == null) {
      OwnerThread = Thread.currentThread() ;
      lockCount = 1; return true;
   }
   if (OwnerThread = Thread.currentThread()){
      lockCount ++; return true;
   }
   return false;
}
public synchronized Thread getMutexOwner() {
   return OwnerThread;
}
public synchronized void freeMyMutex() {
   if (OwnerThread == Thread.currentThread()) {
      if (--lockCount == 0) {
         OwnerThread = null;
         notify();
      }
   }
}
```

Fig. 6.36 Implementation of the synchronization class MyMutex with wait() and notify()
avoiding active waiting.

Additionally, the modified implementation realizes a nested locking mechanism
which allows multiple locking of a synchronization object by the same thread.
The number of locks is counted in the variable lockCount; this variable is initialized to 0 and is incremented or decremented by each call of getMyMutex() or
freeMyMutex(), respectively. In Fig. 6.36, the method getMyMutex() is now also
declared synchronized. With the implementation in Fig. 6.30, this would lead to
a deadlock. But in Fig. 6.36, no deadlock can occur, since the activation of wait()
releases the implicit mutex variable before the thread is suspended and inserted into
the waiting queue of the object.

6.4.3.3 Barrier Synchronization

A barrier synchronization is a synchronization point at which each thread waits until all participating threads have reached this synchronization point. Only then the threads proceed with their execution. A barrier synchronization can be implemented in Java using wait() and notify(). This is shown in Fig. 6.37 for a class Barrier, see also [162]. The Barrier class contains a constructor which initializes a Barrier object with the number of threads to wait for (t2w4). The actual synchronization is provided by the method waitForRest(). This method must be called by each thread at the intended synchronization point. Within the method, each thread decrements t2w4 and calls wait() if t2w4 is > 0. This blocks each arriving thread within the Barrier object. The last arriving thread wakes up all waiting threads using notifyAll().

Objects of the Barrier class can be used only once, since the synchronization counter t2w4 is decremented to 0 during the synchronization process. An example for the use of the Barrier class for the synchronization of a multi-phase computation is given in Fig. 6.38, see also [162]. The program illustrates an algorithm with three phases (doPhase1(), doPhase2(), doPhase3()) which are separated from each other by a barrier synchronization using Barrier objects bp1, bp2, and bpEnd. Each of the threads created in the constructor of ProcessIt executes the three phases

```java
public class Barrier() {
  private int t2w4;
  private InterruptedException e;
  public Barrier(int n) {
    this.t2w4 = n;
  }
  public synchronized int waitForRest()
      throws InterruptedException {
    int nThreads = --t2w4;
    if (e != null) throw e;
    if (t2w4 <= 0) {
      notifyAll(); return nThreads;
    }
    while (t2w4 > 0) {
      if (e != null) throw e;
      try { wait(); }
      catch(InterruptedException e) { this.e = e; notifyAll(); }
    }
    return nThreads;
  }
}
```

Fig. 6.37 Implementation of a barrier synchronization in Java with wait() and notify().

```
public class ProcessIt implements Runnable {
  String is[];
  Barrier bpStart, bp1, bp2, bpEnd;
  public ProcessIt(String sources[]) {
    is = sources;
    bpStart = new Barrier(sources.length);
    bp1 = new Barrier(sources.length);
    bp2 = new Barrier(sources.length);
    bpEnd = new Barrier(sources.length);
    for (int i = 0; i < sources.length; i++)
      new Thread(this).start();
  }
  public void run() {
    try {
      int i = bpStart.waitForRest();
      doPhase1(is[i]);
      bp1.waitForRest();
      doPhase2(is[i]);
      bp2.waitForRest();
      doPhase3(is[i]);
      bpEnd.waitForRest();
    }
    catch (InterruptedException e)
  }
  public static void main(String args[]) {
    ProcessIt pi = new ProcessIt(args);
  }
}
```

Fig. 6.38 Use of the Barrier class for the implementation of a multi-phase algorithm.

6.4.3.4 Condition Variables

The mechanism provided by wait() and notify() in Java has some similarities to the synchronization mechanism of condition variables in Pthreads, see Section 6.1.3, page 325. The main difference lies in the fact that wait() and notify() are provided by the general Object class. Thus, the mechanism is implicitly bound to the internal mutex variable of the object for which wait() and notify() are activated. This facilitates the use of this mechanism by avoiding the explicit association of a mutex variable as needed when using the corresponding mechanism in Pthreads. But the fixed binding of wait() and notify() to a specific mutex variable also reduces flexibility, since it is not possible to combine an arbitrary mutex variable with the waiting queue of an object.

When calling wait() or notify(), a Java thread must be the owner of the mutex variable of the corresponding object; otherwise an exception IllegalMonitorStateException is raised. With the mechanism of wait() and notify() it is not possible to use the same mutex variable for the synchronization of the waiting queues of different objects. This would be useful, e.g., for the implementation of producer and consumer threads with a common data buffer, see, e.g., Fig. 6.20. But wait() and notify() can be used for the realization of a new class which mimics the mechanism of condition variables in Pthreads. Fig. 6.39 shows an implementation of such a class CondVar, see also [162, 142]. The class CondVar provides the methods cvWait(), cvSignal() and cvBroadcast() which mimic the behavior of pthread_cond_wait(), pthread_cond_signal() and pthread_cond_broadcast(), respectively. These methods allow the use of an arbitrary mutex variable for the synchronization. This mutex variable is provided as a parameter of type MyMutex for each of the methods, see Fig. 6.39.

Thus, a single mutex variable of type MyMutex can be used for the synchronization of several condition variables of type CondVar. When calling cvWait(), a thread will be blocked and put in the waiting queue of the corresponding object of type CondVar. The internal synchronization within cvWait() is performed with the internal mutex variable of this object. The class CondVar also allows a simple porting of Pthreads programs with condition variables to Java programs.

Fig. 6.40 shows as example the realization of a buffer mechanism with producer and consumer threads by using the new class CondVar, see also [142]. A producer thread can insert objects into the buffer by using the method put(). A consumer thread can remove objects from the buffer by using the method take(). The condition objects notFull and notEmpty of type CondVar use the same mutex variable mutex for synchronization.

6.4.4 Extended Synchronization Patterns

The synchronization mechanisms provided by Java can be used to implement more complex synchronization patterns which can then be used in parallel applicaton programs. This will be demonstrated in the following for the example of a semaphore mechanism, see page 159.

A semaphore mechanism can be implemented in Java by using wait() and notify(). Figure 6.41 shows a simple implementation, see also [142, 162]. The method acquire() waits (if necessary), until the internal counter of the semaphore object has reached at least the value 1. As soon as this is the case, the counter is decremented. The method release() increments the counter and uses notify() to wake up a waiting thread that has been blocked in acquire() by calling wait(). A waiting thread can only exist, if the counter has had the value 0 before incrementing it. Only in this case, it can be blocked in acquire(). Since the counter is only incremented by one, it is sufficient to wake up a single waiting thread. An alternative would be to use notifyAll(), which wakes up all waiting threads. Only one of these threads would succeed in decrementing the counter, which would then have the value 0 again. Thus, all other threads that had been woken up would be blocked again by calling wait.

The semaphore mechanism shown in Fig. 6.41 can be used for the synchronization of producer and consumer threads. A similar mechanism has already been im-

```
public class CondVar {
  private MyMutex syncVar; /* use MyMutex for synchronization */
  public CondVar() {
    this(new MyMutex());
  }
  public CondVar(MyMutex sv) {
    syncVar = sv;
  }
  public void cvWait() throws InterruptedException {
    cvWait(syncVar, 0);
  }
  public void cvWait(MyMutex sv) throws InterruptedException {
    cvWait(sv, 0);
  }
  public void cvWait(int millis) throws InterruptedException {
    cvWait(syncVar, millis);
  }
  public void cvWait(MyMutex sv, int millis)
      throws InterruptedException {
    int i = 0;
    InterruptedException exception;
    synchronized (this) {
      if (sv.getMutexOwner() != Thread.currentThread())
        throw new IllegalMonitorStateException ("thread not owner");
      while (sv.getMutexOwner() == Thread.currentThread()) {
        i++; sv.freeMyMutex();
      }
      try { if (millis == 0) wait(); else wait(millis); }
      catch (InterruptedException e) { exception = e; }
    }
    for (; i > 0; i--) sv.getMyMutex();
    if (exception != null) throw exception;
  }
  public void cvSignal() {
    cvSignal(syncVar);
  }
  public synchronized void cvSignal(MyMutex sv) {
    if (sv.getMutexOwner() != Thread.currentThread())
      throw new IllegalMonitorStateException ("thread not owner");
    notify();
  }
  public void cvBroadcast() {
    cvBroadcast(syncVar);
  }
  public synchronized void cvBroadcast(MyMutex sv) {
    if (sv.getMutexOwner() != Thread.currentThread())
      throw new IllegalMonitorStateException("thread not owner");
    notifyAll();
  }
}
```

Fig. 6.39 Class CondVar for the implementation of the Pthreads condition variable mechanism using the Java signaling mechanism.

```
class PThreadsStyleBuffer {
    private final MyMutex mutex = new MyMutex();
    private final CondVar notFull = new CondVar(mutex);
    private final CondVar notEmpty = new CondVar(mutex);
    private int count = 0;
    private int takePtr = 0;
    private int putPtr = 0;
    private final Object[] array;

    public PThreadsStyleBuffer(int capacity) {
        array = new Object[capacity];
    }

    public void put(Object x) throws InterruptedException {
        mutex.getMyMutex();
        try {
            while (count == array.length)
                notFull.cvWait();

            array[putPtr] = x;
            putPtr = (putPtr + 1) % array.length;
            ++count;
            notEmpty.cvSignal();
        }
        finally {
            mutex.freeMyMutex();
        }
    }

    public Object take() throws InterruptedException {
        Object x = null;
        mutex.getMyMutex();
        try {
            while (count == 0)
                notEmpty.cvWait();

            x = array[takePtr];
            array[takePtr] = null;
            takePtr = (takePtr + 1) % array.length;
            --count;
            notFull.cvSignal();
        }
        finally {
            mutex.freeMyMutex();
        }
        return x;
    }
}
```

Fig. 6.40 Implementation of a buffer mechanism for producer and consumer threads.

```
public class Semaphore {
  private long counter;
  public Semaphore(long init) {
    counter = init;
  }
  public void acquire()
      throws InterruptedException {
    if (Thread.interrupted())
      throw new InterruptedException();
    synchronized (this) {
      try {
        while (counter <= 0) wait();
        counter--;
      }
      catch (InterruptedException ie) {
        notify(); throw ie;
      }
    }
  }
  public synchronized void release() {
    counter++;
    notify();
  }
}
```

Fig. 6.41 Implementation of a semaphore mechanism.

plemented in Fig. 6.34 by using wait() and notify() directly. Fig. 6.43 shows
an alternative implementation with semaphores, see [142]. The producer stores the
objects generated into a buffer of fixed size, the consumer retrieves objects from
this buffer for further processing. The producer can only store objects in the buffer,
if the buffer is not full. The consumer can only retrieve objects from the buffer, if
the buffer is not empty. The actual buffer management is done by a separate class
BufferArray which provides methods insert() and extract() to insert and retrieve objects, see Fig. 6.42. Both methods are synchronized, so multiple threads
can access objects of this class without conflicts. The class BufferArray does not
provide a mechanism to control buffer overflow.

The class BoundedBufferSema in Fig. 6.43 provides the methods put() and
take() to store and retrieve objects in a buffer. Two semaphores putPermits
and takePermits are used to control the buffer management. At each point in
time, these semaphores count the number of permits to store (producer) and retrieve
(consumer) objects. The semaphore putPermits is initialized to the buffer size,
the semaphore takePermits is initialized to 0. When storing an objects by using
put(), the semaphore putPermits is decremented with acquire(); if the buffer
is full, the calling thread is blocked when doing this. After an object has been stored
in the buffer with insert(), a waiting consumer thread (if present) is woken up

```
public class BufferArray {
  private final Object[] array;
  private int putptr = 0;
  private int takeptr = 0;
  public BufferArray (int n) {
    array = new Object[n];
  }

  public synchronized void insert (Object obj) {
    array[putptr] = obj;
    putptr = (putptr +1) % array.length;
  }

  public synchronized Object extract() {
    Object x = array[takeptr];
    array[takeptr] = null;
    takeptr = (takeptr +1) % array.length;
    return x;
  }
}
```

Fig. 6.42 Class BufferArray for buffer management.

by calling release() for the semaphore takePermits. Retrieving an object with take() works similarly with the role of the semaphores exchanged.

In comparion to the implementation in Fig. 6.34, the new implementation in Fig. 6.43 uses two separate objects (of type Semaphore) for buffer control. Depending on the specific situation, this can lead to a reduction of the synchronization overhead: in the implementation from Fig. 6.34 *all* waiting threads are woken up in put() and take(). But only one of these can proceed and retrieve an object from the buffer (consumer) or store an object into the buffer (producer). All other threads are blocked again. In the implementation from Fig. 6.43, only one thread is woken up.

6.4.5 Thread Scheduling in Java

A Java program may consist of several threads which can be executed on one or several of the processors of the execution platform. The threads which are ready for execution compete for execution on a free processor. The programmer can influence the mapping of threads to processors by assigning priorities to the threads. The minimum, maximum and default priorities for Java threads are specified in the following fields of the Thread class:

```
pulic class BoundedBufferSema {
  private final BufferArray buff;
  private final Semaphore putPermits;
  private final Semaphore takePermits;

  public BoundedBufferSema(int capacity)
    throws IllegalArgumentException {
        if (capacity <= 0)
          throw new IllegalArgumentException();
        buff = new BufferArray(capacity);
        putPermits = new Semaphore(capacity);
        takePermits = new Semaphore(0);
  }

  public void put(Object x)
        throws InterruptedException {
    putPermits.acquire();
    buff.insert(x);
    takePermits.release();
  }

  public Object take()
        throws InterruptedException {
    takePermits.acquire();
    Object x = buff.extract();
    putPermits.release();
    return x;
  }
}
```

Fig. 6.43 Buffer management with semaphores.

```
public static final int MIN_PRIORITY     // normally 1
public static final int MAX_PRIORITY     // normally 10
public static final int NORM_PRIORITY    // normally 5
```

A *large* priority value corresponds to a *high* priority. The thread which executes the main() method of a class has by default the priority Thread.NORM_PRIORITY. A newly created thread has by default the same priority as the generating thread. The current priority of a thread can be retrieved or dynamically changed by using the methods

```
public int getPriority();
public int setPriority(int prio);
```

of the Thread class. If there are more executable threads than free processors, a thread with a larger priority is usually favored by the scheduler of the JVM. The exact mechanism for selecting a thread for execution may depend on the implementation of a specific JVM. The Java specification does not define an exact scheduling mechanism to increase flexibility for the implementation of the JVM on different operating systems and different execution platforms. For example, the scheduler might always bring the thread with the largest priority to execution, but it could also integrate an aging mechanism to ensure that threads with a lower priority will be mapped to a processor from time to time to avoid starvation and implement fairness.

Since there is no exact specification for the scheduling of threads with different priorities, priorities cannot be used to replace synchronization mechanisms. Instead, priorities can only be used to express the relative importance of different threads to bring the most important thread to execution in case of doubt.

When using threads with different priorities, the problem of **priority inversion** can occur, see also Section 6.3.3, page 361. A priority inversion happens if a thread with a high priority is blocked to wait for a thread with a low priority, e.g., because this thread has locked the same mutex variable that the thread with the high priority tries to lock. The thread with a low priority can be inhibited from proceeding its execution and releasing the mutex variable as soon as a thread with a medium priority is ready for execution. In this constellation, the thread with high priority can be prevented from execution in favor of the thread with a medium priority.

The problem of priority inversion can be avoided by using **priority inheritance**, see also Section 6.3.3: if a thread with high priority is blocked, e.g., because of an activation of a synchronized method, then the priority of the thread that currently controls the critical synchronization object will be increased to the high priority of the blocked thread. Then, no thread with medium priority can inhibit the thread with high priority from execution. Many JVMs use this method, but this is not guaranteed by the Java specification.

6.4.6 *Package* java.util.concurrent

The java.util.concurrent package provides additional synchronization mechanisms and classes which are based on the standard synchronisation mechanisms described in the previous section, like synchronized blocks, wait() and notify(). The package is available for Java platforms starting with the Java2 platform (Java2 Standard Edition 5.0, J2SE 5.0).

The additional mechanisms provide more abstract and flexible synchronization operations, including atomic variables, lock variables, barrier synchronization, condition variables, and semaphores, as well as different thread-safe data structures like queues, hash-maps, or array lists. The additional classes are similar to those described in [142]. In the following, we give a short overview of the package and refer to [84] for a more detailed description.

6.4.6.1 Semaphore mechanism

The class `Semaphore` provides an implementation of a counting semaphore, which is similar to the mechanism given in Fig. 6.19 for Pthreads. Internally, a `Semaphore` object maintains a counter which counts the number of permits. The most important methods of the `Semaphore` class are:

```
void acquire();
void release();
boolean tryAcquire()
boolean tryAcquire(int permits, long timeout,
                   TimeUnit unit)
```

The method `acquire()` asks for a permit and blocks the calling thread if no permit is available. If a permit is currently available, the internal counter for the number of available permits is decremented and control is returned to the calling thread.

The method `release()` adds a permit to the semaphore by incrementing the internal counter. If another thread is waiting for a permit of this semaphore, this thread is woken up. The method `tryAcquire()` asks for a permit to a semaphore object. If a permit is available, a permit is acquired by the calling thread and control is returned immediately with return value `true`. If no permit is available, control is also returned immediately, but with return value `false`; thus, in contrast to `acquire()`, the calling thread is not blocked. There exist different variants of the method `tryAcquire()` with varying parameters allowing the additional specification of a number of permits to acquire (parameter `permits`), a waiting time (parameter `timeout`) after which the attempt of acquiring the specified number of permits is given up with return value `false`, as well as a time unit (parameter `unit`) for the waiting time. If not enough permits are available when calling a timed `tryAcquire()`, the calling thread is blocked until one of the following events occurs:

- the number of requested permits becomes available because other threads call `release()` for this semaphore; in this case, control is returned to the calling thread with return value `true`;
- the specified waiting time elapses; in this case, control is returned with return value `false`; no permit is acquired in this case, also if some of the requested permits would have been available.

6.4.6.2 Barrier synchronization

The class `CyclicBarrier` provides an implementation of a barrier synchronization. The prefix *cyclic* refers to the fact that an object of this class can be re-used again after all participating threads have passed the barrier. The constructors of the class

```
public CyclicBarrier (int n);
public CyclicBarrier (int n, Runnable action);
```

allow the specification of a number n of threads that must pass the barrier before execution continues after the barrier. The second constructor allows the additional specification of an operation action that is executed as soon as all threads have passed the barrier. The most important methods of CyclicBarrier are await() and reset(). By calling await() a thread waits at the barrier until the specified number of threads have reached the barrier. A barrier object can be reset into its original state by calling reset().

6.4.6.3 Lock Mechanisms

The package java.util.concurrent.locks contains interfaces and classes for locks and for waiting for the occurrence of conditions. The interface Lock defines locking mechanisms which go beyond the standard synchronized methods and blocks and are not limited to the synchronization with the implicit mutex variables of the objects used. The most important methods of Lock are

```
void lock();
boolean tryLock();
boolean tryLock(long time, TimeUnit unit);
void unlock();
```

The method lock() tries to lock the corresponding lock object. If the lock has already been set by another thread, the executing thread is blocked until the locking thread releases the lock by calling unlock(). If the lock object has not been set by another thread when calling lock(), the executing thread becomes owner of the lock without waiting.

The method tryLock() also tries to lock a lock object. If this is successful, the return value is true. If the lock object is already set by another thread, the return value is false; in contrast to lock(), the calling thread is not blocked in this case. For the method tryLock(), additional parameters can be specified to set a waiting time after which control is resumed also if the lock is not available, see tryAcquire() of the class Semaphore. The method unlock() releases a lock which has previously been set by the calling thread.

The class ReentrantLock() provides an implementation of the interface Lock. The constructors of this class

```
public ReentrantLock();
public ReentrantLock(boolean fairness);
```

allow the specification of an additional fairness parameter fairness. If this is set to true, the thread with the longest waiting time can access the lock object if several threads are waiting concurrently for the same lock object. If the fairness parameter is not used, no specific access order can be assumed. Using the fairness parameter can lead to an additional management overhead and hence to a reduced throughput. A typical usage of the class ReentrantLock is illustrated in Fig. 6.44.

```
import java.util.concurrent.locks.*;
pulic class NewClass {
  private ReentrantLock lock = new ReentrantLock();
  //...
  public void method() {
    lock.lock();
    try {
        //...
    } finally { lock.unlock(); }
  }
}
```

Fig. 6.44 Illustration of the use of ReentrantLock objects.

6.4.6.4 Signal mechanism

The interface Condition from the package java.util.concurrent.lock defines a signal mechanism with condition variables which allows a thread to wait for a specific condition. The occurrence of this condition is shown by a signal of another thread, similar to the functionality of condition variables in Pthreads, see Section 6.1.3, page 325. A condition variable is always bound to a lock object, see interface Lock. A condition variable to a lock object can be created by calling the method

Condition newCondition().

This method is provided by all classes which implement the interface Lock. The condition variable returned by the method is bound to the lock object for which the method newCondition() has been called. For condition variables, the following methods are available:

```
void await();
void await(long time, TimeUnit unit);
void signal();
void signalAll();
```

The method await() blocks the executing thread until it is woken up by another thread by signal(). Before blocking, the executing thread releases the lock object as an atomic operation. Thus, the executing thread has to be the owner of the lock object before calling await(). After the blocked thread is woken up again by a signal() of another thread, it first must try to set the lock object again. Only after this is successful, the thread can proceed with its computations.

There is a variant of await() which allows the additional specification of a waiting time. If this variant is used, the calling thread is woken up after the time interval has elapsed, and if no signal() of another thread has arrived in the meantime. By calling signal(), a thread can wake up another thread which is waiting for a condition variable. By calling signalAll(), *all* waiting threads of the condition variable are woken up. The use of condition variables for the realization of a buffer mechanism is illustrated in Fig. 6.45. The condition variables are used in a similar way as the semaphore objects in Fig. 6.43.

```java
import java.util.concurrent.locks.*;
pulic class BoundedBufferCondition {
  private Lock lock = new ReentrantLock();
  private Condition notFull = lock.newCondition();
  private Condition notEmpty = lock.newCondition();
  private Object[] items = new Object[100];
  private int putptr=0, takeptr=0, count=0;
  public void put (Object x)
        throws InterruptedException {
    lock.lock();
    try {
        while (count == items.length)
          notFull.await();
        items[putptr] = x;
        putptr = (putptr +1) % items.length;
        ++count;
        notEmpty.signal();
    } finally { lock.unlock(); }
  }
  public Object take()
        throws InterruptedException {
    lock.lock();
    try {
        while (count == 0)
          notEmpty.await();
        Object x = items[takeptr];
        takeptr = (takeptr +1) % items.length;
        --count;
        notFull.signal();
        return x;
    } finally {lock.unlock();}
  }
}
```

Fig. 6.45 Implementation of a buffer mechanism by using condition variables.

6.4.6.5 Atomic Operations

The package `java.util.concurrent.atomic` provides atomic operations for simple data types, allowing a lock-free access to single variables. An example is the class `AtomicInteger` which comprises the following methods

```java
boolean compareAndSet (int expect, int update);
int getAndIncrement();
```

The first method sets the value of the variable to the value update, if the variable previously had the value expect. In this case, the return value is true. If the variable has not the expected value, the return value is false; no operation is performed. The operation is performed *atomically*, i.e., during the execution, the operation cannot be interrupted.

The second method increments the value of the variable atomically and returns the previous value of the variable as a result. The class AtomicInteger provides plenty of similar methods.

6.4.6.6 Task-based execution of programs

The package java.util.concurrent also provides a mechanism for a task-based formulation of programs. A task is a sequence of operations of the program which can be executed by an arbitrary thread. The execution of tasks is supported by the interface Executor:

```
public interface Executor {
   void execute (Runnable command);
}
```

where command is the task which is brought to execution by calling execute(). A simple implementation of the method execute() might merely activate the method command.run() in the current thread. More sophisticated implementations may queue command for execution by one of a set of threads. For multicore processors, several threads are typically available for the execution of tasks. These threads can be combined in a thread pool where each thread of the pool can execute an arbitrary task.

Compared to the execution of each task by a separate thread, the use of task pools typically leads to a smaller management overhead, particularly if the tasks consist of only a few operations. For the organization of thread pools, the class Executors can be used. This class provides methods for the generation and management of thread pools. Important methods are:

```
static ExecutorService newFixedThreadPool(int n);
static ExecutorService newCachedThreadPool();
static ExecutorService newSingleThreadExecutor();
```

The first method generates a thread pool which creates new threads when executing tasks until the maximum number n of threads has been reached. The second method generates a thread pool for which the number of threads is dynamically adapted to the number of tasks to be executed. Threads are terminated, if they are not used for a specific amount of time (60 s). The third method generates a single thread which executes a set of tasks. To support the execution of task-based programs the interface ExecutorService is provided. This interface inherits from the interface Executor and comprises methods for the termination of thread pools. The most important methods are

```
                void shutdown();
                List<Runnable> shutdownNow();
```

The method shutdown() has the effect, that the thread pool does not accept further tasks for execution. Tasks which have already been submitted are still executed before the shutdown. In contrast, the method shutdownNow() additionally stops the tasks which are currently in execution; the execution of waiting tasks is not started. The set of waiting tasks is provided in form of a list as return value. The class ThreadPoolExecutor is an implementation of the interface ExecutorService.

```java
import java.io.IOException;
import java.net.*;
import java.util.concurrent.*;

pulic class TaskWebServer {
  static class RunTask implements Runnable {
    private Socket myconnection;
    public RunTask (Socket connection) {
      myconnection = connection;
    }
    public void run() {
      // handleRequest(myconnection);
    }
  }
  public static void main (String[] args)
      throws IOException {
    ServerSocket s = new ServerSocket(80);
    ExecutorService pool =
      Executors.newFixedThreadPool(10);
    try {
      while (true) {
        Socket connection = s.accept();
        Runnable task = new RunTask(connection)
        pool.execute(task);
      }
    } catch (IOException ex) {
      pool.shutdown();
    }
  }
}
```

Fig. 6.46 Program fragment of a task-based web server.

Fig. 6.46 illustrates the use of a thread pool for the implementation of a web server which waits for connection requests of clients at a ServerSocket object. If a client request arrives, it is computed as a separate task by submitting this task with execute() to a thread pool. Each task is generated as a Runnable object. The operation handleRequest() to be executed for the request is specified as run() method. The maximum size of the thread pool is set to 10.

6.5 OpenMP

OpenMP is a portable standard for the programming of shared memory systems. The OpenMP API (application program interface) provides a collection of compiler directives, library routines, and environmental variables. The compiler directives can be used to extend the sequential languages Fortran, C, and C++ with single program multiple data (SPMD) constructs, tasking constructs, work-sharing constructs, and synchronization constructs. The use of shared and private data is supported. The library routines and the environmental variable control the OpenMP runtime system.

The OpenMP standard was originally designed in 1997 and is owned and maintained by the OpenMP Architecture Review Board (ARB). Since then many vendors have included the OpenMP standard in their compilers. Currently most compilers support at least Version 5.0 from 2018. The most recent update of the OpenMP standard is Version 5.2 from November 2021 [164]. Version 6.0 is expected to be released in 2023. Information about the OpenMP standard with all specifications as well as information about supporting resources such as compilers, books, forums, presentations, videos and tutorials can be found on the OpenMP web site www.openmp.org. There are also several books with a detailed description of OpenMP, including [30, 31, 152].

The programming model of OpenMP is based on cooperating **threads** running simultaneously on multiple processors or cores. Threads are created and destroyed in a **fork-join** pattern. The execution of an OpenMP program begins with a single thread, the initial thread, which executes the program sequentially until a first `parallel` construct is encountered. At the parallel construct the initial thread creates a team of threads consisting of a certain number of new threads and the initial thread itself. The initial thread becomes the master thread of the team. This fork operation is performed implicitly. The program code inside the parallel construct is called a **parallel region** and is executed in parallel by all threads of the team. The parallel execution mode can be an SPMD style; but an assignment of different tasks to different threads is also possible. OpenMP provides directives for different execution modes, which will be described below. At the end of a parallel region there is an implicit barrier synchronization, and only the master thread continues its execution after this region (implicit join operation). Parallel regions can be nested and each thread encountering a parallel construct creates a team of threads as described above.

The memory model of OpenMP distinguishes between shared memory and private memory. All OpenMP threads of a program have access to the same shared memory. To avoid conflicts, race conditions, or deadlocks, synchronisation mechanisms have to be employed, for which the OpenMP standard provides appropriate library routines. In addition to shared variables, the threads can also use private variables in the *threadprivate* memory, which cannot be accessed by other threads.

An OpenMP program needs to include the header file <omp.h>. The compilation with appropriate options translates the OpenMP source code into multithreaded code. This is supported by several compilers. The Version 4.2 of GCC and newer versions support OpenMP; the option -fopenmp has to be used. Intel's C++ compiler Version 8 and newer versions also support the OpenMP standard and provide additional Intel-specific directives. A compiler supporting OpenMP defines the variable _OPENMP if the OpenMP option is activated.

An OpenMP program can also be compiled into sequential code by a translation without the OpenMP option. The translation ignores all OpenMP directives. However, for the translation into correct sequential code special care has to be taken for some OpenMP runtime functions. The variable _OPENMP can be used to control the translation into sequential or parallel code.

6.5.1 Compiler directives

In OpenMP, parallelism is controlled by compiler directives. For C and C++, OpenMP directives are specified with the #pragma mechanism of the C and C++ standards. The general form of an OpenMP directive is

```
#pragma omp directive [clauses [ ] ...]
```

written in a single line. The clauses are optional and are different for different directives. Clauses are used to influence the behavior of a directive. In C and C++, the directives are case sensitive and apply only to the next code line or to the block of code (written within brackets { and }) immediately following the directive.

6.5.1.1 Parallel region

The most important directive is the parallel construct mentioned before with syntax

```
#pragma omp parallel [clause [clause] ... ]
{ // structured block ... }
```

The parallel construct is used to specify a program part that should be executed in parallel. Such a program part is called a *parallel region*. A team of threads is created to execute the parallel region in parallel. Each thread of the team is assigned a unique thread number, starting from zero for the master thread up to the number of threads minus one. The parallel construct ensures the creation of the team but does not distribute the work of the parallel region among the threads of the team. If there is no further explicit distribution of work (which can be done by other directives), all threads of the team execute the same code on possibly different data in an SPMD mode. One usual way to execute on different data is to employ the thread number also called *thread id*. The user-level library routine

```
int omp_get_thread_num()
```

returns the thread id of the calling thread as integer value. The number of threads remains unchanged during the execution of one parallel region but may be different for another parallel region. The number of threads can be set with the clause

```
num_threads(expression)
```

The user-level library routine

```
int omp_get_num_threads()
```

returns the number of threads in the current team as integer value, which can be used in the code for SPMD computations. At the end of a parallel region there is an implicit barrier synchronisation and the master thread is the only thread which continues the execution of the subsequent program code.

The clauses of a parallel directive include clauses which specify whether data will be private for each thread or shared among the threads executing the parallel region. Private variables of the threads of a parallel region are specified by the private clause with syntax

```
private(list_of_variables)
```

where list_of_variables is an arbitrary list of variables declared before. The private clause has the effect that for each private variable a new version of the original variable with the same type and size is created in the memory of each thread belonging to the parallel region. The private copy can be accessed and modified only by the thread owning the private copy. Shared variables of the team of threads are specified by the shared clause with the syntax

```
shared(list_of_variables)
```

where list_of_variables is a list of variables declared before. The effect of this clause is that the threads of the team access and modify the same original variable in the shared memory. The default clause can be used to specify whether variables in a parallel region are *shared* or *private* by default. The clause

```
default(shared)
```

causes all variables referenced in the construct to be shared except the private variables which are specified explicitly. The clause

```
default(none)
```

requires each variable in the construct to be specified explicitly as *shared* or *private*. The following example shows a first OpenMP program with a parallel region, in which multiple threads perform an SPMD computation on shared and private data.

Example: The program code in Fig. 6.47 uses a parallel construct for a parallel SPMD execution on an array x. The input values are read in the function initialize() by the master thread. Within the parallel region the variables x and

npoints are specified as *shared* and the variables iam, np and mypoints are specified as *private*. All threads of the team of threads executing the parallel region store
the number of threads in the variable np and their own thread id in the variable iam.
The private variable mypoints is set to the number of points assigned to a thread.
The function compute_subdomain() is executed by each thread of the team using
its own private variables iam and mypoints. The actual computations are performed
on the *shared* array x. □

```
#include <stdio.h>
#include <omp.h>

int npoints, iam, np, mypoints;
double *x;

int main() {
  scanf("%d", &npoints);
  x = (double *) malloc(npoints * sizeof(double));
  initialize();
  #pragma omp parallel shared(x,npoints) private(iam,np,mypoints)
  {
    np = omp_get_num_threads();
    iam = omp_get_thread_num();
    mypoints = npoints / np;
    compute_subdomain(x, iam, mypoints);
  }
}
```

Fig. 6.47 OpenMP program with parallel construct.

A nesting of parallel regions by calling a parallel construct within a parallel
region is possible. However, the default execution mode assigns only one thread to
the team of the inner parallel region. The library function

void omp_set_nested(int nested)

with a parameter nested ≠ 0 can be used to change the default execution mode to
more than one thread for the inner region. The actual number of threads assigned to
the inner region depends on the specific OpenMP implementation.

6.5.1.2 Parallel loops

OpenMP provides constructs which can be used within a parallel region to distribute
the work across threads that already exist in the team of threads executing the parallel region. The loop construct causes a distribution of the iterates of a **parallel loop**
and has the syntax

```
#pragma omp for [clause [clause] ... ]
for (i = lower_bound; i op upper_bound; incr_expr) {
  { // loop iterate ... }
}
```

The use of the for construct is restricted to loops which are parallel loops, in which the iterates of the loop are independent of each other and for which the total number of iterates is known in advance. The effect of the for construct is that the iterates of the loop are assigned to the threads of the parallel region and are executed in parallel. The index variable i should not be changed within the loop and is considered as private variable of the thread executing the corresponding iterate. The expressions lower_bound and upper_bound are integer expressions, whose values should not be changed during the execution of the loop. The operator op is a boolean operator from the set { <, <=, >, >= }. The increment expression incr_expr can be of the form

```
++i, i++, --i, i--, i += incr, i -= incr,
i = i + incr, i = incr + i, i = i - incr,
```

with an integer expression incr that remains unchanged within the loop. The parallel loop of a for construct should not be finished with a break command. The parallel loop ends with an implicit synchronization of all threads executing the loop, and the program code following the parallel loop is only executed if all threads have finished the loop. The nowait clause given as clause of the for construct can be used to avoid this synchronization.

The specific distribution of iterates to threads is done by a scheduling strategy. OpenMP supports different scheduling strategies specified by the schedule parameters of the following list:

- schedule(static, block_size) specifies a static distribution of iterates to threads which assigns blocks of size block_size in a *round-robin* fashion to the threads available. When block_size is not given, blocks of almost equal size are formed and assigned to the threads in a blockwise distribution.
- schedule(dynamic, block_size) specifies a dynamic distribution of blocks to threads. A new block of size block_size is assigned to a thread as soon as the thread has finished the computation of the previously assigned block. When block_size is not provided, blocks of size one, i.e., consisting of only one iterate, are used.
- schedule(guided, block_size) specifies a dynamic scheduling of blocks with decreasing size. For the parameter value block_size =1, the new block assigned to a thread has a size which is the quotient of the number of iterates not assigned yet and the number of threads executing the parallel loop. For a parameter value block_size = k > 1, the size of the blocks is determined in the same way, but a block never contains fewer than *k* iterates (except for the last block which may contain fewer than k iterates). When no block_size is given, the blocks consist of one iterate each.

- schedule(auto) delegates the scheduling decision to the compiler and/or run-time system. Thus, any possible mapping of iterates to threads can be chosen.
- schedule(runtime) specifies a scheduling at runtime. At runtime the environ-mental variable OMP_SCHEDULE, which has to contain a character string describ-ing one of the formats given above, is evaluated. Examples are

```
setenv OMP_SCHEDULE "dynamic, 4"
setenv OMP_SCHEDULE "guided"
```

When the variable OMP_SCHEDULE is not specified, the scheduling used depends on the specific implementation of the OpenMP library.

A for construct without any schedule parameter is executed according to a default scheduling method also depending on the specific implementation of the OpenMP library. The use of the for construct is illustrated with the following ex-ample coding a matrix multiplication.

Example: The code fragment in Fig. 6.48 shows a multiplication of 100×100 matrix MA with a 100×100 matrix MB resulting in a matrix MC of the same dimension. The parallel region specifies MA, MB, MC as shared variables and the indices row, col,i as private. The two parallel loops use static scheduling with blocks of row. The first parallel loop initializes the result matrix MC with 0. The second parallel loop

```
#include <omp.h>

double MA[100][100], MB[100][100], MC[100][100];
int i, row, col, size = 100;

int main() {
  read_input(MA, MB);
  #pragma omp parallel shared(MA,MB,MC,size) private(row,col,i)
  {
    #pragma omp for schedule(static)
    for (row = 0; row < size; row++) {
      for (col = 0; col < size; col++)
        MC[row][col] = 0.0;
    }
    #pragma omp for schedule(static)
    for (row = 0; row < size; row++) {
      for (col = 0; col < size; col++)
        for (i = 0; i < size; i++)
          MC[row][col] += MA[row][i] * MB[i][col];
    }
  }
  write_output(MC);
}
```

Fig. 6.48 OpenMP program for a parallel matrix multiplication using a parallel region with two inner for constructs.

performs the matrix multiplication in a nested for loop. The for construct applies
to the first for loop with iteration variable row and, thus, the iterates of the parallel
loop are the nested loops of the iteration variables col and i. The static scheduling
leads to a row-blockwise computation of the matrix MC. The first loop ends with an
implicit synchronization. Since it is not clear that the first and the second parallel
loop have exactly the same assignment of iterates to threads, a nowait clause should
be avoided to guarantee that the initialization is finished before the multiplication
starts.   □

The nesting of the for-construct within the same parallel construct is not
allowed. The nesting of parallel loops can be achieved by nesting parallel constructs so that each parallel construct contains exactly one for construct. This is
illustrated in the following example.

```
#include <omp.h>

double MA[100][100], MB[100][100], MC[100][100];
int i, row, col, size = 100;

int main() {
  read_input(MA, MB);
  #pragma omp parallel private(row,col,i)
  {
    #pragma omp for schedule(static)
    for (row = 0; row < size; row++) {
      #pragma omp parallel shared(MA, MB, MC, size)
      {
        #pragma omp for schedule(static)
        for (col = 0; col < size; col++) {
          MC[row][col] = 0.0;
          for (i = 0; i < size; i++)
            MC[row][col] += MA[row][i] * MB[i][col];
        }
      }
    }
  }
  write_output(MC);
}
```

Fig. 6.49 OpenMP programm for a parallel matrix multiplikation with nested parallel loops.

Example: The program code in Fig. 6.49 shows a modified version of the matrix
multiplication in the last example. Again, the for-construct applies to the for loop
with the iteration index row. The iterates of this parallel loop start with another
parallel construct which contains a second for-construct applying to the loop
with iteration index col. This leads to a parallel computation, in which each entry

of MC can be computed by a different thread. There is no need for a synchronization
between initialization and computation.   □

The OpenMP program in Fig. 6.49 implements the same parallelism as the
Pthreads program for matrix multiplication in Fig. 6.1, see page 319. A difference
between the two programs is that the Pthreads program starts the threads explicitly.
The thread creation in the OpenMP program is done implicitly by the OpenMP library which deals with the implementation of the nested loop and guarantees the
correct execution. Another difference is that there is a limitation for the number of
threads in the Pthreads program. The matrix size 8×8 in the Pthreads program from
Fig. 6.1 leads to a correct program. A matrix size 100×100, however, would lead
to the start of 10000 threads, which is too large for most Pthreads implementation.
There is no such limitation in the OpenMP program.

6.5.1.3 Non-iterativ Work-Sharing Constructs

The OpenMP library provides the sections construct to distribute non-iterative
tasks to threads. Within the sections construct different code blocks are indicated
by the section construct as tasks to be distributed. The syntax for the use of a
sections construct is the following.

```
#pragma omp sections [clause [clause] ... ]
{
  [#pragma omp section]
    { // structured block ... }
  [#pragma omp section
    { // structured block ... }
    ⋮
  ]
}
```

The section constructs denote structured blocks which are independent of each
other and can be executed in parallel by different threads. Each structured block
starts with #pragma omp section which can be omitted for the first block. The
sections construct ends with an implicit synchronization unless a nowait clause
is specified.

6.5.1.4 Single excution

The single construct is used to specify that a specific structured block is executed
by only one thread of the team, which is not necessarily the master thread. This can
be useful for tasks like control messages during a parallel execution. The single
construct has the syntax

```
#pragma omp single [Parameter [Parameter] ... ]
{ // structured block ... }
```

and can be used within a parallel region. The single construct also ends with an implicit synchronization unless a nowait clause is specified. The execution of a structured block within a parallel region by the master thread only is specified by

```
#pragma omp master
  { // structured block ... }
```

All other threads ignore the construct. There is no implicit synchronization of the master threads and the other threads of the team.

6.5.1.5 Syntactic abbreviations

OpenMP provides abbreviated syntax for parallel regions containing only one for construct or only one sections construct. A parallel region with one for construct can be specified as

```
#pragma omp parallel for [clause [clause] ··· ]
  for (i = lower_bound; i op upper_bound; incr_expr) {
    { // loop body ... }
}
```

All clauses of the parallel construct or the for construct can be used. A parallel region with only one sections construct can be specified as

```
#pragma omp parallel sections [clause [clause] ··· ]
{
  [#pragma omp section]
    { // structured block ... }
  [#pragma omp section
    { // structured block ... }
    ⋮
  ]
}
```

6.5.2 Execution environment routines

The OpenMP library provides several execution environment routines that can be used to query and control the parallel execution environment. We present a few of them. The function

```
void omp_set_dynamic (int dynamic_threads)
```

can be used to set a dynamic adjustment of the number of threads by the runtime system and is called outside a parallel region. A parameter value `dynamic_threads` $\neq$ 0 allows the dynamic adjustment of the number of threads for the subsequent parallel region. However, the number of threads within the same parallel region remains constant. The parameter value `dynamic_threads` = 0 disables the dynamic adjustment of the number of threads. The default case depends on the specific OpenMP implementation. The routine

```
int omp_get_dynamic (void)
```

returns information about the current status of the dynamic adjustment. The return value 0 denotes that no dynamic adjustment is set; a return value $\neq$ 0 denotes that the dynamic adjustment is set. The number of threads can be set with the routine

```
void omp_set_num_threads (int num_threads)
```

which has to be called outside a parallel region and influences the number of threads in the subsequent parallel region (without a `num_threads` clause). The effect of this routine depends on the status of the dynamic adjustment. If the dynamic adjustment is set, the value of the parameter `num_threads` is the maximum number of threads to be used. If the dynamic adjustment is not set, the value of `num_threads` denotes the number of threads to be used in the subsequent parallel region. The routine

```
void omp_set_nested (int nested)
```

influences the number of threads in nested parallel regions. The parameter value `nested` = 0 means that the execution of the inner parallel region is executed by one thread sequentially. This is the default. A parameter value `nested` $\neq$ 0 allows a nested parallel execution and the runtime system can use more than one thread for the inner parallel region. The actual behavior depends on the implementation. The routine

```
int omp_get_nested (void)
```

returns the current status of the nesting strategy for nested parallel regions.

6.5.3 Coordination and synchronization of threads

A parallel region is executed by multiple threads accessing the same shared data, so that there is need for synchronization in order to protect critical regions or avoid race condition, see also Chapter 3. OpenMP offers several constructs which can be used for synchronization and coordination of threads within a parallel region. The `critical` construct specifies a **critical region** which can be executed only by a single thread at a time. The syntax is

```
#pragma omp critical [(name)]
    structured block
```

An optional name can be used to identify a specific critical region. When a thread encounters a critical construct, it waits until no other thread executes a critical region of the same name name and then executes the code of the critical region. Unnamed critical regions are considered to be one critical region with the same unspecified name. The barrier construct with syntax

```
#pragma omp barrier
```

can be used to synchronize the threads at a certain point of execution. At such an explicit barrier construct all threads wait until all other threads of the team have reached the barrier and only then they continue the execution of the subsequent program code. The atomic construct can be used to specify that a single assignment statement is an **atomic operation**. The syntax is

```
#pragma omp atomic
statement
```

and can contain statements of the form

```
x binop= E,
x++, ++x, x--, --x,
```

with an arbitrary variable x, a scalar expression E not containing x, and a binary operator binop $\in \{+, -, *, /, \&, \hat{\ }, |, <<, >>\}$. The atomic construct ensures that the storage location x addressed in the statement belonging to the construct is updated atomically, which means that the load and store operations for x are atomic but not the evaluation of the expression E. No interruption is allowed between the load and the store operation for variable x. However, the atomic construct does not enforce exclusive access to x with respect to a critical region specified by a critical construct. An advantage of the atomic construct over the critical construct is that also parts of an array variable can be specified as being atomically updated. The use of a critical construct would protect the entire array.

Example: The following example shows an atomic update of a single array element a[index[i]] += b.

```
extern float a[], *p=a, b; int index[];
#pragma omp atomic
  a[index[i]] += b;
#pragma omp atomic
  p[i] -= 1.0;
```

$\square$

A typical calculation which needs to be synchronized is a **global reduction** operation performed in parallel by the threads of a team. For this kind of calculation OpenMP provides the reduction clause, which can be used for parallel, sections and for constructs. The syntax of the clause is

```
reduction (op: list)
```

where op $\in \{+, -, *, \&, \hat{}, |, \&\&, ||\}$ is a reduction operator to be applied and list is a list of reduction variables which have to be declared as shared. For each of the variables in list, a private copy is created for each thread of the team. The private copies are initialized with the neutral element of the operation op and can be updated by the owning thread. At the end of the region for which the reduction clause is specified, the local values of the reduction variables are combined according to the operator op and the result of the reduction is written into the original shared variable. The OpenMP compiler creates efficient code for executing the global reduction operation. No additional synchronization, such as the critical construct, has to be used to guarantee a correct result of the reduction. The following example illustrates the accumulation of values.

Example: Figure 6.50 shows the accumulation of values in a for construct with the results written into the variables a, y and am. Local reduction operations are performed by the threads of the team executing the for construct using private copies of a, y and am for the local results. It is possible that a reduction operation is performed within a function, such as the function sum used for the accumulation onto y. At the end of the for loop, the values of the private copies of a, y and am are accumulated according to + or ||, respectively, and the final values are written into the original shared variables a, y and am. □

```
#pragma omp parallel for reduction (+: a,y) reduction (||: am)
for (i=0; i<n; i++) {
    a += b[i];
    y = sum (y, c[i]);
    am = am || b[i] == c[i];
}
```

Fig. 6.50 Program fragment for the use of the reduction clause.

The shared memory model of OpenMP might also require to coordinate the memory view of the threads. OpenMP provides the flush construct with the syntax

```
#pragma omp flush [(list)]
```

to produce a consistent view of the memory, where list is a list of variables whose values should be made consistent. For pointers in the list list only the pointer value is updated. If no list is given, all variables are updated. An inconsistent view can occur since modern computers provide memory hierarchies. Updates are usually done in the faster memory parts, like registers or caches, which are not immediately visible to all threads. OpenMP has a specific relaxed-consistency shared memory in which updated values are written back later. But to make sure at a specific program point that a value written by one thread is actually read by another thread, the flush construct has to be used. It should be noted that no synchronization is provided if several threads execute the flush construct.

Example: Figure 6.51 shows an example adopted from the OpenMP specification [163]. Two threads i ($i = 0, 1$) compute work[i] of array work which is written back to memory by the flush construct. The following update on array sync[iam] indicates that the computation of work[iam] is ready and written back to memory. The array sync is also written back by a second flush construct. In the while loop a thread waits for the other thread to have updated its part of sync. The array work is then used in the function combine() only after both threads have updated their elements of work.   □

```
#pragma omp parallel private (iam, neighbor) shared (work, sync)
{
    iam = omp_get_thread_num();
    sync[iam] = 0;
    #pragma omp barrier
    work[iam] = do_work();
    #pragma omp flush (work)
    sync[iam] = 1;
    #pragma omp flush (sync)
    neighbor = (iam != 0) ? (iam - 1) : (omp_get_num_threads() - 1);
    while (sync[neighbor] == 0) {
        #pragma omp flush (sync)
        { }
    }
    combine (work[iam], work[neighbor]);
}
```

Fig. 6.51 Program fragment for the use of the flush construct.

Besides the explicit flush construct there is an implicit flush at several points of the program code, which are:

- a barrier construct,
- entry to and exit from a critical region,
- at the end of a parallel region,
- at the end of a for, sections or single construct without nowait clause,
- entry and exit of lock routines (which will be introduced below).

6.5.3.1 Locking mechanism

The OpenMP runtime system also provides runtime library functions for a synchronization of threads with the **locking mechanism**. The locking mechanism has been described in Section 4.4 and in this Section 6 for Pthreads and Java threads. The specific locking mechanism of the OpenMP library provides two kinds of lock variables on which the locking runtime routines operate. *Simple locks* of type omp_lock_t can be locked only once. *Nestable locks* of type omp_nest_lock_t can be locked multiple times by the same thread. OpenMP lock variables should be accessed only by OpenMP locking routines. A lock variable is initialized by one of the following initialization routines

```
void omp_init_lock (omp_lock_t *lock);
void omp_init_nest_lock (omp_nest_lock_t *lock);
```

for simple and nestable locks, respectively. A lock variable is removed with the routines

```
void omp_destroy_lock (omp_lock_t *lock);
void omp_destroy_nest_lock (omp_nest_lock_t *lock);
```

An initialized lock variable can be in the states *locked* or *unlocked*. At the beginning, the lock variable is in the state *unlocked*. A lock variable can be used for the synchronization of threads by locking and unlocking. To lock a lock variable the functions

```
void omp_set_lock (omp_lock_t *lock)
void omp_set_nest_lock (omp_nest_lock_t *lock)
```

are provided. If the lock variable is available, the thread calling the lock routine locks the variable. Otherwise, the calling thread blocks. A simple lock is available when no other thread has locked the variable before without unlocking it. A nestable lock variable is available when no other thread has locked the variable without unlocking it or when the calling thread has locked the variable, i.e., multiple locks for one nestable variable by the same thread are possible counted by an internal counter. When a thread uses a lock routine to lock a variable successfully, this thread is said to *own* the lock variable. A thread owning a lock variable can unlock this variable with the routines

```
void omp_unset_lock (omp_lock_t *lock)
void omp_unset_nest_lock (omp_nest_lock_t *lock)
```

For a nestable lock, the routine omp_unset_nest_lock () decrements the internal counter of the lock. If the counter has the value 0 afterwards, the lock variable is in the state *unlocked*. The locking of a lock variable without a possible blocking of the calling thread can be performed by one of the routines

```
void omp_test_lock (omp_lock_t *lock)
void omp_test_nest_lock (omp_nest_lock_t *lock)
```

for simple and nestable lock variables, respectively. When the lock is available, the routines lock the variable or increment the internal counter and return a result value $\neq 1$. When the lock is not available, the test routine returns 0 and the calling thread is not blocked.

Example: Figure 6.52 illustrates the use of nestable lock variables, see [163]. A data structure pair consists of two integer a and b and a nestable lock variable 1

which is used to synchronize the updates of a, b or the entire pair. It is assumed
that the lock variable l has been initialized before calling f(). The increment function incr_a() for incrementing a, incr_b() for incrementing b, and incr_pair()
for incrementing both integer variables are given. The function incr_a() is only
called from incr_pair() and does not need an additional locking. The functions
incr_b() and incr_pair() are protected by the lock since they can be called concurrently.   □

```c
#include <omp.h>

typedef struct {
  int a, b;
  omp_nest_lock_t l;
} pair;

void incr_a (pair *p, int a) {
  p->a += a;
}

void incr_b (pair *p, int b) {
  omp_set_nest_lock (&p->l);
  p->b += b;
  omp_unset_nest_lock (&p->l);
}

void incr_pair (pair *p, int a, int b) {
  omp_set_nest_lock (&p->l);
  incr_a (p, a);
  incr_b (p, b);
  omp_unset_nest_lock (&p->l);
}

void f (pair *p) {
  extern int work1(), work2(), work3();
  #pragma omp parallel sections
  {
    #pragma omp section
      incr_pair (p, work1(), work2());
    #pragma omp section
      incr_b (p,work3());
  }
}
```

Fig. 6.52 Program fragment illustrating the use of nestable lock variables.

6.5.4 OpenMP task model

In Version 3.0, OpenMP has been extended by a task model to support the efficient execution of algorithms with a more irregular structure of parallelism or a dynamic creation of work. Such situations cannot easily be captured by regular parallel loops which are provided by OpenMP for loops. The OPenMP task model uses an implicit task pool into which work units (called tasks) can dynamically be inserted during the execution of a program. In contrast, an implementation of an explicit task pool using Pthreads has been presented in Sect. 6.2.1. A task captures a structured block of statements which are executed sequentially, see [18] for the design of OpenMP tasks. Tasks are specified by the programmer using the following pragma:

```
#pragma omp task [clauses[[,]clauses] ...]
    // structured block
```

The task pragma must be positioned within a surrounding parallel region. The execution of the pragma leads to the creation of a new task containing the specified structured block of statements. The newly created task can be executed immediately by the thread executing the pragma, but the execution of the task can also be deferred. If the task execution is deferred, the task is inserted into a pool of tasks. Any thread of the surrounding parallel region can take out the task at a later time and can execute it. There is no guaranteed order for the execution of the generated tasks, i.e., it should not be assumed that tasks are executed in the order in which they are generated and this should be taken into consideration for a correct behavior of the program. Thus, the program should be designed such that it behaves correctly for any order of task execution. Without further specification, all tasks generated have the same priority for execution. The `priority (integer)` clause can be used to give a hint to the runtime system that some tasks are more important than other tasks. The priority-value is a non-negative integer value that defines the task priority. Without a priority specification, all generated tasks have priority 0, which is the smallest priority. Specifying a larger priority value for a task tells the runtime system that this task should be executed first, if all other tasks have a smaller priority.

Figure 6.53 shows a program fragment demonstrating the use of tasks for the traversal of a linked list. A team of threads is created by the `parallel` pragma. Using the `single` pragma inside ensures that a single thread of the team traverses the linked list sequentially and creates a task for each list node. The created task performs some computations on the corresponding list node, which is captured by the function `DoWork()`. After creation, the tasks can be executed by any thread of the team of threads. Omitting the `single` pragma in this example would specify that each thread of the team traverses the linked list and multiple tasks would be created for each node.

An immediate execution of a newly created task can be obtained by using an `if` clause of the form

```
    if (expression)
```

in the task specification. If the specified scalar expression evaluates to false, the executing thread suspends the execution of its current task and immediately exe-

```
int main() {
    struct node* p;

// create linked list and let p point to head of this list
    ♯pragma omp parallel
        ♯pragma omp single
            while (p != NULL) {
                ♯pragma omp task firstprivate(p)
                    DoWork(p)
                p = p->next;
            }
}
```

Fig. 6.53 Program fragment illustrating the use of the task construct in OpenMP for the traversal of a linked list with head p.

cutes the specified task until completion. This does not include the execution of descendents of the specified task, if there are any. The usage of an if clause with an immediate execution of a task is reasonable, if the granularity of the task is small such that the overhead for a deferred execution including an insertion into the task pool and a later extraction out of the task pool would be too large.

If the execution of the task is deferred, the task can be executed by an arbitrary thread of the team to which the creating thread belongs. Per default, tasks are tied, i.e., the thread that starts the execution of the task, executes the task until its final completion. If the execution of the task is suspended, e.g., due to the creation of an immediate task using an if clause, only the same thread that started the execution of the task can resume the execution. Using the untied clause for the task creation, the programmer can specify that any thread of the team can resume the execution of such an untied task.

The clauses specified for a task definition can be used to define variables for the task, cutoff strategies, and dependencies between tasks. Clauses to define variables for the newly created tasks are

- private: a new copy for each of the variables specified is created; the new copies are not initialized;
- firstprivate: a new copy for each of the variables specified is created and the variables are initialized with the content of the original variables at task creation time;
- shared: no new copy of the variables listed is created; references to the variables listed refer to the original copy of the variable in the context of the task creation pragma.

Variables that are not listed are implicitly treated as firstprivate. This behavior can be changed with a default clause, e.g., defining
```
default (shared)
```

During the execution of a task, the task pragma can be used to generate further tasks. The newly generated tasks are called **child tasks**. A task can wait for the completion of all its child tasks that have been generated so far by using the taskwait construct:

♯pragma omp taskwait [clauses[[,]clauses] ...]

When executing a taskwait construct, the thread executing the task containing the taskwait suspends the execution of the current task until all immediate child tasks have been terminated. After the suspension of the task execution, the thread can execute another task. To wait also for the descendants of the child tasks, the taskgroup construct can be used:

♯pragma omp taskgroup [clauses[[,]clauses] ...]
    // structured block

The execution of the taskgroup pragma suspends the execution of the current task until all its child tasks that are created in the structured block and all descendents of these child tasks have been finished.

```
void traverse(struct node* p) {
    if (p->left != NULL)
♯pragma omp task //p is firstprivate by default
        traverse(p->left);
    if (p->right != NULL)
♯pragma omp task // p is firstprivate by default
        traverse(p->right);
♯pragma omp taskwait
    make_computation(p->data);
}

int main(int argc, char **argv) {
    struct node* root;
    construct_tree(root);
♯pragma omp parallel
    {
♯pragma omp master
        traverse(root);
    }
}
```

Fig. 6.54 Program fragment illustrating the use of the task and taskwait constructs in OpenMP for the recursive traversal of a binary tree with root p.

The following example illustrates the use of tasks for the recursive traversal of a binary tree, see Fig. 6.54. The main program specifies a parallel region within which the master thread calls the function traverse for the root of the tree to be traversed. In the function traverse, a new task is created for each child node encountered. The task creation is recursive, i.e., during the execution of a task, new tasks are created for the child nodes. This is done by the master thread for the child nodes of the root node. For inner nodes, the task creation can be performed by any other thread of the team of threads. After the generation of the tasks for the child nodes of a node, the function make_computation() is used to make some computation with the data of the node. The use of the taskwait pragma in traverse ensures that the child node computations are finished before the computation of the function make_computation() is started. This results in a postorder traversal of the tree, i.e., for each node the computations for the child nodes are executed before the computations of the parent node. Without the taskwait pragma, any traversal order could result.

```
void traverse(struct node* p) {
♯pragma omp taskgroup
{
    if (p->left != NULL)
    ♯pragma omp task //p is firstprivate by default
        traverse(p->left);
    if (p->right != NULL)
    ♯pragma omp task // p is firstprivate by default
        traverse(p->right);
}
make_computation(p->data);
}
```

Fig. 6.55 Alternative implementation of a recursive traversal of a binary tree with root p using taskgroup instead of taskwait as in Fig. 6.54.

An alternative implementation for the traverse() function is given in Fig. 6.55. The same main program as in Fig. 6.54 can be used. However, in Fig. 6.55 a taskgroup pragma is used for the synchronization instead of the taskwait in Fig. 6.54. The taskgroup pragma ensures that all direct and indirect child tasks have been finished before the execution of the code after the taskgroup construct is continued, i.e., the function make_computation() for a node p is executed after the recursive traversal of the left and the right subtree of p has been finished. This also ensures a postorder traversal of the tree.

Another method to define tasks in OpenMP is the taskloop pragma. The taskloop pragma can be used to partition the iterates of a loop into chunks, see Sect. 4.7.1, and to assign the chunks to tasks for a parallel execution by the threads of the team of the surrounding parallel region:

```
♯pragma omp taskloop [clauses[[,]clauses] ...]
    // structured for loop
```

How many iterations are put into one task is decided by the OpenMP runtime system. The number of iterations to be put into one task can also be defined by the programmer by using the grainsize (int) clause with an integer value. The grainsize clause is a hint to the runtime system that the grain size should be at least as large as the integer value specified. Alternatively, the num_tasks clause can be used to specify the number of tasks to be generated by the taskloop pragma. The iterates of the loop are distributed among these tasks, i.e., the num_tasks clause leads to an implicit specification of the chunk size.

```c
#include <stdio.h>
#include <omp.h>
#include <stdlib.h>

int main(void) {
    const int N = 600;
    //double MA[N][N], MB[N][N], MC[N][N];
    double *MA, *MB, *MC;
    MA = (double*)malloc(sizeof(double)*N*N);
    MB = (double*)malloc(sizeof(double)*N*N);
    MC = (double*)malloc(sizeof(double)*N*N);
    int i, row, col, size = N;
    initialize_matrix(MA);
    initialize_matrix(MB);
    #pragma omp parallel default ( shared ) private ( row , col , i )
    {
        #pragma omp single
        {
            #pragma omp taskloop shared ( MC ) num_tasks (50)
                for ( row = 0; row < size ; row ++)
                    for ( col = 0; col < size ; col ++)
                        MC [ row *N+ col ] = 0.0;
            #pragma omp taskloop grainsize(10) shared(MA,MB,MC)
                for ( row = 0; row < size; row++)
                    for ( col = 0; col < size; col++)
                        for ( i = 0; i < size; i++)
                            MC[row*N+col] += MA[row*N+i] * MB[i*N+col];
        } // single
    } // parallel
} // main
```

Fig. 6.56 Program fragment illustrating the use of the OpenMP taskloop construct for the multiplication of two matrices MA and MB.

Figure 6.56 shows the use of the taskloop pragma for the multiplication of two $N \times N$ matrices MA and MB. The result matrix is stored in matrix MC. The three matrices are dynamically allocated by the main thread as arrays with $N \cdot N$ double entries using malloc(). The matrices MA and MB are initialized with the function initialize_matrix(), which is not shown in the figure. The parallel pragma creates a parallel region with a certain number of worker threads. Then, the single pragma is used to ensure that only one of these threads generates the tasks for the iterates of the parallel loops. The initialization of the result matrix MC with 0.0 is performed using a taskloop pragma with the clause num_tasks (50), i.e., 50 tasks are created. The actual matrix multiplication is performed with a separate taskloop pragma using the clause grainsize(10), i.e., the tasks generated should comprise the computation of 10 rows of the result matrix. The matrices MA, MB, and MC are shared between the threads that have been created by the surrounding parallel region. There is an implicit barrier synchronization between the two taskloop constructs, ensuring that all tasks that have been generated for the matrix initialization are finished before the tasks for the actual matrix multiplication are generated.

For the taskgroup and taskloop pragmas, reduction operations can be specified using the clause

reduction(op: ListOfVariables)

where op is a reduction operator and ListOfVariables is a list of variables for which the reduction operation is performed, see also Section 6.5.3. The reduction clause has been introduced in Version 5.0 of OpenMP. Each of the tasks created participates in the reduction operation.

Figure 6.57 demonstrates the use of the taskloop pragma for the computation of a scalar product of two vectors x and y. The grainsize (100) clause is used to specify that the tasks generated comprise at least 100 iterations of the for loop succeeding the taskloop pragma. The reduction clause is used to accumulate the partial results contributed by each task to the final result of the scalar product in variable res.

```
double dotprod(int n, double *x, double *y) {
double res = 0.0;
int i;
#pragma omp taskloop grainsize(100) reduction(+: res)
for (i = 0; i < n; i++)
    res += x[i] * y[i];
return res;
}
```

Fig. 6.57 Program fragment illustrating the use of the OpenMP taskloop construct with a reduction clause for the computation of a scalar product of two vectors x and y.

More information about the usage of tasks in OpenMP can be found in [18, 164].

6.6 Exercises for Chapter 6

Exercise 6.1. Modify the matrix multiplication program from Fig. 6.1 on page 319 so that a fixed number of threads is used for the multiplication of matrices of arbitrary size. For the modification, let each thread compute the number of rows of the result matrix instead of a single entry. Compute the rows that each thread must compute such that each thread has about the same number of rows to compute. Is there any synchronization required in the program?

Exercise 6.2. Use the task pool implementation from Section 6.2.1 on page 333 to implement a parallel matrix multiplication. To do so, use the function `thread_mult()` from Fig. 6.1 to define a task as the computation of one entry of the result matrix and modify the function if necessary so that it fits to the requirements of the task pool. Modify the main program so that all tasks are generated and inserted into the task pool before the threads to perform the computations are started. Measure the resulting execution time for different numbers of threads and different matrix sizes and compare the execution time with the execution time of the implementation of the last exercise.

Exercise 6.3. Consider the r/w lock mechanism in Fig. 6.5. The implementation given does not provide operations that are equivalent to the function `pthread_mutex_trylock()`. Extend the implementation from Fig. 6.5 by specifying functions `rw_lock_rtrylock()` and `rw_lock_wtrylock()` which return EBUSY if the requested read or write permit cannot be granted.

Exercise 6.4. Consider the r/w lock mechanism in Fig. 6.5. The implementation given favors read request over write requests in the sense that a thread will get a write permit only if no other threads request a read permit, but read permits are given without waiting also in the presence of other read permits. Change the implementation such that write permits have priority, i.e., as soon as a write permit arrives, no more read permits are granted until the write permit has been granted and the corresponding write operation is finished. To test the new implementation write a program which starts three threads, two read-threads and one write-thread. The first read-thread requests five read permits one after another. As soon as it gets the read permits it prints a control message and waits for two seconds (use `sleep(2)`) before requesting the next read permit. The second read-thread does the same except that it only waits one second after the first read permit and two seconds otherwise. The write-thread first waits five seconds and then requests a write permit and prints a control message after it has obtained the write permit; then the write permit is released again immediately.

Exercise 6.5. An r/w lock mechanism allows multiple readers to access a data structure concurrently, but only a single writer is allowed to access the data structures at a time. We have seen a simple implementation of r/w-locks in Pthreads in Fig. 6.5. Transfer this implementation to Java threads by writing a new class RWlock with entries num_r and num_w to count the current number of read and write permits given. The class `RWlock` provides methods similar to the functions in Fig. 6.5 to request or release a read or write permit.

Exercise 6.6. Consider the pipelining programming pattern and its Pthreads implementation in Section 6.2.2. In the example given, each pipeline stage adds 1 to the integer value received from the predecessor stage. Modify the example such that pipeline stage i adds the value i to the value received from the predecessor. In the modification, there should still be only one function `pipe_stage()` expressing the computations of a pipeline stage. This function must receive an appropriate parameter for the modification.

Exercise 6.7. Use the task pool implementation from Section 6.2.1 to define a parallel loop pattern. The loop body should be specified as function with the loop variable as parameter. The iteration space of the parallel loop is defined as the set of all values that the loop variable can have. To execute a parallel loop, all possible indices are stored in a parallel data structure similar to a task pool which can be accessed by all threads. For the access, a suitable synchronization must be used.

(a) Modify the task pool implementation accordingly such that functions for the definition of a parallel loop and for retrieving an iteration from the parallel loop are provided. The thread function should also be provided.
(b) The parallel loop pattern from (a) performs a dynamic load balancing since a thread can retrieve the next iteration as soon as its current iteration is finished. Modify this operation such that a thread retrieves a chunk of iterations instead of a single operation to reduce the overhead of load balancing for fine-grained iterations.
(c) Include guided self-scheduling (GSS) in your parallel loop pattern. GSS adapts the number of iterations retrieved by a thread to the total number of iterations that are still available. if n threads are used and there are R_i remaining iterations, the next thread retrieves

$$x_i = \lceil \frac{R_i}{n} \rceil$$

iterations. For the next retrieval, $R_{i+1} = R_i - x_i$ iterations remain. R_1 is the initial number of iterations to be executed.
(d) Use the parallel loop pattern to express the computation of a matrix multiplication where the computation of each matrix entry can be expressed as an iteration of a parallel loop. Measure the resulting execution time for different matrix sizes. Compare the execution time for the two load balancing schemes (standard and GSS) implemented.

Exercise 6.8. Consider the client-server pattern and its Pthreads implementation in Section 6.2.3. Extend the implementation given in this section by allowing a cancellation with deferred characteristics. To be cancellation-safe, mutex variables that have been locked must be released again by an appropriate cleanup handler. When a cancellation occurs, allocated memory space should also be released. In the server function `tty_server_routine()`, the variable `running` should be reset when a cancellation occurs. Note that this may create a concurrent access. If a cancellation request arrives during the execution of a synchronous request of a client, the client thread should be informed that a cancellation has occurred. For a cancellation in the function `client_routine()`, the counter `client_threads` should be kept consistent.

Exercise 6.9. Consider the taskpool pattern and its implementation in Pthreads in Section 6.2.1. Implement a Java class `TaskPool` with the same functionality. The task pool should accept each object of a class which implements the interface `Runnable` as task. The tasks should be stored in an array `final Runnable tasks[]`. A constructor `TaskPool(int p, int n)` should be implemented which allocates a task array of size n and creates p threads which access the task pool. The methods `run()` and `insert(Runnable w)` should be implemented according to the Pthreads functions `tpool_thread()` and `tpool_insert()` from Fig. 6.7. Additionally, a method `terminate()` should be provided to terminate the threads that have been started in the constructor. For each access to the task pool. a thread should check whether a termination request has been set.

Exercise 6.10. Transfer the pipelining pattern from Section 6.2.2 for which Figs. 6.10 - 6.13 give an implementation in Pthreads to Java. For the Java implementation, define classes for a pipeline stage as well as for the entire pipeline which provide the appropriate methods to perform the computation of a pipeline stage, to send data into the pipeline, and to retrieve a result from the last stage of the pipeline.

Exercise 6.11. Consider the client-server pattern given as Pthreads implementation in Figs. 6.15 - 6.18. Provide a similar implementation of the client-server pattern using Java threads. Define a suitable class that can be used to store requests for the server. Explain which synchronization operations are needed in the server implementation and give reasons that these synchronization operations are able to avoid deadlock situations.

Exercise 6.12. The program in Fig. 6.34 showing the implementation of a thread-safe buffer mechanism using the `wait()` and `notify()` methods of Java uses `notifyAll()` to wake up all produces or consumer thread. Does the program also work correctly, if `notify()` is used instead? Explain possible effects when using `notify()`.

Exercise 6.13. Consider the program in Fig. 6.54 implementing a postorder traversal of a binary tree. Modify the program in the following ways:

1. Add a suitable function `construct_tree()`.
2. Remove the `taskwait` pragma in the function `traverse()`. Add statements to print the visiting order of the nodes. Run the program several times. What can be observed about the visiting order?

3. Modify the program such that a preorder traversal of the tree results. This means that the computation for a node is executed before the child nodes are visited. Where must the `taskwait` pragma be positioned?
4. Modify the program such that an inorder traversal of the tree results. This means that for each node, first the left child node is visited, then the computation for the node is executed, then the right child node is visited.

Exercise 6.14. Consider the following OpenMP program piece:

```
int x=0;
int y=0;

void foo1() {
#pragma omp critical (x)
    { foo2(); x+=1; }
}
void foo2() {
#pragma omp critical(y)
    { y+=1; }
}
void foo3() {
#pragma omp critical(y)
    { y-=1; foo4(); }
}
void foo4() {
#pragma omp critical(x)
    { x-=1; }
}
int main(int argx, char **argv) {
    int x;
#pragma omp parallel private(i) {
    for (i=0; i<10; i++)
        { foo1(), foo3(); }
}
printf(''%d %d \n'', x,y )
}
```

We assume that two threads execute this piece of code on two cores of a multi-core processors. Can a deadlock situation occur? If so, describe the execution order which leads to the deadlock. If not, give reasons why a deadlock is not possible.

Chapter 7
General Purpose GPU Programming

About this Chapter

Graphics Processing Units (GPUs) are available in almost all current hardware platforms, from standard desktops to computer clusters and, thus, provide easily accessible and low-cost parallel hardware to a broad community. This chapter gives an overview of the architecture of GPUs and introduces the CUDA (Compute Unified Device Architecture) programming model including synchronization concepts and shared memory organization, CUDA thread scheduling, and efficient memory access and tiling techniques.

Graphics Processing Units (GPUs) have originally been designed and used for graphics applications. However, today there is an increasing importance for parallelizing applications from scientific computing and scientific simulations. The usage of a GPU for this purpose is often referred to as General Purpose GPU (GPGPU). Especially for data parallel programs there can be a considerable increase of efficiency due to the inherent SIMD programming model of GPUs. This efficiency improvement is mainly caused by the specific hardware design of GPUs, which has been optimized for big data of graphics applications and a high throughput of floating point operations to be executed by a large number of threads. Nowadays, GPUs may comprise thousands of cores executing these threads. A brief overview of the current architecture design of GPUs is provided in the first Sect. 7.1.

The first programmable GPUs were introduced in 2001 by NVIDIA. In the beginning, the programming effort for implementing general non-graphics applications and simulations on a GPU was extremely high, since programming environments, such as DirectX or OpenGL, designed for graphics applications had to be used. More recent programming environments for GPUs are CUDA (Compute Unified Device Architecture) and OpenCL (Open Computing Language). CUDA is a more generic parallel programming environment supported by NVIDIA GPUs for new generations since 2007 [38] and can also be emulated on CPUs. OpenCL has jointly been developed as a standardized programming model for GPUs by several industrial partners, including Apple, Intel, AMD/ATI and NVIDIA. This chapter provides an introduction to parallel programming with CUDA according to [39, 130, 196, 1] in Sect. 7.2 - 7.5 followed by a brief introduction to OpenCL written from a CUDA perspective according to [91, 156] in Sect. 7.6.

© The Author(s), under exclusive license to Springer Nature Switzerland AG 2023   423
T. Rauber, G. Rünger, *Parallel Programming*, https://doi.org/10.1007/978-3-031-28924-8_7

7.1 The Architecture of GPUs

GPUs have evolved from graphics accelerators with the main purpose to do graphics applications well. This explains why the architecture of manycore GPUs has been developed quite independently from the architecture of general CPUs. Because of the high potential of data parallelism in graphics applications, the design of GPU architectures has used several specialized processor cores much earlier than this has been done for the architecture of CPUs. Today, a single GPU may comprise several thousands of compute cores, which is a much higher number of cores than used in current CPU technology.

GPUs are provided in two basic types, which are integrated GPUs (iGPU) and discrete GPUs. Integrated GPUs are embedded into the processor chip as a separate unit. There is no separate memory for the iGPU. Instead, the iGPU shares its memory with the CPU. This facilitates data transfer between CPU and iGPU [122]. Discrete GPUs are built on a separate chip. Therefore, they have their own memory that is separate from the CPU memory. In this case, data to be processed by the GPU must first be transferred from the main memory of the CPU into the GPU memory, which may cause a significant overhead if the data size is large.

The memory for discrete GPUs is often based on the GDDR (Graphics Double Data Rate) technology, which is optimized for high throughput in comparison to the DDR technology, see Sect. 2.3, which is optimized for low latency [1]. The specifications for the GDDR technology come also from JDEC. In 2022, the newest GPU memory technology available are GDDR6 and GDDR7. The announced data rate of GDDR7 is about 1700 GByte/s compared to 51.2 GByte/s for the DDR5 specifications, see Sect. 2.3. Figure 7.1 illustrates the different memory organizations of discrete GPUs and iGPUs. A detailed description of the integration of iGPUs onto the processor chip and the interaction with the CPU as well as the architecture of the Intel Gen11 iGPU is given in [118]. In the following discrete GPUs are considered in more detail.

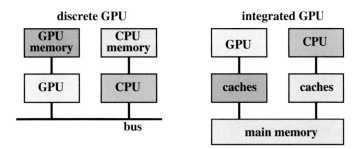

Fig. 7.1 Memory organization of discrete GPUs (left) and iGPUs (right) according to [1].

In addition to graphics processing, GPUs can also be employed for general nongraphics applications. This is useful if the potential of data parallelism of the nongraphics application is large enough to fully utilize the high number of compute cores in a GPU. Scientific simulations internally processing numerical calculations, such as those given in Chapter 8, often have this property. Numerous example implementations have shown that those applications can benefit from the GPU architecture resulting in a much better compute performance than on a CPU. Several supercomputers on the Top500 list, see Sect. 2.10.2 use GPUs as accelerators. The trend to use GPUs for general numerical applications has inspired GPU manufacturers, such as NVIDIA, to develop the programming environment CUDA and OpenCL. These parallel programming environments combine the effort to support non-graphics applications on GPUs on the one hand and to provide a specific programming model which is adequate for the architecture of GPUs.

Both CUDA and OpenCL separate a program into a CPU program (the host program), which include all I/O operations or operation for user interaction, and a GPU program (the device program), which contains all computations to be executed on the GPU. The simplest case of interaction between host program and device program is implemented by a host program that first copies data into the global memory of the GPU and then calls device functions to initiate the processing on the GPU. These device functions take the parallel architecture of the GPU into account to appropriately map the data parallel computations to compute cores. Important aspects are also the memory organization of the GPU and the specific data transfer between CPU and GPU, which will be discussed later in this chapter. This section is devoted to the basic concepts of the architecture of GPUs and concentrates specifically on NVIDIA-GPUs. More detailed architecture descriptions are given in [137, 16, 40, 41, 104].

A GPU comprises several multi-threaded SIMD processors, each processing independent sequences of compute instructions. These SIMD processors are called **streaming multiprocessors (SM)** by NVIDIA. The actual number of SIMD processors incorporated in a GPU depends on the specific GPU model. For example, the NVIDIA H100 GPU belonging to the family of the Hopper architecture has up to 132 independent SIMD processors, depending on the configuration. Each of the SIMD processors has several SIMD function units which can execute the same SIMD instruction on different data. The data to be processed has to be available in the local registers associated to the SIMD function unit. Specific transfer operations are provided to initiate the data transfer from the GPU memory into the registers. The actual transfer may require several machine cycles because the data may reside in the global memory of the GPU (off-chip) from which they have to be fetched. Recent GPU architectures contain a cache memory hierarchy, so that some data transfer operations can use cached data and are faster than data accesses to non-cached data.

At first glance, it seems to be sufficient to provide one thread of control (SIMD thread) to one SIMD processor, since parallelism results because each SIMD function unit executes the same instruction on different data. However, the data transfer operations will cause waiting times of uncertain length. To hide these waiting times, SIMD processors are able to execute several independent SIMD threads. A SIMD thread scheduler picks a SIMD thread ready for execution and initiates the execution of the next SIMD instruction of this thread. Each of the SIMD threads uses an independent set of registers. This method is a special form of multithreading, see Sect. 2.3.

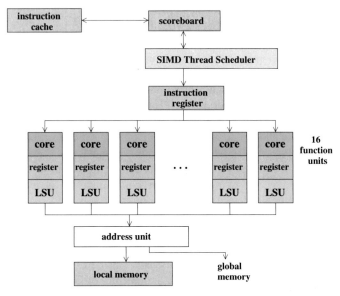

Fig. 7.2 Block diagram of an SIMD processor according to [103]. The SIMD processor has 16 function units (FU), each of which has a separate set of registers and a separate Load-Store-Unit (LSU).

In each execution step, the SIMD thread scheduler can select a different SIMD thread, since the SIMD threads are independent of each other. To support the selection, the SIMD thread scheduler uses a scoreboard. For each SIMD thread, the scoreboard contains the current instruction to be executed and the information whether the operands of this instruction reside in registers or not. The maximum number of SIMD threads to be supported depends on the size of the scoreboard. The scoreboard size is an architectural feature and a typical size is 32 [103]. The actual number of independent SIMD threads is determined by the application program as described later in this chapter.

The number of function units of an SIMD processor depends on the specific GPU model. Figure 7.2 illustrates a simplified internal organization of a single SIMD processor. A total number 32768 of 32-Bit registers is provided so that each of the 16 function units owns 2048 physical registers. Two neighboring physical register can be used as one 64-Bit register. The 2048 registers of one function unit are distributed among the SIMD threads available in the current program execution. In case of 32 SIMD threads, each of these threads can use 64 separate registers of each function unit for the storage of its data. The registers are dynamically assigned to the SIMD threads when they are created. A local memory (on-chip) is available for each SIMD processor, providing fast access for the SIMD threads running on this SIMD processor. The global memory (off-chip) is shared by all SIMD processors and can be accesses also by the CPU. Access to the global memory is much slower than access to the local memories of the SIMD processors.

The actual design of a GPU of the NVIDIA Hopper architecture is more complex than the illustration in Fig. 7.2. Figure 7.3 which shows a block diagram of a single SIMD processor of the Hopper architecture. Each SIMD processor is organized into quadrants. Each quadrant has a separate SIMD thread scheduler and dispatch unit for launching SIMD instructions to the functional units. The dispatch unit can handle 32 threads per cycle. Each quadrant has a separate set of 16384 32-bit registers to store the data for the different threads. There is also combined shared memory and L1 data cache of size 256 KB. Moreover, each quadrant has an L0 instruction cache and there is also a global L1 instruction cache for all four quadrants of an SIMD processor. There is also a combined L1 data cache and shared memory of size 256 KB for each SIMD processor. Corresponding to the quadrant organization, there are four separate sets of function units. The functional units provided are of different type as described in the following. Together, the four quadrants provide

- 128 32-bit floating-point cores; these cores support 32-bit floating-point operations (FP32), 64-bit FP operations (FP64), 16-bit FP operations (FP16), and 8-bit FP operations (FP8);
- four so-called Tensor cores for fast matrix multiply and accumulate operations; again, different data formats (FP8, FP16, FP32, FP64) are supported;
- 64 32-bit integer cores; 16-bit integer (INT16) and 8-bit integer (INT8) operations are also supported;
- 32 transfer units (load-store units, LD/ST) for the transfer of data between registers and memory system,
- four special function units (SFU) that are dedicated to the execution of special functions, such as sinus, cosinus, square root or reciprocal value.

Thus, each SIMD processor exhibits a heterogeneous design with functional units of different functionality integrated. This corresponds to the recent development towards heterogeneous designs as described in Section 2.6.3. Figure 7.3 shows the block diagram of a SIMD processor of the Hopper architecture according to [41, 16].

The peak performance of the NVIDIA Hopper H100 GPU is 66.9 TFlops for FP32 and 33.5 TFlops for FP64, see Table 7.1 for a comparison of different NVIDIA GPUs. There is also a version of the NVIDIA Hopper architecture for the consumer segment, the NVIDIA Ada Lovelace architecture. A detailed performance comparison between CPUs and GPUs is given in [104].

A new feature of the NVIDIA Hopper H100 GPU is the introduction of a Thread Block Cluster architecture that combines a group of Thread Blocks to a cluster, see Fig. 7.4 for an illustration [41]. A single Thread Block contains multiple threads that run concurrently on a single SIMD processor. These threads can easily exchange data using the shared memory of the SIMD processors. A Thread Block Cluster is a group of Thread Blocks that are guaranteed to be concurrently scheduled onto a group of SIMD processors. To support these Thread Block Clusters, the H100 architecture uses the concept of GPU Processing Clusters (GPCs), where each GPC

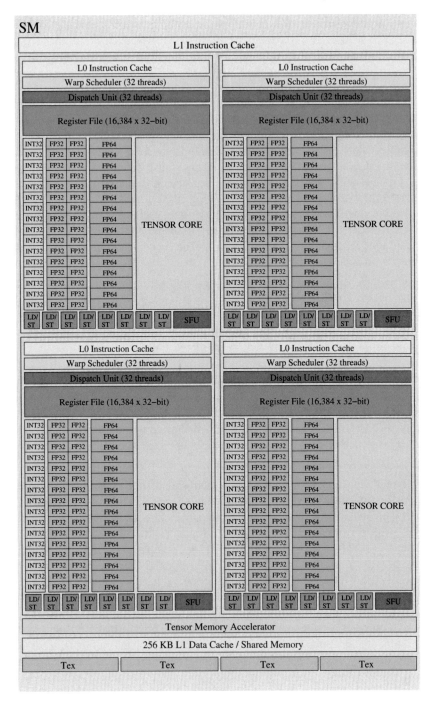

Fig. 7.3 Block diagram of an SIMD processor of the NVIDIA Hopper architecture according to [41]. The entire GPU consists of up to 132 of these SIMD processor.

combines a group of SIMD processors that are always physically close together. The goal of GPCs is to enable efficient cooperation of threads across multiple SIMD processors. The hardware provides a dedicated network for a fast data exchange between threads in a cluster. To support a fast data exchange between SIMD processors, the Hopper architecture supports distributed shared memory (DSM), see Sect. 2.5.2, by combining the shared memory of the different SIMD processors to a single virtual address space, thus enabling direct data exchange between the SIMD processors without the need to involve the global memory into the data exchange.

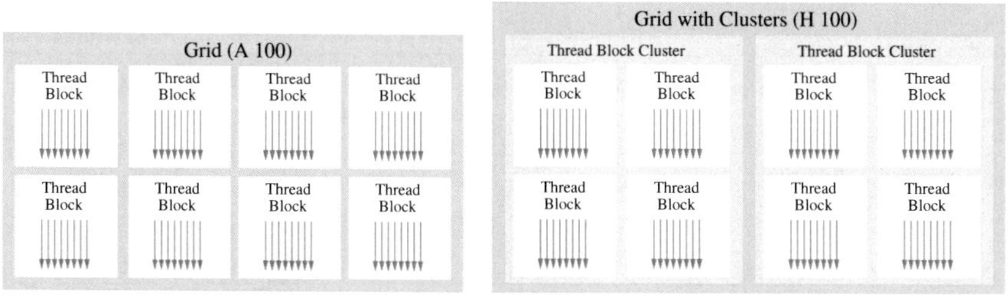

Fig. 7.4 Comparison between the original CUDA programming model (left) and the extended CUDA programming model with additional clusters introduced by the NVIDIA Hopper architecture (right) according to [41].

The use of the parallel units of GPU architectures can result in an efficient execution if the program is coded so that the SIMD processors can be fully employed. Especially, the application program has to provide an appropriate number of SIMD threads so that the SIMD thread scheduler of each SIMD processor has enough threads available for switching between them so that it is possible to hide the latency of memory accesses. For efficiency purposes, a good organization of the memory layout of the application data is also important. This can be explicitly planned by the application programmer using appropriate commands in CUDA or OpenCL and can be supported by specific parallel programming techniques, as described in the following sections of this chapter.

7.2 Introduction to CUDA programming

The coarse structure of a CUDA program is built up of phases which are either executed on a host or on one of the devices. The host is a traditional central processing unit (CPU) and the program phases for the host are C programs which can be compiled by a standard C compiler. The devices are massively parallel processors with a large number of execution units, such as GPUs, for processing massively data parallel program phases. The device code is written in C and CUDA specific

Table 7.1 Summary of important characteristics of several NVIDIA GPUs of different architectures according to [137]: NVidia Tesla P100 (Pascal architecture), NVidia Tesla V100 (Volta architecture), NVidia A100 (Ampere architecture), and NVidia H100 SXMX (Hopper architecture), see also [137, 16] and [104]. The TFlops values, denoted as TF in the table, are given for floating point operations of single precision (32 bit) and are the maximum reachable values. The consumption in Watt is the thermal design power (TDP) of the manufacturers, which is actually the value for the cooling system but can also serve as indication for the maximum power consumption.

Nvidia GPU	P100 (Pascal)	V100 (Volta)	A100 (Ampere)	H100 (Hopper)
number of transistors	$15.3 \cdot 10^9$	$21.1 \cdot 10^9$	$54.2 \cdot 10^9$	$80 \cdot 10^9$
GPU die size	610 mm2	828 mm2	815 mm2	814 mm2
SIMD processors	56	80	108	132
SIMD cores per SIMD processor	64	64	64	128
total number of SIMD cores	3584	5120	6912	16896
L2-Cache	4096 KB	6144 KB	40960 KB	50 MB
Performance	10.6 TF	15.7 TF	19.5 TF	60 TF
memory size	16 GB	16/32 GB	40 GB	80 GB
shared memory size per SM	64 KB	up to 96 KB	up to 154 KB	up to 228
register file size per SM	256 KB	256 KB	256 KB	256 KB
register file size GPU	14336 KB	20480 KB	27648 KB	33792 KB
Bandwidth memory	720 GB/sec	900 GB/sec	1555 GB/sec	3352 GB/sec
power consumption (TDP)	300 W	300 W	400 W	700 W

extensions for specifying data parallel executions. A data parallel function to be executed on the device is called a kernel function or simply kernel. The parallelism of a CUDA program resides in the kernel functions, which typically generate a large number of CUDA threads. CUDA threads are light weight threads and require only a few cycles for generating and scheduling threads, in contrast to CPU threads which need thousands of cycles for those tasks.

A CUDA program given in a file *.cu consisting of a host program and kernel functions is compiled by the NVIDIA-C-compiler (nvcc), which separates both program parts. The kernel functions are translated into PTX (Parallel Thread Execution) assembler code. PTX is the NVIDIA assembler language which, similar to the X86 assembler language, provides a set of machine commands that ensure the compatibility of different generations of NVIDIA GPUs. A detailed description of PTX is given in [104]. Calls of kernel functions in the host program are translated into CUDA runtime system calls for starting the corresponding function on the GPU. The host program is then translated by a standard C compiler.

The execution of a CUDA program starts by executing the host program which calls kernel functions to be processed in parallel on the GPU. The call of a kernel function initiates the generation of a set of CUDA threads, called a grid of threads. A grid is terminated as soon as all threads of the grid have finished their part of the kernel function. After invoking a kernel function the CPU continues to process the host program which might lead to the call of other kernel functions. CUDA extends the C function declaration syntax to distinguish host and kernel functions. The CUDA-specific keyword __global__ indicates that the function is a kernel that can be called from a host function to be executed on a GPU. The keyword __device__ indicates that the function is a kernel to be called from another kernel or device function. Recursive function calls or indirect function calls through pointers are not allowed in these device functions. A host function is declared using the keyword __host__ and is simply a traditional C function that is executed on the host and can only be called by another host function. All function without any keyword are host function by default.

In order to execute a kernel function on a device, the data have to reside in the device memory. Thus, a CUDA program typically contains data transfer operations from the host to the GPU memory and also from the GPU to the CPU memory to transfer data results back to the host. These data transfer operation are explicitly coded in CUDA programs using CUDA-specific functions.

Before performing the data transfer from the host to the GPU memory, an appropriate amount of memory has to be allocated. This is done by the function

$$\texttt{cudaMalloc(void **, size\_t)}$$

which is called by the host program to allocate memory in the global memory of the GPU. The function cudaMalloc() has two parameters: The first parameter is a pointer to the memory to be allocated and the second parameter specifies the size in Bytes of the memory allocated. The function

$$\texttt{cudaFree(void *)}$$

is called to free the storage space of the data objects given in the parameter after the computation is done. A data transfer from host to GPU is requested by calling the copy function

$$\texttt{cudaMemcpy(void *, const void *, size\_t, enum cudaMemcpyKind)}$$

with four parameters:

- a pointer to the destination of the transfer operation,
- a pointer to the source of the transfer operation,
- the number of bytes to be copied, and
- a predefined symbolic constant specifying the type of the memory operation to be used in the transfer operations, e.g. from host to device.

Types provided for the data transfer are cudaMemcpyHostToDevice for the transfer from host to device and cudaMemcpyDeviceToHost for the transfer from device to host. Copy operations from host to host or from device to device are also possible. However, a copy operation between two different GPUs is not possible.

After having initiated the data transfer into the global memory of the GPU, the host program can invoke a kernel function on the GPU working on those data. The invocation of a kernel function contains an execution configuration which specifies the grid organization of the threads to be generated for the execution of that kernel function. Threads in a grid are organized as a two-level hierarchy. At the first level, each grid consists of several thread blocks all of which have the same number of threads. The second level is the organization of the threads within each thread block, which is identical for all thread blocks of the same grid.

Blocks of a grid have a two-dimensional structure, in which each thread block has a unique two-dimensional coordinate given in the CUDA specific keywords `blockIdx.x` and `blockIdx.y`. This holds for CUDA Versions 2 and older; since CUDA Version 3, a three-dimensional block structure is supported. The threads within a thread block are, in turn, organized in a three-dimensional structure and the unique three-dimensional coordinate of each thread is given in the CUDA specific variables `threadIdx.x`, `threadIdx.y` and `threadIdx.z`. The maximum number of threads in a thread block is 512 threads (up to Version 2) and 1024 threads (since Version 3). According to this two-level hierarchy, each thread of a grid is uniquely identified by the block coordinates of the thread block it belongs to and its thread coordinates within this block. These coordinate values can be used in kernel functions to distinguish the parallel threads resulting in a data parallel execution.

The size of the thread grid and the thread blocks generated for the execution of a specific kernel invocation are defined in the execution configuration mentioned before. For the specification of an execution configuration, two `struct` variables of type `dim3` are declared; `dim3` is an integer vector-type built up from the vector-type `unit3` and initiated with the value 1, if not specified otherwise. These `struct` parameters describe the two- or three-dimensional organization of the thread grid and the thread blocks and are included in the kernel call syntax surrounded by $<<<$ and $>>>$ as illustrated in the following example:

```
// execution configuration
dim3 gsize(gx, gy);
dim3 bsize(bx, by, bz);
// call to a kernel function
KernelFct <<<gsize, bsize>>>(...);
```

The two-dimensional grid structure for the execution of kernel `KernelFct` is given in `gsize` and has size gx × gy. The three-dimensional structure of the block is given in `bsize` and has size bx × by × bz. The sizes of the current grid and its blocks are stored in the CUDA specific variables `gridDim` and `blockDim`. More precisely, the sizes of the grid are given by `gridDim.x` and `gridDim.y` and have the values gx and gy, respectively, in the example given above. Analogously, the size of the thread blocks are stored in `blockDim.x`, `blockDim.y` and `blockDim.z` and contain the values bx, by and bz, respectively, in the example.

Figure 7.5 illustrates a specific execution configuration for a grid of size `dim3 gsize(3,2)` and thread blocks of size `dim3 bsize(4,3)`. The third dimension is not given any size and, thus, the size of this dimension in 1. The invocation of a kernel function `KernelFct` with this execution configuration results in the assignment `gridDim.x=3`, `gridDim.y=2`, `blockDim.x=4`, `blockDim.y=3`, and `blockDim.z=1`. Typical grids contain many more blocks and the values for

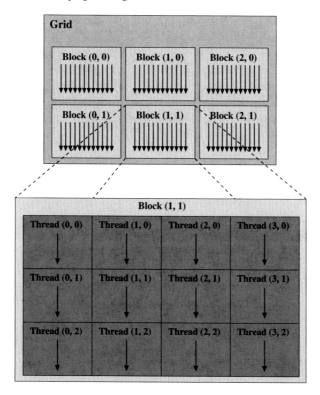

Fig. 7.5 Illustration of a CUDA execution configuration for grid dimension dim3 gsize(3, 2) and block dimension dim3 bsize(4, 3).

gridDim.x and gridDim.y can be between 0 and 65535. Correspondingly, also the possible value for the block identifier blockIdx.x and blockIDx.y are limited and can be between 0 and gridDim.x-1 and 0 and gridDim.y-1, respectively. It has to be noticed that the pair of identifiers (blockIdx.x, blockIdx.y) denote the column in the first component and the row in the second, which is the reverse order than in the specification of field elements.

The example execution configuration in Fig. 7.5 generates 3*2=6 thread blocks with 4*3=12 threads each, resulting in a total of 72 threads. If only a one-dimensional grid with one-dimensional thread blocks is needed for a kernel function call, the size of the grid and the blocks can be used directly in the execution configuration. For example, the call

```
KernelFct <<<8, 16>>>(...);
```

is identical to the call

```
dim3 gsize(8, 1);
dim3 bsize(16, 1);
KernelFct <<<gsize, bsize>>>(...);
```

As mentioned before, the invocation of a kernel function initiates the generation of a grid of threads according to the given execution configuration, and the sizes of the grid and the thread blocks as well as the identifiers are stored in the variables gridDim, blockDim, blockIdx and threadIdx. The threads generated can use these variables to execute the kernel function in an SIMD programming model, which is illustrated by the following example vector addition in Fig. 7.6. This CUDA program also illustrates the cooperation between host program and kernel function as well as the use of the execution configuration.

Example: The CUDA program in Fig. 7.6 implements the addition of two integer vectors a and b of length N and stores the result in integer vector c of length N. The host program starts with the declaration of the arrays a, b, c of length N on the CPU and then allocates memory of size N * sizeof(int) for corresponding vectors ad, bd, cd in the global memory of the GPU, using the CUDA function cudaMalloc(). Next, the input vectors a and b are copied into the vectors ad and bd in the GPU memory using the CUDA function cudaMemcpy() with constant cudaMemcpyHostToDevice. The Kernel function vecadd() is called using the configuration $<<< 10, 16 >>>$, i.e., a grid with 10 blocks having 16 threads each is generated. The result vector c is copied back to the CPU memory using cudaMempcy with constant cudaMempcyDeciceToHost. The function write_out(c) is meant to output the result. Finally, the host programs free the data structures ad, bd, and cd in the GPU memory by calls to cudaFree(). The kernel function vecadd has three parameters of type pointer to int. Each of the threads generated in the thread grid of size 10×16 executes the function vecadd() in the SIMD programming model, see Section 3.3.2. The actual execution of the thread differ because of different thread identifier. A program specific thread identifier tid is declared and has the value threadIdx.x + blockId.x * blockDim.x. This results in a single one-dimensional array of program specific thread identifiers.

The computations to be executed are the additions of two values a[i] and b[i], $i \in \{0, \dots, N-1\}$. In each step of the for loop, the threads process consecutive elements of the vectors a and b in parallel, starting with the element a[i] and b[i] with i=tid. If the length N of the vectors is larger than the number of threads, i.e. blockDim.x * gridDim.x = 10 * 16, each threads executes N/(blockDim.x * gridDim.x) additions, assuming N is divisible by blockDim.x * gridDim.x. The assignment of threads to additions implements a cyclic data distribution for the one-dimensional arrays a, b and c, see Sect. 3.5.

It should be noticed that the thread generation in CUDA is implicit in contrast to an explicit thread generation in Pthreads for example. The thread generation in CUDA is achieved by just providing an execution configuration.   □

```
#include <stdio.h>
#define N 1600

__global__ void vecadd (int *a, int *b, int *c) {
    int tid = threadIdx.x + blockIdx.x * blockDim.x;
    int i;
    for (i=tid; i<N; i += blockDim.x * gridDim.x)
        c[i] = a[i] + b[i];
}

int main( void ) {
    int a[N], b[N], c[N];
    int *ad, *bd, *cd;

    cudaMalloc ((void **) &ad, N * sizeof(int));
    cudaMalloc ((void **) &bd, N * sizeof(int));
    cudaMalloc ((void **) &cd, N * sizeof(int));
    read_in (a); read_in (b);
    cudaMemcpy (ad, a, N * sizeof(int), cudaMemcpyHostToDevice);
    cudaMemcpy (bd, b, N * sizeof(int), cudaMemcpyHostToDevice);
    vecadd<<<10,16>>> (ad, bd, cd);
    cudaMemcpy (c, cd, N * sizeof(int), cudaMemcpyDeviceToHost);
    write_out (c);
    cudaFree (ad);
    cudaFree (bd);
    cudaFree (cd);
}
```

Fig. 7.6 A simple CUDA program for the addition of two vectors a and b.

7.3 Synchronization and Shared Memory

The threads generated for the parallel execution of a kernel function can be synchronized using the barrier synchronization operation
    __syncthreads();
A barrier synchronization involves all threads of the same thread block. Threads
of different blocks of the same grid of threads, however, cannot be synchronized
by a barrier synchronization. The barrier synchronization of one block of threads
has the effect that the threads wait at the synchronization call until all other
threads of the block also reach this program point. It is important that the function
__synthreads() is actually called by all threads of the block, since the execution
can only proceed if all threads of the block synchronize at this function call. This
constraint requires a careful coding of if-then-else constructs in a program with
SIMD threads such that all threads are able to reach the synchronization point.

  If a call to __synthreads() would reside in the then part of an if-then-else
construct, then only those threads executing the then part would reach the synchronization point and would wait for other threads of the block. However, if some of
the thread do not execute the then part, the waiting threads can never continue their

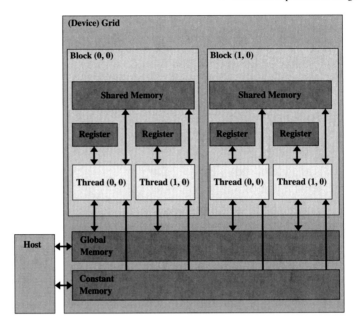

Fig. 7.7 Memory organization of the CUDA programming model with the global memory, the constant memory, the shared memories assigned to blocks, and registers assigned to single threads.

execution, since some of the threads of the block can never reach the synchronization point. Even if an `if-then-else` construct is used and both parts, the `then` part and the `else` part, contain a call to the function `__synthreads()`, these two calls define different synchronization points, each of which is reached only by a subset of the threads of the blocks so that the execution cannot be continued.

The design of an execution configuration, especially the subdivision of a grid into thread blocks is essential for the potential to synchronize, as it has been explained above. Thus, we have already introduced two reasons why different execution configurations lead to different executions for the same kernel: these are the increase of the number of threads in a grid for a potential SIMD parallelism and the definition of a potential synchronization structure. Another important property to be exploited by the thread block structure is the memory organization of the CUDA programming model. This is introduced next.

The CUDA programming model provides the CPU memory and the GPU memory. The GPU memory is organized in a hierarchy of different memory types, as illustrated in Fig. 7.7. The global memory and the constant memory can be accessed by the CPU for reading or writing, as already introduced for the function `cudaMemcpy()`. The global memory can also be written and read by the GPU. The constant memory supports read-only access by the GPU, which has a short latency. The registers and the shared memory are on-chip memories and have short access times. Registers are assigned to individual threads and a thread can only access its own registers. The data stored in registers are declared as private variables.

The shared memory is assigned to an entire block of threads and the shared data within the shared memory can be accessed by all threads of the block. The threads of the same thread block can exchange data through the shared variables. The entire shared memory is broken up into fragments so that each block has its own shared memory. Figure 7.7 shows two thread blocks (0, 0) and (1, 0) and two shared memories, one for each block.

The storage of data in the different memories of the CUDA memory organization is declared explicitly in a CUDA program by a CUDA specific declaration system. All scalar variables declared in a kernel or a device function are private variables to be stored in registers if possible. There is a copy of such a private variable for each thread. The lifetime of a private variable is within a kernel invocation and there is no way to access it after the kernel function has been finished and its threads are terminated. Private array variables are also declared within a kernel or a device function and a separate copy of the variable in stored for each of the threads in the global memory.

A variable declaration preceded by the keyword __shared__ declares a shared variable, for which several versions are created, one for each block. All threads of a block can access the variable copy of their block. The lifetime of shared variables is also restricted to the execution of the kernel function. Since the access to shared variables is very fast, these variables should be used for data that are accessed frequently in the computation.

The keyword __constant__ is used for the declaration of variables in the constant memory. These variables are declared outside any function and their lifetime is the entire execution time of the program. The size of the constant memory is currently limited to 65536 Bytes. Variables in the global memory are declared using the keyword __global__. Global variables can be accessed by all threads during the entire execution time of the program. The access to global variables cannot be synchronized in CUDA, i.e., there are no further synchronization mechanisms, such as locking methods. The only way to synchronize threads is the barrier synchronization introduced above. There are, however, several atomic operations for integer variables available and also an atomic addition for floating point variables that allow a synchronized access. For the atomic operations, the variables can reside in the global memory or, since CUDA Version 2, in the shared memory. The next example implementing the multiplication of two vectors, i.e. a scalar product, illustrates the usage of the shared memory and the block-oriented synchronization.

Example: Figure 7.8 shows a CUDA program for the multiplication of two vectors a and b of length N. Such a multiplication of two vectors is also called scalar product, see Sect. 3.7. The host program for the scalar product first allocates memory for the vectors a and b in the CPU memory as well as memory for the corresponding vectors ad and bd in the global memory of the GPU. Additionally, memory space is allocated for the vector part_c of length n_blocks on the GPU. The length n_blocks denotes the number of thread blocks in the execution configuration of the kernel

```
#include <stdio.h>
const int N = 32 * 1024;
const int threadsPerBlock = 256;
const int n_blocks = N / threadsPerBlock;

__global__ void scal_prod (float *a, float *b, float *c) {
    __shared__ float part_prod[threadsPerBlock];
    int tid = threadIdx.x + blockIdx.x * blockDim.x;
    int i,size, thread_index = threadIdx.x;
    float part_res = 0;

    for(i=tid; i<N; i += blockDim.x * gridDim.x)
        part_res += a[i] * b[i];
    part_prod[thread_index] = part_res;
    __syncthreads();
    size = blockDim.x/2;
    while (size != 0) {
        if(thread_index < size)
            part_prod[thread_index] += part_prod[thread_index + size];
        __syncthreads();
        size = size/2;
    }
    if (thread_index == 0)
        c[blockIdx.x] = part_prod[0];
}

int main (void) {
    float *a, *b, c, *part_c;
    float *ad, *bd, *part_cd;
    a = (float *) malloc (N*sizeof(float));
    b = (float *) malloc (N*sizeof(float));
    part_c = (float*) malloc( n_blocks*sizeof(float));
    cudaMalloc ((void **) &ad, N*sizeof(float));
    cudaMalloc ((void **) &bd, N*sizeof(float));
    cudaMalloc ((void **) &part_cd, n_blocks*sizeof(float));
    read_in (a); read_in (b);
    cudaMemcpy (ad, a, N*sizeof(float), cudaMemcpyHostToDevice);
    cudaMemcpy (bd, b, N*sizeof(float), cudaMemcpyHostToDevice);
    scal_prod<<<n_blocks,threadsPerBlock>>>( ad, bd, part_cd);
    cudaMemcpy (part_c, part_cd,
                n_blocks*sizeof(float), cudaMemcpyDeviceToHost);
    c = 0;
    for (int i=0; i<n_blocks; i++) c += part_cd[i];
    write_out (c);
    cudaFree (ad);
    cudaFree (bd);
    cudaFree (part_cd);
}
```

Fig. 7.8 CUDA program for the multiplication of two vectors a und b.

function scal_prod() and the vector part_cd is intended to store intermediate results of those thread blocks. After reading in the vectors a and b, their content is copied to the vectors ad and bd on the GPU using the function cudaMemcpy(). This is analogous to the program shown earlier in Fig. 7.6.

The scalar product is computed by the kernel function scal_prod() using an execution configuration with a one-dimensional block organization in the grid and a one-dimensional organization of the threads within each block. It is assumed that the number threadsPerBlock of threads in a block is a power of two and the length N of the input vectors is divisible by this number of threads per block. The array part_prod is a shared variable and is declared in the kernel function so that a private copy is created for each block.

The scalar product is calculated in three computation phases. The first two of these phases are executed by the kernel function scal_prod() on the GPU and the third phase is executed on the CPU providing the final result. The first phase on the GPU starts with the calculation of intermediate results of the scalar product. Each thread computes the scalar product of a part of the vectors ad and bd according to a data cyclic distribution of these vectors and stores the intermediate results at position thread_index in array part_prod. This calculation is performed in a for-loop whose index starts with the specific thread identifier tid and has a step size blockDim.x * gridDim.x resulting in the cyclic distribution. For the specific case in Fig. 7.8, each thread computes only one element, since the number of threads is identical to the length of the input array, i.e. N = 32*1024. The first phase is completed by a synchronization with __syncthreads().

The second phase on the GPU adds up the intermediate results in each block. This is done in a while-loop, in which the number of values to be added is halved in each iteration step: In the first iteration step, halve of the threads in a block add two intermediate results each and store the result back into the array part_prod in the position of the first operand. In the second iteration step, a quarter of the threads of a block add two values of part_prod and store the result back. After $\log_2(\text{blockDim.x})$ steps of the while-loop, a single thread adds the last two values resulting in the final result of the scalar product of the thread block and stores it into part_prod[0]. The threads to be involved actively in an iteration step of the while loop are chosen according to their thread identifiers thread_index in the block. For a correct computation, each iteration step is completed by a synchronization of the threads in the block.

At the end of the second phase, there is a final result in the shared variable part_prod[0], i.e. there is one final value for each block. These values are copied back into the CPU memory where they are stored in array part_c of length n_blocks. These values in part_c are accumulated on the CPU and provide the final result of the scalar product in the scalar variable c. Figure 7.8 shows a CUDA program implementing the three computation phases described. Figure 7.9 illustrates the computation phases for example data structures of specific sizes. □

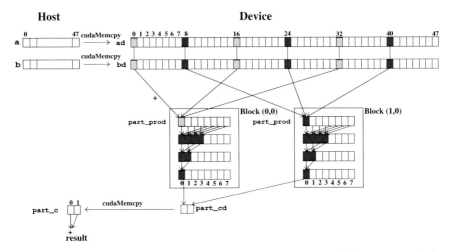

Fig. 7.9 Illustration of the computations and data structures of the CUDA program implementing a scalar product in Fig. 7.8 with a vector size N=3*16 and an execution configuration
<<< 2, 8 >>>, i.e. 2*8=16 threads, in the call of kernel function scal_prod. The figure shows
the computations of the host on the left side and the computations of the GPU on the right
side. The vectors a and b of length 48 are processed by 16 threads organized in the two blocks
(0, 0) and (1, 0). Each of the threads computes its unique thread identification number tid
as a number between 0 and 15. As an example it is illustrated that the thread with tid=0
computes the partial scalar product of the vector elements 0, 16 and 32 and stores the result
in the global memory of the GPU and that the thread with tid=8 processes vector elements 8,
24 and 40. After the computations of the partial scalar products by all threads, the resulting
values are accumulated in each of the blocks separately in $log_2\ 8 = 3$ steps. The result is stored
in array part_cd, which has one component for each block. After copying this array back into
the CPU memory, the CPU computes the final result of the scalar product by accumulating
the two values in array part_c.

7.4 CUDA Thread Scheduling

In typical kernel function executions, the number of threads generated for that kernel
exceeds the number of compute units, i.e. the number of streaming multiprocessor or
SIMD function units, respectively, see Sect. 7.1. Therefore, there has to be a method-
ology to assign threads to compute units. This is implemented by the CUDA thread
scheduling. The CUDA thread scheduling exploits that different thread blocks of a
grid are independent of each other and can be executed in any execution order. The
scheduling of thread blocks can assign as many executable thread blocks to SIMD
processors as possible for the specific hardware.

Basically, the threads within the same thread block are also independent of each
other if there is no barrier synchronization. However, the CUDA system does not
schedule individual threads but uses an additional concept to divide blocks into
smaller sets of threads, called warps, and schedules these warps. For current GPU architectures, the size of warps is 32. Threads of a warp are assigned and processed together and the scheduling manages and determines the execution order. The CUDA

thread scheduling is part of the CUDA architecture and programming model and cannot be influenced directly by the application programmer.

The division of thread blocks into warps is based on the thread index `threadIdx`. For a one-dimensional block, 32 threads with consecutive values of `threadIdx.x` are combined to form one warp. If the number of threads is not divisible by 32, then the last warp is filled with virtual threads. A two-dimensional block is first linearized in a row-wise manner and then consecutive threads are combined for a warp. Analogously, a three-dimensional block is linearized and subdivided into warps. Thus, depending on the organization of a grid into blocks the same set of threads can result in different divisions into warps and a different scheduling.

Warps are executed in the CUDA computation model SIMT (single instruction, multiple threads). In this model, the hardware executes the same instruction for all threads of a warp and only then proceeds to the next instruction. The SIMT computation model is suited to exploit the hardware efficiently, see Sect. 7.1, especially if all threads of the warp have the same flow of control. If threads of the same warp have different control flow paths in an `if-then-else` construct, since some of the threads execute the `then` part and the other threads execute the `else` part, then the threads have different control flow paths, which have to be executed one after another in the SIMT model. This leads to longer execution times and should be avoided by a more suitable definition of threads. An example is the reduction operation accumulating the elements of an array in parallel. Such as reduction operation has been used for the second phase in the scalar product in Fig. 7.9.

In the loop implementing the reduction operation, the number of threads executing an iteration is reduced in every iteration step so that only half of the threads continue the computation in the next iteration. The other threads do not execute any operation. This leads to the situation that the threads have different control flow paths, which are either an addition operation or no operation. If the reduction operation is implemented such that each thread adds neighboring array elements, then each warp contains threads with two different control flow paths. In contrast, if the addition of array elements is organized as shown in the reduction operation of Phase 2 of the scalar product, see the illustration in Fig. 7.8, then all threads with smaller `threadIdx.x`-values perform an addition while threads with larger `threadIdx.x`-values do not perform any operation. For large numbers of threads this means that in the earlier iterations of the reduction the warps containing 32 threads each will contain either only threads performing an addition or only threads performing no operation. When the number of threads executing an addition decreases towards the end of the reduction and is finally smaller than 32, then this final warp has threads with different control flow paths.

7.5 Efficient Memory Access and Tiling Technique

The execution of a kernel function usually requires to access a lot of data from the global memory, caused by the data parallel execution model of the threads. Since these global memory accesses are expensive, data should be copied into the shared memory or the registers, which have a much faster access time. To support copy operations from the global memory into the shared memory or registers, CUDA provides a technique that combines neighboring data and copies them together, resulting in fast copy operations. This technique is called memory coalescing. The coalescing technique exploits the fact that at any point in time threads of a warp execute the same instruction. If this instruction is a load operation, then the hardware can detect whether the parallel load operations address neighboring memory locations in the global memory. The hardware then combines the load operations of neighboring memory locations to only one memory access, which is much faster than several single memory accesses.

To exploit the coalescing technique for efficient memory access, the application programmer should organize the data in the CUDA program in such a way that threads with neighboring thread identifiers to be combined in the same warp access neighboring elements of the arrays used in the program. This means that n threads $T_0, T_1, \ldots, T_{n-1}$ should access consecutive memory locations $M, M + 1, \ldots, M + n - 1$, where M denotes the first address of an array.

For the access to a two-dimensional array, threads with consecutive thread identifiers should access neighboring element in the rows. The rowwise storage of two-dimensional arrays then enables the hardware to combine the data access to one copy operation. In contrast, if threads in the same warp would access neighboring elements in the columns of an array, then a coalescing of the data accesses would not be possible, since neighboring column elements are stored in different places in the memory. More precisely, for a two-dimensional array with m columns, the elements of the same column are stored in memory locations with a distance of m, which avoids coalescing.

The data layout of CUDA programs should be designed such that coalescing is possible so that efficient programs can result. A well-known programming technique, which can be used for this purpose, is the tiling technique. In the tiling technique, a two-dimensional array is decomposed into smaller two-dimensional arrays of the same size, which are called tiles. The algorithm using the array has then to be modified such that the program execution fits to the tile structure. This means expecially that nested loops that access the two-dimensional data structure are modified so that smaller data sets are processed. The tiling techniques is illustrated in the following example program, which is a matrix matrix multiplication. First, a CUDA program without tiling technique is given in Fig. 7.10, and then the tiling technique is introduced in Fig. 7.13.

Example: The CUDA program `matmult.cu` in Fig. 7.10 contains a host program and the kernel function `MatMulKernel()` implementing a matrix multiplication of two matrices A and B of size N×N. The result is stored in matrix C. The host function `MatMul()` captures the interaction of the CPU with the GPU using copy operations and the call of the kernel function `MatMulKernel()`. In the parallel implementation, each element of the result matrix C is computed by a separate thread, i.e. for computing an element of C in row i and column j, this thread computes a scalar product of the i-th row of A and the j-th column of B. Thus, the number of threads generated for the execution of `MatMulKernel` corresponds to the number of elements in C. The execution configuration specifies two-dimensional blocks of size 32×32, i.e., the configuration `dim3 dBlock(32, 32)` is chosen. The size of the two-dimensional grid is defined such that the number of threads generated fits the number N×N of matrix elements, i.e., the grid size is `dim3 dGrid (N/32, N/32)`.

The arrays Ad, Bd, and Cd to be used on the GPU are allocated in the global memory of the GPU from which they are accessed by the threads. For the assignment of threads to the computation of an element of the result matrix, a specific row index `row` and a specific column index `col` are used based on the thread identifier. The variables `row` and `column` are private variables so that there exists a separate copy for each thread. The program uses a linear storage of the arrays in row-major order. The computation of one scalar product is implemented in a `for` loop with the loop body

`Cval= Cval+A [row*N+e]*B[e*N+col]`.

The value of the private variable `Cval` is then copied into the global array C at the appropriate position `row * N + col`.   □

The matrix multiplication given in Fig. 7.10 accesses the data directly in the global memory, which results in a high data access time. This data access time can be hidden by a high number of threads so that always some threads can be executed while others access data. But still the bandwidth for accessing the global memory may limit the overall performance. As an example, we consider the NVIDIA GTX 680, which has a maximum memory bandwidth of 192 GB/sec, i.e. in each second, $192 * 10^9$ Bytes or equivalently $48 * 10^9$ floating point values of single precision (32 bit) can be loaded from the global memory. The loop body of the `for` loop in the kernel function `MatMulKernel()` in Fig. 7.10 requires two data accesses and performs two arithmetic operations. Since only $48 * 10^9$ floating point values can be loaded per second, this means that a maximum of $48 * 10^9$ floating point operations can be performed per second. However, the NVIDIA GTX 680 has a maximum performance of 3090 GFLOPS, which means that only about 1,6 % of the maximum performance is exploited.

The efficiency of a CUDA program can be increased by first loading data into the shared memory and then accessing them from this faster memory. This is advantageous if data are used several times by the threads of the same block. For the matrix multiplication, there is a multiple use of data, since the threads of a block of size 32×32 access 32 rows of matrix A and 32 columns of matrix B several times, see also [130]. The access pattern is illustrated in Fig. 7.11 for matrices of size 12×12 with 16 thread blocks of size 3×3. The thread blocks (0,0) and (1,2) are highlighted. As example, we first consider thread block (0,0). Thread (0,0) of this block computes C[0][0] and accesses row 0 of A and column 0 of B for this computation. These accesses are illustrated by the arrows in the figure. The arrows also indicate the access order to the rows of *A* and the columns of *B*. Thread (1,0) computing C[0,1] also accesses row 0 of A but column 1 of B, which is a multiple use of row 0 of A. Similarly, thread (2,0) also accesses row 0 of A for its computation of C[0,2].

```
#include <stdio.h>
typedef float * Matrix;
const int N = 32 * 32;
__global__ void MatMulKernel(const Matrix A,const Matrix B,Matrix C){
  float Cval = 0;
  int row = blockIdx.y * blockDim.y + threadIdx.y;
  int col = blockIdx.x * blockDim.x + threadIdx.x;
  for (int e = 0; e < N; ++e)
    Cval += A[row * N + e] * B[e * N + col];
  C[row * N + col] = Cval;
}
void MatMul (const Matrix A, const Matrix B, Matrix C){
  int size = N * N * sizeof(float);
  Matrix Ad,Bd,Cd;
  cudaMalloc (&Ad, size); cudaMalloc(&Bd, size);
  cudaMemcpy (Ad, A, size, cudaMemcpyHostToDevice);
  cudaMemcpy (Bd, B, size, cudaMemcpyHostToDevice);
  cudaMalloc (&Cd, size);
  dim3 dBlock(32,32);
  dim3 dGrid(N/32,N/32);
  MatMulKernel<<<dGrid,dBlock>>> (Ad,Bd,Cd);
  cudaMemcpy (C,Cd,size, cudaMemcpyDeviceToHost);
  cudaFree (Ad); cudaFree (Bd); cudaFree (Cd);
}
int main() {
  Matrix A, B, C;
  A = (float *) malloc(N * N * sizeof(float));
  B = (float *) malloc(N * N * sizeof(float));
  C = (float *) malloc(N * N * sizeof(float));
  read_in(A); read_in(B);
  MatMul (A,B,C);
  write_out (C);
}
```

Fig. 7.10 CUDA program implementing a matrix multiplication of two $N \times N$ matrices A and B. The matrix size N is assumed to be a multiple of 32.

Thus, for the computation of an entire block with nine elements of the result matrix C, the nine threads of thread block $(0,0)$ need to access only six vectors, three rows of A (indicated by X) and three columns of B (indicated by Y). Similar accesses are performed for thread block $(1,2)$ as well as the other thread blocks. Notice that the array indices are denoted using the standard matrix indices, but the indices of the threads in the thread blocks use the CUDA specific numbering, as given in Fig 7.5.

To take advantage of this data reuse for the matrix multiplication using 32×32 thread blocks, sub-arrays of size $32 \times N$ of A and of size $N \times 32$ of B could be pre-loaded into the shared memory so that all data accesses of a thread block address the shared memory and, thus, the memory access time is reduced. However, since the size of the shared memory is limited, a sub-array of size $32 \times N$ might be too large to fit into this memory. In such cases, the sub-array to be reloaded should be smaller

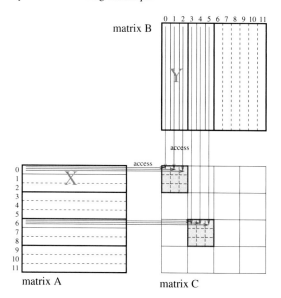

Fig. 7.11 Illustration of the access pattern of a matrix-matrix multiplication with input matrices A and B and result matrix C of size 12×12 and 16 thread blocks of size 3×3.

while being still large enough to provide data accesses from the shared memory. This can be achieved by the tiling programming technique, see Sect. 4.7.

For the matrix multiplication from Fig. 7.10, the tiling technique can be implemented by subdividing the sub-array of size $32\times N$ further into tiles of size 32×32, which are pre-loaded into the shared memory one after another. To process such a tile before other data of the same sub-array are needed, the matrix multiplication program has to be re-structured in such a way that data in the pre-loaded tile are accessed consecutively, i.e., locality of data accesses is required, see Sect. 2.9. This is achieved by dividing the scalar product to be computed by each tile into sub-scalar products using only the vector of length 32 loaded with the tile. An appropriate ordering of elements to be loaded can even exploit the coalescing of the hardware to further increase the efficiency.

Figure 7.12 illustrates the tiling technique for 12×12 matrices using 3×3 tiles, i.e., there are four tiles in each dimension of the matrices. In the figure, tile X of matrix A and tile Y of matrix B are used for the sub-computation of tile T of the output matrix C. Tile X of matrix A is also used for the sub-computation of tile S of matrix C. However, these are not all tiles that are needed for the computation of T and S. For tile R of matrix C, all tiles of matrix A and of matrix B that are needed for the complete computation of tile R are highlighted. As shown in the figure, these are the tiles 1,2,3, and 4 of A and of B.

Example: The CUDA kernel function in Fig. 7.13 implements the tiling technique for the matrix multiplication by pre-loading tiles Ads and Bds into the shared memory, see [130]. The tiles Ads and Bds have size TILE_WIDTH×TILE_WIDTH. Each

matrix B

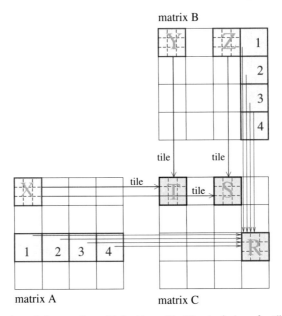

Fig. 7.12 Illustration of the matrix multiplication with tiling technique for tiles of size 3×3.

thread is responsible for the computation of one element of the result matrix C. The indices of the element to be computed by a thread is determined according to its block and thread identifier. Since these values are accessed often during the computation, they are loaded into private variables bx, by, tx, and ty, for which each thread has private copies to be stored in its registers.

The loop for calculating the scalar product needed for each element of the result matrix is now broken up into several phases where each phase processes the data of one tile of A and B. These phases are implemented in the outer loop over m=0, ..., N/TILE_WIDTH-1. Each phase first loads two tiles Ads and Bds, one from each input matrix, into the shared memory. These tiles are loaded together by all threads of a block, i.e., thread (tx,ty) of a block loads elements into Ads[ty][tx] and Bds[ty][tx]. Notice that each thread block has its own copies of Ads and Bds and loads different elements from the matrices A and B in the global memory. The elements to be loaded are identified using the values Row and Col as well as the size N of the matrices and the size of a tile TILE_WIDTH. The matrices Ads and Bds in the shared memories will contain different values in each iteration phase.

The transfer of data into the shared memory is finalized by a synchronization, which is necessary, since each thread explicitly loads only one element of Ads and Bds but uses values loaded by other threads of the same block. The computation of a partial scalar product based on the data tiles loaded is implemented in a for loop over k=0, ..., TILE_WIDTH-1. The intermediate result is accumulated into the private variable Cval. The computation of the scalar product is also synchronized for all threads of the block. This is necessary, since the next outer iteration phase

```
#define TILE_WIDTH 32
__global__ void MatMulTileKernel (Matrix A, Matrix B, Matrix C) {
    __shared__ float Ads[TILE_WIDTH][TILE_WIDTH];
    __shared__ float Bds[TILE_WIDTH][TILE_WIDTH];
    int bx = blockIdx.x; int by = blockIdx.y;
    int tx = threadIdx.x; int ty = threadIdx.y;
    int Row = by * TILE_WIDTH + ty;
    int Col = bx * TILE_WIDTH + tx;
    float Cval = 0.0;
    for (int m = 0; m < N/TILE_WIDTH; m++) { /* loop over tiles */
        Ads[ty][tx] = A[Row*N + (m*TILE_WIDTH + tx)];
        Bds[ty][tx] = B[(m*TILE_WIDTH + ty)*N + Col];
        __syncthreads();
        for (int k = 0; k < TILE_WIDTH; k++) /* loop within tile */
            Cval += Ads[ty][k] * Bds[k][tx];
        __syncthreads();
    }
    C[Row * N + Col] = Cval; /* write back to global memory */
}
```

Fig. 7.13 CUDA kernel function for the multiplication of two N × N matrices A and B using tiles according to [130]. The host program from Fig. 7.10 can be used.

reloads the tiles, which should be done only after the computations on the current values have been finished for all threads. After the N/TILE_WIDTH iteration phases, all threads of a block have jointly finished the computation of that part of the result matric C for which the block is responsible. Each thread then copies its value Cval into the global array C at position Row * N + Col.   □

7.6 Introduction to OpenCL

Another application programming interface for compute environments including GPUs is OpenCL. OpenCL is a standardized, cross-platform programming interface proposed in 2008 which supports the use of compute environments consisting of CPUs, GPUs, and other processing units. Similar to CUDA, OpenCL is based on the programming language C. The programming model is also very similar to that of CUDA and we briefly introduce OpenCL by explaining the similarities and differences to CUDA, see [130]. OpenCL provides methods to explicitly address components of a heterogeneous compute platform and to assign tasks to those components. Thus, the programming with OpenCL is somehow more complex than programming with CUDA. However, with OpenCL it is possible to exploit very different hardware components, such as laptops or nodes of a supercomputer.

An OpenCL platform is a heterogeneous platform, comprising a single host and one or more OpenCL devices, called compute devices. The host is responsible for the interaction with the external environment, such as user interactions or I/O, and

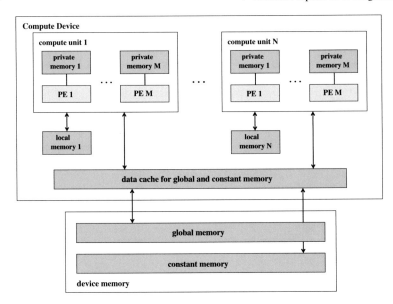

Fig. 7.14 Illustration of the OpenCL memory hierarchy for N compute units and M processing elements (PE) in each compute unit. The host is not shown.

the devices are used for parallel computations. An OpenCL application program consists of a host program and a set of kernels implemented in the OpenCL-C programming language. The kernels are called by the host program to be executed on one of the devices. When a kernel is called, the OpenCL runtime system generates a global index space called NDRanges and for each point in the index space, called work item, an instance of the kernel is executed. The work items correspond to CUDA threads; the difference is that work items can be directly addressed by their global indices in NDRanges. The name NDRanges is an abbreviation for N-dimensional index space; currently N can be 1, 2, or 3.

The work items can be grouped into work groups, which have the same dimension as NDRanges. In each dimension, the sizes of NDRanges has to be divisible by the number of work groups. The work groups have a work group identifier, and the work items of a work group have an additional local identifier in their group, so that a work item can be identified by its work group identifier and its local identifier. Thus, in this respect an index space NDRanges resembles the grid in CUDA and the work groups resemble the blocks in CUDA. The contrast to CUDA is that work items have two ways to be identified, first by the combination of group identifier and local identifier and second by the global identifier in NDRanges, which is not possible in CUDA. The barrier synchronization `barrier()`, however, is only referring to all work items of the same work group, so that groups are necessary for synchronization purposes. The parallel computing model of OpenCL is SIMD or SPMD, which means that all work items execute the same operation on different data.

```
__kernel void vectoradd (__global const float * a,
    __global const float * b, __global const float * c) {
    int id = get_global_id(0);
    c[id] = a[id] + b[id];
}
```

Fig. 7.15 OpenCL kernel for a vector addition of two vectors a and b with result vector c.

Data of an OpenCL program can be stored in five different memory types, very similar to CUDA. The host memory on the CPU is only accessible by the host program. On a device, there are the global memory, the constant memory, the local memory, and the private memory, see Fig. 7.14. The host program can dynamically allocate space in the global memory of the GPU, and the host as well as the device program can access this memory type. It corresponds to the CUDA global memory. The constant memory can be written and read by the host program and can be read by the device program. In contrast to CUDA, the host program can dynamically allocate memory space in the constant memory. The size of the constant memory is not limited to 64 KB and the actual size can be different for different devices. The local memory can only be written and read by the work items of one work group, and not by other work groups or the host, and corresponds to the CUDA shared memory. The private memory is assigned to one work item only.

The kernels in OpenCL correspond to CUDA kernel functions, however they are declared using the keyword __kernel, and not __global__ as in CUDA. Figure 7.15 shows an OpenCL kernel vectoradd() implementing an addition of vectors a and b with the result delivered in vector c. Each work item is addressed using the global identifier get_global_id(0) and computes exactly one element of the result vector.

The heterogeneous compute environment to be used for the execution of an OpenCL program is defined in the context containing all devices. The assignment of computations to devices is specified in a command queue. Such a command queue can contain calls of kernels, memory allocation operations, copy operations, or synchronizations, which are executed by the device one after another. Due to the concept of command queues, it is possible to specify task parallelism by specifying several command queues. The correct interaction has to be guaranteed. OpenCL provides the concept of events, which are initiated by the commands of a command queue. The programming technique using commands and events is more complex and more details can be found in [91, 156].

7.7 Exercises for Chapter 7

Exercise 7.1. Consider the program in Fig. 7.8 computing the dot product of two vectors a and b. Assume that the program is executed on an Nvidia GTX 680 GPU. Answer the following questions:

(a) Specify the total number of threads used for the execution of the program.
(b) Specify the number of threads in a warp.
(c) Specify the number of threads in a block.

Exercise 7.2. Consider the illustration of a dot product of two vectors a and b in Fig. 7.9. Adapt the figure to the execution of the program in Fig. 7.8 for $N = 4 \cdot 16$ and execution configuration `scal_prod <<<4,4>>>`. Draw a corresponding illustration for this execution configuration. How many threads are used in total? Which entries of a and b does the thread with tid $= 4$ access? How many steps are required to compute the partial dot product for each block?

Exercise 7.3. Write a complete CUDA program for the addition of two vectors a and b. The result should be provided in vector c. The vectors a and b should be initialized in the host program. The host program also allocates vectors ad, bd, and cd in the global memory of the GPU using `cudaMalloc()` and uses `cudaMemcpy()` to initialize ad and bd with the content of vector a and b, respectively. The kernel function should be designed such that a separate thread is used for adding two elements of ad and bd. After the completion of the kernel function, the host program copies the result, which has been computed in cd, back to the host and prints it out.

Exercise 7.4. Write a complete CUDA program for the computation of a matrix-vector product, see Section 3.7. $A \in \mathbb{R}^{n \times m}$ is an $n \times m$ matrix and $b \in \mathbb{R}^m$ is a vector of size m. The result vector $c \in \mathbb{R}^n$ of length n is computed as $c = A \cdot b$. The host program initializes the input matrix A and the input vector b and transfers them to the global memory of the GPU. After the call of the kernel function, the result vector c is transferred back to the host and is printed out. For the kernel function performing the actual computation $c_i = \sum_{j=1}^{m} k_{ij} \cdot b_j$ for $i = 1, \ldots, n$, two versions should be provided:

(a) The global-memory version of the kernel function performs all computations in the global memory of the GPU. A separate thread is used for each entry of the result vector c. Each thread computes the dot product of a row of A with vector b and writes the result at the corresponding position in vector c.

(b) The shared-memory version of the kernel function exploits the fact that the same vector b is used for the computation of all dot products.
To exploit this reuse, a vector bds of size BLOCK_SIZE is defined in the shared memory. Similar to the tiling technique, an outer loop is used to iterate over the blocks of the input vector b, and in each iteration of the outer loop, a block of b is processed by the different threads. For the processing, each thread loads an appropriate element of b into the shared vector bds and then computes a partial dot product using the block of vector b in bds. Note that each thread of a thread block loads a different element of b into bds. Note that a synchronization is required to ensure that the loading of bds is finished before the actual computation starts.

Chapter 8
Algorithms for Systems of Linear Equations

About this Chapter

The solution of a system of simultaneous linear equations is a fundamental algorithm in numerical linear algebra and is a basic ingredient of many scientific simulations. This chapter presents several standard methods for solving systems of linear equations such as Gaussian elimination, direct solution methods for linear systems with tridiagonal structure or banded matrices, in particular cyclic reduction and recursive doubling, iterative solution methods, Cholesky factorization for sparse matrices, and the conjugate gradient method. All presentations mainly concentrate on the parallel aspects.

Examples for the usage of algorithms for solving linear equation systems are scientific or engineering problems modeled by ordinary or differential equations. The numerical solution is often based on discretization methods leading to a system of linear equations. In this chapter, we present several standard methods for solving systems of linear equations of the form

$$Ax = b \tag{8.1}$$

where $A \in \mathbb{R}^{n \times n}$ is an $(n \times n)$-matrix of real numbers, $b \in \mathbb{R}^n$ is a vector of size n, and $x \in \mathbb{R}^n$ is an unknown solution vector of size n specified by the linear system (8.1) to be determined by a solution method. There exists a solution x for equation (8.1) if the matrix A is non-singular, which means that a matrix A^{-1} with $A \cdot A^{-1} = I$ exists; I denotes the n-dimensional identity matrix and $\cdot$ denotes the matrix product. Equivalently, the determinant of matrix A is not equal to zero. For the exact mathematical properties we refer to a standard book for linear algebra [85]. The emphasis of the presentation in this chapter is on parallel implementation schemes for linear system solvers.

The solution methods for linear systems are classified as direct and iterative. **Direct solution methods** determine the exact solution (except rounding errors) in a fixed number of steps depending on the size n of the system. Elimination methods and factorization methods are considered in the following. **Iterative solution methods** determine an approximation of the exact solution. Starting with a start vector, a sequence of vectors is computed which converges to the exact solution. The compu-

T. Rauber, G. Rünger, *Parallel Programming*, https://doi.org/10.1007/978-3-031-28924-8_8

tation is stopped if the approximation has an acceptable precision. Often, iterative solution methods are faster than direct methods and their parallel implementation is straightforward. On the other hand, the system of linear equations needs to fulfill some mathematical properties in order to guarantee the convergence to the exact solution. For sparse matrices, in which many entries are zeros, there is an advantage for iterative methods since they avoid a fill-in of the matrix with non-zero elements.

This chapter starts with a presentation of Gaussian elimination in Section 8.1. The Gaussian elimination is a direct solver, and its parallel implementation with different data distribution patterns is examined. In Section 8.2, direct solution methods for linear systems with tridiagonal structure or banded matrices, in particular cyclic reduction and recursive doubling, are discussed. Section 8.3 is devoted to iterative solution methods, Section 8.5 presents the Cholesky factorization, and Section 8.4 introduces the conjugate gradient method. The emphasis lies on the potential parallelism of these methods and the exploitation in parallel programs.

8.1 Gaussian Elimination

For the well-known Gaussian elimination, we briefly present the sequential method and then discuss parallel implementations with different data distributions. The section closes with an analysis of the parallel runtime of the Gaussian elimination with double-cyclic data distribution.

8.1.1 Gaussian Elimination and LU Decomposition

Written out in full the linear system $Ax = b$ has the form

$$a_{11}x_1 + a_{12}x_2 + \ldots + a_{1n}x_n = b_1$$
$$\vdots \qquad \vdots \qquad \qquad \vdots \quad \vdots$$
$$a_{i1}x_1 + a_{i2}x_2 + \ldots + a_{in}x_n = b_i$$
$$\vdots \qquad \vdots \qquad \qquad \vdots \quad \vdots$$
$$a_{n1}x_1 + a_{n2}x_2 + \ldots + a_{nn}x_n = b_n$$

The Gaussian elimination consists of two phases, the forward elimination and the backward substitution. The forward elimination transforms the linear system (8.1) into a linear system $Ux = b'$ with an $(n \times n)$ matrix U in upper triangular form. The transformation employs reordering of equations or operations that add a multiple of one equation to another. Hence, the solution vector remains unchanged. In detail, the **forward elimination** performs the following $n - 1$ steps. The matrix $A^{(1)} := A = (a_{ij})$ and the vector $b^{(1)} := b = (b_i)$ are subsequently transformed into matrices $A^{(2)}, \ldots, A^{(n)}$ and vectors $b^{(2)}, \ldots, b^{(n)}$, respectively. The linear equation systems $A^{(k)}x = b^{(k)}$ have the same solution as equation (8.1) for $k = 2, \ldots, n$. The matrix $A^{(k)}$ computed in step $k - 1$ has the form

$$A^{(k)} = \begin{bmatrix} a_{11} & a_{12} & \cdots & a_{1,k-1} & a_{1k} & \cdots & a_{1n} \\ 0 & a_{22}^{(2)} & \cdots & a_{2,k-1}^{(2)} & a_{2k}^{(2)} & \cdots & a_{2n}^{(2)} \\ \vdots & & \ddots & \ddots & \vdots & & \vdots \\ \vdots & & & \ddots & a_{k-1,k-1}^{(k-1)} & a_{k-1,k}^{(k-1)} & \cdots & a_{k-1,n}^{(k-1)} \\ \vdots & & & 0 & a_{kk}^{(k)} & \cdots & a_{kn}^{(k)} \\ \vdots & & & \vdots & \vdots & \ddots & \vdots \\ 0 & \cdots & \cdots & 0 & a_{nk}^{(k)} & \cdots & a_{nn}^{(k)} \end{bmatrix}.$$

The first $k - 1$ rows are identical to the rows in matrix $A^{(k-1)}$. In the first $k - 1$ columns, all elements below the diagonal element are zero. Thus, the last matrix $A^{(n)}$ has upper triangular form. The matrix $A^{(k+1)}$ and the vector $b^{(k+1)}$ are calculated from $A^{(k)}$ and $b^{(k)}$, $k = 1, \ldots, n - 1$, by subtracting suitable multiples of row k of $A^{(k)}$ and element k of $b^{(k)}$ from the rows $k+1, k+2, \ldots, n$ of A and elements $b_{k+1}^{(k)}, b_{k+2}^{(k)} \ldots, b_n^{(k)}$, respectively. The elimination factors for row i are

$$l_{ik} = a_{ik}^{(k)}/a_{kk}^{(k)}, \qquad i = k+1, \ldots, n. \tag{8.2}$$

They are chosen such that the coefficient of x_k of the unknown vector x is eliminated from equations $k+1, k+2, \ldots, n$. The rows of $A^{(k+1)}$ and the entries of $b^{(k+1)}$ are calculated according to

$$a_{ij}^{(k+1)} = a_{ij}^{(k)} - l_{ik}a_{kj}^{(k)} \tag{8.3}$$

$$b_i^{(k+1)} = b_i^{(k)} - l_{ik}b_k^{(k)} \tag{8.4}$$

for $k < j \leq n$ and $k < i \leq n$. Using the equation system $A^{(n)}x = b^{(n)}$, the result vector x is calculated in the **backward substitution** in the order $x_n, x_{n-1}, \ldots, x_1$ according to

$$x_k = \frac{1}{a_{kk}^{(n)}} \left(b_k^{(n)} - \sum_{j=k+1}^{n} a_{kj}^{(n)} x_j \right) \tag{8.5}$$

Figure 8.1 shows a program fragment in C for a sequential Gaussian elimination. The inner loop computing the matrix elements is iterated approximately k^2 times so that the entire loop has runtime $\sum_{k=1}^{n} k^2 = \frac{1}{6}n(n+1)(2n+1) \approx n^3/3$ which leads to an asymptotic runtime $O(n^3)$.

```
double *gauss_sequential (double **a, double *b)
{
  double *x, sum, l[MAX_SIZE];
  int i,j,k,r;

  x = (double *) malloc(n * sizeof(double));
  for (k = 0; k < n-1; k++) { /* Forward elimination */
    r = max_col(a,k);
    if (k != r) exchange_row(a,b,r,k);
    for (i=k+1; i < n; i++) {
      l[i] = a[i][k]/a[k][k];
      for (j=k+1; j < n; j++)
        a[i][j] = a[i][j] - l[i] * a[k][j];
      b[i] = b[i] - l[i] * b[k];
    }
  }
  for (k = n-1; k >= 0; k--) { /* Backward substitution */
    sum = 0.0;
    for (j=k+1; j < n; j++)
      sum = sum + a[k][j] * x[j];
    x[k] = 1/a[k][k] * (b[k] - sum);
  }
  return x;
}
```

Fig. 8.1 Program fragment in C notation for a sequential Gaussian elimination of the linear system $Ax = b$. The matrix A is stored in array a, the vector b is stored in array b. The indices start with 0. The function max_col(a,k) and exchange_row(a,b,r,k) implement pivoting. The function max_col(a,k) returns the index r with $|a_{rk}| = \max_{k \leq s \leq n}(|a_{sk}|)$. The function exchange_row(a,b,r,k) exchanges the rows r and k of A and the corresponding elements b_r and b_k of the right hand side.

8.1.1.1 LU decomposition and triangularization

The matrix A can be represented as the matrix product of an upper triangular matrix $U := A^{(n)}$ and a lower triangular matrix L which consists of the elimination factors (8.2) in the following way

$$
L = \begin{bmatrix}
1 & 0 & 0 & \cdots & & 0 \\
l_{21} & 1 & 0 & \cdots & & 0 \\
l_{31} & l_{32} & 1 & & & 0 \\
\vdots & \vdots & \ddots & & \ddots & 0 \\
l_{n1} & l_{n2} & l_{n3} & \cdots & l_{n,n-1} & 1
\end{bmatrix}
$$

The matrix representation $A = L \cdot U$ is called triangularization or LU decomposition. When only the LU decomposition is needed, the right side of the linear system does not have to be transformed. Using the LU decomposition, the linear system $Ax = b$ can be rewritten as

$$Ax = LA^{(n)}x = Ly = b \quad \text{with} \quad y = A^{(n)}x \tag{8.6}$$

and the solution can be determined in two steps. In the first step, the vector y is obtained by solving the triangular system $Ly = b$ by forward substitution. The forward substitution corresponds to the calculation of $y = b^{(n)}$ from equation (8.4). In the second step, the vector x is determined from the upper triangular system $A^{(n)}x = y$ by backward substitution. The advantage of the LU factorization over the elimination method is that the factorization into L and U is done only once but can be used to solve several linear systems with the same matrix A and different right hand side vectors b without repeating the elimination process.

8.1.1.2 Pivoting

Forward elimination and LU decomposition require the division by $a_{kk}^{(k)}$ and so these methods can only be applied when $a_{kk}^{(k)} \neq 0$. That is, even if $\det A \neq 0$ and the system $Ax = y$ is solvable, there does not need to exist a decomposition $A = LU$ when $a_{kk}^{(k)}$ is a zero element. However, for a solvable linear system, there exists a matrix resulting from permutations of rows of A, for which an LU decomposition is possible, i.e., $BA = LU$ with a permutation matrix B describing the permutation of rows of A. The permutation of rows of A, if necessary, is included in the elimination process. In each elimination step, a **pivot element** is determined to substitute $a_{kk}^{(k)}$. A pivot element is needed when $a_{kk}^{(k)} = 0$ and when $a_{kk}^{(k)}$ is very small which would induce an elimination factor, which is very large leading to imprecise computations. Pivoting strategies are used to find an appropriate pivot element. Typical strategies are column pivoting, row pivoting and total pivoting.

    Column pivoting considers the elements $a_{kk}^{(k)} \dots a_{nk}^{(k)}$ of column k and determines the element $a_{rk}^{(k)}, k \leq r \leq n$, with the maximum absolute value. If $r \neq k$, the rows r and k of matrix $A^{(k)}$ and the values $b_k^{(k)}$ and $b_r^{(k)}$ of the vector $b^{(k)}$ are exchanged. **Row pivoting** determines a pivot element $a_{kr}^{(k)}, k \leq r \leq n$, within the elements $a_{kk}^{(k)} \dots a_{kn}^{(k)}$ of row k of matrix $A^{(k)}$ with the maximum absolute value. If $r \neq k$, the columns k and r of $A^{(k)}$ are exchanged. This corresponds to an exchange of the enumeration of the unknowns x_k and x_r of vector x. **Total pivoting** determines the element with the maximum absolute value in the matrix $\tilde{A}^{(k)} = (a_{ij}^{(k)}), k \leq i, j \leq n$, and exchanges columns and rows of $A^{(k)}$ depending on $i \neq k$ and $j \neq k$. In practice, row or column pivoting is used instead of total pivoting, since they have smaller computation time and total pivoting may also destroy special matrix structures like banded structures.

    The implementation of pivoting avoids the actual exchange of rows or columns in memory and uses index vectors pointing to the current rows of the matrix. The indexed access to matrix elements is more expensive but in total the indexed access is usually less expensive than moving entire rows in each elimination step. When supported by the programming language, a dynamic data storage in form of separate vectors for rows of the matrix, which can be accessed through a vector pointing to the rows, may lead to more efficient implementations. The advantage is that matrix elements can still be accessed with a two-dimensional index expression but the exchange of rows corresponds to a simple exchange of pointers.

8.1.2 Parallel Row-Cyclic Implementation

A parallel implementation of the Gaussian elimination is based on a data distribution of matrix A and of the sequence of matrices $A^{(k)}$, $k = 2, \ldots, n$, which can be a row-oriented, a column-oriented or a checkerboard distribution, see Section 3.5. In this subsection, we consider a row-oriented distribution.

From the structure of the matrices $A^{(k)}$ it can be seen that a blockwise row-oriented data distribution is not suitable because of load imbalances: For a blockwise row-oriented distribution processor P_q, $1 \leq q \leq p$, owns the rows $(q-1) \cdot n/p + 1, \ldots, q \cdot n/p$ so that after the computation of $A^{(k)}$ with $k = q \cdot n/p + 1$ there is no computation left for this processor and it becomes idle. For a row-cyclic distribution, there is a better load balance, since processor P_q, $1 \leq q \leq p$, owns the rows q, $q + p$, $q + 2p \ldots$, i.e., it owns all rows i with $1 \leq i \leq n$, and $q = ((i-1) \bmod p) + 1$. The processors begin to get idle only after the first $n - p$ stages, which is reasonable for $p \ll n$. Thus, we consider a parallel implementation of the Gaussian elimination with a row-cyclic distribution of matrix A and a column-oriented pivoting. One step of the forward elimination computing $A^{(k+1)}$ and $b^{(k+1)}$ for given $A^{(k)}$ and $b^{(k)}$ performs the following computation and communication phases:

1. **Determination of the local pivot element:** Each processor considers its local elements of column k in the rows $k, \ldots, n$ and determines the element (and its position) with the largest absolute value.
2. **Determination of the global pivot element:** The global pivot element is the local pivot element which has the largest absolute value. A single-accumulation operation with the maximum operation as reduction determines this global pivot element. The root processor of this global communication operation sends the result to all other processors.
3. **Exchange of the pivot row:** If $k \neq r$ for a pivot element $a_{rk}^{(k)}$, the row k owned by processor P_q and the pivot row r owned by processor $P_{q'}$ have to be exchanged. When $q = q'$, the exchange can be done locally by processor P_q. When $q \neq q'$, then communication with single-transfer operations is required. The elements b_k and b_r are exchanged accordingly.
4. **Distribution of the pivot row:** Since the pivot row (now row k), is required by all processors for the local elimination operations, processor P_q sends the elements $a_{kk}^{(k)}, \ldots, a_{kn}^{(k)}$ of row k and the element $b_k^{(k)}$ to all other processors.
5. **Computation of the elimination factors:** Each processor locally computes the elimination factors l_{ik} for which it owns the row i according to Formula (8.2).

6. **Computation of the matrix elements:** Each processor locally computes the elements of $A^{(k+1)}$ and $b^{(k+1)}$ using its elements of $A^{(k)}$ and $b^{(k)}$ according to Formulas (8.3) and (8.4).

The computation of the solution vector x in the backward substitution is inherently sequential since the values x_k, $k = n, \ldots, 1$, depend on each other and are computed one after another. In step k, processor P_q owning row k computes the value x_k according to Formula (8.5) and sends the value to all other processors by a single-broadcast operation.

A program fragment implementing the computation phases 1–6 and the backward substitution is given in Figure 8.2. The matrix A and the vector b are stored in a two- and a one-dimensional array a and b, respectively. Some of the local functions are already introduced in the program in Figure 8.1. The SPMD program uses the variable me to store the individual processor number. This processor number, the current value of k and the pivot row are used to distinguish between different computations for the single processors. The global variables n and p are the system size and the number of processors executing the parallel program. The parallel algorithm is implemented in the program in Figure 8.2 in the following way:

1. **Determination of the local pivot element:** The function max_col_loc(a,k) determines the row index r of the element a[r][k], which has the largest local absolute value in column k for the rows $\geq k$. When a processor has no element of column k for rows $\geq k$, the function returns -1.

2. **Determination of the global pivot element:** The global pivoting is performed by an MPI_Allreduce() operation, implementing a single-accumulation with a subsequent single-broadcast. The MPI reduction operation MPI_MAXLOC for data type MPI_DOUBLE_INT consisting of one double value and one integer value is used. The MPI operations have been introduced in Section 5.2. The MPI_Allreduce() operation returns y with the pivot element in y.val and the processor owning the corresponding row in y.node. Thus, after this step all processors know the global pivot element and the owner for possible communication.

3. **Exchange of the pivot row:** Two cases are considered:

   - If the owner of the pivot row is the processor also owning row k (i.e. k%p == y.node), the rows k and r are exchanged locally by this processor for $r \neq k$. Row k is now the pivot row. The functions copy_row(a,b,k,buf) copies the pivot row into the buffer buf, which is used for further communication.
   - If different processors own the row k and the pivot row r, row k is sent to the processor y.node owning the pivot row with MPI_Send and MPI_Recv operations. Before the send operation, the function copy_row(a,b,k,buf) copies row k of array a and element k of array b into a common buffer buf so that only one communication operation needs to be applied. After the communication, the processor y.node finalizes its exchange with the pivot row. The function copy_exchange_row(a,b,r,buf,k) exchanges the row r (still the pivot row) and the buffer buf. The appropriate row index r is known from the

```
double *gauss_cyclic (double **a, double *b)
{
  double *x, l[MAX_SIZE], *buf;
  int i,j,k,r, tag=42;
  MPI_Status status;
  struct { double val; int node; } z,y;
  x = (double *) malloc(n * sizeof(double));
  buf = (double *) malloc((n+1) * sizeof(double));
  for (k=0 ; k<n-1 ; k++) { /* Forward elimination */
    r = max_col_loc(a,k);
    z.node = me;
    if (r != -1) z.val = fabs(a[r][k]); else z.val = 0.0;
    MPI_Allreduce(&z,&y,1,MPI_DOUBLE_INT,MPI_MAXLOC,MPI_COMM_WORLD);
    if (k % p == y.node){ /* Pivot row and row k are on the same processor */
      if (k % p == me) {
        if (a[k][k] != y.val) exchange_row(a,b,r,k);
        copy_row(a,b,k,buf);
      }
    }
    else /* Pivot row and row k are owned by different processors */
      if (k % p == me) {
        copy_row(a,b,k,buf);
        MPI_Send(buf+k,n-k+1,MPI_DOUBLE,y.node,tag,
                            MPI_COMM_WORLD);
      }
      else if (y.node == me) {
        MPI_Recv(buf+k,n-k+1,MPI_DOUBLE,MPI_ANY_SOURCE,
                            tag,MPI_COMM_WORLD,&status);
        copy_exchange_row(a,b,r,buf,k);
      }
    MPI_Bcast(buf+k,n-k+1,MPI_DOUBLE,y.node,MPI_COMM_WORLD);
    if ((k % p != y.node) && (k % p == me)) copy_back_row(a,b,buf,k);
    i = k+1; while (i % p != me) i++;
    for ( ; i<n; i+=p) {
      l[i] = a[i][k] / buf[k];
      for (j=k+1; j<n; j++)
        a[i][j] = a[i][j] - l[i]*buf[j];
      b[i] = b[i] - l[i]*buf[n];
    }
  }
  for (k=n-1; k>=0; k--) { /* Backward substitution */
    if (k % p == me) {
      sum = 0.0;
      for (j=k+1; j < n; j++) sum = sum + a[k][j] * x[j];
      x[k] = 1/a[k][k] * (b[k] - sum); }
    MPI_Bcast(&x[k],1,MPI_DOUBLE,k%p,MPI_COMM_WORLD);
  }
  return x;
}
```

Fig. 8.2 Program fragment with C-notation and MPI operations for the Gaussian elimination with row-cyclic distribution.

former local determination of the pivot row. Now the former row k is the row r and the buffer buf contains the pivot row.

Thus, in the both cases the pivot row is stored in buffer buf.

4. **Distribution of the pivot row:** Processor y.node sends the buffer buf to all other processors by an MPI_Bcast() operation. For the case that the pivot row was owned by a different processor than the owner of row k, the content of buf is copied into row k by this processor using copy_back_row().

5. and 6. **Computation of the elimination factors and the matrix elements:** The computation of the elimination factors and the new arrays a and b is done in parallel. Processor P_q starts this computation with the first row $i > k$ with i mod $p = q$.

For a row-cyclic implementation of the Gaussian elimination, an alternative way of storing array a and vector b can be used. The alternative data structure consists of a one-dimensional array of pointers and n one-dimensional arrays of length $n+1$ each containing one row of a and the corresponding element of b. The entries in the pointer-array point to the row-arrays. This storage scheme not only facilitates the exchange of rows but is also convenient for a distributed storage. For a distributed memory, each processor P_q stores the entire array of pointers but only the rows i with i mod $p = q$; all other pointers are NULL-pointers. Figure 8.3 illustrates this storage scheme for $n = 8$. The advantage of storing an element of b together with a is that the copy operation into a common buffer can be avoided. Also the computation of the new values for a and b is now only one loop with $n + 1$ iterations. This implementation variant is not shown in Figure 8.2.

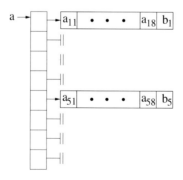

Fig. 8.3 Data structure for the Gaussian elimination with $n = 8$ and $p = 4$ showing the rows stored by processor P_1. Each row stores $n + 1$ elements consisting of one row of the matrix a and the corresponding element of b.

8.1.3 Parallel Implementation with Checkerboard Distribution

A parallel implementation using a block-cyclic checkerboard distribution for matrix A can be described with the parameterized data distribution introduced in Section 3.5. The parameterized data distribution is given by a distribution vector

$$((p_1,b_1),(p_2,b_2)) \tag{8.7}$$

with a $p_1 \times p_2$ virtual processor mesh with p_1 rows, p_2 columns and $p_1 \cdot p_2 = p$ processors. The numbers b_1 and b_2 are the sizes of a block of data with b_1 rows and b_2 columns. The function $\mathcal{G} : P \to \mathbb{N}^2$ maps each processor to a unique position in the processor mesh. This leads to the definition of p_1 **row groups**

$$R_q = \{Q \in P \mid \mathcal{G}(Q) = (q,\cdot)\}$$

with $|R_q| = p_2$ for $1 \le q \le p_1$ and p_2 **column groups**

$$C_q = \{Q \in P \mid \mathcal{G}(Q) = (\cdot,q)\}$$

with $|C_q| = p_1$ for $1 \le q \le p_2$. The row groups as well as the column groups are a partition of the entire set of processors, i.e.,

$$\bigcup_{q=1}^{p_1} R_q = \bigcup_{q=1}^{p_2} C_q = P$$

and $R_q \cap R_{q'} = \emptyset = C_q \cap C_{q'}$ for $q \ne q'$. Row i of the matrix A is distributed across the local memories of the processors of only one row group, denoted $Ro(i)$ in the following. This is the row group R_k with $k = \left\lfloor \frac{i-1}{b_1} \right\rfloor \bmod p_1 + 1$. Analogously, column j is distributed within one column group, denoted as $Co(j)$, which is the column group C_k with $k = \left\lfloor \frac{j-1}{b_2} \right\rfloor \bmod p_2 + 1$.

Example: For a matrix of size 12×12 (i.e. $n = 12$) p=4 processors $\{P_1, P_2, P_3, P_4\}$ and distribution vector $((p_1,b_1),(p_2,b_2)) = ((2,2),(2,3))$, the virtual processor mesh has size 2×2 and the data blocks have size 2×3. There are two row groups and two column groups:

$$R_1 = \{Q \in P \mid \mathcal{G}(Q) = (1,j), j = 1,2\}$$
$$R_2 = \{Q \in P \mid \mathcal{G}(Q) = (2,j), j = 1,2\}$$
$$C_1 = \{Q \in P \mid \mathcal{G}(Q) = (j,1), j = 1,2\}$$
$$C_2 = \{Q \in P \mid \mathcal{G}(Q) = (j,2), j = 1,2\}.$$

The distribution of matrix A is shown in Figure 8.4. It can be seen that row 5 is distributed in row group R_1 and that column 7 is distributed in column group C_1. □

Using a checkerboard distribution with distribution vector (8.7), the computation of $A^{(k)}$ has the following implementation, which has a different communication pat-

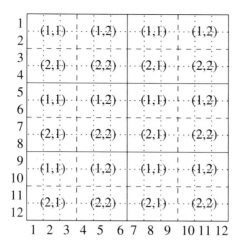

Fig. 8.4 Illustration of a checkerboard distribution for a 12×12 matrix. The tuples denote the position of the processors in the processor mesh owning the data block.

tern than the previous implementation. Figure 8.5 illustrates the communication and computation phases of the Gaussian elimination with checkerboard distribution.

1. **Determination of the local pivot element:** Since column k is distributed across the processors of column group $Co(k)$, only these processors determine the element with the largest absolute value for row $\geq k$ within their local elements of column k.

2. **Determination of the global pivot element:** The processors in group $Co(k)$ perform a single-accumulation operation within this group, for which each processor in the group provides its local pivot element from phase 1. The reduction operation is the maximum operation also determining the index of the pivot row (and not the number of the owning processor as before). The root processor of the single-accumulation operation is the processor owning the element $a_{kk}^{(k)}$. After the single-accumulation, the root processor knows the pivot element $a_{rk}^{(k)}$ and its row index. This information is sent to all other processors.

3. **Exchange of the pivot row:** The pivot row r containing the pivot element $a_{rk}^{(k)}$ is distributed across row group $Ro(r)$. Row k is distributed across the row group $Ro(k)$, which may be different from $Ro(r)$. If $Ro(r) = Ro(k)$, the processors of $Ro(k)$ exchange the elements of the rows k and r locally within the columns they own. If $Ro(r) \neq Ro(k)$, each processor in $Ro(k)$ sends its part of row k to the corresponding processor in $Ro(r)$; this is the unique processor which belongs to the same column group.

4. **Distribution of the pivot row:** The pivot row is needed for the recalculation of matrix A, but each processor needs only those elements with column indices for which it owns elements. Therefore, each processor in $Ro(r)$ performs a group-

oriented single-broadcast operation within its column group sending its part of
the pivot row to the other processors.

5. **Computation of the elimination factors:** The processors of column group
   $Co(k)$ locally compute the elimination factors l_{ik} for their elements i of column k
   according to Formula (8.2).

5a. **Distribution of the elimination factors:** The elimination factors l_{ik} are needed
   by all processors in the row group $Ro(i)$. Since the elements of row i are distributed across the row group $Ro(i)$, each processor of column group $Co(k)$
   performs a group-oriented single-broadcast operation in its row group $Ro(i)$ to
   broadcast its elimination factors l_{ik} within this row group.

6. **Computation of the matrix elements:** Each processor locally computes the elements of $A^{(k+1)}$ and $b^{(k+1)}$ using its elements of $A^{(k)}$ and $b^{(k)}$ according to Formulas (8.3) and (8.4).

The backward substitution for computing the n elements of the result vector x is
done in n consecutive steps where each step consists of the following computations.

1. Each processor of the row group $Ro(k)$ computes that part of $\sum_{j=k+1}^{n} a_{kj}^{(n)} x_j$ which
   contains its local elements of row k.
2. The entire sum $\sum_{j=k+1}^{n} a_{kj}^{(n)} x_j$ is determined by the processors of row group $Ro(k)$
   by a group-oriented single-accumulation operation with the processor P_q as root
   which stores the element $a_{kk}^{(n)}$. Addition is used as reduction operation.
3. Processor P_q computes the value of x_k according to Formula (8.5).
4. Processor P_q sends the value of x_k to all other processors by a single-broadcast
   operation.

A pseudocode for an SPMD program in C notation with MPI operations implementing the Gaussian elimination with checkerboard distribution of matrix A is given in
Figure 8.6. The computations correspond to those given in the pseudocode for the
row-cyclic distribution in Figure 8.2, but the pseudocode additionally uses several
functions organizing the computations on the groups of processors. The functions
`Co(k)` and `Ro(k)` denote the column or row groups owning column k or row k, respectively. The function `member(me,G)` determines whether processor me belongs
to group G. The function `grp_leader()` determines the first processor in a group.
The functions `Cop(q)` and `Rop(q)` determine the column or row group, respectively, to which a processor q belongs. The function `rank(q,G)` returns the local
processor number (rank) of a processor in a group G.

1. **Determination of the local pivot element:** The determination of the local pivot
   element is performed only by the processors in column group `Co(k)`.
2. **Determination of the global pivot element:** The global pivot element is again
   computed by an `MPI_MAXLOC` reduction operation, but in contrast to Figure 8.2
   the index of the row of the pivot element is calculated and not the processor
   number owning the pivot element. The reason is that all processors which own a
   part of the pivot row need to know that some of their data belongs to the current
   pivot row; this information is used in further communication.

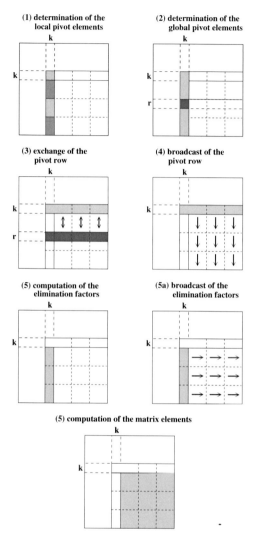

Fig. 8.5 Computation phases of the Gaussian elimination with checkerboard distribution.

3. **Exchange of the pivot row:** For the exchange and distribution of the pivot row
   r, the cases $\text{Ro}(\text{k}) == \text{Ro}(\text{r})$ and $\text{Ro}(\text{k})! = \text{Ro}(\text{r})$ are distinguished.

   - When the pivot row and the row k are stored by the same row group, each
     processor of this group exchanges its data elements of row k and row r locally
     using the function $\texttt{exchange\_row\_loc()}$ and copies the elements of the pivot
     row (now row k) into the buffer $\texttt{buf}$ using the function $\texttt{copy\_row\_loc()}$.
     Only the elements in column k or higher are considered.

- When the pivot row and the row k are stored by different row groups, communication is required for the exchange of the pivot row. The function `compute_partner(Ro(r),me)` computes the communicaton partner for the calling processor me, which is the processor $q \in Ro(r)$ belonging to the same column group as me. The function `compute_size(n,k,Ro(k))` computes the number of elements of the pivot row, which is stored for the calling processor in columns greater than k; this number depends on the size of the row group `Ro(k)`, the block size and the position k. The same function is used later to determine the number of elimination factors to be communicated.

4. **Distribution of the pivot row:** For the distribution of the pivot row r, a processor takes part in a single-broadcast operation in its column group. The roots of the broadcast operation performed in parallel are the processors $q \in Ro(r)$. The participants of a broadcast are the processors $q' \in Cop(q)$, either as root when $q' \in Ro(r)$ or as recipient otherwise.

5. **Computation of the elimination factors:** The function `compute_elim_fact_loc()` is used to compute the elimination factors l_{ik} for all elements a_{ik} owned by the processor. The elimination factors are stored in buffer `elim_buf`.

5a. **Distribution of the elimination factors:** A single-broadcast operation is used to send the elimination factors to all processors in the same row group `Rop(q)`; the corresponding communicator `comm(Rop(q))` is used. The number (rank) of the root processor q for this broadcast operation in a group G is determined by the function `rank(q,G)`.

6. **Computation of the matrix elements:** The computation of the matrix elements by `compute_local_entries()` and the backward substitution performed by `backward_substitution()` are similar to the pseudocode in Figure 8.2.

The main difference to the program in Figure 8.2 are that more communication is required and that almost all collective communication operations are performed on subgroups of the set of processors and not on the entire set of processors.

8.1.4 Analysis of the Parallel Execution Time

The analysis of the parallel execution time of the Gaussian elimination uses functions expressing the computation and communication time depending on the characteristics of the parallel machine, see also Section 4.5. The function describing the parallel execution time of the program in Figure 8.6 additionally contains the parameters p_1, p_2, b_1 and b_2 of the parameterized data distribution in Formula (8.7). In the following, we model the parallel execution time of the Gaussian elimination with checkerboard distribution, neglecting pivoting and backward substitution for simplicity, see also [181]. These are the phases 4, 5, 5a, and 6 of the Gaussian elimination. For the derivation of functions reflecting the parallel execution time, these four SPMD computation phases can be considered separately, since there is a barrier synchronization after each phase.

```
double * gauss_double_cyclic (double **a, double *b)
{
 double *x, *buf, *elim_buf;
 int i,j,k,r,q, ql, size, buf_size, elim_size, psz;
 struct { double val; int pvtline; } z,y;
 MPI_Status status;

 x = (double *) malloc(n * sizeof(double));
 buf = (double *) malloc((n+1) * sizeof(double));
 elim_buf = (double *) malloc((n+1) * sizeof(double));
 for (k=0; k<n-1; k++) {
   if (member(me, Co(k))) {
    r = max_col_loc(a,k);
    z.pvtline = r; z.val = fabs(a[r][k]);
    MPI_Reduce(&z,&y,1,MPI_DOUBLE_INT,MPI_MAXLOC,
               grp_leader(Co(k)),comm(Co(k)));
   }
   MPI_Bcast(&y,1,MPI_DOUBLE_INT,grp_leader(Co(k)),MPI_COMM_WORLD);
   r = y.pvtline;
   if(Ro(k) == Ro(r)){
           /*pivot row and row k are in the same row group */
     if (member(me, Ro(k))) {
         if (r != k) exchange_row_loc(a,b,r,k);
         copy_row_loc(a,b,k,buf); }   }
   else /* pivot row and row k are in different row groups */
     if (member(me, Ro(k))) {
         copy_row_loc(a,b,k,buf);
         q = compute_partner(Ro(r),me);
         psz = compute_size(n,k,Ro(k));
         MPI_Send(buf+k,psz,MPI_DOUBLE,q,tag,MPI_COMM_WORLD); }
     else if (member(me,Ro(r))) {
         /* executing processor contains a part of the pivot row */
         q = compute_partner(Ro(k),me);
         psz = compute_size(n,k,Ro(r));
         MPI_Recv(buf+k,psz,MPI_DOUBLE,q,tag,MPI_COMM_WORLD,&status);
         exchange_row_buf(a,b,r,buf,k);
     }
   for (q=0; q<p; q++) /* broadcast of pivot row */
     if (member(q,Ro(r)) && member(me,Cop(q))) {
         ql = rank(q,Cop(q)); buf_size = compute_size(n,k,Ro(k));
         MPI_Bcast(buf+k,buf_size,MPI_DOUBLE,ql,comm(Cop(q)));}
   if ((Ro(k) != Ro(r)) && (member(me,Ro(k)))
     copy_row_loc(a,b,buf,k);
   if (member(me,Co(k))) elim_buf = compute_elim_fact_loc(a,b,k,buf);
   for (q=0; q<p; q++) /* broadcast of elimination factors */
       if (member(q,Co(k)) && member(me,Rop(q))) {
         ql = rank(q,Rop(q)); elim_size = compute_size(n,k,Co(k));
         MPI_Bcast(elim_buf,elim_size,MPI_DOUBLE,ql,comm(Rop(q))); }
   compute_local_entries(a,b,k,elim_buf,buf); }
 backward_substitution(a,b,x);
 return x;
}
```

Fig. 8.6 Program of the Gaussian elimination with checkerboard distribution.

For a communication phase, a formula describing the time of a collective communication operation is used which describes the communication time as a function of the number of processors and the message size. For the Gaussian elimination (without pivoting), the phases 4 and 5a implement communication with a single-broadcast operation. The communication time for a single-broadcast with p processors and message size m is denoted as $T_{sb}(p,m)$. We assume that independent communication operations on disjoint processor sets can be done in parallel. The values for p and m have to be determined for the specific situation. These parameters depend on the data distribution and the corresponding sizes of row and column groups as well as on the step k, $k = 1,\ldots,n$, of the Gaussian elimination since messages get smaller for increasing k.

Also, the modeling of the computation times of phases 5 and 6 depend on the step number k, since less elimination factors or matrix elements have to be computed for increasing k and thus the number of arithmetic operations decreases with increasing k. The time for an arithmetic operation is denoted as t_{op}. Since the processors perform an SPMD program, the processor computing the most arithmetic operations determines the computation time for the specific computation phase. The following modeling of communication and computation time for one step k uses the index sets

$$I_q = \{(i,j) \in \{1...n\} \times \{1...n\} \mid P_q \text{ owns } a_{ij}\},$$

which contain the indices of the matrix elements stored locally in the memory of processor P_q.

- The broadcasting of the pivot row k in phase 4. of step k sends the elements of row k with column index $\geq k$ to the processors needing the data for computations in rows $\geq k$. Since the pivot row is distributed across the processors of the row group $Ro(k)$, all the processors of $Ro(k)$ send their part of row k. The amount of data sent by one processor $q \in Ro(k)$ is the number of elements of row k with column indices $\geq k$. (i.e., with indices $((k,k),\ldots,(k,n)))$ owned by processor q. This is the number

$$N_q^{row \geq k} := \#\{(k,j) \in I_q \mid j \geq k\}. \tag{8.8}$$

(The symbol $\#X$ for a set X denotes the number of elements of this set X.) The processor $q \in Ro(k)$ sends its data to those processors owning elements in the rows with row index $\geq k$ which have the same column indices as the elements of processor q. These are the processors in the column group $Cop(q)$ of the processor q and thus these processors are the recipients of the single-broadcast operation of processor q. Since all column groups of the processors $q \in Ro(k)$ are disjoint, the broadcast operation can be done in parallel and the communication time is:

$$\max_{q \in Ro(k)} T_{sb}(\#Cop(q), N_q^{row \geq k}).$$

- In phase 5 of step k, the elimination factors using the elements $a_{kk}^{(k)}$ and the elements $a_{ik}^{(k)}$ for $i > k$ are computed by the processors owning these elements of column k, i.e., by the processors $q \in Co(k)$, according to Formula (8.2). Each of the processors computes the elimination factors of its part, which are

$$N_q^{col>k} := \#\{(i,k) \in I_q | i > k\} \tag{8.9}$$

elimination factors for processor $q \in Co(k)$. Since the computations are done in parallel this results in the computation time

$$\max_{q \in Co(k)} N_q^{col>k} \cdot t_{op} \, .$$

- In phase 5a the elimination factors are sent to all processors which recalculate the matrix elements with indices $(i,j), i > k, j > k$. Since the elimination factors $l_{ik}^{(k)}$, $l = k+1,\ldots,n$, are needed within the same row i, a row-oriented single-broadcast operation is used to send the data to the processors owning parts of row i. A processor $q \in Co(k)$ sends its data to the processors in its row group $Rop(q)$. These are the data elements computed in the previous phase, i.e., $N_q^{col>k}$ data elements, and the communication time is:

$$\max_{q \in Co(k)} T_{sb}(\#Rop(q), N_q^{col>k}) \, .$$

- In phase 6 of step k, all matrix elements in the lower right rectangular area are recalculated. Each processor q recalculates the entries it owns; these are the number of elements per column for rows with indices $> k$ (i.e. $N_q^{col>k}$) multiplied by the number of elements per row for columns with indices $> k$ (i.e. $N_q^{row>k}$). Since two arithmetic operations are performed for one entry according to Formula (8.4), the computation time is

$$\max_{q \in P} N_q^{col>k} \cdot N_q^{row>k} \cdot 2t_{op} \, .$$

In total, the parallel execution for all phases and all steps is:

$$T(n,p) = \sum_{k=1}^{n-1} \Big\{ \max_{q \in Ro(k)} T_{sb}(\#Cop(q), N_q^{row \geq k})$$
$$+ \max_{q \in Co(k)} N_q^{col>k} \cdot t_{op} \tag{8.10}$$
$$+ \max_{q \in Co(k)} T_{sb}(\#Rop(q), N_q^{col>k})$$
$$+ \max_{q \in P} N_q^{col>k} \cdot N_q^{row>k} \cdot 2t_{op} \Big\}$$

This parallel execution time can be expressed in terms of the parameters of the data distribution $((p_1,b_1),(p_2,b_2))$, the problem size n and the step number k by estimating the sizes of messages and the number of arithmetic operations. For the estimation, larger blocks of data, called **superblocks**, are considered. Superblocks consist of $p_1 \times p_2$ consecutive blocks of size $b_1 \times b_2$, i.e., it has p_1b_1 rows and p_2b_2 columns. There are $\left\lceil \frac{n}{p_1b_1} \right\rceil$ superblocks in the row direction and $\left\lceil \frac{n}{p_2b_2} \right\rceil$ in the column direction. Each of the p processors owns one data block of size $b_1 \times b_2$ of a superblock. The two-dimensional matrix A is covered by these superblocks and from this covering, it can be estimated how many elements of smaller matrices $A^{(k)}$ are owned by a specific processor.

The number of elements owned by a processor q in row k for column indices $\geq k$ can be estimated by

$$N_q^{row \geq k} \leq \left\lceil \frac{n-k+1}{p_2b_2} \right\rceil b_2 \leq \left(\frac{n-k+1}{p_2b_2} + 1 \right) b_2 = \frac{n-k+1}{p_2} + b_2 \qquad (8.11)$$

where $\left\lceil \frac{n-k+1}{p_2b_2} \right\rceil$ is the number of superblocks covering row k for column indices $\geq k$, which are $n - k + 1$ indices, and b_2 is the number of column elements that each processor of $Ro(k)$ owns in a complete superblock. For the covering of one row, the number of columns p_2b_2 of a superblock is needed. Analogously, the number of elements owned by a processor q in column k for row indices $> k$ can be estimated by

$$N_q^{col > k} \leq \left\lceil \frac{n-k}{p_1b_1} \right\rceil b_1 \leq \left(\frac{n-k}{p_1b_1} + 1 \right) b_1 = \frac{n-k}{p_1} + b_1 \qquad (8.12)$$

where $\left\lceil \frac{n-k}{p_1b_1} \right\rceil$ is the number of superblocks covering column k for row indices $> k$, which are $n - k$ row indices, and b_1 is the number of row elements that each processor of $Co(k)$ owns in a complete superblock. Using these estimations, the parallel execution time in Formula (8.10) can be approximated by

$$T(n,p) \approx \sum_{k=1}^{n-1} \left(T_{sb}(p_1, \frac{n-k+1}{p_2} + b_2) + (\frac{n-k}{p_1} + b_1) \cdot t_{op} \right.$$
$$\left. + \ T_{sb}(p_2, \frac{n-k}{p_1} + b_1) + (\frac{n-k}{p_1} + b_1)(\frac{n-k}{p_2} + b_2) \cdot 2t_{op} \right).$$

Suitable parameters leading to a good performance can be derived from this modeling. For the communication time of a single-broadcast operation, we assume a communication time

$$T_{sb}(p,m) = \log p \cdot (\tau + m \cdot t_c),$$

with a startup time τ and a transfer time t_c. This formula models the communication time in many interconnection networks, like a hypercube. Using the summation formula $\sum_{k=1}^{n-1}(n-k+1) = \sum_{k=2}^{n}k = (\sum_{k=1}^{n}k) - 1 = \frac{n(n+1)}{2} - 1$ the communication time in phase 4. results in

$$\sum_{k=1}^{n-1} T_{sb}\left(p_1, \frac{n-k+1}{p_2} + b_2\right)$$

$$= \sum_{k=1}^{n-1} \log p_1 \left(\left(\frac{n-k+1}{p_2} + b_2\right) t_c + \tau\right)$$

$$= \log p_1 \left(\left(\frac{n(n+1)}{2} - 1\right) \frac{1}{p_2} t_c + (n-1)b_2 t_c + (n-1)\tau\right).$$

For the second and the third term the summation formula $\sum_{k=1}^{n-1}(n-k) = \frac{n(n-1)}{2}$ is used, so that the computation time

$$\sum_{k=1}^{n-1}\left(\frac{n-k}{p_1} + b_1\right) \cdot t_{op} = \left(\frac{n(n-1)}{2p_1} + (n-1)b_1\right) \cdot t_{op}$$

and the communication time

$$\sum_{k=1}^{n-1} T_{sb}\left(p_2, \frac{n-k}{p_1} + b_1\right)$$

$$= \sum_{k=1}^{n-1} \log p_2 \left(\left(\frac{n-k}{p_1} + b_1\right) t_c + \tau\right)$$

$$= \log p_2 \left(\left(\frac{n(n-1)}{2}\right) \frac{1}{p_1} t_c + (n-1)b_1 t_c + (n-1)\tau\right).$$

result. For the last term, the summation formula $\sum_{k=1}^{n-1}\frac{n-k}{p_1} \cdot \frac{n-k}{p_2} = \frac{1}{p}\sum_{k=1}^{n-1}k^2 = \frac{1}{p}\frac{n(n-1)(2n-1)}{6}$ is used. The total parallel execution time is

$$T(n,p) = \log p_1 \left(\left(\frac{n(n+1)}{2} - 1\right)\frac{t_c}{p_2} + (n-1)b_2 t_c + (n-1)\tau\right)$$

$$+ \left(\frac{n(n-1)}{2}\frac{1}{p_1} + (n-1)b_1\right)t_{op}$$

$$+ \log p_2 \left(\left(\frac{n(n-1)}{2}\frac{t_c}{p_1}\right) + (n-1)b_1 t_c + (n-1)\tau\right)$$

$$+ \left(\frac{n(n-1)(2n-1)}{6p} + \frac{n(n-1)}{2}\left(\frac{b_2}{p_1} + \frac{b_1}{p_2}\right) + (n-1)b_1 b_2\right)2t_{op}.$$

The block sizes b_i, $1 \le b_i \le n/p_i$, for $i = 1, 2$ are contained in the execution time as factors and, thus, the minimal execution time is achieved for $b_1 = b_2 = 1$. In the resulting formula the terms $(\log p_1 + \log p_2)((n-1)(\tau + t_c)) = \log p((n-1)(\tau + t_c))$, $(n-1) \cdot 3t_{op}$ and $\frac{n(n-1)(2n-1)}{3p} \cdot t_{op}$ are independent of the specific choice of p_1 and p_2 and need not be considered. The terms $\frac{n(n-1)}{2}\frac{1}{p_1}t_{op}$ and $\frac{t_c}{p_2}(n-1)\log p_1$ are asymmetric in p_1 and p_2. For simplicity we ignore these terms in the analysis, which is justified since these terms are small compared to the remaining terms; the first term has t_{op} as operand, which is usually small, and the second term with t_c as operand has a factor only linear in n. The remaining terms of the execution time are symmetric in p_1 and p_2 and have constants quadratic in n. Using $p_2 = p/p_1$ this time can be expressed as

$$T_S(p_1) = \frac{n(n-1)}{2} \left(\frac{p_1 \log p_1}{p} + \frac{\log p - \log p_1}{p_1} \right) t_c + \frac{n(n-1)}{2} \left(\frac{1}{p_1} + \frac{p_1}{p} \right) 2t_{op} .$$

The first derivative is

$$T'_S(p_1) = \frac{n(n-1)}{2} \left(\frac{1}{p \cdot \ln 2} + \frac{\log p_1}{p} - \frac{\log p}{p_1^2} + \frac{\log p_1}{p_1^2} - \frac{1}{p_1^2 \cdot \ln 2} \right) t_c$$
$$+ \frac{n(n-1)}{2} \left(\frac{1}{p} - \frac{1}{p_1^2} \right) 2t_{op} .$$

For $p_1 = \sqrt{p}$ it is $T'_S(p_1) = 0$ since $\frac{1}{p} - \frac{1}{p_1^2} = \frac{1}{p} - \frac{1}{p} = 0$, $\frac{1}{p \ln 2} - \frac{1}{p_1^2 \ln 2} = 0$, and $\frac{\log p_1}{p} - \frac{\log p}{p_1^2} + \frac{\log p_1}{p_1^2} = 0$. The second derivation $T''(p_1)$ is positive for $p_1 = \sqrt{p}$ and, thus, there is a minimum at $p_1 = p_2 = \sqrt{p}$.

In summary, the analysis of the most influential parts of the parallel execution time of the Gaussian elimination has shown that $p_1 = p_2 = \sqrt{p}$, $b_1 = b_2 = 1$ is the best choice. For an implementation, the values for p_1 and p_2 have to be adapted to integer values.

8.2 Direct Methods for Linear Systems with Banded Structure

Large linear systems with banded structure often arise when discretizing partial differential equations. The coefficient matrix of a banded system is sparse with non-zero elements in the main diagonal of the matrix and a few further diagonals. As a motivation, we first present the discretization of a two-dimensional Poisson equation resulting in such a banded system in Sect. 8.2.1. In Sect. 8.2.2, the solution methods recursive doubling and cyclic reduction are applied for the solution of tridiagonal systems, i.e., banded systems with only three non-zero diagonals, and the parallel implementation is discussed. General banded matrices are treated with cyclic reduction in Subsection 8.2.3 and the discretized Poisson equation is used as an example in Subsection 8.2.4.

8.2.1 Discretization of the Poisson Equation

As a typical example of an elliptic partial differential equation we consider the Poisson equation with Dirichlet boundary conditions. This equation is often called the **model problem** since its structure is simple but the numerical solution is very similar for many other more complicated partial differential equations, see [73, 96, 209].

The two-dimensional Poisson equation has the form

$$-\Delta u(x,y) = f(x,y) \quad \text{for all } (x,y) \in \Omega \tag{8.13}$$

with domain $\Omega \subset \mathbb{R}^2$.

The function $u : \mathbb{R}^2 \to \mathbb{R}$ is the unknown solution function and the function $f : \mathbb{R}^2 \to \mathbb{R}$ is the right-hand side, which is continuous in Ω and its boundary. The operator Δ is the two-dimensional **Laplace operator**,

$$\Delta = \frac{\partial^2}{\partial x^2} + \frac{\partial^2}{\partial y^2}$$

containing the second partial derivatives with respect to x or y. ($\partial/\partial x$ and $\partial/\partial y$ denote the first partial derivatives with respect to x or y, and $\partial^2/\partial x^2$ and $\partial^2/\partial y^2$ denote the second partial derivatives with respect to x or y, respectively.) Using this notation, the Poisson equation (8.13) can also be written as

$$-\frac{\partial^2 u}{\partial x^2} - \frac{\partial^2 u}{\partial y^2} = f(x,y) \, .$$

The model problem (8.13) uses the unit square $\Omega = (0,1) \times (0,1)$ and assumes a Dirichlet boundary condition

$$u(x,y) = \varphi(x,y) \quad \text{for all } (x,y) \in \partial\Omega \, , \tag{8.14}$$

where φ is a given function and $\partial\Omega$ is the boundary of domain Ω, which is $\partial\Omega = \{(x,y) \mid 0 \le x \le 1, y=0 \text{ or } y=1\} \cup \{(x,y) \mid 0 \le y \le 1, x=0 \text{ or } x=1\}$. The boundary condition uniquely determines the solution u of the model problem. Figure 8.7 (left), illustrates the domain and the boundary of the model problem.

An example of the Poisson equation from electrostatics is the equation

$$\Delta u = -\frac{\rho}{\varepsilon_0} \, ,$$

where ρ is the charge density, ε_0 is a constant and u is the unknown potential to be determined [119].

For the numerical solution of equation $-\Delta u(x,y) = f(x,y)$, the method of finite differences can be used, which is based on a discretization of the domain $\Omega \cup \partial\Omega$ in both directions. The discretization is given by a regular mesh with $N+2$ mesh points in x-direction and in y-direction, where N points are in the inner part and 2 points are on the boundary. The distance between points in the x- or y-direction is $h = \frac{1}{N+1}$. The mesh points are

$$(x_i, y_j) = (ih, jh) \text{ for } i, j = 0, 1, \cdots, N+1 \, .$$

Poisson equation   mesh for the unit square

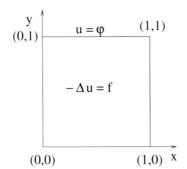
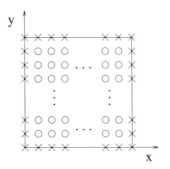

× boundary values
○ inner mesh points

Fig. 8.7 Left: Poisson equation with Dirichlet boundary condition on the unit square $\Omega =$ $(0,1) \times (0,1)$. Right: The numerical solution discretizes the Poisson equation on a mesh with equidistant mesh points with distance $1/(N+1)$. The mesh has N^2 inner mesh points and additional mesh points on the boundary.

The points on the boundary are the points with $x_0 = 0, y_0 = 0, x_{N+1} = 1$ or $y_{N+1} = 1$. The unknown solution function u is determined at the points (x_i, y_j) of this mesh, which means that values $u_{ij} := u(x_i, y_j)$ for $i, j = 0, 1, \cdots, N+1$ are to be found.

For the inner part of the mesh, these values are determined by solving a linear equation system with N^2 equations which is based on the Poisson equation in the following way. For each mesh point (x_i, y_j), $i, j = 1, \cdots, N$, a Taylor expansion is used for the x or y-direction. The Taylor expansion in x-direction is

$$u(x_i + h, y_j) = u(x_i, y_j) + h \cdot u_x(x_i, y_j) + \frac{h^2}{2} u_{xx}(x_i, y_j)$$
$$+ \frac{h^3}{6} u_{xxx}(x_i, y_j) + O(h^4) \,,$$
$$u(x_i - h, y_j) = u(x_i, y_j) - h \cdot u_x(x_i, y_j) + \frac{h^2}{2} u_{xx}(x_i, y_j)$$
$$- \frac{h^3}{6} u_{xxx}(x_i, y_j) + O(h^4) \,,$$

where u_x denotes the partial derivative in x-direction (i.e., $u_x = \partial u / \partial x$) and u_{xx} denotes the second partial derivative in x-direction (i.e., $u_{xx} = \partial^2 u / \partial x^2$). Adding these two Taylor expansions results in

$$u(x_i + h, y_j) + u(x_i - h, y_j) = 2u(x_i, y_j) + h^2 u_{xx}(x_i, y_j) + O(h^4) \,.$$

Analogously, the Taylor expansion for the y-direction can be used to get

$$u(x_i, y_j + h) + u(x_i, y_j - h) = 2u(x_i, y_j) + h^2 u_{yy}(x_i, y_j) + O(h^4) \,.$$

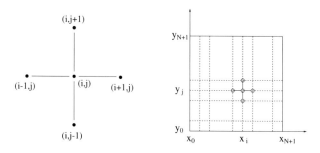

Fig. 8.8 Five-point stencil resulting from the discretization of the Laplace operator with a finite difference scheme. The computation at one mesh point uses values at the four neighbor mesh points.

From the last two equations, an approximation for the Laplace operator $\Delta u = u_{xx} + u_{yy}$ at the mesh points can be derived

$$\Delta u(x_i, y_j) = -\frac{1}{h^2}(4u_{ij} - u_{i+1,j} - u_{i-1,j} - u_{i,j+1} - u_{i,j-1}),$$

where the higher order terms $O(h^4)$ are neglected. This approximation uses the mesh point (x_i, y_j) itself and its four neighbor points; see Fig. 8.8. This pattern is known as **five-point stencil**. Using the approximation of Δu and the notation $f_{ij} := f(x_i, y_j)$ for the values of the right hand side, the **discretized Poisson equation** or **Five-point formula** results:

$$\frac{1}{h^2}(4u_{ij} - u_{i+1,j} - u_{i-1,j} - u_{i,j+1} - u_{i,j-1}) = f_{ij} \tag{8.15}$$

for $1 \leq i, j \leq N$. For the points on the boundary, the values of u_{ij} result from the boundary condition (8.14) and are given by

$$u_{ij} = \varphi(x_i, y_j) \tag{8.16}$$

for $i = 0, N+1$ and $j = 0, \ldots, N+1$ or $j = 0, N+1$ and $i = 0, \ldots, N+1$. The inner mesh points which are immediate neighbors of the boundary, i.e., the mesh points with $i = 1, i = N, j = 1$ or $j = N$, use the boundary values in their five-point stencil; the four mesh points in the corners use two boundary values and all other points use one boundary value. For all points with $i = 1, i = N, j = 1$ or $j = N$, the values of u_{ij} in the formulas (8.15) are replaced by the values (8.16). For the mesh point (x_1, y_1) for example the equation

$$\frac{1}{h^2}(4u_{11} - u_{21} - u_{12}) = f_{11} + \frac{1}{h^2}\varphi(0, y_1) + \frac{1}{h^2}\varphi(x_1, 0)$$

results. The five-point formula (8.15) including boundary values represents a linear equation system with N^2 equations, N^2 unknown values and a coefficient matrix

$A \in \mathbb{R}^{N^2 \times N^2}$. In order to write the equation system (8.15) with boundary values (8.16) in matrix form $Az = d$, the N^2 unknowns u_{ij}, $i, j = 1, \ldots, N$, are arranged in row-oriented order in a one-dimensional vector z of size $n = N^2$ which has the form

$$z = \left(u_{11}, u_{21}, \ldots, u_{N1}, u_{12}, u_{22}, \ldots, u_{N2}, \ldots, u_{1N}, u_{2N}, \ldots, u_{NN} \right).$$

The mapping of values u_{ij} to vector elements z_k is

$$z_k := u_{ij} \text{ with } k = i + (j-1)N \text{ for } i, j = 1, \ldots, N.$$

Using the vector z, the five-point formula has the form

$$\frac{1}{h^2} \left(4z_{i+(j-1)N} - z_{i+1+(j-1)N} - z_{i-1+(j-1)N} - z_{i+jN} - z_{i+(j-2)N} \right)$$
$$= d_{i+(j-1)N}$$

with $d_{i+(j-1)N} = f_{ij}$ and a corresponding mapping of the values f_{ij} to a one-dimensional vector d. Replacing the indices by $k = 1, \ldots, n$ with $k = i + (j-1)N$ results in

$$\frac{1}{h^2} \left(4z_k - z_{k+1} - z_{k-1} - z_{k+N} - z_{k-N} \right) = d_k. \tag{8.17}$$

Thus, the entries in row k of the coefficient matrix contain five entries which are $a_{kk} = 4$ and $a_{k,k+1} = a_{k,k-1} = a_{k,k+N} = a_{k,k-N} = -1$.

The building of the vector d and the coefficient matrix $A = (a_{ij})$, $i, j = 1, \ldots, N^2$, can be performed by the following algorithm, see [96]. The loops over i and j, $i, j = 1, \ldots, N$ visit the mesh points (i, j) and build one row of the matrix A of size $N^2 \times N^2$. When (i, j) is an inner point of the mesh, i.e., $i, j, \neq 1, N$, the corresponding row of A contains five elements at the position $k, k+1, k-1, k+N, k-N$ for $k = i + (j-1)N$. When (i, j) is at the boundary of the inner part, i.e., $i = 1, j = 1, i = N$ or $j = N$, the boundary values for φ are used.

```
/* Algorithm for building the matrix A and the vector d */
Initialize all entries of A with 0;
for (j = 1; j <= N; j++)
    for (i = 1; i <= N; i++) {
        /* Build dk and row k of A with k = i + (j-1)N */
        k = i + (j-1)·N;
        a_{k,k} = 4/h²;
        d_k = f_{ij};
        if (i > 1) a_{k,k-1} = -1/h² else d_k = d_k + 1/h² φ(0,y_j);
        if (i < N) a_{k,k+1} = -1/h² else d_k = d_k + 1/h² φ(1,y_j);
        if (j > 1) a_{k,k-N} = -1/h² else d_k = d_k + 1/h² φ(x_i,0);
        if (j < N) a_{k,k+N} = -1/h² else d_k = d_k + 1/h² φ(x_i,1);
    }
```

The linear equation system resulting from this algorithm has the structure:

$$\frac{1}{h^2} \begin{pmatrix} B & -I & & 0 \\ -I & B & \ddots & \\ & \ddots & \ddots & -I \\ 0 & & -I & B \end{pmatrix} \cdot z = d \, , \tag{8.18}$$

where I denotes the $N \times N$ unit matrix, which has the value 1 in the diagonal elements and the value 0 in all other entries. The matrix B has the structure

$$B = \begin{pmatrix} 4 & -1 & & 0 \\ -1 & 4 & \ddots & \\ & \ddots & \ddots & -1 \\ 0 & & -1 & 4 \end{pmatrix} . \tag{8.19}$$

Figure 8.9 illustrates the two-dimensional mesh with five-point stencil (above) and the sparsity structure of the corresponding coefficient matrix A of Formula (8.17).

In summary, the Formulas (8.15) and (8.17) represent a linear equation system with a sparse coefficient matrix, which has non-zero elements in the main diagonal and its direct neighbors as well as in the diagonals in distance N. Thus, the linear equation system resulting from the Poisson equation has a banded structure, which should be exploited when solving the system. In the following, we present solution methods for linear equation systems with banded structure and start the description with tridiagonal systems. These systems have only three non-zero diagonals in the main diagonal and its two neighbors. A tridiagonal system results, for example, when discretizing the one-dimensional Poisson equation.

8.2.2 Tridiagonal Systems

For the solution of a linear equation system $Ax = y$ with a banded or tridiagonal coefficient matrix $A \in \mathbb{R}^{n \times n}$ specific solution methods can exploit the sparse matrix structure. A matrix $A = (a_{ij})_{i,j=1,\ldots,n} \in \mathbb{R}^{n \times n}$ is called banded when its structure takes the form of a band of non-zero elements around the principal diagonal. More precisely, this means a matrix A is a **banded matrix** if there exists $r \in \mathbb{N}$, $r \leq n$, with

$$a_{ij} = 0 \text{ for } |i - j| > r \, .$$

The number r is called the **semi-bandwidth** of A. For $r = 1$ a banded matrix is called **tridiagonal matrix**. We first consider the solution of tridiagonal systems which are linear equation systems with tridiagonal coefficient matrix.

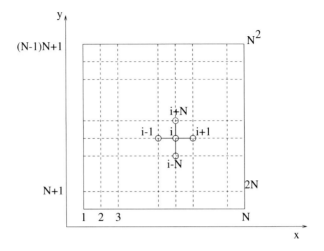

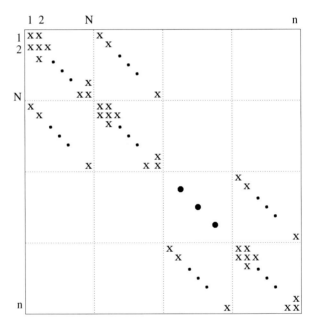

Fig. 8.9 Rectangular mesh in the x-y-plane of size $N \times N$ and the $n \times n$ coefficient matrix with $n = N^2$ of the corresponding linear equation system of the five-point formula. The sparsity structure of the matrix corresponds to the adjacency relation of the mesh points. The mesh can be considered as adjacency graph of the non-zero elements of the matrix.

8.2.2.1 Gaussian elimination for tridiagonal systems

For the solution of a linear equation system $Ax = y$ with tridiagonal matrix A, the Gaussian elimination can be used. Step k, $k < n$, of the forward elimination (without pivoting) results in the following computations, see also Section 8.1:

1. Compute $l_{ik} := a_{ik}^{(k)} / a_{kk}^{(k)}$ for $i = k+1, \ldots, n$.
2. Subtract l_{ik} times the kth row from the rows $i = k+1, \ldots, n$, i.e., compute

$$a_{ij}^{(k+1)} = a_{ij}^{(k)} - l_{ik} \cdot a_{kj}^{(k)} \quad \text{for } k \leq j \leq n \text{ and } k < i \leq n .$$

The vector y is changed analogously.

Because of the tridiagonal structure of A, all matrix elements a_{ik} with $i \geq k+2$ are zero elements, i.e., $a_{ik} = 0$. Thus, in each step k of the Gaussian elimination only one elimination factor $l_{k+1} := l_{k+1,k}$ and only one row with only one new element have to be computed. Using the notation

$$A = \begin{pmatrix} b_1 & c_1 & & & 0 \\ a_2 & b_2 & c_2 & & \\ & a_3 & b_3 & \ddots & \\ & & \ddots & \ddots & c_{n-1} \\ 0 & & & a_n & b_n \end{pmatrix} \tag{8.20}$$

for the matrix elements and starting with $u_1 = b_1$, these computations are:

$$l_{k+1} = a_{k+1}/u_k \tag{8.21}$$
$$u_{k+1} = b_{k+1} - l_{k+1} \cdot c_k .$$

After $n - 1$ steps an LU decomposition $A = LU$ of the matrix (8.20) with

$$L = \begin{pmatrix} 1 & & & 0 \\ l_2 & 1 & & \\ & \ddots & \ddots & \\ 0 & & l_n & 1 \end{pmatrix} \quad \text{and} \quad U = \begin{pmatrix} u_1 & c_1 & & 0 \\ & \ddots & \ddots & \\ & & u_{n-1} & c_{n-1} \\ 0 & & & u_n \end{pmatrix}$$

results. The right hand side y is transformed correspondingly according to

$$\tilde{y}_{k+1} = y_{k+1} - l_{k+1} \cdot \tilde{y}_k .$$

The solution x is computed from the upper triangular matrix U by a backward substitution, starting with $x_n = \tilde{y}_n/u_n$ and solving the equations $u_i x_i + c_i x_{i+1} = \tilde{y}_i$ one after another resulting in

$$x_i = \frac{\tilde{y}_i}{u_i} - \frac{c_i}{u_i} x_{i+1} \quad \text{for } i = n-1, \ldots, 1 .$$

The computational complexity of the Gaussian elimination is reduced to $O(n)$ for tridiagonal systems. However, the elimination phase computing l_k and u_k according to Eq. (8.21) is inherently sequential, since the computation of l_{k+1} depends on u_k and the computation of u_{k+1} depends on l_{k+1}. Thus, in this form the Gaussian elimination or LU decomposition has to be computed sequentially and is not suitable for a parallel implementation.

8.2.2.2 Recursive doubling for tridiagonal systems

An alternative approach for solving a linear equation system with tridiagonal matrix is the method of **recursive doubling** or **cyclic reduction**. The methods of recursive doubling and cyclic reduction also use elimination steps but contain potential parallelism [86, 85]. Both techniques can be applied if the coefficient matrix is either symmetric and positive definite or diagonally dominant [144]. The elimination steps in both methods are applied to linear equation systems $Ax = y$ with the tridiagonal matrix structure shown in (8.20), i.e.,

$$
\begin{aligned}
b_1\, x_1 + c_1\, x_2 &= y_1 \\
a_i\, x_{i-1} + b_i\, x_i + c_i\, x_{i+1} &= y_i \quad \text{for } i = 2, \ldots, n-1 \\
a_n\, x_{n-1} + b_n\, x_n &= y_n \ .
\end{aligned}
$$

The method, which was first introduced by Hockney and Golub in [112], uses two equations $i-1$ and $i+1$ to eliminate the variables x_{i-1} and x_{i+1} from equation i. This results in a new equivalent equation system with a coefficient matrix with three non-zero diagonals where the diagonals are moved to the outside. Recursive doubling and cyclic reduction can be considered as two implementation variants for the same numerical idea of the method of Hockney and Golub. The implementation of recursive doubling repeats the elimination step which finally results in a matrix structure in which only the elements in the principal diagonal are non-zero and the solution vector x can be computed easily. Cyclic reduction is a variant of recursive doubling which also eliminates variables using neighboring rows. But in each step the elimination is only applied to half of the equations and, thus, less computations are performed. On the other hand, the computation of the solution vector x requires a substitution phase.

We would like to mention that the terms recursive doubling and cyclic reduction are used in different ways in the literature. Cyclic reduction is sometimes used for the numerical method of Hockney and Golub in both implementation variants, see [73, 144]. On the other hand recursive doubling (or full recursive doubling) is sometimes used for a different method, the method of Stone [211]. This method applies the implementation variants sketched above in Equation (8.21) resulting from the Gaussian elimination, see [74, 220]. In the following, we start the description of recursive doubling for the method of Hockney and Golub according to [74] and [13].

Recursive doubling considers three neighboring equations $i-1, i, i+1$ of the equation system $Ax = y$ with tridiagonal coefficient matrix A in the form (8.20) for $i = 3, 4, \ldots, n-2$. These equations are:

$$
\begin{aligned}
a_{i-1} x_{i-2} &+ b_{i-1} x_{i-1} &+ c_{i-1} x_i & & & = y_{i-1} \\
&\quad\ a_i x_{i-1} &+ b_i x_i & + c_i x_{i+1} & & = y_i \\
& & a_{i+1} x_i & + b_{i+1} x_{i+1} & + c_{i+1} x_{i+2} & = y_{i+1}
\end{aligned}
$$

Equation $i-1$ is used to eliminate x_{i-1} from the ith equation and equation $i+1$ is used to eliminate x_{i+1} from the ith equation. This is done by reformulating equation $i-1$ and $i+1$ such that x_{i-1} and x_{i+1} are expressed by the following formulas:

$$
x_{i-1} = \frac{y_{i-1}}{b_{i-1}} - \frac{a_{i-1}}{b_{i-1}} x_{i-2} - \frac{c_{i-1}}{b_{i-1}} x_i ,
$$

$$
x_{i+1} = \frac{y_{i+1}}{b_{i+1}} - \frac{a_{i+1}}{b_{i+1}} x_i - \frac{c_{i+1}}{b_{i+1}} x_{i+2} ,
$$

and inserting those formulas of x_{i-1} and x_{i+1} into equation i. The resulting new equation i is

$$
a_i^{(1)} x_{i-2} + b_i^{(1)} x_i + c_i^{(1)} x_{i+2} = y_i^{(1)} \tag{8.22}
$$

with coefficients

$$
\begin{aligned}
a_i^{(1)} &= \alpha_i^{(1)} \cdot a_{i-1} \\
b_i^{(1)} &= b_i + \alpha_i^{(1)} \cdot c_{i-1} + \beta_i^{(1)} \cdot a_{i+1} \\
c_i^{(1)} &= \beta_i^{(1)} \cdot c_{i+1} \\
y_i^{(1)} &= y_i + \alpha_i^{(1)} \cdot y_{i-1} + \beta_i^{(1)} \cdot y_{i+1}
\end{aligned} \tag{8.23}
$$

and

$$
\alpha_i^{(1)} := -a_i / b_{i-1} ,
$$

$$
\beta_i^{(1)} := -c_i / b_{i+1} ,
$$

where the upper index (1) indicates that the coefficients are newly computed in the first step. For the special cases $i = 1, 2, n-1, n$, the coefficients are given by:

$$
\begin{aligned}
b_1^{(1)} &= b_1 + \beta_1^{(1)} \cdot a_2 & y_1^{(1)} &= y_1 + \beta_1^{(1)} \cdot y_2 \\
b_n^{(1)} &= b_n + \alpha_n^{(1)} \cdot c_{n-1} & y_n^{(1)} &= b_n + \alpha_n^{(1)} \cdot y_{n-1} \\
a_1^{(1)} &= a_2^{(1)} = 0 \text{ and} & c_{n-1}^{(1)} &= c_n^{(1)} = 0 .
\end{aligned}
$$

The values for $a_{n-1}^{(1)}, a_n^{(1)}, b_2^{(1)}, b_{n-1}^{(1)}, c_1^{(1)}, c_2^{(1)}, y_2^{(1)}$ and $y_{n-1}^{(1)}$ are defined as in equation (8.23). The equations (8.22) form a linear equation system $A^{(1)} x = y^{(1)}$ with a coefficient matrix having a new main diagonal with non-zero elements and two new diagonals with non-zero elements in distance 2 from the main diagonal:

$$A^{(1)} = \begin{pmatrix} b_1^{(1)} & 0 & c_1^{(1)} & & & & 0 \\ 0 & b_2^{(1)} & 0 & c_2^{(1)} & & & \\ a_3^{(1)} & 0 & b_3^{(1)} & \ddots & \ddots & & \\ & a_4^{(1)} & \ddots & & \ddots & \ddots & c_{n-2}^{(1)} \\ & & \ddots & \ddots & \ddots & 0 \\ 0 & & & a_n^{(1)} & 0 & b_n^{(1)} \end{pmatrix}$$

Comparing the structure of $A^{(1)}$ with the structure of A, it can be seen that the diagonals are moved to the outside and now have a distance 2 to the main diagonal.
In the second step, this method is applied to the equations $i-2, i, i+2$ of the equation system $A^{(1)}x = y^{(1)}$ for $i = 5, 6, \dots, n-4$. Equation $i-2$ is used to eliminate
x_{i-2} from the ith equation and equation $i+2$ is used to eliminate x_{i+2} from the ith
equation. This results in a new ith equation

$$a_i^{(2)} x_{i-4} + b_i^{(2)} x_i + c_i^{(2)} x_{i+4} = y_i^{(2)} ,$$

which contains the variables x_{i-4}, x_i and x_{i+4}. The cases $i = 1, \dots, 4, n-3, \dots, n$ are
treated separately as shown for the first elimination step. Altogether a next equation
system $A^{(2)}x = y^{(2)}$ results in which the diagonals are further moved to the outside
and now have a distance 4 to the main diagonal. The structure of $A^{(2)}$ is:

$$A^{(2)} = \begin{pmatrix} b_1^{(2)} & 0 & 0 & 0 & c_1^{(2)} & & & & 0 \\ 0 & b_2^{(2)} & & & & c_2^{(2)} & & \\ 0 & & \ddots & & & & \ddots & \\ 0 & & & \ddots & & & & c_{n-4}^{(2)} \\ a_5^{(2)} & & & & \ddots & & & 0 \\ & a_6^{(2)} & & & & \ddots & & 0 \\ & & \ddots & & & & \ddots & 0 \\ 0 & & & a_n^{(2)} & 0 & 0 & 0 & b_n^{(2)} \end{pmatrix} .$$

Again, the upper index (2) indicates the newly computed coefficients in this
transformation step. The following steps of the recursive doubling algorithm apply the same method to the modified equation system of the last step. Step k transfers the side-diagonals $2^k - 1$ positions away from the main diagonal, compared

to the original coefficient matrix. This is achieved by considering the equations $i - 2^{k-1}, i, i + 2^{k-1}$:

$$a_{i-2^{k-1}}^{(k-1)} x_{i-2^k} + b_{i-2^{k-1}}^{(k-1)} x_{i-2^{k-1}} + c_{i-2^{k-1}}^{(k-1)} x_i \qquad\qquad = y_{i-2^{k-1}}^{(k-1)}$$

$$a_i^{(k-1)} x_{i-2^{k-1}} + b_i^{(k-1)} x_i + c_i^{(k-1)} x_{i+2^{k-1}} \qquad\qquad = y_i^{(k-1)}$$

$$a_{i+2^{k-1}}^{(k-1)} x_i + b_{i+2^{k-1}}^{(k-1)} x_{i+2^{k-1}} + c_{i+2^{k-1}}^{(k-1)} x_{i+2^k} = y_{i+2^{k-1}}^{(k-1)}$$

Equation $i - 2^{k-1}$ is used to eliminate $x_{i-2^{k-1}}$ from the ith equation and equation $i + 2^{k-1}$ is used to eliminate $x_{i+2^{k-1}}$ from the ith equation. Again, the elimination is performed by computing the coefficients for the next equation system. These coefficients are:

$$a_i^{(k)} = \alpha_i^{(k)} \cdot a_{i-2^{k-1}}^{(k-1)} \quad \text{for } i = 2^k + 1, \dots, n, \text{ and } a_i^{(k)} = 0 \text{ otherwise },$$

$$c_i^{(k)} = \beta_i^{(k)} \cdot c_{i+2^{k-1}}^{(k-1)} \quad \text{for } i = 1, \dots, n - 2^k, \text{ and } c_i^{(k)} = 0 \text{ otherwise}, \qquad (8.24)$$

$$b_i^{(k)} = \alpha_i^{(k)} \cdot c_{i-2^{k-1}}^{(k-1)} + b_i^{(k-1)} + \beta_i^{(k)} \cdot a_{i+2^{k-1}}^{(k-1)} \quad \text{for } i = 1, \dots, n,$$

$$y_i^{(k)} = \alpha_i^{(k)} \cdot y_{i-2^{k-1}}^{(k-1)} + y_i^{(k-1)} + \beta_i^{(k)} \cdot y_{i+2^{k-1}}^{(k-1)} \quad \text{for } i = 1, \dots, n,$$

with

$$\alpha_i^{(k)} := -a_i^{(k-1)} / b_{i-2^{k-1}}^{(k-1)} \quad \text{for } i = 2^{k-1} + 1, \dots, n, \qquad (8.25)$$

$$\beta_i^{(k)} := -c_i^{(k-1)} / b_{i+2^{k-1}}^{(k-1)} \quad \text{for } i = 1, \dots, n - 2^{k-1}.$$

The modified equation i results by multiplying equation $i - 2^{k-1}$ from step $k - 1$ with $\alpha_i^{(k)}$, multiplying equation $i + 2^{k-1}$ from step $k - 1$ with $\beta_i^{(k)}$, and adding both to equation i. The resulting ith equation is

$$a_i^{(k)} x_{i-2^k} + b_i^{(k)} x_i + c_i^{(k)} x_{i+2^k} = y_i^{(k)} \qquad (8.26)$$

with the coefficients (8.24). The cases $k = 1, 2$ are special cases of this formula. The initialization for $k = 0$ is the following:

$$a_i^{(0)} = a_i \qquad \text{for } i = 2, \dots, n,$$

$$b_i^{(0)} = b_i \qquad \text{for } i = 1, \dots, n,$$

$$c_i^{(0)} = c_i \qquad \text{for } i = 1, \dots, n - 1,$$

$$y_i^{(0)} = y_i \qquad \text{for } i = 1, \dots, n,$$

and $a_1^{(0)} = 0$, $c_n^{(0)} = 0$. Also, for the steps $k = 0, \dots, \lceil \log n \rceil$ and $i \in \mathbb{Z} \setminus \{1, \dots, n\}$ the values

$$a_i^{(k)} = c_i^{(k)} = y_i^{(k)} = 0,$$

$$b_i^{(k)} = 1,$$

$$x_i = 0,$$

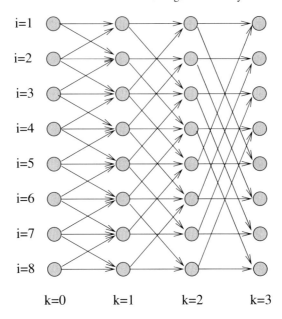

Fig. 8.10 Dependence graph for the computation steps of the recursive doubling algorithm in the case of three computation steps and eight equations. The computations of step k are shown in column k of the illustration. Column k contains one node for each equation i, thus representing the computation of all coefficients needed in step k. Column 0 represents the data of the coefficient matrix of the linear system. An edge from a node i in step k to a node j in step $k+1$ means that the computation at node j needs at least one coefficient computed at node i.

are set. In each transformation step, the distance of the side diagonals to the main diagonal is doubled. After $N = \lceil \log n \rceil$ steps, the original matrix A is transformed into a diagonal matrix $A^{(N)}$

$$A^{(N)} = \text{diag}(b_1^{(N)}, \ldots, b_n^{(N)}),$$

in which only the main diagonal contains non-zero elements. The solution x of the linear equation system can be directly computed using this matrix $A^{(N)}$ and the correspondingly modified vector $y^{(N)}$ in the following way:

$$x_i = y_i^{(N)} / b_i^{(N)} \text{ for } i = 1, 2, \ldots, n.$$

In summary, the recursive doubling algorithm consists of two main phases:

Recursive doubling algorithm:

1. Elimination phase: compute the values $a_i^{(k)}$, $b_i^{(k)}$, $c_i^{(k)}$ and $y_i^{(k)}$ for $k = 1, \dots, \lceil \log n \rceil$ and $i = 1, \dots, n$ according to Eqs. (8.24) and (8.25).
2. Solution phase: compute $x_i = y_i^{(N)}/b_i^{(N)}$ for $i = 1, \dots, n$ with $N = \lceil \log n \rceil$.

The first phase consists of $\lceil \log n \rceil$ steps where in each step $O(n)$ values are computed. The sequential asymptotic runtime of the algorithm is therefore $O(n \cdot \log n)$ which is asymptotically slower than the $O(n)$ runtime for the Gaussian elimination approach described earlier in Subsection 8.2.2.1. The advantage of the recursive doubling is that the computations in each step of the elimination and the substitution phases are independent and can be performed in parallel. Figure 8.10 illustrates the computations of the recursive doubling algorithm and the data dependencies between different steps by a dependence graph in which nodes represent the computations and arrows between nodes show a data dependence.

8.2.2.3 Cyclic reduction for tridiagonal systems

The recursive doubling algorithm offers a large degree of potential parallelism but has a larger computational complexity than the Gaussian elimination caused by computational redundancy. The **cyclic reduction** algorithm is a modification of recursive doubling which reduces the amount of computations to be performed. In each step, only half of the variables in the equation system are eliminated which means that only half of the values $a_i^{(k)}$, $b_i^{(k)}$, $c_i^{(k)}$ and $y_i^{(k)}$ are computed. As a consequence, a computationally more expensive substitution phase is needed to compute the solution vector x. The elimination and the substitution phases of cyclic reduction are described by the following two phases.

Cyclic reduction algorithm:

1. Elimination phase: For $k = 1, \dots, \lfloor \log n \rfloor$ compute $a_i^{(k)}$, $b_i^{(k)}$, $c_i^{(k)}$ and $y_i^{(k)}$ with $i = 2^k, \dots, n$ and step-size 2^k. In each step, the number of equations of the form (8.26) is reduced by a factor of $1/2$. In step $k = \lfloor \log n \rfloor$, there is only one equation left for $i = 2^N$ with $N = \lfloor \log n \rfloor$.
2. Substitution phase: For $k = \lfloor \log n \rfloor, \dots, 0$ compute x_i according to Eq. (8.26) for $i = 2^k, \dots, n$ with step-size 2^{k+1}, resulting in:

$$x_i = \frac{y_i^{(k)} - a_i^{(k)} \cdot x_{i-2^k} - c_i^{(k)} \cdot x_{i+2^k}}{b_i^{(k)}} . \tag{8.27}$$

Implicitly, the cyclic reduction algorithm transforms the tridiagonal matrix in Eq. (8.20) $\log n$ times into the equation systems $A_{cyc}^{(i)} \cdot x = y_{cyc}^{(i)}$, $i = 1, 2, \dots, \log n$. For $N = \log n$ with $8 = n = 2^N$, the implicit coefficient matrices $A_{cyc}^{(1)}$, $A_{cyc}^{(2)}$, $A_{cyc}^{(3)}$ of the intermediate equation systems of cyclic reduction have the form given in Fig. 8.11.

After first transformation step of cyclic reduction:

$$
A_{cyc}^{(1)} =
\begin{pmatrix}
b_1 & c_1 & & & & & & \\
& b_2^{(1)} & & c_2^{(1)} & & & & \\
& a_3 & b_3 & c_3 & & & & \\
& a_4^{(1)} & & b_4^{(1)} & & c_4^{(1)} & & \\
& & & a_5 & b_5 & c_5 & & \\
& & & a_6^{(1)} & & b_6^{(1)} & & c_6^{(1)} \\
& & & & & a_7 & b_7 & c_7 \\
& & & & & a_8^{(1)} & & b_8^{(1)}
\end{pmatrix}
\qquad
\text{righthand side}\quad y_{cyc}^{(1)} =
\begin{pmatrix}
y_1 \\
y_2^{(1)} \\
y_3 \\
y_4^{(1)} \\
y_5 \\
y_6^{(1)} \\
y_7 \\
y_8^{(1)}
\end{pmatrix}
$$

After second transformation step of cyclic reduction:

$$
A_{cyc}^{(2)} =
\begin{pmatrix}
b_1 & c_1 & & & & & & \\
& b_2^{(1)} & & c_2^{(1)} & & & & \\
& a_3 & b_3 & c_3 & & & & \\
& & & b_4^{(2)} & & & c_4^{(2)} & \\
& & & a_5 & b_5 & c_5 & & \\
& & & a_6^{(1)} & & b_6^{(1)} & & c_6^{(1)} \\
& & & & & a_7 & b_7 & c_7 \\
& & & a_8^{(2)} & & & & b_8^{(2)}
\end{pmatrix}
\qquad
\text{righthand side}\quad y_{cyc}^{(2)} =
\begin{pmatrix}
y_1 \\
y_2^{(1)} \\
y_3 \\
y_4^{(2)} \\
y_5 \\
y_6^{(1)} \\
y_7 \\
y_8^{(2)}
\end{pmatrix}
$$

After third transformation step of cyclic reduction:

$$
A_{cyc}^{(3)} =
\begin{pmatrix}
b_1 & c_1 & & & & & & \\
& b_2^{(1)} & & c_2^{(1)} & & & & \\
& a_3 & b_3 & c_3 & & & & \\
& & & b_4^{(2)} & & & c_4^{(2)} & \\
& & & a_5 & b_5 & c_5 & & \\
& & & a_6^{(1)} & & b_6^{(1)} & & c_6^{(1)} \\
& & & & & a_7 & b_7 & c_7 \\
& & & & & & & b_8^{(3)}
\end{pmatrix}
\qquad
\text{righthand side}\quad y_{cyc}^{(3)} =
\begin{pmatrix}
y_1 \\
y_2^{(1)} \\
y_3 \\
y_4^{(2)} \\
y_5 \\
y_6^{(1)} \\
y_7 \\
y_8^{(3)}
\end{pmatrix}
$$

Fig. 8.11 Structure of the matrices $A_{cyc}^{(1)}$, $A_{cyc}^{(2)}$, $A_{cyc}^{(3)}$ and the right hand side vectors $v_{cyc}^{(1)}$, $y_{cyc}^{(2)}$, $y_{cyc}^{(3)}$ of the equation systems $A_{cyc}^{(i)} \cdot x = y_{cyc}^{(i)}$. The upper index at the matrix and vector elements indicates the transformation step of its calculation. The matrices and the vectors are usually implicit and only the newly computed elements are stored.

The upper index at a matrix element indicates the transformation step in which the value is computed. All matrix elements not shown are zero elements.

Figure 8.12 illustrates the computations of the elimination and the substitution phase of cyclic reduction represented by nodes and their dependencies represented by arrows. In each computation step k, $k = 1, \ldots, \lfloor \log n \rfloor$, of the elimination phase, there are $\frac{n}{2^k}$ nodes representing the computations for the coefficients of one equation. This results in

$$
\frac{n}{2} + \frac{n}{4} + \frac{n}{8} + \cdots + \frac{n}{2^N} = n \cdot \sum_{i=1}^{\lfloor \log n \rfloor} \frac{1}{2^i} \leq n
$$

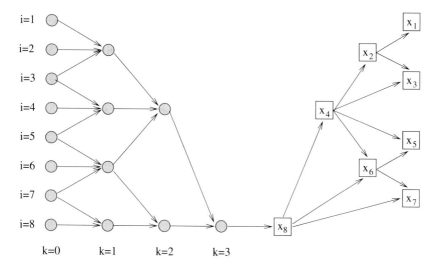

Fig. 8.12 Dependence graph illustrating the dependencies between neighboring computation steps of the cyclic reduction algorithm for the case of three computation steps and eight equations in analogy to the representation in Figure 8.10. The first four columns represent the computations of the coefficients. The last columns in the graph represent the computation of the solution vector x in the second phase of the cyclic reduction algorithm, see (8.27).

computation nodes with $N = \lfloor \log n \rfloor$ and the execution time of cyclic reduction is $O(n)$. Thus, the computational complexity is the same as for the Gaussian elimination, however, the cyclic reduction offers potential parallelism which can be exploited in a parallel implementation as described in the following.

The computations of the numbers $\alpha_i^{(k)}$, $\beta_i^{(k)}$ require a division by $b_i^{(k)}$ and, thus, cyclic reduction as well as recursive doubling is not possible if any number $b_i^{(k)}$ is zero. This can happen even when the original matrix is invertible and has non-zero diagonal elements or when the Gaussian elimination can be applied without pivoting. However, for many classes of matrices it can be shown that a division by zero is never encountered. Examples are matrices A which are symmetric and positive definite or invertible and diagonally dominant, see [74] or [144] (using the name odd-even reduction). (A matrix A is symmetric if $A = A^T$ and positive definite if $x^T A x > 0$ for all x. A matrix is diagonally dominant if in each row the absolute value of the diagonal element exceeds the sum of the absolute values of the other elements in the row without the diagonal in the row.)

8.2.2.4 Parallel implementation of cyclic reduction

We consider a parallel algorithm for the cyclic reduction for p processors. For the description of the phases we assume $n = p \cdot q$ for $q \in \mathbb{N}$ and $q = 2^Q$ for $Q \in \mathbb{N}$. Each processor stores a block of rows of size q, i.e., processor P_i stores the rows of A

with the numbers $(i-1)q+1,\ldots,i\cdot q$ for $1 \leq i \leq p$. We describe the parallel algorithm with data exchange operations that are needed for an implementation with a distributed address space. As data distribution a row-blockwise distribution of the matrix A is used to reduce the interaction between processors as much as possible. The parallel algorithm for the cyclic reduction comprises three phases: the elimination phase stopping earlier than described above, an additional recursive doubling phase, and a substitution phase.

Phase 1: Parallel reduction of the cyclic reduction in $\log q$ **steps:** Each processor computes the first $Q = \log q$ steps of the cyclic reduction algorithm, i.e., processor P_i computes for $k = 1,\ldots,Q$ the values

$$a_j^{(k)}, b_j^{(k)}, c_j^{(k)}, y_j^{(k)}$$

for $j = (i-1)\cdot q + 2^k,\ldots,i\cdot q$ with step-size 2^k. After each computation step, processor P_i receives four data values from P_{i-1} (if $i > 1$) and from processor P_{i+1} (if $i < n$) computed in the previous step. Since each processor owns a block of rows of size q, no communication with any other processor is required. The size of data to be exchanged with the neighboring processors is a multiple of 4 since four coefficients $(a_j^{(k)}, b_j^{(k)}, c_j^{(k)}, y_j^{(k)})$ are transferred. Only one data block is received per step and so there are at most $2Q$ messages of size 4 for each step.

Phase 2: Parallel recursive doubling for tridiagonal systems of size p: Processor P_i is responsible for the ith equation of the following p-dimensional tridiagonal system

$$\tilde{a}_i \tilde{x}_{i-1} + \tilde{b}_i \tilde{x}_i + \tilde{c}_i \tilde{x}_{i+1} = \tilde{y}_i \qquad \text{for } i = 1,\ldots,p$$

with

$$\left. \begin{aligned} \tilde{a}_i &= a_{i\cdot q}^{(Q)} \\ \tilde{b}_i &= b_{i\cdot q}^{(Q)} \\ \tilde{c}_i &= c_{i\cdot q}^{(Q)} \\ \tilde{y}_i &= y_{i\cdot q}^{(Q)} \\ \tilde{x}_i &= x_{i\cdot q} \end{aligned} \right\} \quad \text{for } i = 1,\ldots,p\,.$$

For the solution of this system, we use recursive doubling. Each processor is assigned one equation. Processor P_i performs $\lceil \log p \rceil$ steps of the recursive doubling algorithm. In step k, $k = 1,\ldots,\lceil \log p \rceil$, processor P_i computes

$$\tilde{a}_i^{(k)}, \tilde{b}_i^{(k)}, \tilde{c}_i^{(k)}, \tilde{y}_i^{(k)}$$

for which the values of

$$\tilde{a}_j^{(k-1)}, \tilde{b}_j^{(k-1)}, \tilde{c}_j^{(k-1)}, \tilde{y}_j^{(k-1)}$$

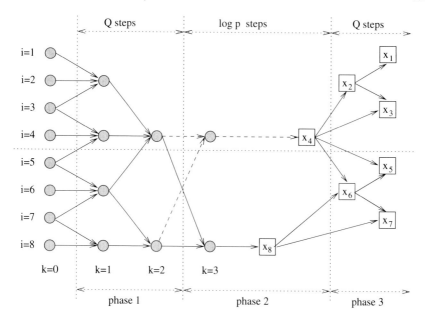

Fig. 8.13 Illustration of the parallel algorithm for the cyclic reduction for $n = 8$ equations and $p = 2$ processors. Each of the processors is responsible for $q = 4$ equations; we have $Q = 2$. The first and the third phase of the computation have $\log q = 2$ steps. The second phase has $\log p = 1$ step. As recursive doubling is used in the second phase, there are more components of the solution to be computed in the second phase compared with the computation shown in Figure 8.12.

from the previous step computed by a different processor are required. Thus, there is a communication in each of the $\lceil \log p \rceil$ steps with a message size of four values. After step $N' = \lceil \log p \rceil$ processor P_i computes

$$\tilde{x}_i = \tilde{y}_i^{(N')}/\tilde{b}_i^{(N')}. \tag{8.28}$$

Phase 3: Parallel substitution of cyclic reduction: After the second phase, the values $\tilde{x}_i = x_{i\cdot q}$ are already computed. In this phase, each processor P_i, $i = 1, \ldots, p$, computes the values x_j with $j = (i-1)q + 1, \ldots, iq - 1$ in several steps according to Eq. (8.27). In step k, $k = Q - 1, \ldots, 0$, the elements x_j, $j = 2^k, \ldots, n$, with step size 2^{k+1} are computed. Processor P_i computes x_j with j div $q + 1 = i$ for which the values $\tilde{x}_{i-1} = x_{(i-1)q}$ and $\tilde{x}_{i+1} = x_{(i+1)q}$ computed by processors P_{i-1} and P_{i+1} are needed. Figure 8.13 illustrates the parallel algorithm for $p = 2$ and $n = 8$.

8.2.2.5 Parallel execution time

The execution time of the parallel algorithm can be modeled by the following run-time functions.

Phase 1 executes

$$Q = \log q = \log \frac{n}{p} = \log n - \log p$$

steps where in step k with $1 \leq k \leq Q$ each processor computes at most $q/2^k$ coefficient blocks of 4 values each. Each coefficient block requires 14 arithmetic operations according to Eq. (8.23). The computation time of phase 1 can therefore be estimated as

$$T_1(n,p) = 14 t_{op} \cdot \sum_{k=1}^{Q} \frac{q}{2^k} \leq 14 \frac{n}{p} \cdot t_{op} .$$

Moreover, each processor exchanges in each of the Q steps two messages of 4 values each with its two neighboring processors by participating in single transfer operations. Since in each step the transfer operations can be performed by all processors in parallel without interference, the resulting communication time is

$$C_1(n,p) = 2Q \cdot t_{s2s}(4) = 2 \cdot \log \frac{n}{p} \cdot t_{s2s}(4) ,$$

where $t_{s2s}(m)$ denotes the time of a single-transfer operation with message size m. Phase 2 executes $\lceil \log p \rceil$ steps. In each step, each processor computes 4 coefficients requiring 14 arithmetic operations. Then the value $\tilde{x}_i = x_{i \cdot q}$ is computed according to Eq. (8.28) by a single arithmetic operation. The computation time is therefore

$$T_2(n,p) = 14 \lceil \log p \rceil \cdot t_{op} + t_{op} .$$

In each step, each processor sends and receives 4 data values from other processors, leading to a communication time

$$C_2(n,p) = 2 \lceil \log p \rceil \cdot t_{s2s}(4) .$$

In each step k of phase 3, $k = 0, \ldots, Q-1$, each processor computes 2^k components of the solution vector according to Eq. (8.27). For each component, five operations are needed. Altogether, each processor computes $\sum_{k=0}^{Q-1} 2^k = 2^Q - 1 = q - 1$ components with one component already computed in phase 2. The resulting computation time is

$$T_3(n,p) = 5 \cdot (q-1) \cdot t_{op} = 5 \cdot \left(\frac{n}{p} - 1 \right) \cdot t_{op} .$$

Moreover, each processor exchanges one data value with each of its neighboring processors; the communication time is therefore

$$C_3(n,p) = 2 \cdot t_{s2s}(1) \, .$$

The resulting total computation time is

$$
\begin{aligned}
T(n,p) &= \left(14\frac{n}{p} + 14 \cdot \lceil \log p \rceil + 5\frac{n}{p} - 4 \right) \cdot t_{op} \\
&\simeq \left(19\frac{n}{p} + 14 \cdot \log p \right) \cdot t_{op} \, .
\end{aligned}
$$

The communication overhead is

$$
\begin{aligned}
C(n,p) &= \left[2 \cdot \log\frac{n}{p} + 2\lceil \log p \rceil \right] t_{s2s}(4) + 2 \cdot t_{s2s}(1) \\
&\simeq 2 \cdot \log n \cdot t_{s2s}(4) + 2 \cdot t_{s2s}(1) \, .
\end{aligned}
$$

Compared to the sequential algorithm, the parallel implementation leads to a small computational redundancy of $14 \cdot \log p$ operations. The communication overhead increases logarithmically with the number of rows, whereas the computation time increases linearly.

8.2.3 Generalization to Banded Matrices

The cyclic reduction algorithm can be generalized to banded matrices with semi-bandwidth $r > 1$. For the description we assume $n = s \cdot r$. The matrix is represented as a block-tridiagonal matrix of the form

$$
\begin{pmatrix}
B_1^{(0)} & C_1^{(0)} & & & & 0 \\
A_2^{(0)} & B_2^{(0)} & C_2^{(0)} & & & \\
& \ddots & \ddots & \ddots & & \\
& & A_{s-1}^{(0)} & B_{s-1}^{(0)} & C_{s-1}^{(0)} & \\
0 & & & A_s^{(0)} & B_s^{(0)} &
\end{pmatrix}
\begin{pmatrix}
X_1 \\ X_2 \\ \vdots \\ X_{s-1} \\ X_s
\end{pmatrix}
=
\begin{pmatrix}
Y_1^{(0)} \\ Y_2^{(0)} \\ \vdots \\ Y_{s-1}^{(0)} \\ Y_s^{(0)}
\end{pmatrix}
$$

where

$$
\begin{aligned}
A_i^{(0)} &= (a_{lm})_{l \in I_i, m \in I_{i-1}} \quad \text{for } i = 2, \ldots, s \, , \\
B_i^{(0)} &= (a_{lm})_{l \in I_i, m \in I_i} \quad \text{for } i = 1, \ldots, s \, , \\
C_i^{(0)} &= (a_{lm})_{l \in I_i, m \in I_{i+1}} \quad \text{for } i = 1, \ldots, s-1 \, ,
\end{aligned}
$$

are sub-matrices of A. The index sets are for $i = 1, \ldots, s$

$$I_i = \{ j \in \mathbb{N} \mid (i-1)r < j \le ir \} .$$

The vectors $X_i, Y_i^{(0)} \in \mathbb{R}^r$ are

$$X_i = (x_l)_{l \in I_i} \text{ and } Y_i^{(0)} = (y_l)_{l \in I_i} \text{ for } i = 1, \ldots, s$$

The algorithm from above is generalized by applying the described computation steps for elements according to Eq. (8.23) to blocks and using matrix operations instead of operations on single elements. In the first step, three consecutive matrix equations $i-1, i, i+1$ for $i = 3, 4, \ldots, s-2$ are considered:

$$
\begin{aligned}
A_{i-1}^{(0)} X_{i-2} + B_{i-1}^{(0)} X_{i-1} + C_{i-1}^{(0)} X_i &= Y_{i-1}^{(0)} \\
A_i^{(0)} X_{i-1} + B_i^{(0)} X_i + C_i^{(0)} X_{i+1} &= Y_i^{(0)} \\
A_{i+1}^{(0)} X_i + B_{i+1}^{(0)} X_{i+1} + C_{i+1}^{(0)} X_{i+2} &= Y_{i+1}^{(0)}
\end{aligned}
$$

Equation $(i-1)$ is used to eliminate subvector X_{i-1} from equation i and equation $(i+1)$ is used to eliminate subvector X_{i+1} from equation i. The algorithm starts with the following initializations:

$$
A_1^{(0)} := 0 \in \mathbb{R}^{r \times r}, \ C_s^{(0)} := 0 \in \mathbb{R}^{r \times r}
$$

and for $k = 0, \ldots, \lceil \log s \rceil$ and $i \in \mathbb{Z} \setminus \{1, \ldots, s\}$

$$
A_i^{(k)} = C_i^{(k)} := 0 \in \mathbb{R}^{r \times r},
$$

$$
B_i^{(k)} := I \in \mathbb{R}^{r \times r},
$$

$$
Y_i^{(k)} := 0 \in \mathbb{R}^r .
$$

In step $k = 1, \ldots, \lceil \log s \rceil$ the following sub-matrices

$$
\alpha_i^{(k)} := -A_i^{(k-1)} \left(B_{i-2^{k-1}}^{(k-1)} \right)^{-1},
$$

$$
\beta_i^{(k)} := -C_i^{(k-1)} \left(B_{i+2^{k-1}}^{(k-1)} \right)^{-1},
$$

$$
\begin{aligned}
A_i^{(k)} &= \alpha_i^{(k)} \cdot A_{i-2^{k-1}}^{(k-1)} \\
C_i^{(k)} &= \beta_i^{(k)} \cdot C_{i+2^{k-1}}^{(k-1)} \\
B_i^{(k)} &= \alpha_i^{(k)} C_{i-2^{k-1}}^{(k-1)} + B_i^{(k-1)} + \beta_i^{(k)} A_{i+2^{k-1}}^{(k-1)}
\end{aligned}
\tag{8.29}
$$

and the vector

$$
Y_i^{(k)} = \alpha_i^{(k)} Y_{i-2^{k-1}}^{(k-1)} + Y_i^{(k-1)} + \beta_i^{(k)} Y_{i+2^{k-1}}^{(k-1)}
\tag{8.30}
$$

are computed. The resulting matrix equations are

$$
A_i^{(k)} X_{i-2^k} + B_i^{(k)} X_i + C_i^{(k)} X_{i+2^k} = Y_i^{(k)}
\tag{8.31}
$$

for $i = 1, \ldots, s$. In summary, the method of cyclic reduction for banded matrices comprises the following two phases:

1. Elimination phase: For $k = 1, \ldots, \lceil \log s \rceil$ compute the matrices $A_i^{(k)}, B_i^{(k)}, C_i^{(k)}$ and the vector $Y_i^{(k)}$ for $i = 2^k, \ldots, s$ with step-size 2^k according to (8.29) and (8.30).
2. Substitution phase: For $k = \lfloor \log s \rfloor, \ldots, 0$ compute subvector X_i for $i = 2^k, \ldots, s$ with step-size 2^{k+1} by solving the linear equation system (8.31), i.e.,

$$B_i^{(k)} X_i = Y_i^{(k)} - A_i^{(k)} X_{i-2^k} - C_i^{(k)} X_{i+2^k} .$$

The computation of $\alpha_i^{(k)}$ and $\beta_i^{(k)}$ requires a matrix inversion or the solution of a dense linear equation system with a direct method requiring $O(r^3)$ computations, i.e., the computations increase with the bandwidth cubically. The first step requires the computation of $O(s) = O(n/r)$ sub-matrices, the asymptotic runtime for this step is therefore $O(nr^2)$. The second step solves a total number of $O(s) = O(n/r)$ linear equation systems, also resulting in a asymptotic runtime of $O(nr^2)$.

For the parallel implementation of the cyclic reduction for banded matrices, the parallel method described for tridiagonal systems with its three phases can be used. The main difference is that arithmetic operations in the implementation for tridiagonal systems are replaced by matrix operations in the implementation for banded systems, which increases the amount of computations for each processor. The computational effort for the local operations is now $O(r^3)$. Also, the communication between the processors exchanges larger messages. Instead of single numbers entire matrices of size $r \times r$ are exchanged so that the message size is $O(r^2)$. Thus, with growing semi-bandwidth r of the banded matrix the time for the computation increases faster than the communication time. For $p \ll s$ an efficient parallel implementation can be expected.

8.2.4 Solving the Discretized Poisson Equation

The cyclic reduction algorithm for banded matrices presented in Sect. 8.2.3 is suitable for the solution of the discretized two-dimensional Poisson equation. As shown in Section 8.2.1, this linear equation system has a banded structure with semi-bandwidth N where N is the number of discretization points in the x- or y-dimension of the two-dimensional domain, see Figure 8.9. The special structure has only four non-zero diagonals and the band has a sparse structure. The use of the Gaussian elimination method would not preserve the sparse banded structure of the matrix, since the forward elimination for eliminating the two lower diagonals leads to fill-ins with non-zero elements between the two upper diagonals. This induces a higher computational effort which is needed for banded matrices with a dense band of semi-bandwidth N. In the following, we consider the method of cyclic reduction for banded matrices, which preserves the sparse banded structure.

The blocks of the discretized Poisson equation $Az = d$ for a representation as blocked tridiagonal matrix are given by (8.18) and (8.19). Using the notation for the banded system, we get.

$$B_i^{(0)} := \frac{1}{h^2}B \quad \text{for } i = 1,\ldots,N,$$

$$A_i^{(0)} := -\frac{1}{h^2}I \text{ and } C_i^{(0)} := -\frac{1}{h^2}I \quad \text{for } i = 1,\ldots,N.$$

The vector $d \in \mathbb{R}^n$ consists of N subvectors $D_j \in \mathbb{R}^N$, i.e.,

$$d = \begin{pmatrix} D_1 \\ \vdots \\ D_N \end{pmatrix} \text{ with } D_j = \begin{pmatrix} d_{(j-1)N+1} \\ \vdots \\ d_{jN} \end{pmatrix}.$$

Analogously, the solution vector consists of N subvectors Z_j of length N each, i.e.,

$$z = \begin{pmatrix} Z_1 \\ \vdots \\ Z_N \end{pmatrix} \text{ mit } Z_j = \begin{pmatrix} z_{(j-1)N+1} \\ \vdots \\ z_{j \cdot N} \end{pmatrix}.$$

The initialization for the cyclic reduction algorithm is given by:

$$B^{(0)} := B,$$
$$D_j^{(0)} := D_j \quad \text{for } j = 1,\ldots,N,$$
$$D_j^{(k)} := 0 \quad \text{for } k = 0,\ldots,\lceil \log N \rceil, \ j \in \mathbb{Z}\setminus\{1,\ldots,N\},$$
$$Z_j := 0 \quad \text{for } j \in \mathbb{Z}\setminus\{1,\ldots,N\}.$$

In step k of the cyclic reduction, $k = 1,\ldots,\lfloor \log N \rfloor$, the matrices $B^{(k)} \in \mathbb{R}^{N \times N}$ and the vectors $D_j^{(k)} \in \mathbb{R}^N$ for $j = 1,\ldots,N$ are computed according to

$$B^{(k)} = (B^{(k-1)})^2 - 2I,$$
$$D_j^{(k)} = D_{j-2^{k-1}}^{(k-1)} + B^{(k-1)}D_j^{(k-1)} + D_{j+2^{k-1}}^{(k-1)}. \tag{8.32}$$

For $k = 0,\ldots,\lfloor \log N \rfloor$ equation (8.31) has the special form:

$$-Z_{j-2^k} + B^{(k)}Z_j - Z_{j+2^k} = D_j^{(k)} \qquad \text{for } j = 1,\ldots,n. \tag{8.33}$$

Together Eq. (8.32) and (8.33) represent the method of cyclic reduction for the discretize Poisson equation, which can be seen by induction. For $k = 0$ the Eq. (8.33) is the initial equation system $Az = d$. For $0 < k < \lceil \log N \rceil$ and $j \in \{1,\ldots,N\}$ the three equations

$$
\begin{aligned}
-Z_{j-2^{k+1}} + B^{(k)}Z_{j-2^k} - Z_j & = D^{(k)}_{j-2^k} \\
- Z_{j-2^k} + B^{(k)}Z_j - Z_{j+2^k} & = D^{(k)}_j \\
- Z_j + B^{(k)}Z_{j+2^k} - Z_{j+2^{k+1}} & = D^{(k)}_{j+2^k}
\end{aligned}
\tag{8.34}
$$

are considered. The multiplication of (8.33) with $B^{(k)}$ from the left results in

$$
-B^{(k)}Z_{j-2^k} + B^{(k)}B^{(k)}Z_j - B^{(k)}Z_{j+2^k} = B^{(k)}D^{(k)}_j .
\tag{8.35}
$$

Adding Eq. (8.35) with the first part in (8.34) and the third part in (8.34) results in

$$
-Z_{j-2^{k+1}} - Z_j + B^{(k)}B^{(k)}Z_j - Z_j - Z_{j+2^{k+1}} = D^{(k)}_{j-2^k} + B^{(k)}D^{(k)}_j + D^{(k)}_{j+2^k} ,
$$

which shows that Formula (8.32) for $k+1$ is derived. In summary, the cyclic reduction for the discretized two-dimensional Poisson equation consists of the following two steps:

1. Elimination phase: For $k = 1, \ldots, \lfloor \log N \rfloor$, the matrices $B^{(k)}$ and the vectors $D^{(k)}_j$ are computed for $j = 2^k, \ldots, N$ with step-size 2^k according to Eq. (8.32).
2. Substitution phase: For $k = \lfloor \log N \rfloor, \ldots, 0$, the linear equation system

$$
B^{(k)}Z_j = D^{(k)}_j + Z_{j-2^k} + Z_{j+2^k}
$$

for $j = 2^k, \ldots, N$ with step-size 2^{k+1} is solved.

In the first phase, $\lfloor \log N \rfloor$ matrices and $O(N)$ subvectors are computed. The computation of each matrix includes a matrix multiplication with time $O(N^3)$. The computation of a subvector includes a matrix-vector multiplication with complexity $O(N^2)$. Thus, the first phase has a computational complexity of $O(N^3 \log N)$. In the second phase, $O(N)$ linear equation systems are solved. This requires time $O(N^3)$ when the special structure of the matrices $B^{(k)}$ is not exploited. In [74] it is shown how to reduce the time by exploiting this structure. A parallel implementation of the discretized Poisson equation can be done in an analogeous way as shown in the previous section.

8.3 Iterative Methods for Linear Systems

In this section, we introduce classical iteration methods for solving linear equation systems, including the Jacobi iteration, the Gauss-Seidel iteration and the SOR method (successive over-relaxation), and discuss their parallel implementation. Direct methods as presented in the previous sections involve a factorization of the coefficient matrix. This can be impractical for large and sparse matrices, since fill-ins with non-zero elements increase the computational work. For banded matrices, special methods can be adapted and used as discussed in Section 8.2. Another possibility is to use iterative methods as presented in this section.

Iterative methods for solving linear equation systems $Ax = b$ with coefficient matrix $A \in \mathbb{R}^{n \times n}$ and right-hand side $b \in \mathbb{R}^n$ generate a sequence of approximation vectors $\{x^{(k)}\}_{k=1,2,\dots}$ that converges to the solution $x^* \in \mathbb{R}^n$. The computation of an approximation vector essentially involves a matrix-vector multiplication with the iteration matrix of the specific problem. The matrix A of the linear equation system is used to build this iteration matrix. For the evaluation of an iteration method it is essential how quickly the iteration sequence converges. Basic iteration methods are the Jacobi- and the Gauss-Seidel methods, which are also called **relaxation methods** historically, since the computation of a new approximation depends on a combination of the previously computed approximation vectors. Depending on the specific problem to be solved, relaxation methods can be faster than direct solution methods. But still these methods are not fast enough for practical use. A better convergence behavior can be observed for methods like the SOR method, which has a similar computational structure. The practical importance of relaxation methods is their use as preconditioner in combination with solution methods like the conjugate gradient method or the multigrid method. Iterative methods are a good first example to study parallelism typical also for more complex iteration methods. In the following, we describe the relaxation methods according to [25], see also [85, 209]. Parallel implementations are considered in [73, 74, 86, 194].

8.3.1 Standard Iteration Methods

Standard iteration methods for the solution of a linear equation system $Ax = b$ are based on a splitting of the coefficient matrix $A \in \mathbb{R}^{n \times n}$ into

$$A = M - N \quad \text{with } M, N \in \mathbb{R}^{n \times n} ,$$

where M is a non-singular matrix for which the inverse M^{-1} can be computed easily, e.g. a diagonal matrix. For the unknown solution x^* of the equation $Ax = b$ we get

$$Mx^* = Nx^* + b .$$

This equation induces an iteration of the form $Mx^{(k+1)} = Nx^{(k)} + b$, which is usually written as

$$x^{(k+1)} = Cx^{(k)} + d \tag{8.36}$$

with iteration matrix $C := M^{-1}N$ and vector $d := M^{-1}b$. The iteration method is called *convergent* if the sequence $\{x^{(k)}\}_{k=1,2,\dots}$ converges towards x^* independently from the choice of the start vector $x^{(0)} \in \mathbb{R}^n$, i.e., $\lim_{k \to \infty} x^{(k)} = x^*$ or $\lim_{k \to \infty} \|x^{(k)} - x^*\| = 0$. When a sequence converges then the vector x^* is uniquely defined by $x^* =$

$Cx^* + d$. Subtracting this equation from equation (8.36) and using induction leads to the equality $x^{(k)} - x^* = C^k(x^{(0)} - x^*)$, where C^k denotes the matrix resulting from k multiplications of C. Thus, the convergence of (8.36) is equivalent to

$$\lim_{k \to \infty} C^k = 0.$$

A result from linear algebra shows the relation between the convergence criteria and the spectral radius $\rho(C)$ of the iteration matrix C. (The spectral radius of a matrix is the eigenvalue with the largest absolute value, i.e., $\rho(C) = \max_{\lambda \in EW} |\lambda|$ with $EW = \{\lambda \mid Cv = \lambda v, v \neq 0\}$.) The following properties are equivalent, see [209]:

(1) Iteration (8.36) converges for every $x^{(0)} \in \mathbb{R}^n$,
(2) $\lim\limits_{k \to \infty} C^k = 0$,
(3) $\rho(C) < 1$.

Well-known iteration methods are the Jacobi, the Gauss-Seidel, and the SOR method.

8.3.1.1 Jacobi iteration

The Jacobi iteration is based on the splitting $A = D - L - R$ of the matrix A with $D, L, R \in \mathbb{R}^{n \times n}$. The matrix D holds the diagonal elements of A, $-L$ holds the elements of the lower triangular of A without the diagonal elements and $-R$ holds the elements of the upper triangular of A without the diagonal elements. All other elements of D, L, R are zero. The splitting is used for an iteration of the form

$$Dx^{(k+1)} = (L+R)x^{(k)} + b$$

which leads to the iteration matrix $C_{Ja} := D^{-1}(L+R)$ or

$$C_{Ja} = (c_{ij})_{i,j=1,\dots,n} \text{ with } c_{ij} = \begin{cases} -a_{ij}/a_{ii} & \text{for } j \neq i, \\ 0 & \text{otherwise} \end{cases}.$$

The matrix form is used for the convergence proof, not shown here. For the practical computation, the equation written out with all its components is more suitable:

$$x_i^{(k+1)} = \frac{1}{a_{ii}} \left(b_i - \sum_{j=1, j \neq i}^{n} a_{ij} x_j^{(k)} \right), \quad i = 1, \dots, n. \tag{8.37}$$

The computation of one component $x_i^{(k+1)}$, $i \in \{1, \dots, n\}$, of the $(k+1)$th approximation requires all components of the kth approximation vector x^k. Considering a sequential computation in the order $x_1^{(k+1)}, \dots, x_n^{(k+1)}$, it can be observed that the values $x_1^{(k+1)}, \dots, x_{i-1}^{(k+1)}$ are already known when $x_i^{(k+1)}$ is computed. This information is exploited in the Gauss-Seidel iteration method.

8.3.1.2 Gauss-Seidel iteration

The Gauss-Seidel iteration is based on the same splitting of the matrix A as the Jacobi iteration, i.e., $A = D - L - R$, but uses the splitting in a different way for an iteration

$$(D - L)x^{(k+1)} = Rx^{(k)} + b .$$

Thus, the iteration matrix of the Gauss-Seidel method is $C_{Ga} := (D - L)^{-1}R$; this form is used for numerical properties like convergence proofs, not shown here. The component form for the practical use is

$$x_i^{(k+1)} = \frac{1}{a_{ii}} \left(b_i - \sum_{j=1}^{i-1} a_{ij} x_j^{(k+1)} - \sum_{j=i+1}^{n} a_{ij} x_j^{(k)} \right) , \quad i = 1, \ldots, n . \tag{8.38}$$

It can be seen that the components of $x_i^{(k+1)}$, $i \in \{1, \ldots, n\}$, uses the new information $x_1^{(k+1)}, \ldots, x_{i-1}^{(k+1)}$ already determined for that approximation vector. This is useful for a faster convergence in a sequential implementation but the potential parallelism is now restricted.

8.3.1.3 Convergence criteria

For the Jacobi and the Gauss-Seidel iteration the following convergence criteria based on the structure of A is often helpful. The Jacobi and the Gauss-Seidel iteration converge if the matrix A is **strongly diagonal dominant**, i.e.,

$$|a_{ii}| > \sum_{j=1, j \neq i}^{n} |a_{ij}| , \quad i = 1, \ldots, n .$$

When the absolute values of the diagonal elements are large compared to the sum of the absolute values of the other row elements, this often leads to a better convergence. Also, when the iteration methods converge, the Gauss-Seidel iteration often converges faster than the Jacobi iteration, since always the most recently computed vector components are used. Still the convergence is usually not fast enough for practical use. Therefore, an additional relaxation parameter is introduced to speed up the convergence.

8.3.1.4 JOR method

The JOR method or Jacobi over-relaxation is based on the splitting $A = \frac{1}{\omega}D - L - R - \frac{1-\omega}{\omega}D$ of the matrix A with a **relaxation parameter** $\omega \in \mathbb{R}$. The component form of this modification of the Jacobi method is

$$x_i^{(k+1)} = \frac{\omega}{a_{ii}} \left(b_i - \sum_{j=1, j\neq i}^{n} a_{ij}x_j^{(k)} \right) + (1-\omega)x_i^{(k)} , \quad i = 1,\ldots,n . \tag{8.39}$$

More popular is the modification with a relaxation parameter for the Gauss-Seidel method.

8.3.1.5 SOR method

The SOR method or (**successive over-relaxation**) is a modification of the Gauss-Seidel iteration that speeds up the convergence of the Gauss-Seidel method by introducing a relaxation parameter $\omega \in \mathbb{R}$. This parameter is used to modify the way in which the combination of the previous approximation $x^{(k)}$ and the components of the current approximation $x_1^{(k+1)},\ldots,x_{i-1}^{(k+1)}$ are combined in the computation of $x_i^{(k+1)}$. The $(k+1)$th approximation computed according to the Gauss-Seidel iteration (8.38) is now considered as intermediate result $\hat{x}^{(k+1)}$ and the next approximation $x^{(k+1)}$ of the SOR method is computed from both vectors $\hat{x}^{(k+1)}$ and $x^{(k+1)}$ in the following way:

$$\hat{x}_i^{(k+1)} = \frac{1}{a_{ii}} \left(b_i - \sum_{j=1}^{i-1} a_{ij}x_j^{(k+1)} - \sum_{j=i+1}^{n} a_{ij}x_j^{(k)} \right) , \quad i = 1,\ldots,n , \tag{8.40}$$

$$x_i^{(k+1)} = x_i^{(k)} + \omega(\hat{x}_i^{(k+1)} - x_i^{(k)}) , \quad i = 1,\ldots,n . \tag{8.41}$$

Substituting Eq. (8.40) into Eq. (8.41) results in the iteration:

$$x_i^{(k+1)} = \frac{\omega}{a_{ii}} \left(b_i - \sum_{j=1}^{i-1} a_{ij}x_j^{(k+1)} - \sum_{j=i+1}^{n} a_{ij}x_j^{(k)} \right) + (1-\omega)x_i^{(k)} \tag{8.42}$$

for $i = 1,\ldots,n$. The corresponding splitting of the matrix A is $A = \frac{1}{\omega}D - L - R - \frac{1-\omega}{\omega}D$ and an iteration step in matrix form is

$$(D - wL)x^{(k+1)} = (1-\omega)Dx^{(k)} + \omega Rx^{(k)} + \omega b .$$

The convergence of the SOR method depends on the properties of A and the value chosen for the relaxation parameter ω. For example the following property holds: if A is symmetric and positive definite and $\omega \in (0,2)$, then the SOR method converges for every start vector $x^{(0)}$. For more numerical properties see books on numerical linear algebra, e.g., [25, 74, 85, 209].

8.3.1.6 Implementation using matrix operations

The iteration (8.36) computing $x^{(k+1)}$ for a given vector $x^{(k)}$ consists of

- a matrix-vector multiplication of the iteration matrix C with $x^{(k)}$ and

- a vector-vector addition of the result of the multiplication with vector d.

The specific structure of the iteration matrix, i.e., C_{Ja} for the Jacobi iteration and C_{Ga} for the Gauss-Seidel iteration, is exploited. For the Jacobi iteration with $C_{Ja} = D^{-1}(L+R)$ this results in the computation steps:

- a matrix-vector multiplication of $L+R$ with $x^{(k)}$,
- a vector-vector addition of the result with b and
- a matrix-vector multiplication with D^{-1} (where D is a diagonal matrix and thus D^{-1} is easy to compute).

A sequential implementation uses Formula (8.37) and the components $x_i^{(k+1)}$, $i = 1,\ldots,n$, are computed one after another. The entire vector $x^{(k)}$ is needed for this computation. For the Gauss-Seidel iteration with $C_{Ga} = (D-L)^{-1}R$ the computation steps are:

- a matrix-vector multiplication $Rx^{(k)}$ with upper triangular matrix R,
- a vector-vector addition of the result with b and
- the solution of a linear system with lower triangular matrix $(D-L)$.

A sequential implementation uses formula (8.38). Since the most recently computed approximation components are always used for computing a value $x_i^{(k+1)}$, the previous value $x_i^{(k)}$ can be overwritten. The iteration method stops when the current approximation is close enough to the exact solution. Since this solution is unknown, the relative error is used for error control and after each iteration step the convergence is tested according to

$$\|x^{(k+1)} - x^{(k)}\| \le \varepsilon \|x^{(k+1)}\| \tag{8.43}$$

where ε is a predefined error value and $\|.\|$ is a vector norm such as $\|x\|_\infty = \max_{i=1,\ldots,n} |x|_i$ or $\|x\|_2 = \left(\sum_{i=1}^n |x_i|^2\right)^{\frac{1}{2}}$.

8.3.2 Parallel implementation of the Jacobi Iteration

In the Jacobi iteration (8.37), the computations of the components $x_i^{(k+1)}$, $i = 1,\ldots,n$, of approximation $x^{(k+1)}$ are independent of each other and can be executed in parallel. Thus, each iteration step has a maximum degree of potential parallelism of n and $p = n$ processors can be employed. For a parallel system with distributed memory, the values $x_i^{(k+1)}$ are stored in the individual local memories. Since the computation of one of the components of the next approximation requires all components of the previous approximation, communication has to be performed to create a replicated distribution of $x^{(k)}$. This can be done by multi-broadcast operation.

When considering the Jacobi iteration built up of matrix and vector operations, a parallel implementation can use the parallel implementations introduced in Section

3.7. The iteration matrix C_{Ja} is not built up explicitly but matrix A is used without its diagonal elements. The parallel computation of the components of $x^{(k+1)}$ corresponds to the parallel implementation of the matrix-vector product using the parallelization with scalar products, see Section 3.7. The vector addition can be done after the multi-broadcast operation by each of the processors or before the multi-broadcast operation in a distributed way. When using the parallelization of the linear combination from Section 3.7, the vector addition takes place after the accumulation operation. The final broadcast operation is required to provide $x^{(k+1)}$ to all processors also in this case.

Figure 8.14 shows a parallel implementation of the Jacobi iteration using C notation and MPI operations from [167]. For simplicity it is assumed that the matrix size n is a multiple of the number of processors p. The iteration matrix is stored in a row blockwise way so that each processor owns n/p consecutive rows of matrix A which are stored locally in array local_A. The vector b is stored in a corresponding blockwise way. This means that the processor me, $0 \leq$ me $< p$, stores the rows me $\cdot n/p+1,\ldots,($me $+ 1) \cdot n/p$ of A in local_A and the corresponding components of b in local_b. The iteration uses two local arrays x_old and x_new for storing the previous and the current approximation vectors. The symbolic constant GLOB_MAX is the maximum size of the linear equation system to be solved. The result of the local matrix-vector multiplication is stored in local_x; local_x is computed according to (8.37). An MPI_Allgather() operation combines the local results so that each processor stores the entire vector x_new. The iteration stops when a predefined number max_it of iteration steps has been performed or when the difference between x_old and x_new is smaller than a predefined value tol. The function distance() implements a maximum norm and the function output(x_new,global_x) returns array global_x which contains the last approximation vector to be the final result.

8.3.3 Parallel Implementation of the Gauss-Seidel Iteration

The Gauss-Seidel iteration (8.38) exhibits data dependencies, since the computation of the component $x_i^{(k+1)}$, $i \in \{1,\ldots,n\}$, uses the components $x_1^{(k+1)},\ldots,x_{i-1}^{(k+1)}$ of the same approximation and the components of $x^{(k+1)}$ have to be computed one after another. Since for each $i \in \{1,\ldots,n\}$ the computation (8.38) corresponds to a scalar product of the vector

$$\left(x_1^{(k+1)},\ldots,x_{i-1}^{(k+1)},0,x_{i+1}^{(k)},\ldots,x_n^{(k)}\right)$$

and the ith row of A, this means the scalar products have to be computed one after another. Thus, parallelism is only possible within the computation of each single scalar product: each processor can compute a part of the scalar product, i.e., a local scalar product, and the results are then accumulated. For such an implementation a column blockwise distribution of matrix A is suitable. Again, we assume that n is a multiple of the numbers of processors p. The approximation vectors are distributed

```
int Parallel_jacobi(int n, int p, int max_it, float tol)
{
  int i_local, i_global, j, i;
  int n_local, it_num;
  float x_temp1[GLOB_MAX], x_temp2[GLOB_MAX], local_x[GLOB_MAX];
  float *x_old, *x_new, *temp;

  n_local = n/p; /* local blocksize */
  MPI_Allgather(local_b, n_local, MPI_FLOAT, x_temp1, n_local,
                MPI_FLOAT, MPI_COMM_WORLD);
  x_new = x_temp1;
  x_old = x_temp2;
  it_num = 0;
  do {
    it_num ++;
    temp = x_new; x_new = x_old; x_old = temp;
    for (i_local = 0; i_local < n_local; i_local++) {
      i_global = i_local + me * n_local;
      local_x[i_local] = local_b[i_local];
      for (j = 0; j < i_global; j++)
        local_x[i_local] = local_x[i_local] -
                           local_A[i_local][j] * x_old[j];
      for (j = i_global+1 ; j < n; j++)
        local_x[i_local] = local_x[i_local] -
                           local_A[i_local][j] * x_old[j];
      local_x[i_local] = local_x[i_local]/ local_A[i_local][i_global];
    }
    MPI_Allgather(local_x, n_local, MPI_FLOAT, x_new, n_local,
                  MPI_FLOAT, MPI_COMM_WORLD);
  } while ((it_num < max_it) && (distance(x_old,x_new,n) >= tol));
  output(x_new,global_x);
  if (distance(x_old, x_new, n) < tol ) return 1;
  else return 0;
}
```

Fig. 8.14 Program fragment in C notation and with MPI communication operations for a parallel implementation of the Jacobi iteration. The arrays local_x, local_b and local_A are declared globally. The dimension of local_A is n_local × n. A pointer-oriented storage scheme as shown in Fig. 8.3 is not used here so that the array indices in this implementation differ from the indices in a sequential implementation. The computation of local_x[i_local] is performed in two loops with loop index j; the first loop corresponds to the multiplication with array elements in row i_local to the left of the main diagonal of A and the second loop corresponds to the multiplication with array elements in row i_local to the right of the main diagonal of A. The result is divided by local_A[i_local][i_global] which corresponds to the diagonal element of that row in the global matrix A.

correspondingly in a blockwise way. Processor P_q, $1 \leq q \leq p$, computes that part of the scalar product for which it owns the columns of A and the components of the approximation vector $x^{(k)}$. This is the computation

$$s_{qi} = \sum_{\substack{j=(q-1)\cdot n/p+1 \\ j<i}}^{q\cdot n/p} a_{ij}x_j^{(k+1)} + \sum_{\substack{j=(q-1)\cdot n/p+1 \\ j>i}}^{q\cdot n/p} a_{ij}x_j^{(k)} . \tag{8.44}$$

The intermediate results s_{qi} computed by processors P_q, $q = 1,\ldots,p$, are accumulated by a single-accumulation operation with the addition as reduction operation and the value $x_i^{(k+1)}$ is the result. Since the next approximation vector $x^{(k+1)}$ is expected in a blockwise distribution, the value $x_i^{(k+1)}$ is accumulated at the processor owning the ith component, i.e., $x_i^{(k+1)}$ is accumulated by processor P_q with $q = \lceil i/(n/p) \rceil$. A parallel implementation of the SOR method corresponds to the parallel implementation of the Gauss-Seidel iteration, since both methods differ only in the additional relaxation parameter of the SOR method.

Figure 8.15 shows a program fragment using C notation and MPI operations of a parallel Gauss-Seidel iteration. Since only the most recently computed components of an approximation vector are used in further computations, the component $x_i^{(k)}$ is overwritten by $x_i^{(k+1)}$ immediately after its computation. Therefore, only one array x is needed in the program. Again, an array local_A stores the local part of matrix A which is a block of columns in this case; n_local is the size of the block. The for-loop with loop index i computes the scalar products sequentially; within the loop body the parallel computation of the inner product is performed according to Formula (8.44). An MPI reduction operation computes the components at differing processors root which finalizes the computation.

8.3.4 Gauss-Seidel Iteration for Sparse Systems

The potential parallelism for the Gauss-Seidel iteration or the SOR method is limited because of data dependencies so that a parallel implementation is only reasonable for very large equation systems. Each data dependency in Formula (8.38) is caused by a coefficient (a_{ij}) of matrix A, since the computation of $x_i^{(k+1)}$ depends on the value $x_j^{(k+1)}$, $j < i$, when $(a_{ij}) \neq 0$. Thus, for a linear equation system $Ax = b$ with sparse matrix $A = (a_{ij})_{i,j=1,\ldots,n}$ there is a larger degree of parallelism caused by less data dependencies. If $a_{ij} = 0$, then the computation of $x_i^{(k+1)}$ does not depend on $x_j^{(k+1)}$, $j < i$. For a sparse matrix with many zero elements the computation of $x_i^{(k+1)}$ only needs a few $x_j^{(k+1)}$, $j < i$. This can be exploited to compute components of the $(k+1)$th approximation $x^{(k+1)}$ in parallel.

In the following, we consider sparse matrices with a banded structure like the discretized Poisson equation, see Eq. (8.13) in Section 8.2.1. The computation of

```
n_local = n/p;
do {
    delta_x = 0.0;
    for (i = 0; i < n; i++) {
        s_k = 0.0;
        for (j = 0; j < n_local; j++)
            if (j + me * n_local != i)
                s_k = s_k + local_A[i][j] * x[j];
        root = i/n_local;
        i_local = i % n_local;
        MPI_Reduce(&s_k, &x[i_local], 1, MPI_FLOAT, MPI_SUM, root,
                   MPI_COMM_WORLD);
        if (me == root) {
            x_new = (b[i_local] - x[i_local]) / local_A[i][i_local];
            delta_x = max(delta_x, abs(x[i_local] - x_new));
            x[i_local] = x_new;
        }
    }
    MPI_Allreduce(&delta_x, &global_delta, 1, MPI_FLOAT,
                  MPI_MAX, MPI_COMM_WORLD);
} while(global_delta > tol);
```

Fig. 8.15 Program fragment in C notation and using MPI operations for a parallel Gauss-Seidel iteration for a dense linear equation system. The components of the approximation s are computed one after another according to Formula (8.38), but each of these computations is done in parallel by all processors. The matrix is stored in a column blockwise way in the local arrays local_A. The vectors x and b are also distributed blockwise. Each processor computes the local error and stores it in delta_x. An MPI_Allreduce() operation compute the global error global_delta from these values so that each processor can perform the convergence test global_delta > tol.

$x_i^{(k+1)}$ uses the elements in the ith row of A, see Fig. 8.9, which has non-zero elements a_{ij} for $j = i - \sqrt{n}, i - 1, i, i + 1, i + \sqrt{n}$. Formula (8.38) of the Gauss-Seidel iteration for the discretized Poisson equation has the specific form

$$x_i^{(k+1)} = \frac{1}{a_{ii}} \left(b_i - a_{i,i-\sqrt{n}} \cdot x_{i-\sqrt{n}}^{(k+1)} - a_{i,i-1} \cdot x_{i-1}^{(k+1)} - a_{i,i+1} \cdot x_{i+1}^{(k)} \right.$$

$$\left. - a_{i,i+\sqrt{n}} \cdot x_{i+\sqrt{n}}^{(k)} \right), \quad i = 1, \ldots, n \,. \tag{8.45}$$

Thus, the two values $x_{i-\sqrt{n}}^{(k+1)}$ and $x_{i-1}^{(k+1)}$ have to be computed before the computation of $x_i^{(k+1)}$. The dependencies of the values $x_i^{(k+1)}$, $i = 1, \ldots, n$, on $x_j^{(k+1)}$, $j < i$, are illustrated in Fig. 8.16 (a) for the corresponding mesh of the discretized physical domain. The computation of $x_i^{(k+1)}$ corresponds to the mesh point i, see also Section 8.2.1. In this mesh, the computation of $x_i^{(k+1)}$ depends on all computations for mesh points which are located in the upper left part of the mesh. On the other hand, computations for mesh points $j > i$ which are located to the right or below mesh point i need value $x_i^{(k+1)}$ and have to wait for its computation.

a) Data dependencies of the SOR method

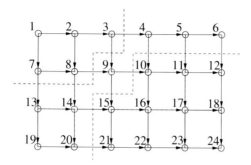

b) Independent computations within the diagonals

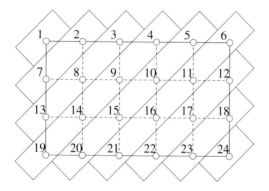

Fig. 8.16 Data dependence of the Gauss-Seidel and the SOR method for a rectangular mesh of size 6×4 in the x-y-plane. a) The data dependencies between the computations of components are depicted as arrows between nodes in the mesh. As an example, for mesh point 9 the set of nodes which have to be computed before point 9 and the set of nodes which depend on mesh point 9 are shown. b) The data dependencies lead to areas of independent computations; these are the diagonals of the mesh from the upper right to the lower left. The computations for mesh points within the same diagonal can be computed in parallel. The length of the diagonals is the degree of potential parallelism which can be exploited.

The data dependencies between computations associated to mesh points are depicted in the mesh by arrows between the mesh points. It can be observed that the mesh points in each diagonal from left to right are independent of each other; these independent mesh points are shown in Fig. 8.16 (b). For a square mesh of size $\sqrt{n} \times \sqrt{n}$ with the same number of mesh points in each dimension, there are at most $\sqrt{n}$ independent computations in a single diagonal and at most $p = \sqrt{n}$ processors can be employed.

A parallel implementation can exploit the potential parallelism in a loop structure with an outer sequential loop and an inner parallel loop. The outer sequential loop visits the diagonals one after another from the upper left corner to the lower right corner. The inner loop exploits the parallelism within each diagonal of the mesh. The number of diagonals is $2\sqrt{n}-1$ consisting of $\sqrt{n}$ diagonals in the upper left triangular mesh and $\sqrt{n}-1$ in the lower triangular mesh. The first $\sqrt{n}$ diagonals $l = 1,\ldots,\sqrt{n}$ contain l mesh points i with

$$i = l + j \cdot (\sqrt{n} - 1) \quad \text{for} \quad 0 \le j < l \,.$$

The last $\sqrt{n}-1$ diagonals $l = 2,\ldots,\sqrt{n}$ contain $\sqrt{n}-l+1$ mesh points i with

$$i = l \cdot \sqrt{n} + j \cdot (\sqrt{n} - 1) \quad \text{for} \quad 0 \le j \le \sqrt{n} - l \,.$$

For an implementation on a distributed memory machine, a distribution of the approximation vector x, the right hand side b and the coefficient matrix A is needed. The elements a_{ij} of matrix A are distributed in such a way that the coefficients for the computation of $x_i^{(k+1)}$ according to Formula (8.45) are locally available. Because the computations are closely related to the mesh, the data distribution is chosen for the mesh and not the matrix form.

The program fragment with C notation in Fig. 8.17 shows a parallel SPMD implementation. The data distribution is chosen such that the data associated to mesh points in the same mesh row are stored on the same processor. A row cyclic distribution of the mesh data is used. The program has two loop nests: the first loop nest treats the upper diagonals and the second loop nest treats the last diagonals. In the inner loops, the processor with name me computes the mesh points which are assigned to it due to the row cyclic distribution of mesh points. The function collect_elements() sends the data computed to the neighboring processor, which needs them for the computation of the next diagonal. The function convergence_test(), not expressed explicitly in this program, can be implemented similarly as in the program in Fig. 8.15 using the maximum norm for $x^{(k+1)} - x^{(k)}$.

The program fragment in Fig. 8.17 uses two-dimensional indices for accessing array elements of array a. For a large sparse matrix, a storage scheme for sparse matrices would be used in practice. Also, for a problem such as the discretized Poisson equation where the coefficients are known it is suitable to code them directly as constants into the program. This saves expensive array accesses but the code is less flexible to solve other linear equation systems.

For an implementation on a shared memory machine, the inner loop is performed in parallel by $p = \sqrt{n}$ processors in an SPMD pattern. No data distribution is needed but the same distribution of work to processors is assigned. Also, no communication is needed to send data to neighboring processors. However, a barrier synchronization is used instead to make sure that the data of the previous diagonal are available for the next one.

A further increase of the potential parallelism for solving sparse linear equation systems can be achieved by the method described in the next subsection.

```
sqn = sqrt(n);
do {
    for (l = 1; l <= sqn; l++) {
        for (j = me; j < l; j+=p) {
            i = l + j * (sqn-1) - 1; /* start numbering with 0 */
            x[i] = 0;
            if (i-sqn >= 0) x[i] = x[i] - a[i][i-sqn] * x[i-sqn];
            if (i > 0) x[i] = x[i] - a[i][i-1] * x[i-1];
            if (i+1 < n) x[i] = x[i] - a[i][i+1] * x[i+1];
            if (i+sqn < n) x[i] = x[i] - a[i][i+sqn] * x[i+sqn];
            x[i] = (x[i] + b[i]) / a[i][i];
        }
        collect_elements(x,l);
    }
    for (l = 2; l <= sqn; l++) {
        for (j = me -1 +1; j <= sqn -1; j+=p) {
            if (j >= 0) {
                i = l * sqn + j * (sqn-1) - 1;
                x[i] = 0;
                if (i-sqn >= 0) x[i] = x[i] - a[i][i-sqn] * x[i-sqn];
                if (i > 0) x[i] = x[i] - a[i][i-1] * x[i-1];
                if (i+1 < n) x[i] = x[i] - a[i][i+1] * x[i+1];
                if (i+sqn < n) x[i] = x[i] - a[i][i+sqn] * x[i+sqn];
                x[i] = (x[i] + b[i]) / a[i][i];
            }
        }
        collect_elements(x,l);
    }
} while(convergence_test() < tol);
```

Fig. 8.17 Program fragment of the parallel Gauss-Seidel iteration for a linear equation system with the banded matrix from the discretized Poisson equation. The computational structure uses the diagonals of the corresponding discretization mesh, see Fig. (8.16).

8.3.5 Red-black Ordering

The potential parallelism of the Gauss-Seidel iteration or the successive over-relaxation for sparse systems resulting from discretization problems can be increased by an alternative ordering of the unknowns and equations. The goal of the reordering is to get an equivalent equation system in which more independent computations exist and, thus, a higher potential parallelism results. The most frequently used reordering technique is the **red-black ordering**. The two-dimensional mesh is regarded as a checkerboard where the points of the mesh represent the squares of the checkerboard and get corresponding colors. The point (i, j) in the mesh is colored according to the value of $i + j$: if $i + j$ is even, then the mesh point is red, and if $i + j$ is odd, then the mesh point is black.

The points in the grid now form two sets of points. Both sets are numbered separately in a rowwise way from left to right. First the red points are numbered by

$1,\ldots,n_R$ where n_R is the number of red points. Then, the black points are numbered by $n_R + 1,\ldots,n_R + n_B$ where n_B is the number of black points and $n = n_R + n_B$. The unknowns associated with the mesh points get the same numbers as the mesh points: There are n_R unknowns associated with the red points denoted as $\hat{x}_1,\ldots,\hat{x}_{n_R}$ and n_B unknowns associated with the black points denoted as $\hat{x}_{n_R+1},\ldots,\hat{x}_{n_R+n_B}$. (The notation $\hat{x}$ is used to distinguish the new ordering from the original ordering of the unknowns x. The unknowns are the same as before but their positions in the system differ.) Figure 8.18 shows a mesh of size 6×4 in its original rowwise numbering in part a) and a red-black ordering with the new numbering in part b).

In a linear equation system using red-black ordering, the equations of red unknowns are arranged before the equations with the black unknown. The equation system $\hat{A}\hat{x} = \hat{b}$ for the discretized Poisson equation has the form:

$$\hat{A} \cdot \hat{x} = \begin{pmatrix} D_R & F \\ E & D_B \end{pmatrix} \cdot \begin{pmatrix} \hat{x}_R \\ \hat{x}_B \end{pmatrix} = \begin{pmatrix} \hat{b}_1 \\ \hat{b}_2 \end{pmatrix}, \tag{8.46}$$

where $\hat{x}_R$ denotes the subvector of size n_R of the first (red) unknowns and $\hat{x}_B$ denotes the subvector of size n_B of the last (black) unknowns. The right-hand side b of the original equation system is reordered accordingly and has subvector $\hat{b}_1$ for the first n_R equations and subvector $\hat{b}_2$ for the last n_B equations. The matrix $\hat{A}$ consists of four blocks $D_R \in \mathbb{R}^{n_R \times n_R}$, $D_B \in \mathbb{R}^{n_B \times n_B}$, $E \in \mathbb{R}^{n_B \times n_R}$ and $F \in \mathbb{R}^{n_R \times n_B}$. The submatrices D_R and D_B are diagonal matrices and the submatrices E and F are sparse banded matrices. The structure of the original matrix of the discretized Poisson equation in Fig. 8.9 in Section 8.2.1 is thus transformed into a matrix $\hat{A}$ with the structure shown in Fig. 8.18 c).

The diagonal form of the matrix D_R and D_B shows that a red unknown $\hat{x}_i$, $i \in \{1,\ldots,n_R\}$, does not depend on the other red unknowns and a black unknown $\hat{x}_j, j \in \{n_R + 1,\ldots,n_R + n_B\}$, does not depend on the other black unknowns. The matrices E and F specify the dependencies between red and black unknowns. The row i of matrix F specifies the dependencies of the red unknowns $\hat{x}_i$ ($i < n_R$) on the black unknowns $\hat{x}_j, j = n_R + 1,\ldots,n_R + n_B$. Analogously, a row of matrix E specifies the dependencies of the corresponding black unknowns on the red unknowns.

The transformation of the original linear equation system $Ax = b$ into the equivalent system $\hat{A}\hat{x} = \hat{b}$ can be expressed by a permutation $\pi : \{1,\ldots,n\} \rightarrow \{1,\ldots,n\}$. The permutation maps a node $i \in \{1,\ldots,n\}$ of the rowwise numbering onto the number $\pi(i)$ of the red-black numbering in the following way:

$$x_i = \hat{x}_{\pi(i)}, \quad b_i = \hat{b}_{\pi(i)}, \quad i = 1,\ldots,n \text{ or } x = P\hat{x} \text{ and } b = P\hat{b}$$

with a permutation matrix $P = (P_{ij})_{i,j=1,\ldots,n}, P_{ij} = \begin{cases} 1 & \text{if} \quad j = \pi(i) \\ 0 & \text{otherwise} \end{cases}$. For the matrices A and $\hat{A}$ the equation $\hat{A} = P^T A P$ holds. Since for a permutation matrix the inverse is equal to the transposed matrix, i.e., $P^T = P^{-1}$, this leads to $\hat{A}\hat{x} =$

a) Mesh in the x–y–plane with rowwise numbering

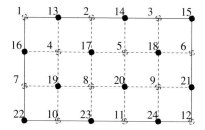

b) Mesh in the x–y–plane with red–black numbering

c) Matrix structure of the discretized Poisson equation with red–black ordering

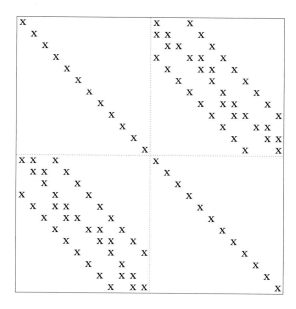

Fig. 8.18 Rectangular mesh in the x-y-plane of size 6×4 with (a) rowwise numbering, (b) red-black numbering and (c) the matrix of the corresponding linear equation system of the five-point formula with red-black numbering.

$P^T A P P^T x = P^T b = \hat{b}$. The easiest way to exploit the red-black ordering is to use an iterative solution method as discussed earlier in this section.

8.3.5.1 Gauss-Seidel iteration for red-black systems

The solution of the linear equation system (8.46) with the Gauss-Seidel iteration is based on a splitting of the matrix $\hat{A}$ of the form $\hat{A} = \hat{D} - \hat{L} - \hat{U}$, $\hat{D}, \hat{L}, \hat{U} \in \mathbb{R}^{n \times n}$,

$$
\hat{D} = \begin{pmatrix} D_R & 0 \\ 0 & D_B \end{pmatrix}, \quad \hat{L} = \begin{pmatrix} 0 & 0 \\ -E & 0 \end{pmatrix}, \quad \hat{U} = \begin{pmatrix} 0 & -F \\ 0 & 0 \end{pmatrix},
$$

with a diagonal matrix $\hat{D}$, a lower triangular matrix $\hat{L}$, and an upper triangular matrix $\hat{U}$. The matrix 0 is a matrix in which all entries are 0. With this notation, iteration step k of the Gauss-Seidel method is given by

$$
\begin{pmatrix} D_R & 0 \\ E & D_B \end{pmatrix} \cdot \begin{pmatrix} x_R^{(k+1)} \\ x_B^{(k+1)} \end{pmatrix} = \begin{pmatrix} b_1 \\ b_2 \end{pmatrix} - \begin{pmatrix} 0 & F \\ 0 & 0 \end{pmatrix} \cdot \begin{pmatrix} x_R^{(k)} \\ x_B^{(k)} \end{pmatrix} \tag{8.47}
$$

for $k = 1, 2, \ldots$. According to equation system (8.46), the iteration vector is split into two subvectors $x_R^{(k+1)}$ and $x_B^{(k+1)}$ for the red and the black unknowns, respectively. (To simplify the notation, we use x_R instead of $\hat{x}_R$ in the following discussion of the red-black ordering.)

The linear equation system (8.47) can be written in vector notation for vectors $x_R^{(k+1)}$ and $x_B^{(k+1)}$ in the form

$$
D_R \cdot x_R^{(k+1)} = b_1 - F \cdot x_B^{(k)} \quad \text{for } k = 1, 2, \ldots , \tag{8.48}
$$

$$
D_B \cdot x_B^{(k+1)} = b_2 - E \cdot x_R^{(k+1)} \quad \text{for } k = 1, 2, \ldots , \tag{8.49}
$$

in which the decoupling of the red subvector $x_R^{(k+1)}$ and the black subvector $x_B^{(k+1)}$ becomes obvious: In Eq. (8.48) the new red iteration vector $x_R^{(k+1)}$ depends only on the previous black iteration vector $x_B^{(k)}$ and in Eq. (8.49) the new black iteration vector $x_B^{(k+1)}$ depends only on the red iteration vector $x_R^{(k+1)}$ computed before in the same iteration step. There is no additional dependence. Thus, the potential degree of parallelism in each of the equations of (8.48) or (8.49) is similar to the potential parallelism in the Jacobi iteration. In each iteration step k, the components of $x_R^{(k+1)}$ according to Eq. (8.48) can be computed independently, since the vector $x_B^{(k)}$ is known, which leads to a potential parallelism with $p = n_R$ processors. Afterwards, the vector $x_R^{(k+1)}$ is known and the components of the vector $x_B^{(k+1)}$ can be computed independently according to (8.49), leading to a potential parallelism of $p = n_R$ processors.

For a parallel implementation, we consider the Gauss-Seidel iteration of the red-black ordering (8.48) and (8.49) written out in a component based form:

$$\left(x_R^{(k+1)}\right)_i = \frac{1}{\hat{a}_{ii}}\left(\hat{b}_i - \sum_{j\in N(i)} \hat{a}_{ij}\cdot(x_B^{(k)})_j\right), \quad i=1,\ldots,n_R,$$

$$\left(x_B^{(k+1)}\right)_i = \frac{1}{\hat{a}_{i+n_R,i+n_R}}\left(\hat{b}_{i+n_R} - \sum_{j\in N(i)} \hat{a}_{i+n_R,j}\cdot(x_R^{(k+1)})_j\right), \quad i=1,\ldots,n_B.$$

The set $N(i)$ denotes the set of adjacent mesh points for mesh point i. According to the red-black ordering, the set $N(i)$ contains only black mesh points for a red point i and vice versa. An implementation on a shared memory machine can employ at most $p = n_R$ or $p = n_B$ processors. There are no access conflicts for the parallel computation of $x_R^{(k)}$ or $x_B^{(k)}$ but a barrier synchronization is needed between the two computation phases. The implementation on a distributed memory machine requires a distribution of computation and data. As discussed before for the parallel SOR method, it is useful to distribute the data according to the mesh structure such that the processor P_q to which the mesh point i is assigned is responsible for the computation or update of the corresponding component of the approximation vector. In a row oriented distribution of a squared mesh with $\sqrt{n}\times\sqrt{n}=n$ mesh points to p processors, $\sqrt{n}/p$ rows of the mesh are assigned to each processor P_q, $q\in\{1,\ldots,p\}$. In the red-black coloring this means that each processor owns $\frac{1}{2}\frac{n}{p}$ red and $\frac{1}{2}\frac{n}{p}$ black mesh points. (For simplicity we assume that $\sqrt{n}$ is a multiple of p.) Thus, the mesh points

$$(q-1)\cdot\frac{n_R}{p}+1,\ldots,q\cdot\frac{n_R}{p} \quad \text{for } q=1,\ldots,p \quad \text{and}$$

$$(q-1)\cdot\frac{n_B}{p}+1+n_R,\ldots,q\cdot\frac{n_B}{p}+n_R \quad \text{for} \quad q=1,\ldots,p$$

are assigned to processor P_q. Figure 8.19 shows an SPMD program implementing the Gauss-Seidel iteration with red-black ordering. The coefficient matrix A is stored according to the pointer-based scheme introduced earlier in Fig. 8.3. After the computation of the red components xr, a function collect_elements(xr) distributes the red vector to all other processors for the next computation. Analogously, the black vector xb is distributed after its computation. The function collect_elements() can be implemented by a multi-broadcast operation.

8.3.5.2 SOR method for red-black systems

An SOR method for the linear equation system (8.46) with relaxation parameter ω can be derived from the Gauss-Seidel computation (8.48) and (8.49) by using the combination of the new and the old approximation vector as introduced in Formula (8.41). One step of the SOR method has then the form:

```
local_nr = nr/p; local_nb = nb/p;
do {
    mestartr = me * local_nr;
    for (i= mestartr; i < mestartr + local_nr; i++) {
        xr[i] = 0;
        for (j ∈ N(i))
            xr[i] = xr[i] - a[i][j] * xb[j];
        xr[i] = (xr[i]+b[i]) / a[i][i] ;
    }
    collect_elements(xr);
    mestartb = me * local_nb + nr;
    for (i= mestartb; i < mestartb + local_nb; i++) {
        xb[i] = 0;
        for (j ∈ N(i))
            xb[i] = xb[i] - a[i+nr][j] * xr[j];
        xb[i]= (xb[i] + b[i+nr]) / a[i+nr][i+nr];
    }
    collect_elements(xb);
} while (convergence_test());
```

Fig. 8.19 Program fragment for the parallel implementation of the Gauss-Seidel method with the red-black ordering. The arrays xr and xb denote the unknowns corresponding to the red or black mesh points. The processor number of the executing processor is stored in me.

$$\tilde{x}_R^{(k+1)} = D_R^{-1} \cdot b_1 - D_R^{-1} \cdot F \cdot x_B^{(k)} \,,$$
$$\tilde{x}_B^{(k+1)} = D_B^{-1} \cdot b_2 - D_B^{-1} \cdot E \cdot x_R^{(k+1)} \,,$$
$$x_R^{(k+1)} = x_R^{(k)} + \omega \left(\tilde{x}_R^{(k+1)} - x_R^{(k)} \right) \,, \tag{8.50}$$
$$x_B^{(k+1)} = x_B^{(k)} + \omega \left(\tilde{x}_B^{(k+1)} - x_B^{(k)} \right) \,, \quad k = 1, 2, \dots \,.$$

The corresponding splitting of matrix $\hat{A}$ is $\hat{A} = \frac{1}{\omega}\hat{D} - \hat{L} - \hat{U} - \frac{1-\omega}{\omega}\hat{D}$ with the matrices $\hat{D}, \hat{L}, \hat{U}$ introduced above. This can be written using block matrices:

$$\begin{pmatrix} D_R & 0 \\ \omega E & D_B \end{pmatrix} \cdot \begin{pmatrix} x_R^{(k+1)} \\ x_B^{(k+1)} \end{pmatrix} \tag{8.51}$$
$$= (1-\omega) \begin{pmatrix} D_R & 0 \\ 0 & D_B \end{pmatrix} \cdot \begin{pmatrix} x_R^{(k)} \\ x_B^{(k)} \end{pmatrix} - \omega \begin{pmatrix} 0 & F \\ 0 & 0 \end{pmatrix} \cdot \begin{pmatrix} x_R^{(k)} \\ x_B^{(k)} \end{pmatrix} + \omega \begin{pmatrix} b_1 \\ b_2 \end{pmatrix}$$

For a parallel implementation the component form of this system is used. On the other hand, for the convergence results the matrix form and the iteration matrix have to be considered. Since the iteration matrix of the SOR method for a given linear equation system $Ax = b$ with a certain order of the equations and the iteration matrix of the SOR method for the red-black system $\hat{A}\hat{x} = \hat{b}$ are different, convergence results cannot be transferred. The iteration matrix of the SOR method with red-black ordering is:

$$\hat{S}_\omega = \left(\frac{1}{\omega}\hat{D} - \hat{L}\right)^{-1}\left(\frac{1-\omega}{\omega}\hat{D} + \hat{U}\right).$$

For a convergence of the method it has to be shown that $\rho(\hat{S}_\omega) < 1$ for the spectral radius of $\hat{S}_\omega$ and $\omega \in \mathbb{R}$. In general, the convergence cannot be derived from the convergence of the SOR method for the original system, since $P^T S_\omega P$ is not identical to $\hat{S}_\omega$, although $P^T A P = \hat{A}$ holds. However, for the specific case of the model problem, i.e., the discretized Poisson equation, the convergence can be shown. Using the equality $P^T A P = \hat{A}$, it follows that $\hat{A}$ is symmetric and positive definite and, thus, the method converges for the model problem, see [74].

Figure 8.20 shows a parallel SPMD implementation of the SOR method for the red-black ordered discretized Poisson equation. The elements of the coefficient matrix are coded as constants. The unknowns are stored in a two-dimensional structure corresponding to the two-dimensional mesh and not as vector so that unknowns appear as x[i][j] in the program. The mesh points and the corresponding computations are distributed among the processors; the mesh points belonging to a specific processor are stored in myregion. The color red or black of a mesh point (i, j) is an additional attribute which can be retrieved by the functions is_red() and is_black(). The value f[i][j] denotes the discretized right-hand side of the Poisson equation as described earlier, see Eq. (8.15). The functions exchange_red_borders() and exchange_black_borders() exchange the red or black data of the red or black mesh points between neighboring processors.

```
do {
    for ((i,j) ∈ myregion) {
        if (is_red(i,j))
            x[i][j] = omega/4 * (h*h*f[i][j] + x[i][j-1] + x[i][j+1]
                     + x[i-1][j] + x[i+1][j]) + (1- omega) * x[i][j];
    }
    exchange_red_borders(x);
    for ((i,j) ∈ myregion) {
        if (is_black(i,j))
            x[i][j] = omega/4 * (h*h*f[i][j] + x[i][j-1] + x[i][j+1]
                     + x[i-1][j] + x[i+1][j]) + (1- omega) * x[i][j];
    }
    exchange_black_borders(x);
} while (convergence_test());
```

Fig. 8.20 Program fragment of a parallel SOR method for a red-black ordered discretized Poisson equation.

8.4 Conjugate Gradient Method

The conjugate gradient method or CG method is a solution method for linear equation systems $Ax = b$ with symmetric and positive definite matrix $A \in \mathbb{R}^{n \times n}$, which has been introduced in [107]. (A is symmetric if $a_{ij} = a_{ji}$ and positive definite if $x^T A x > 0$ for all $x \in \mathbb{R}^n$ with $x \neq 0$.) The CG method builds up a solution $x^* \in \mathbb{R}^n$ in at most n steps in the absence of roundoff errors. Considering roundoff errors more than n steps may be needed to get a good approximation of the exact solution x^*. For sparse matrices a good approximation of the solution can be achieved in less than n steps, also with roundoff errors [188]. In practice, the CG method is often used as preconditioned CG method which combines a CG method with a preconditioner [194]. Parallel implementations are discussed in [86, 165, 166, 194]; [195] gives an overview. In this section, we present the basic CG method and parallel implementations according to [25, 85, 209].

8.4.1 Sequential CG method

The CG method exploits an equivalence between the solution of a linear equation system and the minimization of a function.

More precisely, the solution x^* of the linear equation system $Ax = b$, $A \in \mathbb{R}^{n \times n}$, $b \in \mathbb{R}^n$, is the minimum of the function $\Phi : M \subset \mathbb{R}^n \to \mathbb{R}$ with

$$\Phi(x) = \frac{1}{2} x^T A x - b^T x , \tag{8.52}$$

if the matrix A is symmetric and positive definite. A simple method to determine the minimum of the function Φ is the method of the steepest gradient [85] which uses the negative gradient. For a given point $x_c \in \mathbb{R}^n$ the function decreases most rapidly in the direction of the negative gradient. The method computes the following two steps.

(a) Computation of the negative gradient $d_c \in \mathbb{R}^n$ at point x_c,

$$d_c = - \text{ grad } \Phi(x_c) = - \left(\frac{\partial}{\partial x_1} \Phi(x_c), \dots, \frac{\partial}{\partial x_n} \Phi(x_c) \right) = b - A x_c$$

(b) Determination of the minimum of Φ in the set

$$\{ x_c + t d_c \mid t \geq 0 \} \cap M ,$$

which forms a line in $\mathbb{R}^n$ (line search). This is done by inserting $x_c + t d_c$ into Formula (8.52). Using $d_c = b - A x_c$ and the symmetry of matrix A we get

$$\Phi(x_c + t d_c) = \Phi(x_c) - t d_c^T d_c + \frac{1}{2} t^2 d_c^T A d_c . \tag{8.53}$$

The minimum of this function with respect to $t \in \mathbb{R}$ can be determined using the derivative of this function with respect to t. The minimum is

$$t_c = \frac{d_c^T d_c}{d_c^T A d_c} \tag{8.54}$$

The steps (a) and (b) of the method of the steepest gradient are used to create a sequence of vectors $x_k, k = 0, 1, 2, \ldots$, with $x_0 \in \mathbb{R}^n$ and $x_{k+1} = x_k + t_k d_k$. The sequence $(\Phi(x_k))_{k=0,1,2,\ldots}$ is monotonically decreasing which can be seen by inserting Formula (8.54) into Formula (8.53). The sequence converges towards the minimum but the convergence might be slow [85].

The CG method uses a technique to determine the minimum which exploits orthogonal search directions in the sense of **conjugate** or **A-orthogonal** vectors d_k. For a given matrix A, which is symmetric and non-singular, two vectors $x, y \in \mathbb{R}^n$ are called conjugate or A-orthogonal, if $x^T A y = 0$. If A is positive definite, k pairwise conjugate vectors $d_0, \ldots, d_{n-1}$ (with $d_i \neq 0, i = 0, \ldots, k-1$ and $k \leq n$) are linearly independent [25]. Thus, the unknown solution vector x^* of $Ax = b$ can be represented as a linear combination of the conjugate vectors $d_0, \ldots, d_{n-1}$, i.e.,

$$x^* = \sum_{k=0}^{n-1} t_k d_k . \tag{8.55}$$

Since the vectors are orthogonal, $d_k^T A x^* = \sum_{l=0}^{n-1} d_k^T A t_l d_l = t_k d_k^T A d_k$. This leads to

$$t_k = \frac{d_k A x^*}{d_k^T A d_k} = \frac{d_k^T b}{d_k^T A d_k}$$

for the coefficients t_k. Thus, when the orthogonal vectors are known, the values t_k, $k = 0, \ldots, n-1$, can be computed from the right-hand side b.

The algorithm for the CG method uses a representation

$$x^* = x_0 + \sum_{i=0}^{n-1} \alpha_i d_i, \tag{8.56}$$

of the unknown solution vector x^* as a sum of a starting vector x_0 and a term $\sum_{i=0}^{n-1} \alpha_i d_i$ to be computed. The second term is computed recursively by

$$x_{k+1} = x_k + \alpha_k d_k, \quad k = 1, 2, \ldots, \quad \text{with} \tag{8.57}$$

$$\alpha_k = \frac{-g_k^T d_k}{d_k^T A d_k} \quad \text{and} \quad g_k = A x_k - b . \tag{8.58}$$

The Formulas (8.57) and (8.58) determine x^* according to (8.56) by computing α_i and adding $\alpha_i d_i$ in each step, $i = 1, 2, \ldots$. Thus, the solution is computed after at most n steps. If not all directions d_k are needed for x^*, less than n steps are required.

Algorithms implementing the CG method do not choose the conjugate vectors $d_0, \ldots, d_{n-1}$ before but compute the next conjugate vector from the given gradient

g_k by adding a correction term. The basic algorithm for the CG method is given in Figure 8.21.

Select $x_0 \in \mathbb{R}^n$
Set $d_0 = -g_0 = b - Ax_0$
While $(\|g_k\| > \varepsilon)$ compute for $k = 0, 1, 2, \ldots$
(1) $w_k = Ad_k$
(2) $\alpha_k = \dfrac{g_k^T g_k}{d_k^T w_k}$
(3) $x_{k+1} = x_k + \alpha_k d_k$
(4) $g_{k+1} = g_k + \alpha_k w_k$
(5) $\beta_k = \dfrac{g_{k+1}^T g_{k+1}}{g_k^T g_k}$
(6) $d_{k+1} = -g_{k+1} + \beta_k d_k$

Fig. 8.21 Algorithm of the CG method. (1) and (2) compute the values α_k according to Eq. (8.58). The vector w_k is used for the intermediate result Ad_k. (3) is the computation given in Formula (8.57). (4) computes g_{k+1} for the next iteration step according to Formula (8.58) in a recursive way. $g_{k+1} = Ax_{k+1} - b = A(x_k + \alpha_k d_k) - b = g_k + A\alpha_k d_k$. This vector g_{k+1} represents the error between the approximation x_k and the exact solution. (5) and (6) compute the next vector d_{k+1} of the set of conjugate gradients.

8.4.2 Parallel CG Method

The parallel implementation of the CG method is based on the algorithm given in Figure 8.21. Each iteration step of this algorithm implementing the CG method consists of the following basic vector and matrix operations.

8.4.2.1 Basic operations of the CG algorithm

The basic operations of the CG algorithm are:

(1)   a matrix-vector multiplication Ad_k,
(2)   two scalar products $g_k^T g_k$ and $d_k^T w_k$,
(3)   a so-called *axpy*-operation $x_k + \alpha_k d_k$,
        (The name *axpy* comes from *a x* plus *y* describing the computation.)
(4)   an *axpy*-operation $g_k + \alpha_k w_k$,
(5)   a scalar product $g_{k+1}^T g_{k+1}$, and
(6)   an *axpy*-operation $-g_{k+1} + \beta_k d_k$.

The result of $g_k^T g_k$ is needed in two consecutive steps and so the computation of one scalar product can be avoided by storing $g_k^T g_k$ in the scalar value γ_k. Since there are mainly one matrix-vector product and scalar products, a parallel implementation can be based on parallel versions of these operations.

Like the CG method many algorithms from linear algebra are built up from basic operations like matrix-vector operations or *axpy*-operations and efficient implementations of these basic operations lead to efficient implementations of entire algorithms. The **BLAS** (*Basic Linear Algebra Subroutines*) library offers efficient implementations for a large set of basic operations. This includes many *axpy*-operations which denote that a vector x is multiplied by a scalar value a and then added to another vector y. The prefixes s in saxpy or d daxpy denote axpy operations for *simple precision* and *double precision*, respectively. Introductory descriptions of the BLAS library are given in [55] or [73]. A standard way to parallelize algorithms for linear algebra is to provide efficient parallel implementations of the BLAS operations and to build up a parallel algorithm from these basic parallel operations. This technique is ideally suited for the CG method since it consists of such basic operations.

Here, we consider a parallel implementation based on the parallel implementations for matrix-vector multiplication or scalar product for distributed memory machines as presented in Section 3. These parallel implementations are based on a data distribution of the matrix and the vectors involved. For an efficient implementation of the CG method it is important that the data distributions of different basic operations fit together in order to avoid expensive data re-distributions between the operations. Figure 8.22 shows a data dependence graph in which the nodes correspond to the computation steps (1)–(6) of the CG algorithm in Figure 8.21 and the arrows depict a data dependency between two of these computation steps. The arrows are annotated with data structures computed in one step (outgoing arrow) and needed for another step with incoming arrow. The data dependence graph for one iteration step k is a directed acyclic graph (DAG). There are also data dependences to the previous iteration step $k-1$ and the next iteration step $k+1$, which are shown as dashed arrows.

There are the following dependences in the CG method: The computation (2) needs the result w_k from computation (1) but also the vector d_k and the scalar value γ_k from the previous iteration step $k-1$; γ_k is used to store the intermediate result $\gamma_k = g_k^T g_k$. Computation (3) needs α_k from computation step (2) and the vectors x_k, d_k from the previous iteration step $k-1$. Computation (4) also needs α_k from computation step (2) and vector w_k from computation (1). Computation (5) needs vector g_{k+1} from computation (4) and scalar value γ_k from the previous iteration step $k-1$; Computation (6) needs the scalar value from β_k from computation (5) and vector d_k from iteration step $k-1$. This shows that there are many data dependences between the different basic operations. But it can also be observed that computation (3) is independent from the computations (4)–(6). Thus, the computation sequence (1),(2),(3),(4),(5),(6) as well as the sequence (1),(2),(4),(5),(6),(3) can be used. The independence of computation (3) from computations (4)–(6) is also another source of parallelism, which is a coarse-grain parallelism of two linear algebra operations performed in parallel, in contrast to the fine-grained parallelism exploited for a single basic operation. In the following, we concentrate on the fine-grained parallelism of basic linear algebra operations.

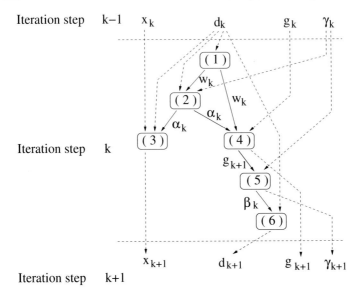

Fig. 8.22 Data dependences between the computation steps (1)–(6) of the CG method in Figure 8.21. Nodes represent the computation steps of one iteration step k. Incoming arrows are annotated by the data required and outgoing arrows are annotated by the data produced. Two nodes have an arrow between them if one of the nodes produces data which are required by the node with the incoming arrow. The data dependences to the previous iteration step $k-1$ or the next iteration step $k+1$ are given as dashed arrows. The data are named in the same way as in Figure 8.21; additionally the scalar γ_k is used for the intermediate result $\gamma_k = g_k^T g_k$ computed in step (5) and required for the computations of α_k and β_k in computation steps (2) and (5) of the next iteration step.

When the basic operations are implemented on a distributed memory machine, the data distribution of matrices and vectors and the data dependences between operations might require data re-distribution for a correct implementation. Thus, the data dependence graph in Figure 8.22 can also be used to study the communication requirements for re-distribution in a message-passing program. Also the data dependences between two iteration steps may lead to communication for data re-distribution.

To demonstrate the communication requirements, we consider an implementation of the CG method in which the matrix A has a row-wise blockwise distribution and the vectors d_k, ω_k, g_k, x_k and r_k have a blockwise distribution. In one iteration step of a parallel implementation, the following computation and communication operations are performed.

8.4.2.2 Parallel CG implementation with blockwise distribution

The parallel CG implementation has to consider data distributions in the following way:

(0) Before starting the computation of iteration step k, the vector d_k computed in the previous step has to be re-distributed from a block-wise distribution of step $k-1$ to a replicated distribution required for step k. This can be done with a multi-broadcast operation.

(1) The matrix-vector multiplication $w_k = Ad_k$ is implemented with a row block-wise distribution of A as described in Section 3.7. Since d_k is now replicated, no further communication is needed. The result vector w_k is distributed in a block-wise way.

(2) The scalar products $d_k^T w_k$ is computed in parallel with the same blockwise distribution of both vectors. (The scalar product $\gamma_k = g_k^T g_k$ is computed in the previous iteration step.) Each processor computes a local scalar product for its local vectors. The final scalar product is then computed by the root processor of a single-accumulation operation with the addition as reduction operation. This processor owns the final result α_k and sends it to all other processors by a single-broadcast operation.

(3) The scalar value α_k is known by each processor and thus the $axpy$-operation $x_{k+1} = x_k + \alpha_k d_k$ can be done in parallel without further communication. Each processor performs the arithmetic operations locally and the vector x_{k+1} results in a blockwise distribution.

(4) The $axpy$-operation $g_{k+1} = g_k + \alpha_k w_k$ is computed analogously to computation step (3) and the result vector g_{k+1} is distributed in a blockwise way.

(5) The scalar products $\gamma_{k+1} = g_{k+1}^T g_{k+1}$ is computed analogously to computation step (2). The resulting scalar value β_k is computed by the root processor of a single-accumulation operation and then broadcasted to all other processors.

(6) The $axpy$-operation $d_{k+1} = -g_{k+1} + \beta_k d_k$ is computed analogously to computation step (3). The result vector d_{k+1} has a blockwise distribution.

8.4.2.3 Parallel execution time

The parallel execution time of one iteration step of the CG method is the sum of the parallel execution times of the basic operations involved. We derive the parallel execution time for p processors; n is the system size. It is assumed that n is a multiple of p. The parallel execution time of one $axpy$-operation is given by

$$T_{axpy} = 2 \cdot \frac{n}{p} \cdot t_{op} ,$$

since each processor computes n/p components and the computation of each component needs one multiplication and one addition. As in earlier sections, the time for one arithmetic operation is denoted by t_{op}. The parallel execution time of a scalar product is

$$T_{scal\_prod} = 2 \cdot \left(\frac{n}{p} - 1 \right) \cdot t_{op} + T_{acc}(+)(p,1) + T_{sb}(p,1) \, ,$$

where $T_{acc}(op)(p,m)$ denotes the communication time of a single-accumulation operation with reduction operation op on p processors and message size m. The computation of the local scalar products with n/p components requires n/p multiplications and $n/p - 1$ additions. The distribution of the result of the parallel scalar product, which is a scalar value, i.e., has size 1, needs the time of a single-broadcast operation $T_{sb}(p,1)$. The matrix-vector multiplication needs time

$$T_{math\_vec\_mult} = 2 \cdot \frac{n^2}{p} \cdot t_{op} \, ,$$

since each processor computes n/p scalar products. The total computation time of the CG method is

$$T_{CG} = T_{mb}(p, \frac{n}{p}) + T_{math\_vec\_mult} + 2 \cdot T_{scal\_prod} + 3 \cdot T_{axpy},$$

where $T_{mb}(p,m)$ is the time of a multi-broadcast operation with p processors and message size m. This operation is needed for the re-distribution of the direction vector d_k from iteration step k .

8.5 Cholesky Factorization for Sparse Matrices

Linear equation systems arising in practice are often large but have sparse coefficient matrices, i.e., they have many zero entries. For sparse matrices with regular structure, like banded matrices, only the diagonals with non-zero elements are stored and the solution methods introduced in the previous sections can be used. For an unstructured pattern of non-zero elements in sparse matrices, however, a more general storage scheme is needed and other parallel solution methods are applied. In this section, we consider the Cholesky factorization as an example of such a solution method. The general sequential factorization algorithm and its variants for sparse matrices are introduced in Section 8.5.1. A specific storage scheme for sparse unstructured matrices is given in Section 8.5.2. In Section 8.5.3, we discuss parallel implementations of sparse Cholesky factorization for shared memory machines.

8.5.1 Sequential Algorithm

The Cholesky factorization is a direct solution method for a linear equation system $Ax = b$. The method can be used if the coefficient matrix $A = (a_{ij}) \in \mathbb{R}^{n \times n}$ is symmetric and positive definite, i.e., if $a_{ij} = a_{ji}$ and $x^T A x > 0$ for all $x \in \mathbb{R}^n$ with $x \neq 0$. For a symmetric and positive definite $n \times n$-matrix $A \in \mathbb{R}^{n \times n}$ there exists a unique triangular factorization

$$A = LL^T \tag{8.59}$$

where $L = (l_{ij})_{i,j=1,\dots,n}$ is a lower triangular matrix, i.e., $l_{ij} = 0$ for $i < j$ and $i, j \in \{1,\dots,n\}$, with positive diagonal elements, i.e., $l_{ii} > 0$ for $i = 1,\dots,n$; L^T denotes the transposed matrix of L, i.e., $L^T = (l_{ij}^T)_{i,j=1,\dots,n}$ with $l_{ij}^T = l_{ji}$ [209]. Using the factorization in Eq. (8.59), the solution x of a system of equations $Ax = b$ with $b \in \mathbb{R}^n$ is determined in two steps by solving the triangular systems $Ly = b$ and $L^T x = y$ one after another. Because of $Ly = LL^T x = Ax = b$, the vector $x \in \mathbb{R}^n$ is the solution of the given linear equation system.

The implementation of the Cholesky factorization can be derived from a column-wise formulation of $A = LL^T$. Comparing the elements of A and LL^T, we obtain:

$$a_{ij} = \sum_{k=1}^{n} l_{ik} l_{kj}^T = \sum_{k=1}^{n} l_{ik} l_{jk} = \sum_{k=1}^{j} l_{ik} l_{jk} = \sum_{k=1}^{j} l_{jk} l_{ik}$$

since $l_{jk} = 0$ for $k > j$ and by exchanging elements in the last summation. Denoting the columns of A as $\tilde{a}_1,\dots,\tilde{a}_n$ and the columns of L as $\tilde{l}_1,\dots,\tilde{l}_n$, this results in an equality for column $\tilde{a}_j = (a_{1j},\dots,a_{nj})$ and columns $\tilde{l}_k = (l_{1k},\dots,l_{nk})$ for $k \leq j$:

$$\tilde{a}_j = \sum_{k=1}^{j} l_{jk} \tilde{l}_k$$

leading to

$$l_{jj} \tilde{l}_j = \tilde{a}_j - \sum_{k=1}^{j-1} l_{jk} \tilde{l}_k \tag{8.60}$$

for $j = 1,\dots,n$. If the columns $\tilde{l}_k, k = 1,\dots,j-1$, are already known, the right-hand side of Formula (8.60) is computable and the column $\tilde{l}_j$ can also be computed. Thus, the columns of L are computed one after another. The computation of column $\tilde{l}_j$ has two cases. For the diagonal element the computation is:

$$l_{jj} l_{jj} = a_{jj} - \sum_{k=1}^{j-1} l_{jk} l_{jk} \quad \text{or} \quad l_{jj} = \sqrt{a_{jj} - \sum_{k=1}^{j-1} l_{jk}^2} \,.$$

For the elements $l_{ij}, i > j$, the computation is:

$$l_{ij} = \frac{1}{l_{jj}} \left(a_{ij} - \sum_{k=1}^{j-1} l_{jk} l_{ik} \right);$$

The elements in the upper triangular of matrix L are $l_{ij} = 0$ for $i < j$.

The Cholesky factorization yields the factorization $A = LL^T$ for a given matrix A [79] by computing $L = (l_{ij})_{i=0,..n-1, j=0,..i}$ from $A = (a_{ij})_{i,j=0,...,n-1}$ column by column from left to right according to the following algorithm, in which the numbering starts with 0:

(I)

$$
\begin{array}{l}
\texttt{for} \quad \texttt{(j=0; j<n; j++) \{} \\[4pt]
\qquad l_{jj} = \sqrt{a_{jj} - \sum_{k=0}^{j-1} l_{jk}^2} \, ; \\[8pt]
\qquad \texttt{for} \quad \texttt{(i=j+1; i<n; i++)} \\[4pt]
\qquad\qquad l_{ij} = \frac{1}{l_{jj}} \left(a_{ij} - \sum_{k=0}^{j-1} l_{jk} l_{ik} \right) ; \\[8pt]
\texttt{\}}
\end{array}
$$

For each column j, first the new diagonal element l_{jj} is computed using the elements in row j; then, the new elements of column j are computed using row j of A and all columns i of L with $i < j$, see Fig. 8.23 (left).

For dense matrices A, the Cholesky factorization requires $O(n^2)$ storage space and $O(n^3/6)$ arithmetic operations [209]. For sparse matrices, drastic reductions in storage and execution time can be achieved by exploiting the sparsity of A, i.e., by storing and computing only the non-zero entries of A.

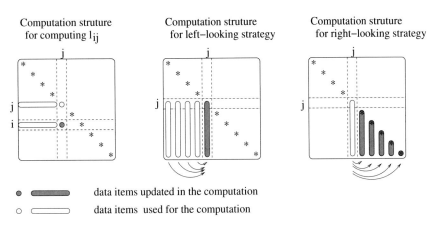

Computation struture for computing l_{ij}

Computation struture for left–looking strategy

Computation struture for right–looking strategy

● ▬▬▬▬ data items updated in the computation

○ ▭▭▭▭ data items used for the computation

Fig. 8.23 Computational structures and data dependences for the computation of L according to the basic algorithm (left), the left-looking algorithm (middle), and the right-looking algorithm (right).

The Cholesky factorization usually causes fill-ins for sparse matrices A which means that the matrix L has non-zeros in positions which are zero in A. The number of fill-in elements can be reduced by reordering the rows and columns of A resulting in a matrix PAP^T with a corresponding permutation matrix P. For Cholesky factorization, P can be chosen without regard to numerical stability, because no pivoting is required [79]. Since PAP^T is also symmetric and positive definite for any permutation matrix P, the factorization of A can be done with the following steps:

1. **Re-ordering:** Find a permutation matrix $P \in \mathbb{R}^{n \times n}$ that minimizes the storage requirement and computing time by reducing fill-ins. The reordered linear equation system is $(PAP^T)(Px) = Pb$.
2. **Storage allocation:** Determine the structure of the matrix L and set up the sparse storage scheme. This is done before the actual computation of L and is called (*symbolic factorization*), see [79].
3. **Numerical factorization:** Perform the factorization $PAP^T = LL^T$.
4. **Triangular solution:** Solve $Ly = Pb$ and $L^T z = y$. Then, the solution of the original system is $x = P^T z$.

The problem of finding an ordering that minimizes the amount of fill-in is NP-complete [224]. But there exist suitable heuristics for reordering. The most popular sequential fill-in reduction heuristic is the minimum degree algorithm [79]. Symbolic factorization by a graph-theoretic approach is described in detail in [79]. In the following, we concentrate on the numerical factorization, which is considered to require by far the most computation time, and assume that the coefficient matrix is already in reordered form.

8.5.1.1 Left-looking algorithms

According to [157], we denote the sparsity structure of column j and row i of L (excluding diagonal entries) by

$$Struct(L_{*j}) = \{k > j | l_{kj} \neq 0\}$$
$$Struct(L_{i*}) = \{k < i | l_{ik} \neq 0\}$$

$Struct(L_{*j})$ contains the row indices of all non-zeros of column j and $Struct(L_{i*})$ contains the column indices of all non-zeros of row i. Using these sparsity structures a slight modification of computation scheme (I) results. The modification uses the following procedures for manipulating columns [157, 192]:

(II)

```
cmod(j,k) =
    for each i ∈ Struct(L*k) with i ≥ j :
        aij = aij − ljklik ;
cdiv(j) =
    ljj = √ajj ;
    for each i ∈ Struct(L*j) :
        lij = aij/ljj ;
```

Procedure $cmod(j,k)$ modifies column j by subtracting a multiple with factor l_{jk} of column k from column j for columns k already computed. Only the non-zero elements of column k are considered in the computation. The entries a_{ij} of the original matrix a are now used to store the intermediate results of the computation of L. Procedure $cdiv(j)$ computes the square root of the diagonal element and divides all entries of column j by this square root of its diagonal entry l_{jj}. Using these two procedures, column j can be computed by applying $cmod(j,k)$ for each $k \in Struct(L_{j*})$ and then completing the entries by applying $cdiv(j)$. Applying $cmod(j,k)$ to columns $k \notin Struct(L_{j*})$ has no effect because $l_{jk} = 0$. The columns of L are computed from left to right and the computation of a column $\tilde{l}_j$ needs all columns $\tilde{l}_k$ to the left of column $\tilde{l}_j$. This results in the following *left-looking* algorithm:

$$(III)$$
```
left_cholesky =
    for  j = 0, ..., n-1 {
        for each  k ∈ Struct(L_j*):
            cmod(j,k);
        cdiv(j);
    }
```

The code in scheme (III) computes the columns one after another from left to right. The entries of column j are modified after all columns *to the left* of j have completely been computed, i.e., the same target column j is used for a number of consecutive $cmod(j,k)$ operations; this is illustrated in Fig. 8.23 (middle).

8.5.1.2 Right-looking algorithm

An alternative way is to use the entries of column j after the complete computation of column j to modify all columns k *to the right* of j, that depend on column j, i.e., to modify all columns $k \in Struct(L_{*j})$ by subtracting l_{kj} times the column j from column k. Because $l_{kj} = 0$ for $k \notin Struct(L_{*j})$ only the columns $k \in Struct(L_{*j})$ are manipulated by column j. Still the columns are computed from left to right. The difference to the left-looking algorithm is that the calls to cmod() for a column j are done earlier. The final computation of a column j then consists only of a call to cdiv(j) after all columns to the left are computed. This results in the following *right-looking* algorithm:

$$(IV)$$
```
right_cholesky =
    for  j = 0, ..., n-1 {
        cdiv(j);
        for each  k ∈ Struct(L_*j):
            cmod(k,j);
    }
```

The code fragment shows that in the right-looking algorithm, successive cmod() operations manipulate different target columns with the same column j. An illustration is given in Fig. 8.23, right.

In both the left-looking and the right-looking algorithms, each non-zero l_{ij} leads to an execution of a cmod() operation. In the left-looking algorithm, the $cmod(j,k)$ operation is used to compute column j. In the right-looking algorithm, the $cmod(k,j)$ operation is used to manipulate column $k \in Struct(L_{*j})$ after the computation of column j. Thus, left-looking and right-looking algorithm use the same number of cmod() operations. They use also the same number of cdiv() operations, since there is exactly one cdiv() operation for each column.

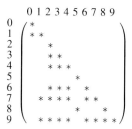

Fig. 8.24 Matrix L with supernodes $I(0) = \{0\}$, $I(1) = \{1\}$, $I(2) = \{2,3,4\}$, $I(5) = \{5\}$, $I(6) = \{6,7\}$, $I(8) = \{8,9\}$. The elimination tree is shown to the right.

8.5.1.3 Supernodes

The supernodal algorithm is a computation scheme for sparse Cholesky factorization that exploits similar patterns of non-zero elements in adjacent columns, see [157, 192]. A supernode is a set

$$I(p) = \{p, p+1, \ldots, p+q-1\}$$

of contiguous columns in L for which for all i with $p \leq i \leq p+q-1$:

$$Struct(L_{*i}) = Struct(L_{*(p+q-1)}) \cup \{i+1, \ldots, p+q-1\} .$$

Thus, a supernode has a dense triangular block above (and including) row $p+q-1$, i.e., all entries are non-zero elements, and an identical sparsity structure for each column below row $p+q-1$, i.e., each column has its non-zero elements in the same rows as the other columns in the supernode. Figure 8.24 shows an example. Because of this identical sparsity structure of the columns, a supernode has the property that each member column modifies the same set of target columns outside its supernode [192]. Thus, the factorization can be expressed in terms of supernodes modifying columns, rather than columns modifying columns.

Using the definitions $first(J) = p$ and $last(J) = p + q - 1$ for a supernode $J = I(p) = \{p, p+1, \ldots, p+q-1\}$, the following additional procedure $smod()$ is defined:

(V)

$$
\begin{aligned}
&\texttt{smod}(j, J) = \\
&\quad r = min\{j - 1, last(J)\}; \\
&\quad \texttt{for } k = first(J), \ldots, r \\
&\quad\quad \texttt{cmod}(j, k);
\end{aligned}
$$

which modifies column j with all columns from supernode J. There are two cases for modifying a column with a supernode: When column j belongs to supernode J, then column j is modified only by those columns of J that are to the left in node J. When column j does not belong to supernode J, then column j is modified by all columns of J. Using the procedure $smod()$, the Cholesky factorization can be performed by the following computation scheme, also called *right-looking supernodal* algorithm:

(VI)

```
supernode_cholesky =
    for each supernode J do from left to right {
        cdiv(first(J));
        for j = first(J) + 1, ..., last(J) {
            smod(j, J);
            cdiv(j);
        }
        for k ∈ Struct(L_*(last(J)))
            smod(k, J);
    }
```

This computation scheme still computes the columns of L from left to right. The difference to the algorithms presented before is that the computations associated to a supernode are combined. On the supernode level, a right-looking scheme is used: For the computation of the first column of a supernode J only one `cdiv()` operation is necessary when the modification with all columns to the left is already done. The columns of J are computed in a left-looking way: After the computation of all supernodes to the left of supernode J and because the columns of J are already modified with these supernodes due the supernodal right-looking scheme, column j is computed by first modifying it with all columns of J to the left of j and then performing a `cdiv()` operation. After the computation of all columns of J, all columns

k to the right of J that depend on columns of J are modified with each column in J, i.e., by the procedure $\mathtt{smod}(k,J)$.

An alternative way would be a right-looking computation of the columns of J. An advantage of the supernodal algorithm lies in an increased locality of memory accesses because each column of a supernode J is used for the modification of several columns to the right of J and because all columns of J are used for the modification of the same columns to the right of J.

8.5.2 Storage Scheme for Sparse Matrices

Since most entries in a sparse matrix are zero, specific storage schemes are used to avoid the storage of zero elements. These compressed storage schemes store the non-zero entries and additional information about the row and column indices to identify its original position in the full matrix. Thus, a compressed storage scheme for sparse matrices needs the space for the non-zero elements as well as space for additional information.

A sparse lower triangular matrix L is stored in a compressed storage scheme of size $O(n+nz)$ where n is the number of rows (or columns) in L and nz is the number of non-zeros. We present the storage scheme of the SPLASH implementation which, according to [145], stores a sparse matrix in a compressed manner similar to [78]. This storage scheme exploits the sparsity structure as well as the supernode structure to store the data. We first describe a simpler version using only the sparsity structure without supernodes. Exploiting the supernode structure is then based on this storage scheme.

The storage scheme uses two arrays Nonzero and Row of length nz and three arrays StartColumn, StartRow, and Supernode of length n. The array Nonzero contains the values of all non-zeros of a triangular matrix $L = (l_{kj})_{k \geq j}$ in column-major order, i.e., the non-zeros are ordered column-wise from left to right in a linear array. Information about the corresponding column indices of non-zero elements is implicitly contained in array StartColumn: Position j of array StartColumn stores the index of array Nonzero in which the first non-zero element of column j is stored, i.e., Nonzero[StartColumn[j]] contains l_{jj}. Because the non-zero elements are stored column-wise, StartColumn[$j+1$] $- 1$ contains the last non-zero element of column j. Thus, the non-zeros of the jth column of L are assigned to the contiguous part of array Nonzero with indices from StartColumn[j] to StartColumn[$j+1$] $- 1$. The size of the contiguous part of non-zeros of column j in array Nonzero is $N_j := $ StartColumn[$j+1$]$-$StartColumn[j]. The array Row contains the row indices of the corresponding elements in Nonzero. In the simpler version without supernodes, Row[r] contains the row index of the non-zero stored in Nonzero[r], $r = 0, \ldots, nz - 1$. Corresponding to the block-wise storage scheme in Nonzero, the indices of the non-zeros of one column are stored in a contiguous block in Row.

When the similar sparsity structure of rows in the same supernode is additionally exploited, row indices of non-zeros are stored in a combination of the arrays Row and StartRow in the following way: StartRow$[j]$ stores the index of Row in which the row index of the first non-zero of column j is stored, i.e., Row[StartRow$[j]$] $= j$ because l_{jj} is the first non-zero. For each column the row indices are still stored in a contiguous block of Row. In contrast to the simpler scheme the blocks for different rows in the same supernode are not disjoint but overlap according to the similar sparsity structure of those columns.

The additional array StartRow can be used for a more compact storage scheme for the supernodal algorithm. When j is the first column of a supernode $I(j) = \{j, j+1, \ldots, j+k-1\}$, then column $j+l$ for $1 \leq l < k$ has the same non-zero pattern as row j for rows greater than or equal to $j+l$, i.e., Row[StartRow$[j]+l$] contains the row index of the first element of column $j+l$. Since this is the diagonal element, Row[StartRow$[j]+l$] $= j+l$ holds. The next entries are the row indices of the other non-zero elements of column $j+l$. Thus, the row indices of column $j+l$ are stored in Row[StartRow$[j]+l$], ..., Row[StartRow$[j]+$StartColumn$[j+1]-$StartColumn$[j-1]$]. This leads to StartRow$[j+l]=$ StartRow$[j]+l$ and thus only the row indices of the first column of a supernode have to be stored to get the full information. A fast access to the sets $Struct(L_{*j})$ is given by:

$$Struct(L_{*j}) =$$
$$\{\text{Row[StartRow}[j]+i] \mid 0 \leq i \leq \text{StartColumn}[j+1] - \text{StartColumn}[j-1]\}.$$

The storage scheme is illustrated in Fig. 8.25. The array Supernode is used for the management of supernodes: If a column j is the first column of a supernode J, then the number of columns of J is stored in Supernode$[j]$.

8.5.3 Implementation for Shared Variables

For a parallel implementation of sparse Cholesky factorization, we consider a shared memory machine. There are several sources of parallelism for sparse Cholesky factorization, including fine-grained parallelism within the single operations cmod(j,k) or cdiv(j) as well as column-oriented parallelism in the left-looking, right-looking, and supernodal algorithm.

The sparsity structure of L may lead to an additional source of parallelism which is not available for dense factorization. Data dependences may be avoided when different columns (and the columns having effect on them) have a disjoint sparsity structure. This kind of parallelism can be described by *elimination trees* that express the specific situation of data dependences between columns using the relation $parent(j)$ [157, 147]. For each column j, $0 \leq j < n$, we define:

$$parent(j) = min\{i \mid i \in Struct(L_{*j})\} \quad \text{if} \quad Struct(L_{*j}) \neq \emptyset,$$

i.e., $parent(j)$ is the row index of the first off-diagonal non-zero of column j. If $Struct(L_{*j}) = \emptyset$, then $parent(j) = j$. The element $parent(j)$ is the first column $i > j$ which depends on j. A column l, $j < l < i$, between them does not depend on j since

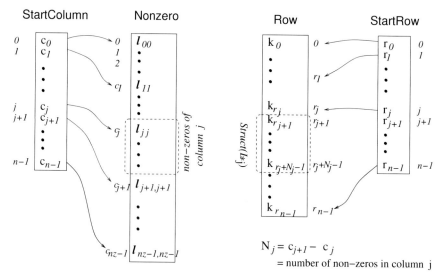

Fig. 8.25 Compressed storage scheme for a sparse lower triangular matrix L. The array `Nonzero` contains the non-zero elements of matrix L and the array `StartColumn` contains the positions of the first elements of columns in `Nonzero`. The array `Row` contains the row indices of elements in `Nonzero`; the first element of a row is given in `StartRow`. For a supernodal algorithm, `Row` can additionally use an overlapping storage (not shown here).

$j \notin Struct(L_{l*})$ and no $cmod(l, j)$ is executed. Moreover we define for $0 \le i < n$

$$children(i) = \{j < i \mid parent(j) = i\},$$

i.e. $children(i)$ contains all columns j that have their first off-diagonal non-zero in row i.

The directed graph $G = (V, E)$ has a set of nodes $V = \{0, \ldots, n-1\}$ with one node for each column and a set of edges E, where $(i, j) \in E$ if $i = parent(j)$ and $i \ne j$. It can be shown that G is a tree, if matrix A is *irreducible*. (A matrix A is called reducible, if A can be permuted such that it is block-diagonal. For a reducible matrix, the blocks can be factorized independently.) In the following, we assume an irreducible matrix. Figure 8.26 shows a matrix and its corresponding elimination tree.

In the following, we denote the subtree with root j by $G[j]$. For sparse Cholesky factorization, an important property of the elimination tree G is that the tree specifies the order in which the columns must be evaluated: the definition of *parent* implies that column i must be evaluated before column j, if $j = parent(i)$. Thus, all the children of column j must be completely evaluated before the computation of j. Moreover, column j does not depend on any column that is not in the subtree $G[j]$. Hence, columns i and j can be computed in parallel, if $G[i]$ and $G[j]$ are disjoint subtrees. Especially, all leaves of the elimination tree can be computed in parallel and the computation does not need to start with column 0. Thus, the sparsity struc-

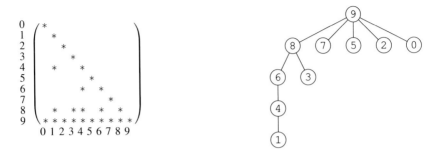

Fig. 8.26 Sparse matrix and corresponding elimination tree.

ture determines the parallelism to be exploited. For a given matrix, elimination trees of smaller height usually represent a larger degree of parallelism than trees of larger height [94].

8.5.3.1 Parallel left-looking algorithms

The parallel implementation of the left-looking algorithm (III) is based on n column tasks $Tcol(0),\ldots,Tcol(n-1)$ where task $Tcol(j)$, $0 \leq j < n$, comprises the execution of $\texttt{cmod}(j,k)$ for all $k \in Struct(L_{j*})$ and the execution of $\texttt{cdiv}(j)$; this is the loop body of the `for`-loop in algorithm (III). These tasks are not independent of each other but have dependences due to the non-zero elements. The parallel implementation uses a task pool for managing the execution of the tasks. The task pool has a central task pool for storing column tasks, which can be accessed by every processor. Each processor is responsible for performing a subset of the column tasks. The assignment of tasks to processors for execution is dynamic, i.e., when a processor is idle, it takes a task from the central task pool.

The dynamic implementation has the advantage that the workload is distributed evenly although the tasks might have different execution times due to the sparsity structure. The concurrent accesses of the processors to the central task pool have to be conflict-free so that the unique assignment of a task to a processor for execution is guaranteed. This can be implemented by a locking mechanism so that only one processor accesses the task pool at a specific time.

There are several parallel implementation variants for the left-looking algorithm differing in the way the column tasks are inserted into the task pool. We consider three implementation variants.

- Variant L1 inserts column task $Tcol(j)$ into the task pool not before all column tasks $Tcol(k)$ with $k \in Struct(L_{j*})$ have been finished. The task pool can be initialized to the leaves of the elimination tree. The degree of parallelism is limited by the number of independent nodes of the tree, since tasks dependent on each other are executed in sequential order. Hence, a processor that has accessed task $Tcol(j)$ can execute the task without waiting for other tasks to be finished.

- Variant L2 allows to start the execution of $Tcol(j)$ without requiring that it can be executed to completion immediately. The task pool is initialized to all column tasks available. The column tasks are accessed by the processors dynamically from left to right, i.e., an idle processor accesses the next column that has not yet been assigned to a processor.

  The computation of column task $Tcol(j)$ is started before all tasks $Tcol(k)$ with $k \in Struct(L_{j*})$ have been finished. In this case, not all operations $\mathtt{cmod}(j,k)$ of $Tcol(j)$ can be executed immediately but the task can perform only those $\mathtt{cmod}(j,k)$ operations with $k \in Struct(L_{j*})$ for which the corresponding tasks have already been executed. Thus, the task might have to wait during its execution for other tasks to be finished.

  To control the execution of a single column task $Tcol(j)$, each column j is assigned a data structure S_j containing all columns $k \in Struct(L_{j*})$ for which $\mathtt{cmod}(j,k)$ can already be executed. When a processor finishes the execution of the column task $Tcol(k)$ (by executing $\mathtt{cdiv}(k)$), it pushes k onto the data structures S_j for each $j \in Struct(L_{*k})$. Because different processors might try to access the same stack at the same time, a locking mechanism has to be used to avoid access conflicts. The processor executing $Tcol(j)$ pops column indices k from S_j and executes the corresponding $\mathtt{cmod}(j,k)$ operation. If S_j is empty, the processor waits for another processor to insert new column indices. When $|Struct(L_{j*})|$ column indices have been retrieved from S_j, the task $Tcol(j)$ can execute the final $\mathtt{cdiv}(j)$ operation.

  Figure 8.27 shows the corresponding implementation. The central task pool is realized implicitly as a parallel loop; the operation $\mathtt{get\_unique\_index()}$ ensures a conflict-free assignment of tasks so that the processors accessing the pool at the same time get different unique loop indices representing column tasks. The loop body of the $\mathtt{while}$-loop implements one task $Tcol(j)$. The data structures $S_1, \ldots, S_n$ are stacks; $\mathtt{pop}(S_j)$ retrieves an element and $\mathtt{push}(j, S_i)$ inserts an element onto the stack.

- Variant L3 is a variation of L2 that takes the structure of the elimination tree into consideration. The columns are not assigned strictly from left to right to the processors, but according to their height in the elimination tree, i.e., the children of a column j in the elimination tree are assigned to processors before their parent j. This variant tries to complete the column tasks in the order in which the columns are needed for the completion of the other columns, thus exploiting the additional parallelism that is provided by the sparsity structure of the matrix.

8.5.3.2 Parallel right-looking algorithm

The parallel implementation of the right-looking algorithm (IV) is also based on a task pool and on column tasks. These column tasks are defined differently than the tasks of the parallel left-looking algorithm: A column task $Tcol(j)$, $0 \leq j < n$,

```
parallel_left_cholesky =
    c = 0;
    while ((j = get_unique_index()) < n) {
        for (i = 0; i < |Struct(L_{j*})|; i++) {
            while (S_j empty) wait();
            k = pop(S_j);
            cmod(j,k);
        }
        cdiv(j);
        for (i ∈ Struct(L_{*j})) : push(j,S_i);
    }
```

Fig. 8.27 Parallel left-looking algorithm according to variant L2. The implicit task pool is implemented in the `while`-loop and the function `get_unique_index()`. The stacks $S_1, \ldots, S_n$ implement the bookkeeping about the dependent columns already finished.

comprises the execution of `cdiv(j)` and `cmod(k, j)` for all $k \in Struct(L_{*j})$, i.e., a column task comprises the final computation for column j and the modifications of all columns $k > j$ right of column j that depend on j. The task pool is initialized with all column tasks corresponding to the leaves of the elimination tree. A task $Tcol(j)$ that is not a leaf is inserted into the task pool as soon as the operations `cmod(j, k)` for all $k \in Struct(L_{j*})$ are executed and a final `cdiv(j)` operation is possible.

Figure 8.28 sketches a parallel implementation of the right-looking algorithm. The task assignment is implemented by maintaining a counter c_j for each column j. The counter is initialized with 0 and is incremented after the execution of each `cmod(j, *)` operation by the corresponding processor using the conflict-free procedure `add_counter()`. For the execution of a `cmod(k, j)` operation of a task $Tcol(j)$, column k must be locked to prevent other tasks from modifying the same column at the same time. A task $Tcol(j)$ is inserted into the task pool, when the counter c_j has reached the value $|Struct(L_{j*})|$.

The difference between this right-looking implementation and the left-looking variant L2 lies in the execution order of the $cmod()$ operations and in the executing processor. In the L2 variant, the operation $cmod(j, k)$ is initiated by the processor computing column k by pushing it on stack S_j, but the operation is executed by the processor computing column j. This execution does not need to be performed immediately after the initiation of the operation. In the right-looking variant, the operation $cmod(j, k)$ is not only initiated, but also executed by the processor that computes column k.

8.5.3.3 Parallel supernodal algorithm

The parallel implementation of the supernodal algorithm uses a partition into *fundamental supernodes*. A supernode $I(p) = \{p, p+1, \ldots, p+q-1\}$ is a fundamental supernode, if for each i with $0 \leq i \leq q-2$, we have $children(p+i+1) = \{p+i\}$,

```
parallel_right_cholesky =
    c = 0;
    initialize_task_pool(TP);
    while ((j = get_unique_index())< n) {
        while (!filled_pool(TP,j)) wait();
        get_column(j);
        cdiv(j);
        for (k ∈ Struct(L*j)) {
            lock(k); cmod(k,j); unlock(k);
            if (add_counter(ck, 1) + 1 == |Struct(Lk*)|) add_column(k);
        }
    }
```

Fig. 8.28 Parallel right-looking algorithm. The column tasks are managed by a task pool TP. Column tasks are inserted into the task pool by add_column() and retrieved from the task pool by get_column(). The function initialize_task_pool() initializes the task pool TP with the leaves of the elimination tree. The condition of the outer while-loop assigns column indices j to processors. The processor retrieves the corresponding column task as soon as the call filled_pool(TP,j) returns that the column task exists in the task pool.

i.e., node $p + i$ is the only child of $p + i + 1$ in the elimination tree [157]. In Fig. 8.24, supernode $I(2) = \{2, 3, 4\}$ is a fundamental supernode whereas supernodes $I(6) = \{6, 7\}$ and $I(8) = \{8, 9\}$ are not fundamental. In a partition into fundamental supernodes, all columns of a supernode can be computed as soon as the first column can be computed and a waiting for the computation of columns outside the supernode is not needed. In the following, we assume that all supernodes are fundamental, which can be achieved by splitting supernodes into smaller ones. A supernode consisting of a single column is fundamental.

The parallel implementation of the supernodal algorithm (VI) is based on supernode tasks $Tsup(J)$ where task $Tsup(J)$ for $0 \leq J < N$ comprises the execution of $\text{smod}(j, J)$ and $\text{cdiv}(j)$ for each $j \in J$ from left to right and the execution of $\text{smod}(k, J)$ for all $k \in Struct(L_{*(last(J))})$; N is the number of supernodes. Tasks $Tsup(J)$ that are ready for execution are held in a central task pool that is accessed by the idle processors. The pool is initialized with supernodes that are ready for completion; these are those supernodes whose first column is a leaf in the elimination tree. If the execution of a $\text{cmod}(k, J)$ operation by a task $Tsup(J)$ finishes the modification of another supernode $K > J$, the corresponding task $Tsup(K)$ is inserted into the task pool.

The assignment of supernode tasks is again implemented by maintaining a counter c_j for each column j of a supernode. Each counter is initialized with 0 and is incremented for each modification that is executed for column j. Ignoring the modifications with columns inside a supernode, a supernode task $Tsup(J)$ is ready for execution, if the counters of the columns $j \in J$ reach the value $|Struct(L_{j*})|$. The implementation of the counters as well as the manipulation of the columns has to be protected, e.g., by a lock-mechanism. For the manipulation of a column $k \notin J$

```
parallel_supernode_cholesky =
     c = 0;
     initialize_task_pool(TP);
     while ((J = get_unique_index()) < N) {
          while (!filled_pool(TP,J)) wait();
          get_supernode(J);
          cdiv(first(J));
          for (j = first(J) + 1; j ≤ last(J); j++) {
               smod(j,J);
               cdiv(j);
          }
          for (k ∈ Struct(L_{*(last(J))})) {
               lock(k); smod(k,J); unlock(k);
               add_counter(c_k, 1);
               if ((k == first(K)) && (c_k + 1 == |Struct(L_{k*})|))
                    add_supernode(K);
          }
     }
```

Fig. 8.29 Parallel supernodal algorithm.

by a $\text{smod}(k,J)$ operation, column k is locked to avoid concurrent manipulation by different processors. Figure 8.29 shows the corresponding implementation.

8.6 Exercises for Chapter 8

Exercise 8.1. For a $n \times m$ matrix A and vectors a and b of length n write a parallel MPI program which computes a rank-1 update $A = A - a \cdot b^\tau$ which can be computed sequentially by

```
for (i=0; i<n; i++)
    for (j=0; j<n; j++)
        A[i][j] = A[i][j]-a[i] · b[j];
```

For the parallel MPI implementation assume that A is distributed among the p processors in a column-cyclic way. The vectors a and b are available at the process with rank 0 only and must be distributed appropriately before the computation. After the update operation, the matrix A should again be distributed in a column-cyclic way.

Exercise 8.2. Implement the rank-1 update in OpenMP. Use a parallel for loop to express the parallel execution.

Exercise 8.3. Extend the program piece in Fig. 8.2 for performing the Gaussian elimination with a row-cycle data distribution to a full MPI program. To do so, all helper functions used and described in the text must be implemented.

Measure the resulting execution times for different matrix sizes and different numbers of processors.

Exercise 8.4. Similar to the previous exercise, transform the program piece in Fig. 8.6 with a total cyclic data distribution to a full MPI program. Compare the resulting execution time for different matrix sizes and different numbers of processors. For which scenarios does a significant difference occur? Try to explain the observed behavior.

Exercise 8.5. Develop a parallel implementation of Gaussian elimination for shared address spaces using OpenMP. The MPI implementation from Fig. 8.2 can be used as an orientation. Explain how the available parallelism is expressed in your OpenMP implementation. Also explain where synchronization is needed when accessing shared data. Measure the resulting execution times for different matrix sizes and different numbers of processors.

Exercise 8.6. Develop a parallel implementation of Gaussian elimination using Java threads. Define a new class `Gaussian` which is structured similar to the Java program in Fig. 6.25 for a matrix multiplication. Explain which synchronization is needed in the program. Measure the resulting execution times for different matrix sizes and different numbers of processors.

Exercise 8.7. Develop a parallel MPI program for Gaussian elimination using a column-cyclic data distribution. An implementation with a row-cyclic distribution has been given in Fig. 8.2. Explain which communication is needed for a column-cyclic distribution and include this communication in your program. Compute the resulting speedup values for different matrix sizes and different numbers of processors.

Exercise 8.8. For $n = 8$ consider the following tridiagonal equation system

$$\begin{pmatrix} 1 & 1 & & & & \\ 1 & 2 & 1 & & & \\ & 1 & 2 & \ddots & & \\ & & \ddots & \ddots & 1 \\ & & & 1 & 2 \end{pmatrix} \cdot x = \begin{pmatrix} 1 \\ 2 \\ 3 \\ \vdots \\ 8 \end{pmatrix}$$

Use the recursive doubling technique from Section 8.2.2, page 478, to solve this equation system.

Exercise 8.9. Develop a sequential implementation of the cyclic reduction algorithm for solving tridiagonal equation systems, see Section 8.2.2, page 478. Measure the resulting sequential execution time for different matrix sizes starting with size $n = 100$ up to size $n = 10^7$.

Exercise 8.10. Transform the sequential implementation of the cyclic reduction algorithm from the last exercise into a parallel implementation for a shared address space using OpenMP. Use an appropriate parallel for loop to express the parallel execution. Measure the resulting parallel execution time for different numbers of processors for the same matrix sizes as in the previous exercise. Compute the resulting speedup values and show the speedup values in a diagram.

Exercise 8.11. Develop a parallel MPI implementation of the cyclic reduction algorithm for a distributed address space based on the description in Section 8.2.2, page 478. Measure the resulting parallel execution time for different numbers of processors and compute the resulting speedup values.

Exercise 8.12. Specify the data dependence graph for the cyclic reduction algorithm for $n = 12$ equations according to Fig. 8.12. For $p = 3$ processors, illustrate the three phases according to Fig. 8.13 and show which dependences lead to communication.

Exercise 8.13. Implement a parallel Jacobi iteration with a pointer-based storage scheme of the matrix A such that global indices are used in the implementation.

Exercise 8.14. Consider the parallel implementation of the Jacobi iteration in Figure 8.14 and provide a corresponding shared memory program using OpenMP operations.

Exercise 8.15. Implement a parallel SOR method for a dense linear equation system by modifying the parallel program in Figure 8.15.

Exercise 8.16. Write a shared memory implementation for solving the discretized Poisson equation with the Gauss-Seidel method.

Exercise 8.17. Develop a shared memory implementation for Cholesky factorization $A = LL^T$ for a dense matrix A using the basic algorithm.

Exercise 8.18. Develop a message-passing implementation for dense Cholesky factorization $A = LL^T$.

Exercise 8.19. Consider a matrix with the following non-zero entries:

$$
\begin{array}{c}
0\ 1\ 2\ 3\ 4\ 5\ 6\ 7\ 8\ 9 \\
\begin{array}{c}
0 \\ 1 \\ 2 \\ 3 \\ 4 \\ 5 \\ 6 \\ 7 \\ 8 \\ 9
\end{array}
\left(
\begin{array}{cccccccccc}
* & & & & & & & & & \\
* & * & & & & & & & & \\
 & & * & & & & & & & \\
 & & * & * & & & & & & \\
 & & & * & * & * & & & & \\
* & * & * & * & * & * & & & & \\
 & & & & & & * & & & \\
 & & * & * & * & & & * & & \\
* & * & & & & & * & & * & * \\
 & & * & * & * & * & * & & * & *
\end{array}
\right)
\end{array}
$$

(a) Specify all supernodes of this matrix.

(b) Consider a supernode J with at least three entries. Specify the sequence of cmod and cdiv operations that are executed for this supernode in the right-looking supernode Cholesky factorization algorithm.

(c) Determine the elimination tree resulting for this matrix.

(d) Explain the role of the elimination tree for a parallel execution.

Exercise 8.20. Derive the parallel execution time of a message-passing program of the CG method for a distributed memory machine with a linear array as interconnection network.

Exercise 8.21. Consider a parallel implementation of the CG method in which the computation step (3) is executed in parallel to the computation step (4). Given a row-blockwise distribution of matrix A and a blockwise distribution of the vector, derive the data distributions for this implementation variant and give the corresponding parallel execution time.

Exercise 8.22. Implement the CG algorithm given in Fig. 8.21 with the blockwise distribution as message-passing program using MPI.

References

1. T.M. Aamodt, W.W.L. Fung, and T.G. Rogers. *General-Purpose Graphics Processor Architectures*. Morgan & Claypool Publishers, 2018.
2. F. Abolhassan, J. Keller, and W.J. Paul. On the Cost–Effectiveness of PRAMs. In *Proc. 3rd IEEE Symposium on Parallel and Distributed Processing*, pages 2–9, 1991.
3. S.V. Adve and K. Gharachorloo. Shared memory consistency models: A tutorial. *IEEE Computer*, 29:66–76, 1995.
4. A. Aggarwal, A.K. Chandra, and M. Snir. On Communication Latency in PRAM Computations. In *Proc. 1989 ACM Symposium on Parallel Algorithms and Architectures (SPAA'89)*, pages 11–21, 1989.
5. A. Aggarwal, A.K. Chandra, and M. Snir. Communication Complexity of PRAMs. *Theoretical Computer Science*, 71:3–28, 1990.
6. A.V. Aho, M.S. Lam, R. Sethi, and J.D. Ullman. *Compilers: Principles, Techniques, and Tools (2nd Edition)*. Addison-Wesley Longman Publishing Co., Inc., Boston, MA, USA, 2006.
7. Y. Ajima, T. Kawashima, T. Okamoto, N. Shida, K. Hirai, T. Shimizu, S. Hiramoto, Y. Ikeda, T. Yoshikawa, U. Uchida, and T. Inoue. The tofu interconnect d. In *2018 IEEE International Conference on Cluster Computing (CLUSTER)*, pages 646–654, 2018.
8. S.G. Akl. *Parallel Computation – Models and Methods*. Prentice Hall, 1997.
9. K. Al-Tawil and C.A. Moritz. LogGP Performance Evaluation of MPI. *High-Performance Distributed Computing, International Symposium on*, page 366, 1998.
10. A. Alexandrov, M. Ionescu, K. E. Schauser, and C. Scheiman. LogGP: Incorporating Long Messages into the LogP Model – One Step Closer towards a Realistic Model for Parallel Computation. In *Proc. 7th ACM Symposium on Parallel Algorithms and Architectures (SPAA'95)*, pages 95–105, Santa Barbara, California, July 1995.
11. Eric Allen, David Chase, Joe Hallett, Victor Luchangco, Jan-Willem Maessen, Sukyoung Ryu, Guy L. Steele, Jr., and Sam Tobin-Hochstadt. *The Fortress Language Specification, version 1.0beta*, March 2007.
12. R. Allen and K. Kennedy. *Optimizing Compilers for Modern Architectures*. Morgan Kaufmann, 2002.
13. S. Allmann, T. Rauber, and G. Rünger. Cyclic Reduction on Distributed Shared Memory Machines. In *Proc. 9th Euromicro Workshop on Parallel and Distributed Processing*, pages 290–297, Mantova, Italien, 2001. IEEE Computer Society Press.
14. G.S. Almasi and A. Gottlieb. *Highly Parallel Computing*. Benjamin Cummings, 1994.
15. G. Amdahl. Validity of the Single Processor Approach to Achieving Large-Scale Computer Capabilities. In *AFIPS Conference Proceedings*, volume 30, pages 483–485, 1967.
16. M. Andersch, G. Palmer, R. Krashinsky, N. Stam, V. Mehta, G. Brito, and S. Ramaswamy. NVIDIA Hopper Architecture In-Depth. https://developer.nvidia.com/blog/nvidia-hopper-architecture-in-depth/, 2022. Nvidia Developer.

© The Author(s), under exclusive license to Springer Nature Switzerland AG 2023
T. Rauber, G. Rünger, *Parallel Programming*, https://doi.org/10.1007/978-3-031-28924-8

17. M. Arafa, B. Fahim, S. Kottapalli, A. Kumar, L. P. Looi, S. Mandava, A. Rudoff, I. M. Steiner, B. Valentine, G. Vedaraman, and S. Vora. Cascade Lake: Next Generation Intel Xeon Scalable Processor. *IEEE Micro*, 39(2):29–36, 2019.

18. E. Ayguade, N. Copty, A. Duran, J. Hoeflinger, Y. Lin, F. Massaioli, X. Teruel, P. Unnikrishnan, and G. Zhang. The Design of OpenMP Tasks. *IEEE Transactions on Parallel and Distributed Systems*, 20(3):404–418, 2009.

19. M. Azimi, N. Cherukuri, D.N. Jayasimha, A. Kumar, P. Kundu, S. Park, I. Schoinas, and A. Vaidya. Integration Challenges and Tradeoffs for Tera-scale Architectures. *Intel Technology Journal*, 11(03), 2007.

20. C.J. Beckmann and C. Polychronopoulos. Microarchitecture Support for Dynamic Scheduling of Acyclic Task Graphs. Technical Report CSRD 1207, University of Illinois, 1992.

21. D.P. Bertsekas and J.N. Tsitsiklis. *Parallel and Distributed Computation*. Athena Scientific, 1997.

22. R. Bird. *Introduction to Functional Programming using Haskell*. Prentice Hall, 1998.

23. P. Bishop and N. Warren. *JavaSpaces in Practice*. Addison Wesley, 2002.

24. F. Bodin, P. Beckmann, D.B. Gannon, S. Narayana, and S. Yang. Distributed C++: Basic Ideas for an Object Parallel Language. In *Proc. Supercomputing'91 Conference*, pages 273–282, 1991.

25. D. Braess. *Finite Elements*. Cambridge, 3rd edition, 2007.

26. P. Brucker. *Scheduling Algorithms*. 4th edition, Springer, 2004.

27. D. R. Butenhof. *Programming with POSIX Threads*. Addison-Wesley, 1997.

28. N. Carriero and D. Gelernter. Linda in Context. *Commun. ACM*, 32(4):444–458, 1989.

29. N. Carriero and D. Gelernter. *How to Write Parallel Programs*. MIT Press, 1990.

30. B. Chapman, G. Jost, and R. van der Pas. *Using OpenMP*. MIT Press, 2007.

31. B. Chapman, G. Jost, and R. van der Pas. *Using OpenMP—The Next Step*. MIT Press, 2017.

32. P. Charles, C. Grothoff, V.A. Saraswat, C. Donawa, A. Kielstra, K. Ebcioglu, C. von Praun, and V. Sarkar. X10: an object-oriented approach to non-uniform cluster computing. In R. Johnson and R.P. Gabriel, editors, *Proceedings of the 20th Annual ACM SIGPLAN Conference on Object-Oriented Programming, Systems, Languages, and Applications (OOPSLA)*, pages 519–538. ACM, October 2005.

33. A. Chin. Complexity Models for All-Purpose Parallel Computation. In *Lectures on Parallel Computation*, chapter 14. Cambridge University Press, 1993.

34. WikiChip Chips and Semi. Cascade Lake - Microarchitectures - Intel. en.wikichip.org/wiki/intel/microarchitectures/cascade_lake. accessed February 2021.

35. M.E. Conway. A Multiprocessor System Design. In *Proc. AFIPS 1963 Fall Joint Computer Conference*, volume 24, pages 139–146. New York: Spartan Books, 1963.

36. T.H. Cormen, C.E. Leiserson, and R.L. Rivest. *Introduction to Algorithms, 2nd edition*. MIT Press, 2003.

37. Intel Corp. Intel Advanced Vector Extensions Programming Reference. Technical report, Intel, 2011.

38. Nvidea Corp. *CUDA programming guide*. NVIDIA, 2007.

39. Nvidea Corp. *NVIDIA CUDA C^{TM} Programming guide, Version 4.1*. NVIDIA CUDA, 2011.

40. Nvidea Corp. *NVIDIA A100 Tensor Core GPU Architecture*, 2020.

41. Nvidea Corp. *NVIDIA H100 Tensor Core GPU Architecture*, 2022.

42. D.E. Culler, A.C. Arpaci-Dusseau, S.C. Goldstein, A. Krishnamurthy, S. Lumetta, T. van Eicken, and K.A. Yelick. Parallel programming in Split-C. In *Proceedings of Supercomputing*, pages 262–273, 1993.

43. D.E. Culler, A.C. Dusseau, R.P. Martin, and K.E. Schauser. Fast Parallel Sorting under LogP: from Theory to Practice. In *Portability and Performance for Parallel Processing*, pages 71–98. Wiley, 1994.

44. D.E. Culler, R. Karp, A. Sahay, K.E. Schauser, E. Santos, R. Subramonian, and T. von Eicken. LogP: Towards a Realistic Model of Parallel Computation. *Proc. 4th ACM SIGPLAN Symp. on Principles and Practice of Parallel Programming (PPoPP'93)*, pages 1–12, 1993.

45. D.E. Culler, J.P. Singh, and A. Gupta. *Parallel Computer Architecture: A Hardware Software Approach*. Morgan Kaufmann, 1999.

46. H.J. Curnov and B.A. Wichmann. A Synthetic Benchmark. *The Computer Journal*, 19(1):43–49, 1976.
47. D. Callahan and B. L. Chamberlain and H. P. Zima. The Cascade High Productivity Language. In *IPDPS*, pages 52–60. IEEE Computer Society, 2004.
48. W.J. Dally and C.L. Seitz. Deadlock-Free Message Routing in Multiprocessor Interconnection Networks. *IEEE Transactions on Computers*, 36(5):547–553, 1987.
49. DEC. The Whetstone Performance. Technical Report, Digital Equipment Corporation, 1986.
50. S. Desrochers, C. Paradis, and V. M. Weaver. A Validation of DRAM RAPL Power Measurements. In *Proceedings of the Second International Symposium on Memory Systems*, MEMSYS '16, page 455–470, New York, NY, USA, 2016. Association for Computing Machinery.
51. E.W. Dijkstra. Cooperating Sequential Processes. In F. Genuys, editor, *Programming Languages*, pages 43–112. Academic Press, 1968.
52. J. Dongarra. Performance of various Computers using Standard Linear Equations Software in Fortran Environment. Technical Report CS-89-85, Computer Science Dept., University of Tennessee, Knoxville, 1990.
53. J. Dongarra and W. Gentzsch, editors. *Computer Benchmarks*. Elsevier, North Holland, 1993.
54. J. J. Dongarra, P. Luszczek, and A. Petitet. The LINPACK Benchmark: past, present and future. *Concurrency and Computation: Practice and Experience*, 15(9):803–820, 2003.
55. J.J Dongarra, I.S. Duff, D.C. Sorenson, and Henk A. van der Vorst. *Solving Linear Systems on Vector and Shared Memory Computers*. SIAM, 1993.
56. J. Duato, S. Yalamanchili, and L. Ni. *Inteconnection Networks - An Engineering Approach*. Morgan Kaufmann, 2003.
57. M. Dubois, C. Scheurich, and F. Briggs. Memory Access Buffering in Multiprocessors. In *Proc. 13th Int. Symp. on Computer Architecture (ISCA'86)*, pages 434–442. ACM, 1986.
58. J. Dümmler, T. Rauber, and G. Rünger. Mixed Programming Models using Parallel Tasks. In J. Dongarra, C.-H. Hsu, K.-C. Li, L.T. Yang, and H. Zima, editors, *Handbook of Research on Scalable Computing Technologies*. Information Science Reference, July 2009.
59. T. El-Ghazawi, W. Carlson, T. Sterling, and K. Yelick. *UPC: Distributed Sahred Memory Programming*. Wiley, 2005.
60. J.R. Ellis. *Bulldog: A Compiler for VLIW Architectures*. MIT Press, 1986.
61. T. Ellis, I. Phillips, and T. Lahey. *Fortran90 Programming*. Addison-Wesley, 1994.
62. J.T. Feo. An Analysis of the Computational and Parallel Complexity of the Livermore Loops. *Parallel Computing*, 7:163–185, 1988.
63. D. Flanagan. *Java in a Nutshell*. O'Reilly, 2005.
64. M.J. Flynn. Some Computer Organizations and their Effectiveness. *IEEE Transactions on Computers*, 21(9):948–960, 1972.
65. S. Fortune and J. Wyllie. Parallelism in Random Access Machines. In *Proc. 10th ACM Symposium on Theory of Computing*, pages 114–118, 1978.
66. High Performance Fortran Forum. High Performance Fortran Language Specification. *Scientific Programming*, 2(1):1–165, 1993.
67. Message Passing Interface Forum. *MPI: A Message-Passing Interface Standard, Version 1.3*. www.mpi-forum.org, 2008.
68. Message Passing Interface Forum. *MPI: A Message-Passing Interface Standard, Version 2.1*. www.mpi-forum.org, 2008.
69. Message Passing Interface Forum. *MPI: A Message-Passing Interface Standard, Version 4.0*. www.mpi-forum.org, 2021.
70. I. Foster. *Designing and Building Parallel Programs*. Addison-Wesley, 1995.
71. I. Foster. Compositional Parallel Programming Languages. *ACM Transactions on Programming Languages and Systems*, 18(4):454–476, 1996.
72. I. Foster. Globus Toolkit Version 4: Software for Service-Oriented Systems. In *Proc. IFIP International Conference on Network and Parallel Computing, Springer LNCS 3779*, pages 2–13, 2006.

73. T. L. Freeman and C. Phillips. *Parallel Numerical Algorithms*. Prentice Hall, 1992.
74. A. Frommer. *Lösung linearer Gleichungssysteme auf Parallelrechnern*. Vieweg, 1990.
75. E. Gabriel, G. E. Fagg, G. Bosilca, T. Angskun, J. J. Dongarra, J. M. Squyres, V. Sahay, P. Kambadur, B. Barrett, A. Lumsdaine, R. H. Castain, D. J. Daniel, R. L. Graham, and T. S. Woodall. Open MPI: Goals, concept, and design of a next generation MPI implementation. In *Proc. 11th European PVM/MPI Users' Group Meeting*, pages 97–104, 2004.
76. M.R. Garey and D.S. Johnson. *Computers and Intractability: A Guide to the Theory of NP-Completeness*. Freeman, 1979.
77. A. Geist, A. Beguelin, J. Dongarra, W. Jiang, R. Manchek, and V. Sunderam. *PVM Parallel Virtual Machine: A User's Guide and Tutorial for Networked Parallel Computing*. MIT Press, Cambridge, MA, 1996.
78. A. George, J. Liu, and E. Ng. User's Guide for SPARSPAK: Waterloo Sparse Linear Equations Package. Technical Report CS-78-30, Department of Computer Science, University of Waterloo, 1980.
79. A. George and J. W-H Liu. *Computer Solution of Large Sparse Positive Definite Systems*. Prentice-Hall, 1981.
80. P.B. Gibbons. A More Practical PRAM Model. In *Proc. 1989 ACM Symposium on Parallel Algorithms and Architectures (SPAA'89)*, pages 158–168, 1989.
81. M.B. Girkar and C. Polychronopoulos. Automatic Extraction of Functional Parallelism from Ordinary Programs. *IEEE Transactions on Parallel and Distributed Systems*, 3(2):166–178, 1992.
82. C.J. Glass and L.M. Li. The Turn Model for Adaptive Routing. In *Proc. 19th Int. Symp. on Computer Architecture (ISCA'92)*, pages 278–287. ACM, 1992.
83. S. Goedecker and A. Hoisie. *Performance Optimization of Numerically Intensive Codes*. Siam, 2001.
84. B. Goetz. *Java Concurrency in Practice*. Addison Wesley, 2006.
85. G. Golub and Ch. Van Loan. *Matrix Computations*. The Johns Hopkins University Press, 3rd. edition, 1996.
86. G. Golub and J. Ortega. *Scientific Computing*. Academic Press, 1993.
87. A. Gottlieb, R. Grishman, C. Kruskal, K. McAuliffe, L. Rudolph, and M. Snir. The NYU Ultracomputer – Designing an MIMD Shared Memory Parallel Computer. *IEEE Transactions on Computers*, 32(2):175–189, February 1983.
88. M.W. Goudreau, J.M. Hill, K. Lang, W.F. McColl, S.D. Rao, D.C. Stefanescu, T. Suel, and T. Tsantilas. A proposal for a BSP Worldwide standard. Technical Report, BSP Worldwide, www.bsp-worldwide.org, 1996.
89. A. Grama, A. Gupta, G. Karypis, and V. Kumar. *Introduction to Parallel Programming*. Addison Wesley, 2003.
90. W. Gropp, E. Lusk, and A. Skjellum. *Using MPI, third edition: Portable Parallel Programming with the Message-Passing Interface*. MIT Press, Camdridge, MA, 2014.
91. Khronos OpenCP Working Group. *The Open CL API Specification Version 3.0*. Khronos Group, 2022.
92. T. Gruber, J. Eitzinger, G. Hager, and G. Wellein. LIKWID 5: Lightweight Performance Tools. https://doi.org/10.5281/zenodo.4275676, 2022. Zenodo.
93. T. Grün, T. Rauber, and J. Röhrig. Support for Efficient Programming on the SB-PRAM. *International Journal of Parallel Programming*, 26(3):209–240, 1998.
94. A. Gupta, G. Karypis, and V. Kumar. Highly scalable parallel algorithms for sparse matrix factorization. *IEEE Trans. Parallel Distrib. Syst.*, 8(5):502–520, 1997.
95. J.L. Gustafson. Reevaluating Amdahl's law. *Commun. ACM*, 31(5):532–533, 1988.
96. W. Hackbusch. *Iterative Solution of Large Sparse Systems of Equations*. Springer, 1994.
97. D. Hackenberg, T. Ilsche, R. Schöne, D. Molka, M. Schmidt, and W. E. Nagel. Power measurement techniques on standard compute nodes: A quantitative comparison. In *2013 IEEE International Symposium on Performance Analysis of Systems and Software (ISPASS)*, pages 194–204, 2013.

98. A. Haidar, H. Jagode, P. Vaccaro, A. YarKhan, S. Tomov, and J. Dongarra. Investigating power capping toward energy-efficient scientific applications. *Concurrency and Computation: Practice and Experience*, 31(6):e4485, 2019. e4485 cpe.4485.

99. K. Hammond and G. Michaelson, editors. *Research Directions in Parallel Functional Programming*. Springer, 1999.

100. J. Handy. *The Cache Memory Book*. Academic Press, 2nd edition, 1998.

101. P.J. Hatcher and M.J. Quinn. *Data-Parallel Programming*. MIT Press, 1991.

102. J. Held, J. Bautista, and S. Koehl. From a Few Cores to Many – A Tera-Scale Computing Research Overview. Intel White Paper, Intel, 2006.

103. J.L. Hennessy and D.A. Patterson. *Computer Architecture - A Quantitative Approach (5. ed.)*. Morgan Kaufmann, 2012.

104. J.L. Hennessy and D.A. Patterson. *Computer Architecture — A Quantitative Approach*. Morgan Kaufmann, 6th edition, 2017.

105. J.L. Hennessy and D.A. Patterson. *Computer Architecture - A Quantitative Approach (6. ed.)*. Morgan Kaufmann, 2019.

106. J. L. Henning. SPEC CPU Suite Growth: An Historical Perspective. *SIGARCH Comput. Archit. News*, 35(1):65–68, mar 2007.

107. M.R. Hestenes and E. Stiefel. Methods of Conjugate Gradients for Solving Linear Systems. *J. Res. Nat. Bur. Stand*, 49:409–436, 1952.

108. Heywood, T. and Ranka, S. A Practical Hierarchical Model of Parallel Computation. *Journal of Parallel and Distributed Computing*, 16:212–249, 1992.

109. J.M.D. Hill, B. McColl, D.C. Stefanescu, M.W. Goudreau, K. Lang, S.B. Rao, T. Suel, T. Tsantilas, and R. Bisseling. BSPlib The BSB Programming Library. Technical Report TR-29-97, Oxford University, May 1997.

110. M. Hill, W. McColl, and D. Skillicorn. Questions and Answers about BSP. *Scientific Programming*, 6(3):249–274, 1997.

111. C.A.R. Hoare. Monitors: An Operating Systems Structuring Concept. *Commun. ACM*, 17(10):549–557, 1974.

112. R. Hockney. A Fast Direct Solution of Poisson's Equation Using Fourier Analysis. *Journal of the ACM*, 12:95–113, 1965.

113. R.W. Hockney. *The Science of Computer Benchmarking*. SIAM, 1996.

114. R. Hoffmann and T. Rauber. Fine-grained task scheduling using adaptive data structures. In *Proc. of Euro-Par*, volume 5168 of *Lecture Notes in Computer Science*, pages 253–262. Springer, 2008.

115. P. Hudak and J. Fasel. A Gentle Introduction to Haskell. *ACM SIGPLAN Notices*, 27, No.5, May 1992.

116. K. Hwang. *Advanced Computer Architecture: Parallelism, Scalability, Programmability*. McGraw-Hill, 1993.

117. F. Ino, N. Fujimoto, and K. Hagihara. LogGPS: a parallel computational model for synchronization analysis. In *PPoPP '01: Proceedings of the eighth ACM SIGPLAN symposium on Principles and practices of parallel programming*, pages 133–142, New York, NY, USA, 2001. ACM.

118. Intel. *Intel Processor Graphics Gen11 Architecture*, 2022. Intel.

119. J.D. Jackson. *Classical Electrodynamics*. Wiley, 3rd edition, 1998.

120. A. Jain and C. Lin. *Cache Replacement Policies*. Morgan & Claypool Publishers, 2019.

121. J. Jájá. *An Introduction to Parallel Algorithms*. Addison-Wesley, 1992.

122. T. Jakobs, L. Reinhardt, and G. Rünger. Performance and energy consumption of a Gram–Schmidt process for vector orthogonalization on a processor integrated GPU. *Sustainable Computing: Informatics and Systems*, September 2020.

123. M. Johnson. *Superscalar Microprocessor Design*. Prentice Hall, 1991.

124. S. Johnsson and C. Ho. Optimum Broadcasting and Personalized Communication in Hypercubes. *IEEE Transactions on Computers*, 38(9):1249–1268, 1989.

125. S. Kaxiras and M. Martonosi. *Computer Architecture Techniques for Power-Efficiency*. Morgan and Claypool Publishers, 2008.

126. J. Keller, C.W. Keßler, and J.L. Träff. *Practical PRAM Programming.* Wiley, 2001.

127. J. Keller, Th. Rauber, and B. Rederlechner. Conservative Circuit Simulation on Shared–Memory Multiprocessors. In *Proc. 10th Workshop on Parallel and Distributed Simulation (PADS'96)*, pages 126–134. ACM, 1996.

128. K. Kennedy, C. Koelbel, and H. Zima. The rise and fall of high performance fortran: an historical object lesson. In *HOPL III: Proceedings of the third ACM SIGPLAN conference on History of programming languages*, pages 7–1–7–22, New York, NY, USA, 2007. ACM.

129. T. Kielmann, H.E. Bal, and K. Verstoep. Fast measurement of logp parameters for message passing platforms. In *IPDPS '00: Proceedings of the 15 IPDPS 2000 Workshops on Parallel and Distributed Processing*, pages 1176–1183, London, UK, 2000. Springer.

130. D.B. Kirk and W.W. Hwu. *Programming Massively Parallel Processors - A Hands-on Approach.* Morgan Kaufmann, 2010.

131. St. Kleiman, D. Shah, and B. Smaalders. *Programming with Threads.* Prentice Hall, 1996.

132. G. Koch. Discovering Multi-Core:Extending the Benefits of Moore's Law. Intel White Paper, Technology@Intel Magazine, 2005.

133. P.M. Kogge. An Exploitation of the Technology Space for Multi-Core Memory/Logic Chips for Highly Scalable Parallel Systems. In *Proceedings of the Innovative Architecture for Future Generation High-Performance Processors and Systems.* IEEE, 2005.

134. Poonacha Kongetira, Kathirgamar Aingaran, and Kunle Olukotun. Niagara: A 32-way multithreaded Sparc processor. *IEEE Micro*, 25(2):21–29, March/April 2005.

135. M. Korch and T. Rauber. A comparison of task pools for dynamic load balancing of irregular algorithms. *Concurrency and Computation: Practice and Experience*, 16:1–47, January 2004.

136. S. Kounev, K.-D. Lange, and J. von Kistowski. *Systems Benchmarking – for Scientists and Engineers.* Springer, 2020.

137. R. Krashinsky, O. Giroux, S. Jones, N. Stam, and S. Ramaswamy. NVIDIA Ampere Architecture In-Depth. https://developer.nvidia.com/blog/nvidia-ampere-architecture-in-depth/, 2020. Nvidia Developer.

138. D. Kuck. Platform 2015 Software-Enabling Innovation in Parallelism for the next Decade. Intel White Paper, TechnologyIntel Magazine, 2005.

139. J. Kurose and K. Ross. *Computer Networking, 8th edition.* Addison Wesley, 2022.

140. L. Lamport. How to Make a Multiprocessor Computer That Correctly Executes Multiprocess Programs. *IEEE Transactions on Computers*, 28(9):690–691, September 1979.

141. J.R. Laurs and R. Rajwar. *Transactional Memory.* Morgan & Claypool Publishers, 2007.

142. D. Lea. *Concurrent Programming in Java: Design Principles and Patterns.* Addison Wesley, 1999.

143. E.A. Lee. The Problem with Threads. *IEEE Computer*, 39(5):33–42, 2006.

144. F.T. Leighton. *Introduction to Parallel Algorithms and Architectures: Arrays, Trees, Hypercubes.* Morgan Kaufmann, 1992.

145. D.E. Lenoski and W. Weber. *Scalable Shared-Memory Multiprocessing.* Morgan Kaufmann, 1995.

146. B. Lewis and D. J. Berg. *Multithreaded Programming with Pthreads.* Prentice Hall, 1998.

147. Joseph W. H. Liu. The Role of Elimination Trees in Sparse Factorization. *SIAM J. Matrix Anal. Appl.*, 11:134–172, 1990.

148. C. Lomont. Introduction to Intel Advanced Vector Extensions. Technical report, Intel, 2011.

149. D.T. Marr, F. Binus, D.L. Hill, G. Hinton, D.A. Konfaty, J.A. Miller, and M. Upton. Hyper-Threading Technology Architecture and Microarchitecture. *Intel Technology Journal*, 6(1):4–15, 2002.

150. Michael Marty. *Cache Coherence Techniques for Multicore Processors.* PhD thesis, University of Wisconsin - Madison, 2008.

151. T. Mattson, B. Sandor, and B. Massingill. *Pattern for Parallel Programming.* Pearson – Addison Wesley, 2005.

152. T.G. Mattson, Y. He, and A.E. Koniges. *The OpenMP Common Core.* MIT Press, 2019.

153. F. McMahon. The Livermore Fortran Kernels: A Computer Test of the Numerical Performance Range. Technical Report UCRL-53745, Lawrence Livermore National Laboratory, Livermore, 1986.

154. M. Metcalf and J. Reid. *Fortran 90/95 explained*. Oxford University Press, 2002.

155. R. Miller and L. Boxer. *Algorithms Sequential and Parallel*. Prentice Hall, 2000.

156. A. Munshi, B.R. Gaster, T.G. Mattson, J. Fung, and D. Ginsburg. *Open CL Programming Guide*. Addison-Wesley, 2012.

157. E.G. Ng and B.W. Peyton. A Supernodal Cholesky Factorization Algorithm for Shared-Memory Multiprocessors. Technical Report, Oak Ridge National Laboratory, 1991.

158. Y. Ngoko and D. Trystram. Scalability in Parallel Processing. In S. K. Prasad, A. Gupta, A. L. Rosenberg, A. Sussman, and C. C. Weems, editors, *Topics in Parallel and Distributed Computing, Enhancing the Undergraduate Curriculum: Performance, Concurrency, and Programming on Modern Platforms*, pages 79–109. Springer, 2018.

159. L.M. Ni and P.K. McKinley. A Survey of Wormhole Routing Techniques in Direct Networks. *IEEE Computer*, pages 62–76, February 1993.

160. B. Nichols, D. Buttlar, and J. Proulx Farrell. *Pthreads Programming*. O'Reilly & Associates, 1997.

161. J. Nieplocha, J. Ju, M.K. Krishnan, B. Palmer, and V. Tipparaju. The Global Arrays User's Manual. Technical Report PNNL-13130, Pacific Northwest National Laboratory, 2002.

162. S. Oaks and H. Wong. *Java Threads*. 3rd edition, O'Reilly, 2004.

163. *OpenMP C and C++ Application Program Interface, Version 1.0*. www.openmp.org, October 1998.

164. *OpenMP Application Programming Interface, Version 5.2*. www.openmp.org, November 2021.

165. J. M. Ortega. *Introduction to Parallel and Vector Solutions of Linear Systems*. Plenum Publishing Corp., 1988.

166. J.M. Ortega and R.G. Voigt. *Solution of Partial Differential Equations on Vector and Parallel Computers*. SIAM, 1985.

167. P.S. Pacheco. *Parallel Programming with MPI*. Morgan Kaufmann, 1997.

168. Dhabaleswar Kumar Panda, Hari Subramoni, Ching-Hsiang Chu, and Mohammadreza Bayatpour. The mvapich project: Transforming research into high-performance mpi library for hpc community. *Journal of Computational Science*, 52:101208, 2021. Case Studies in Translational Computer Science.

169. C.H. Papadimitriou and M. Yannakakis. Towards an Architecture–Independent Analysis of Parallel Algorithms. In *Proc. 20th ACM Symposium on Theory of Computing*, pages 510–513, 1988.

170. D.A. Patterson and J.L. Hennessy. *Computer Organization & Design — The Hardware/Software Interface, 4th edition*. Morgan Kaufmann, 2008.

171. D.A. Patterson and J.L. Hennessy. *Computer Organization & Design — The Hardware/Software Interface, 6th edition*. Morgan Kaufmann, 2020.

172. S. Pelegatti. *Structured Development of Parallel Programs*. Taylor and Francis, 1998.

173. L. Peterson and B. Davie. *Computer Networks - A Systems Approach, 5th edition*. Morgan Kaufmann, 2011.

174. L. Peterson and B. Davie. *Computer Networks - A Systems Approach, 6th edition*. Morgan Kaufmann, 2021.

175. G.F. Pfister. *In Search of Clusters*. Prentice Hall, 2nd edition, 1998.

176. A. Podehl, T. Rauber, and G. Rünger. A Shared-Memory Implementation of the Hierarchical Radiosity Method. *Theoretical Computer Science*, 196(1-2):215–240, 1998.

177. C.D. Polychronopoulos. *Parallel Programming and Compilers*. Kluwer Academic Publishers, 1988.

178. S. Prasad. *Multithreading Programming Techniques*. McGraw-Hill, 1997.

179. S. Ramaswamy, S. Sapatnekar, and P. Banerjee. A Framework for Exploiting Task and Data Parallelism on Distributed-Memory Multicomputers. *IEEE Transactions on Parallel and Distributed Systems*, 8(11):1098–1116, 1997.

180. T. Rauber and G. Rünger. A Transformation Approach to Derive Efficient Parallel Implementations. *IEEE Transactions on Software Engineering*, 26(4):315–339, 2000.
181. T. Rauber and G. Rünger. Deriving Array Distributions by Optimization Techniques. *Journal of Supercomputing*, 15:271–293, 2000.
182. T. Rauber and G. Rünger. Tlib - A Library to Support Programming with Hierarchical Multi-Processor Tasks. *Journal of Parallel and Distributed Computing*, 65(3):347–360, 2005.
183. T. Rauber and G. Rünger. Modeling and Analyzing the Energy Consumption of Fork-Join-based Task Parallel Programs. *Concurrency and Computation: Practice and Experience*, 27(1):211–236, 2015.
184. T. Rauber and G. Rünger. Modeling the effect of application-specific program transformations on energy and performance improvements of parallel ODE solvers. *Journal of Computational Science*, 51, 2021.
185. T. Rauber, G. Rünger, and C. Scholtes. Execution Behavior Analysis and Performance Prediction for a Shared-Memory Implementation of an Irregular Particle Simulation Method. *Simulation: Practice and Theory*, 6:665–687, 1998.
186. T. Rauber, G. Rünger, M. Schwind, H. Xu, and S. Melzner. Energy Measurement, Modeling, and Prediction for Processors with Frequency Scaling. *The Journal of Supercomputing*, 70(3):1451–1476, 2014.
187. T. Rauber, G. Rünger, and M. Stachowski. Performance and energy metrics for multi-threaded applications on DVFS processors. *Sustain. Comput. Informatics Syst.*, 17:55–68, 2018.
188. J.K. Reid. On the Method of Conjugate Gradients for the Solution of Large Sparse Systems of Linear Equations. In *Large Sparse Sets of Linear Equations*, pages 231–254. Academic Press, 1971.
189. G. Rong, F. Xizhou, S. Shuaiwen, C. Hung-Ching, L. Dong, and K.W. Cameron. Power-Pack: Energy Profiling and Analysis of High-Performance Systems and Applications. *IEEE Transactions on Parallel and Distributed Systems*, 21(5):658 –671, may 2010.
190. M. Rosing, R.B. Schnabel, and R.P. Waever. The DINO Parallel Programming language. Technical Report CU-CS-501-90, Computer Science Dept., University of Colorado at Boulder, Boulder, CO, 1990.
191. E. Rotem, A. Naveh, A. Ananthakrishnan, E. Weissmann, and D. Rajwan. Power-management architecture of the intel microarchitecture code-named sandy bridge. *IEEE Micro*, 32(2):20–27, mar 2012.
192. E. Rothberg and A. Gupta. An Evaluation of Left-Looking, Right-Looking and Multifrontal Approaches to Sparse Cholesky Factorization on Hierarchical-Memory Machines. *Int. J. High Speed Computing*, 5(4):537–593, 1993.
193. G. Rünger. Parallel Programming Models for Irregular Algorithms. In *Parallel Algorithms and Cluster Computing*, pages 3–23. Springer Lecture Notes in Computational Science and Engineering, 2006.
194. Y. Saad. *Iterative Methods for Sparse Linear Systems*. International Thomson Publ., 1996.
195. Y. Saad. Krylov Subspace Methods on Supercomputers. *SIAM Journal on Scientific and Statistical Computing*, 10:1200–1332, 1998.
196. J. Sanders and E. Kandrot. *CUDA by Example - An Introduction to General-Purpose GPU programming*. Addison-Wesley, 2011.
197. J.E. Savage. *Models of Computation*. Addison Wesley, 1998.
198. C. Scheurich and M. Dubois. Correct Memory Operation of Cache-Based Multiprocessors. In *Proc. 14th Int. Symp. on Computer Architecture (ISCA'87)*, pages 234–243. ACM, 1987.
199. D. Sima, T. Fountain, and P. Kacsuk. *Advanced Computer Architectures*. Addison-Wesley, 1997.
200. J.P. Singh. *Parallel Hierarchical N-Body Methods and Their Implication for Multiprocessors*. PhD Thesis, Stanford University, 1993.
201. D. Skillicorn and D. Talia. Models and Languages for Parallel Computation. *ACM Computing Surveys*, 30(2):123–169, 1998.
202. B. Smith. Architecture and Applications on the HEP Multiprocessor Computer Systems. *SPIE (Real Time Signal Processing IV)*, 298:241–248, 1981.

203. M. Snir, S. Otto, S. Huss-Ledermann, D. Walker, and J. Dongarra. *MPI: The Complete Reference*. MIT Press, Camdridge, MA, 1996.

204. M. Snir, S. Otto, S. Huss-Ledermann, D. Walker, and J. Dongarra. *MPI: The Complete Reference, Vol.1: The MPI Core*. MIT Press, Camdridge, MA, 1998.

205. D. J. Sorin, M. D. Hill, and D. A. Wood. *A Primer on Memory Consistency and Cache Coherence*. Morgan & Claypool Publishers, 2nd edition, 2020.

206. SPEC Newslett. 2(3):3-4. *SPEC Benchmark Suite Release 1.0*, 1990.

207. W. Stallings. *Computer Organization and Architecture, 9th edition*. Prentice Hall, 2012.

208. R.C. Steinke and G.J. Nutt. A unified theory of shared memory consistency. *Journal of the ACM*, 51(5):800–849, 2004.

209. J. Stoer and R. Bulirsch. *Introduction to Numerical Analysis*. Springer, 2002.

210. H. S. Stone. Parallel Processing with the Perfect Shuffle. *IEEE Transactions on Computers*, 20(2):153–161, 1971.

211. H. S. Stone. An Efficient Parallel Algorithm for the Solution of a Tridiagonal Linear System of Equations. *Journal of the ACM*, 20:27–38, 1973.

212. H. Sutter and J. Larus. Software and the Concurrency Revolution. *2005*, 3(7):54–62, ACM Queue.

213. PAPI team. PAPI's User Guide at `icl.utk.edu/papi`, 2022.

214. D. Terpstra, H. Jagode, H. You, and J. Dongarra. Collecting Performance Data with PAPI-C. In M. S. Müller, M. M. Resch, A. Schulz, and W. E. Nagel, editors, *Tools for High Performance Computing 2009*, pages 157–173, Berlin, Heidelberg, 2010. Springer Berlin Heidelberg.

215. S. Thompson. *Haskell – The Craft of Functional Programming*. Addison Wesley, 1999.

216. D.M. Topkis. Concurrent Broadcast for Information Dissemination. *IEEE Trans. Softw. Eng.*, 11(10):1107–1112, 1985.

217. J. Treibig, G. Hager, and G. Wellein. LIKWID: A Lightweight Performance-Oriented Tool Suite for x86 Multicore Environments. In *39th Int. Conf. on Parallel Processing Workshops*, ICPP '10, pages 207–216. IEEE Computer Society, 2010.

218. L.G. Valiant. A Bridging Model for parallel Computation. *Commun. ACM*, 33(8):103–111, 1990.

219. L.G. Valiant. A Bridging Model for Multi-core Computing. In *Proc. of ESA*, volume 5193, pages 13–28. Springer LNCS, 2008.

220. E.F. van de Velde. *Concurrent Scientific Computing*. Springer, 1994.

221. M. Velten, R. Schöne, T. Ilsche, and D. Hackenberg. Memory performance of AMD EPYC rome and intel cascade lake SP server processors. In D. Feng, S. Becker, N. Herbst, and P. Leitner, editors, *ICPE '22: ACM/SPEC International Conference on Performance Engineering, Bejing, China, April 9 - 13, 2022*, pages 165–175. ACM, 2022.

222. R.P. Weicker. Dhrystone: A Synthetic System Programming Benchmark. *Commun. ACM*, 29(10):1013–1030, 1984.

223. M. Wolfe. *High Performance Compilers for Parallel Computing*. Addison-Wesley, 1996.

224. M. Yannakakis. Computing the Minimum Fill-in is NP-complete. *SIAM J. Algebraic Discrete Methods*, 2:77–79, 1991.

225. S.N. Zheltov and S.V. Bratanov. Measuring HT-Enabled Multi-Core: Advantages of a Thread-Oriented Approach. *Technology & Intel Magazine*, December 2005.

226. J. Zhuo and C. Chakrabarti. Energy-efficient dynamic task scheduling algorithms for DVS systems. *ACM Trans. Embed. Comput. Syst.*, 7(2):1–25, 2008.

227. A.Y.H. Zomaya, editor. *Parallel & Distributed Computing Handbook*. Computer Engineering Series. McGraw-Hill, 1996.

Index

Printed in the United States
by Baker & Taylor Publisher Services